AF315857

Springer

Berlin
Heidelberg
New York
Barcelona
Hong Kong
London
Milan
Paris
Singapore
Tokyo

Numerical Flow Simulation II

CNRS-DFG Collaborative Research Programme
Results 1998–2000

Ernst Heinrich Hirschel (Editor)

 Springer

Prof. Dr. Ernst H. Hirschel
Herzog-Heinrich-Weg 6
85604 Zorneding
e-mail: e.h.hirschel@t-online.de

Library of Congress Cataloging-in-Publication Data applied for

Numerical flow simulation : CNRS DFG collaborative research programme
/ ed. by Ernst Heinrich Hirschel. - Braunschweig ; Wiesbaden : Vieweg;
[Berlin ; Heidelberg ; New York ; Barcelona ; Budapest ; Hongkong ;
London ; Mailand ; Paris ; Singapur ; Tokio] : Springer

2. Results 1998 - 2000. - 2001
  (Notes on numerical fluid mechanics ; 75)
  ISBN 3-540-41608-0

ISSN 0179-9614
ISBN 3-540-41608-0 Springer-Verlag Berlin Heidelberg New York

Springer-Verlag Berlin Heidelberg New York
a member of BertelsmannSpringer Science+Business Media GmbH

http://www.springer.de

The use of general descriptive names, trademarks, etc. in this publication does not imply, even in the absence of a specific statement, that such names are exempt from the relevant protective laws and regulations and therefore free for general use.

Typesetting: Camera ready by authors
Cover design: de'blik, Berlin

Printed on acid-free paper   SPIN: 10796124   62/3020/M – 5 4 3 2 1 0

This volume is dedicated to
Prof. Dr. R. Peyret
on the occasion of his sixtyfifth birthday.

Editorial note

This volume of the Notes on Numerical Fluid Mechanics is the first one to appear at the Springer Verlag, after 74 volumes appeared with Vieweg. The editors of the series wish to thank Vieweg for more than 20 years of a good and fruitful cooperation.

NNFM Editor Addresses

Prof. Dr. Ernst Heinrich Hirschel
(General editor)
Herzog-Heinrich-Weg 6
D-85604 Zorneding
Germany
E-mail: e.h.hirschel@t-online.de

Prof. Dr. Kozo Fujii
Space Transportation Research Division
The Institute of Space and Astronautical Science
3-1-1, Yoshinodai, Sagamihara, Kanagawa,
229-8510
Japan
E-mail: fujii@flab.eng.isas.ac.jp

Dr. Werner Haase
Höhenkirchener Str. 19d
D-85662 Hohenbrunn
Germany
E-mail: wug.haase@t-online.de

Prof. Dr. Bram van Leer
Department of Aerospace Engineering
The University of Michigan
Ann Arbor, MI 48109-2140
USA
E-mail: bram@engin.umich.edu

Prof. Dr. Michael A. Leschziner
Department of Engineering
Queen Mary & Westfield College (QMW)
University of London
Mile End Road
London E1 4NS
Great Britain
E-mail: mike.leschziner@qmw.ac.uk

Prof. Dr. Maurizio Pandolfi
Politecnico di Torino
Dipartimento di Ingegneria Aeronautica e
Spaziale
Corso Duca degli Abruzzi, 24
I - 10129 Torino
Italy
E-mail: pandolfi@polito.it

Prof. Dr. Jaques Periaux
Dassault Aviation
78, Quai Marcel Dassault
F-92552 St. Cloud Cedex
France
E-mail: periaux@rascasse.inria.fr

Prof. Dr. Arthur Rizzi
Department of Aeronautics
KTH Royal Institute of Technology
Teknikringen 8
S-10044 Stockholm
Sweden
E-mail: rizzi@aero.kth.se

Dr. Bernard Roux
IRPHE-IMT
Technopole de Chateau-Gombert
F-13451 Marseille Cedex 20
France
E-mail: broux@irphe.univ-mrs.fr

Foreword

This volume contains nineteen contributions of work, conducted since 1998 in the French – German Research Programme "Numerical Flow Simulation", which was initiated in 1996 by the Centre National de la Recherche Scientifiqué (CNRS) and the Deutsche Forschungsgemeinschaft (DFG).

The main purpose of this second publication on the research programme, is to give an overview over recent progress, and to make the obtained results available to the public. The reports are grouped, like those in the first publication (NNFM 66, 1998), under the four headings "Development of Solution Techniques", "Crystal Growth and Melts", "Flows of Reacting Gases" and "Turbulent Flows". All contributions to this publication were reviewed by a board consisting of T. Alziary de Roquefort (Poitiers, France), H. W. Buggisch (Karlsruhe, Germany), Th. Gallouet (Marseille, France), W. Kordulla (Göttingen, Germany), A. Lerat (Paris, France), R. Rannacher (Heidelberg, Germany), G. Warnecke (Magdeburg, Germany), and the editor. The responsibility for the contents of the reports nevertheless lies with the contributors.

E. H. Hirschel
Editor

Preface

The 8th Joint CNRS - DFG Colloquium on Numerical Flow Simulation was held in the Magnus Haus of the Deutsche Physikalische Gesellschaft, located in the historic center of Berlin, November 5 and 6, 1999. The colloquium was organized by Dr. W. Lachenmeier of the Deutsche Forschungsgemeinschaft in the frame of the French - German Research Program on Numerical Flow Simulation. This program was initiated in 1996 with twenty bilateral collaborative projects of joint French - German laboratory teams under the auspices of Professor Guy Aubert, former Directeur Général of the CNRS, and Professor Wolfgang Frühwald, former President of the DFG. French - German co-operation on numerical methods in fluid mechanics was first proposed by representatives of the Scientific Direction of the Department "Sciences de l´ Ingenieur" of CNRS and of the Reviewing Board of the DFG Priority Research Program "Flow Simulation with High-Performance Computers" in a meeting in January 1991 in Paris with the aim to strengthen both programs by collaborative efforts. In the following years funds for scientific visits and travel were allocated by the CNRS and the DFG. The co-operation grew continuously during the past ten years, resulting in the joint program mentioned above. A brief account of the developments was given in the preface of Volume 66 of the Notes on Numerical Fluid Mechanics, published by Vieweg in 1998 after the 6th CNRS - DFG Colloquium, held in Marseille in November 1997.

The Berlin Colloquium brought over sixty scientists from France and Germany to the now unified capital, just about at the tenth anniversary of the fall of the Berlin wall and of the iron curtain that cut Europe into two isolated parts for 25 years. The participants of the colloquium were greeted by the secretary of state of the Senate of the City of Berlin, Prof. Dr. I. W. Hertel, and by the representative of the President of the DFG, Prof. Dr. Ing., Dr. E. h. mult. O. Mahrenholtz.

The projects for the period 2000 – 2001 were selected by an French - German evaluation committee, consisting of Professors Thierry Alziary de Roquefort, chairman, Hans Buggisch, Thierry Gallouet, Patrick Huerre, Wilhelm Kordulla, Alain Lerat, Rolf Rannacher, and Gerald Warnecke. The selection was based on a scientific program with the 4 major axes: Transition and turbulence, combustion, convection and interfaces, and new solution techniques. The projects proposed had to be based on sound interdisciplinary competency of the three complementary disciplines in CFD: Fundamentals of fluid mechanics and physical flow modeling, applied mathematics, and the computer sciences. The goals of the program are to develop highly accurate and efficient numerical integration methods, adjusted to the architecture of the latest presently available high-performance computers. The combined effort is mainly aimed at studying the interfaces of complex physical flow problems, of aerodynamics and structural dynamics, and of flow problems in internal machines and engines. It is hoped, that thereby CFD research can be promoted on a high level and that collaboration between young scientists using CFD can markedly be intensified.

It is an indicator of the quality of the research of the joint CNRS - DFG scientific program that a large number of contributions were presented at international conferences and published in international archival journals. Moreover, a relatively large number of young scientists who previously were involved in the program, are now occupying academic positions as professors, both in France and in Germany. Among the prices awarded to members of the program were the prestigious Prix Edmond Brun de l´Académie des Sciences, the Ohio Aerospace Institute Distinguished Lecture Award, and the 42nd Ludwig Prandtl Memorial Lecture. Also, the results of the program in many cases enabled its members to obtain support from other funding agencies, including European organizations. This kind of support highly promoted the exchange and collaboration of Ph. D. students.

In the future the CNRS - DFG program will concentrate on larger but fewer collaborative projects, striving even for more intensification of interdisciplinary research than was possible until now. Future scientific and technological developments will require large-scale modeling and simulation, acceleration of advances in the basic sciences and coupling of projects of strategic importance. In Europe, integrated activities for initializing and funding interdisciplinary research are long overdue. It is noteworthy that over the last 15 years Computational Fluid Dynamics has become a pioneering scientific discipline, that, because of the required competencies mentioned earlier influenced other disciplines markedly. It must, however, also be mentioned, that the basic physical understanding of problems to be analyzed and its proper translation into mathematical and computer language will remain the key issue for constructing future solutions of scientifically important problems. There is no doubt, that the previous successful developments in this field in Europe certainly are closely linked to the simultaneous occurrence of interdisciplinary collaborative research programs in the early nineties as the Groupement de Recherche de "Mécanique des Fluides Numérique" at the CNRS in France and the Schwerpunktprogramm "Strömungssimulation mit Hochleistungsrechnern" at the DFG in Germany. It is hoped that future developments will follow in an appropriate manner.

The participants of the CNRS - DFG Program express their gratitude and thank the CNRS, in particular the Direction Scientifique des Sciences de l'Ingénieur; to be mentioned are Profs. M. Champion, P. Le Quéré, D. Vandromme, and R. Peyret. The continuous support of the DFG to the members of the program during recent years is also gratefully acknowledged, and also the sponsoring of this publication. Without Dr. W. Lachenmeier's farsighted planning and procuring the program would not have been so successful. He not only organized the Berlin Colloquium so efficiently, but with great skill he guided the program through the many difficult financial situations. We thank also Prof. E. H. Hirschel for editing this volume.

Aachen and Marseille, September 2000

    E. Krause   P. Bontoux

CONTENTS

I. DEVELOPMENT OF SOLUTION TECHNIQUES

Towards a Parallel Hybrid Highly Accurate
Navier-Stokes Solver

W. Borchers[1], S. Kräutle[1], R. Pasquetti[2], R. Rautmann[3], N. Roß[3], K. Wielage[3], C.J. Xu[4]

[1] Institut für Angewandte Mathematik I, Universität Erlangen-Nürnberg, Martensstr. 3, 91058 Erlangen, Germany

[2] Lab. J.A. Dieudonné, UMR CNRS 6621, Université de Nice-Sophia Antipolis, Parc Valrose, 06108 Nice cedex 2, France

[3] Fachbereich Mathematik/Informatik, Universität-GH Paderborn, Warburgerstr. 100, 33098 Paderborn, Germany

[4] Dpt. of Mathematics, Xiamen University, 361005 Xiamen, China (work done in the Lab. J.A. Dieudonné)

Summary

This paper presents the last progresses obtained in the development of a parallel multi-domain / multi-method Navier-Stokes solver. Three points are addressed: (i) the preconditioning of the iterative procedure which results from using a multi-domain approach, (ii) the semi-Lagrangian method to be used for the transport step and (iii) the first results obtained with finite difference/finite element - spectral Chebyshev couplings.

1   Introduction

Following our previous publication [4], we here present the last progresses of our research concerned with the development of a parallel multi-domain multi-method Navier-Stokes solver. The interest of such an approach may e.g. be found in calculations of wakes behind obstacles. In this case, due to the high aspect ratio of the computational domain, a multi-domain technique is required, at least in the direction of great length. Moreover, in order to capture some phenomena like fluid flow instabilities a highly accurate spectral method is to be prefered. However, the modelling of the obstacle is easier to achieve with a finite element approximation. In addition, special geometries give rise for a natural domain decomposition, for which hybrid solvers are used with advantage.

Let us recall in this introductive part the basic features of the hybrid solver currently developed:

- At a given resolution time, a 3-step method is used to compute the velocity field, namely the 1) transport step, the 2) diffusion step and the 3) projection step. The transport step should be handled with a characteristics method and steps 2) and 3) only require, after the time discretization, efficient multi-domain solvers of elliptic problems.
- The hybrid elliptic solver makes use of the CGBI (Conjugate Gradient Boundary Iteration) method to compute the velocity and the pressure in each sub-domain. The efficiency of this iterative procedure directly results from using efficient preconditioners.
The paper is organized as follows: Section 2 is devoted to the development of efficient preconditioners for the CGBI method. Section 3 discusses the use of the semi-Lagrangian approach instead of the characteristics method in the spectral subdomains. Finally, preliminary results obtained with finite difference/ finite element - spectral Chebyshev couplings are presented in Section 4.

2 Preconditioners for the CGBI method

As explained in [4], CGBI is a domain decomposition method to solve the resolvent equation

$$(\sigma - \Delta)\, u = f, \quad \sigma \geq 0,$$

in a domain Ω with Dirichlet or Neumann boundary conditions

$$
\begin{aligned}
u &= g_D \quad \text{on } \Gamma^D, \\
\frac{\partial u}{\partial \nu} &= g_N \quad \text{on } \Gamma^N,
\end{aligned}
$$

where $\Gamma^D \cap \Gamma^N = \emptyset$, $\Gamma^D \cup \Gamma^N = \partial\Omega$. As also explained in [4], the CGBI method splits this problem into two steps, the second one being solved iteratively. Thus, this section focusses on the second step to develope a preconditioner for CGBI.

2.1 The starting point

Let us consider a domain Ω with Lipschitz boundary[1] which is decomposed into subdomains Ω_i, $i = 1,...,p$, $\Gamma_i = \bar{\Omega}_i \cap \bar{\Omega}_{i+1}$ as suggested by Fig. 1.
Having some Neumann boundary conditions $\varphi := (\varphi_1,...,\varphi_{p-1})$ on the interfaces $\Gamma := (\Gamma_1,...,\Gamma_{p-1})$ between the subdomains, we denote by $u(\varphi) := (u_1(\varphi),...,u_p(\varphi))$ the solutions of the local problems

$$
\begin{aligned}
(\sigma - \Delta)\, u_i &= 0 \quad \text{in } \Omega_i, \\
\frac{\partial u_i}{\partial \nu_i} &= \varphi_i \quad \text{on } \Gamma_i, \\
\frac{\partial u_i}{\partial \nu_i} &= -\varphi_{i-1} \quad \text{on } \Gamma_{i-1}, \\
u_i &= 0 \quad \text{or} \quad \frac{\partial u_i}{\partial \nu_i} = 0 \quad \text{on } \partial\Omega_i \setminus (\Gamma_i \cup \Gamma_{i-1}),
\end{aligned}
$$

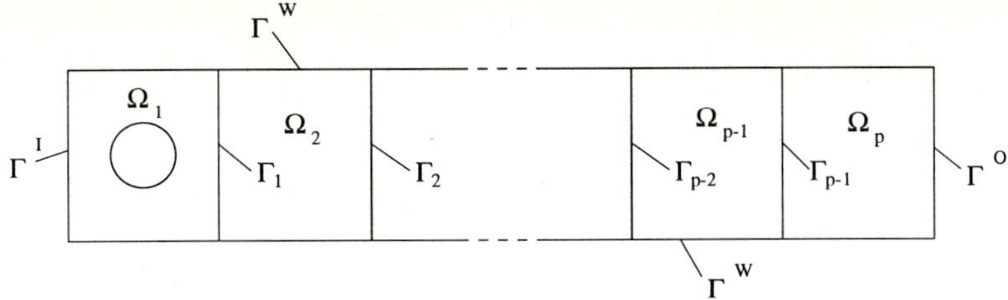

Figure 1   Schema of the computational domain.

where ν_i denotes the outward unit vector normal to Ω_i.

Let us denote the jump of the solution $[u] := ((u_2 - u_1)|_{\Gamma_1},...,(u_p - u_{p-1})|_{\Gamma_{p-1}})$ at the interfaces. Then, the operator for which we have to construct a preconditioner is the mapping

$$A : D(A) \longrightarrow R(A), \quad \varphi \longmapsto [u(\varphi)] \tag{2.1}$$

Let us mention that in the case of Dirichlet boundary conditions on $\partial\Omega$, $\sigma \geq 0$, we have[2] $R(A) = H_{00}^{1/2}(\Gamma) = \underset{i=1}{\overset{p-1}{\times}} H_{00}^{1/2}(\Gamma_i)$, $D(A) = (R(A))^* = H^{-1/2}(\Gamma)$; in the case of Neumann boundary conditions with $\sigma = 0$ we have

$$R(A) = \{\psi = (\psi_1,...,\psi_{p-1}) \in H^{1/2}(\Gamma) \mid \int_{\Gamma_i} \psi_i \, do = 0 \, \forall i\},$$

$$D(A) = \{\varphi \in (H^{1/2}(\Gamma))^* \mid \langle \varphi, \chi_{\Gamma_i} \rangle_{-1/2,1/2} = 0 \, \forall i\},$$

where χ_{Γ_i} denotes the characteristic function of Γ_i.

$u(\varphi)$ is situated in $S = \{u = (u_1,...,u_p) \mid u_i \in H^1(\Omega_i), u|_{\partial\Omega} = 0\}$ in the Dirichlet case and $S = \{u \mid u_i \in H^1(\Omega_i), \int_{\Gamma_i} [u] \, do = 0\}$ in the Neumann case with $\sigma = 0$.

Any Riesz isomorphism

$$C : R(A) \longrightarrow D(A)$$

is a suitable preconditioner [3] [4]; such a preconditioner would lead to the estimate

$$c_1 \langle \varphi, C^{-1}\varphi \rangle_{-1/2,1/2} \leq \langle \varphi, A\varphi \rangle_{-1/2,1/2} \leq c_2 \langle \varphi, C^{-1}\varphi \rangle_{-1/2,1/2} \tag{2.2}$$

with c_1, c_2 depending on the norms of the local trace and extension operators [10]. The constants do *not* depend on the length of the channel Ω and the number of subdomains p. That means that a suitable discretization of C will produce a condition number independent of p, the size of Ω and the discretization parameter N, such that $N+1$ equals the number of grid points on the interface Γ_i.

2.2   Discretization on an equidistant boundary mesh

For the sake of simplicity we may assume that the interfaces Γ_i are straight lines of length 1, which may be identified with the unit interval $(0,1)$. We are considering the square root of the negative Laplacian operator (with homogeneous Dirichlet rsp. Neumann conditions)

---

[1] As pointed out in Section 1, we may have in mind a channel-like domain, e.g. for the calculation of a flow behind an obstacle.

[2] See [12] for the definition of the space $H_{00}^{1/2}$.

$$C\varphi := ((-\Delta)^{1/2}\varphi_1,...,((-\Delta)^{1/2}\varphi_{p-1}) \tag{2.3}$$

on $R(A)$ as a preconditioner. We will follow two different approaches for the discretization of C: A spectral approach and a matrix approach.

The spectral approach

Knowing the values of a $\psi_i = [u(\varphi)]|_{\Gamma_i}$ at the equidistant grid points $y_j = jh$, $j = 0,...,N$, $h = N^{-1}$, we can use the fast Fourier transform (FFT) to perform the decomposition

$$\psi_i(y) = \sum_{k=1}^{\infty} a_{ik} \sin \pi ky \quad \text{rsp.} \quad \psi_i(y) = \sum_{k=1}^{\infty} a_{ik} \cos \pi ky$$

in the Dirichlet/Neumann case. The application of $(-\Delta)^{1/2}$ on each ψ_i just means to multiply each a_{ik} by $k\pi$. This preconditioner requires $O(N \log N)$ operations per processor. This is neglectable compared to the costs of the local solvers on the subdomains which would be at least N^2 for $N \times N$ grid points per subdomain.

The matrix approach

The preconditioning can be performed by application of a (sparse) matrix. Investigating symmetric matrices of the structure

$$\begin{pmatrix} \ddots & \ddots & \ddots & \ddots & \\ \cdots & \alpha_1 & \alpha_0 & \alpha_1 & \alpha_2 & \cdots \\ \cdots & \alpha_2 & \alpha_1 & \alpha_0 & \alpha_1 & \cdots \\ & \ddots & \ddots & \ddots & \ddots \end{pmatrix} \tag{2.4}$$

we could prove [10] that the matrix (2.4), obtained from

$$\tilde{\alpha}_j = \begin{cases} \pi + 1 - (1/2)^{\lfloor \log_2 N \rfloor}, & j = 0 \\ -\frac{\pi}{2}, & j = 1 \\ -\frac{1}{2j}, & j = 2,4,8,16,... \\ 0, & \text{else} \end{cases} \tag{2.5}$$

by odd rsp. even reflection for Dirichlet rsp. Neumann boundary conditions, leads to a bounded condition number. As the spectral approach, it requires $O(N \log N)$ operations.[3]

This matrix approach for the preconditioner can easily be generalized to non-equidistant boundary meshes, as a forthcoming paper will explain.

---

[3] The very simple tridiagonal matrix with $\alpha_0 = \frac{1}{2} + \frac{1}{N}$, $\alpha_1 = -\frac{1}{4}$, $\alpha_i = 0$ else, leads to a condition number $\kappa = O(N^{1/2})$ with only $O(N)$ operations.

2.3 Discretization on the Chebychev-Gauss-Lobatto grid

The use of Chebychev spectral solvers on rectangular subdomains leads to Chebychev-Gauss-Lobatto grids on the interfaces. In this case, the direct application of the two approaches from Section 2.2 seems to be impossible as they require the knowledge of the *equidistant* grid values of ψ_i. Interpolation from the Gauss-Lobatto mesh to an equidistant mesh, however, was observed to destroy the positivity of the preconditioners. A more elegant way is offered [10] by the interpolation theory of weighted Sobolev spaces. Let us restrict to the case of *one* interface at first. The transformation $\widetilde{\cos} := \frac{1}{2}(1+\cos)$ maps the Chebychev-Gauss-Lobatto mesh onto an *equidistant* mesh, and the interpolation theory gives a relation between the $R(A)$-norm of $\psi = \psi_i$ and $\psi \circ \widetilde{\cos}$:
Obviously,

$$
\begin{aligned}
\|\psi\|^2_{L^2(0,1)} &= \frac{1}{2}\|\psi \circ \widetilde{\cos}\|^2_{L^2_{w^{-1}}(0,\pi)}, \\
\|\nabla\psi\|^2_{L^2(0,1)} &= 2\|\nabla(\psi \circ \widetilde{\cos})\|^2_{L^2_w(0,\pi)}
\end{aligned}
\tag{2.6}
$$

where L^2_w is the weighted L^2-norm with the weigth function $w(y) = \frac{1}{\sin y}$ and $\widetilde{\cos} = \frac{1}{2}(1+\cos)$.
In case of Dirichlet boundary conditions, interpolation between the two norms (2.6) leads to the equivalence

$$
\|\psi\|^2_{R(A)} \sim \|\psi \circ \widetilde{\cos}\|_{[L^2_{w^{-1}}(0,\pi),H^1_{0,w}(0,\pi)]_{\frac{1}{2}}}.
\tag{2.7}
$$

Due to a well known result [18], the two weights on the right side of (2.7) extinguish each other, i.e. the $[L^2_{w^{-1}}(0,\pi),H^1_{0,w}(0,\pi)]_{\frac{1}{2}}$ interpolation norm is equivalent to the $[L^2(0,\pi),H^1_0(0,\pi)]_{\frac{1}{2}}$-norm. Finally, integration by parts leads to

$$
\begin{aligned}
\|\psi \circ \widetilde{\cos}\|^2_{[L^2(0,\pi),H^1_0(0,\pi)]_{\frac{1}{2}}} &= \|(-\Delta)^{1/4}(\psi \circ \widetilde{\cos})\|_{L^2(0,\pi)} \\
&= 2\,\langle \bar{C}_{GL}\psi,\psi\rangle_{-1/2,1/2},
\end{aligned}
$$

where

$$
\bar{C}_{GL}\psi := w \cdot (-\Delta)^{1/2}(\psi \circ \widetilde{\cos}) \circ \widetilde{\cos}^{-1}.
$$

So the norms generated by $(-\Delta)^{1/2}$ and $\bar{C}_{GL}$ are equivalent. The application of $\bar{C}_{GL}$ onto a ψ given at the equidistant grid points is easy because $\psi \circ \widetilde{\cos}$ is known at *equidistant* grid points. The result $\bar{C}_{GL}\psi$, again, is known at the Gauss-Lobatto points.
If there is more than one interface we obviously have to use

$$
C_{GL}\psi := (\bar{C}_{GL}\psi_1,...,\bar{C}_{GL}\psi_{p-1}).
$$

2.4 Optimization for rectangular subdomains with singular perturbations $\sigma \to \infty$, $r \to 0$

The preconditioner from Section 2.1 and its realization in Sections 2.2-2.3 leads to a condition number independent of the discretization parameter N, the number of subdomains p and the length of the channel.

If the domain Ω is rectangular (i.e. there is no obstacle in the channel) and if all the subdomains are rectangles of the same size then it is possible to calculate the eigenvalues and eigenfunctions of the operator A (at least in the Dirichlet case) explicitly. This enables us to determine the constants c_1, c_2 in (2.2). Moreover, we can optimize the preconditioner, such that the condition number is independent of σ, of the width of the channel B and finally of the aspect ratio r of the subdomains.

The operator A has the eigenvalues [10]

$$\lambda_{k,m} = \frac{2B}{\sqrt{\sigma B^2 + \pi^2 k^2}} \frac{\cosh r\sqrt{\sigma B^2 + \pi^2 k^2} - \cos\gamma_m}{\sinh r\sqrt{\sigma B^2 + \pi^2 k^2}}, \quad m=1,...,p-1, \, k=1,...,\infty \quad (2.8)$$

where $\gamma_m = \pi m/p$ in the case of Neumann b.c. on the inflow part Γ^I and the outflow part Γ^O of the boundary $\partial\Omega$ and γ_m being the roots of some easy trigonometric equation (depending on k) in the case of Dirichlet b.c. on Γ^I, Γ^O. The eigenvalues are independent of the type of b.c. on the physical part Γ^W of the boundary.

The related eigenfunctions $\varphi_{k,m}$ have, for Neumann b.c. on Γ^I, Γ^O, the shape

$$\varphi_{k,m}|_{\Gamma_i}(y) = a(y)\, b(i),$$

where $a(y) = \sin\pi ky/B$ or $a(y) = \cos\pi ky/B$ for Dirichlet or Neumann b.c. on Γ^W respectively. Furthermore, $b(i) = \sin\pi im/p$. For Dirichlet b.c. on Γ^I, Γ^O we have to replace $b(i)$ by a simple trigonometric expression depending on γ_m.

Some analysis shows that $\lambda_{k,m}$ multiplied by

$$g_k := \frac{\pi}{2B} k + \frac{\sqrt{\sigma}}{2} \tag{2.9}$$

is bounded from above and from below by positive constants independent of k, m, p, B, σ. From this and the definition of C we can conclude directly that the preconditioner[4]

$$C_\sigma := C + \sqrt{\sigma}\, id \tag{2.10}$$

(in the Gauss-Lobatto case we just have to substitute C by C_{GL}) produces a condition number κ independent of N,p,B,σ. It behaves like $O(r^{-2})$ for $r \to 0$. A direct evaluation of $\lambda_{k,m} g_k$ shows that $\kappa \leq 1.5$ for $r=1$ (i.e. square subdomains) and still $\kappa \leq 5$ for $r=0.2$. These theoretical results correspond to our numerical tests (Fig. 3) which showed e.g. an error reduction rate of one full power of 10 for square subdomains.

For very small subdomain aspect ratios r $(r < 0.1)$ the preconditioner C_σ loses efficiency (see Fig. 2). However, it can be proved [10] that every preconditioner which treats each ψ_i separately (i.e. $C(\psi_1,...,\psi_{p-1}) = (C_1(\psi_1),...,C_{p-1}(\psi_{p-1}))$) has the same asymptotic behaviour for $r \to 0$. That means that g_k has to be replaced by $g_{k,m}$ depending on m. In the case of Neumann b.c. on $\partial\Omega$ we have constructed a $g_{k,m}$ such that $\lambda_{k,m} g_{k,m}$ is bounded by constants whose ratio is also independent of r. The related preconditioner is

$$C_T = F^{-1}T\left((-\Delta)^{1/2} + \left(\frac{2}{rB} + \sqrt{\sigma}\right)id\right)T\,F \tag{2.11}$$

where F denotes a Fourier transform acting on the ψ_i with respect to i and T denotes a specific convolution operator. Evaluation of $\lambda_{k,m} g_{k,m}$ shows that now $\kappa < 1.6$ for arbitrary N, p, B, σ and r. Test runs to compare C_σ and C_T for $r \to 0$ are displayed in Fig. 4.

---

[4] For $\sigma > 0$ we may replace C_σ by $C'_\sigma := id + \frac{1}{\sqrt{\sigma}} C$ which converges to the identity operator for $\sigma \to \infty$.

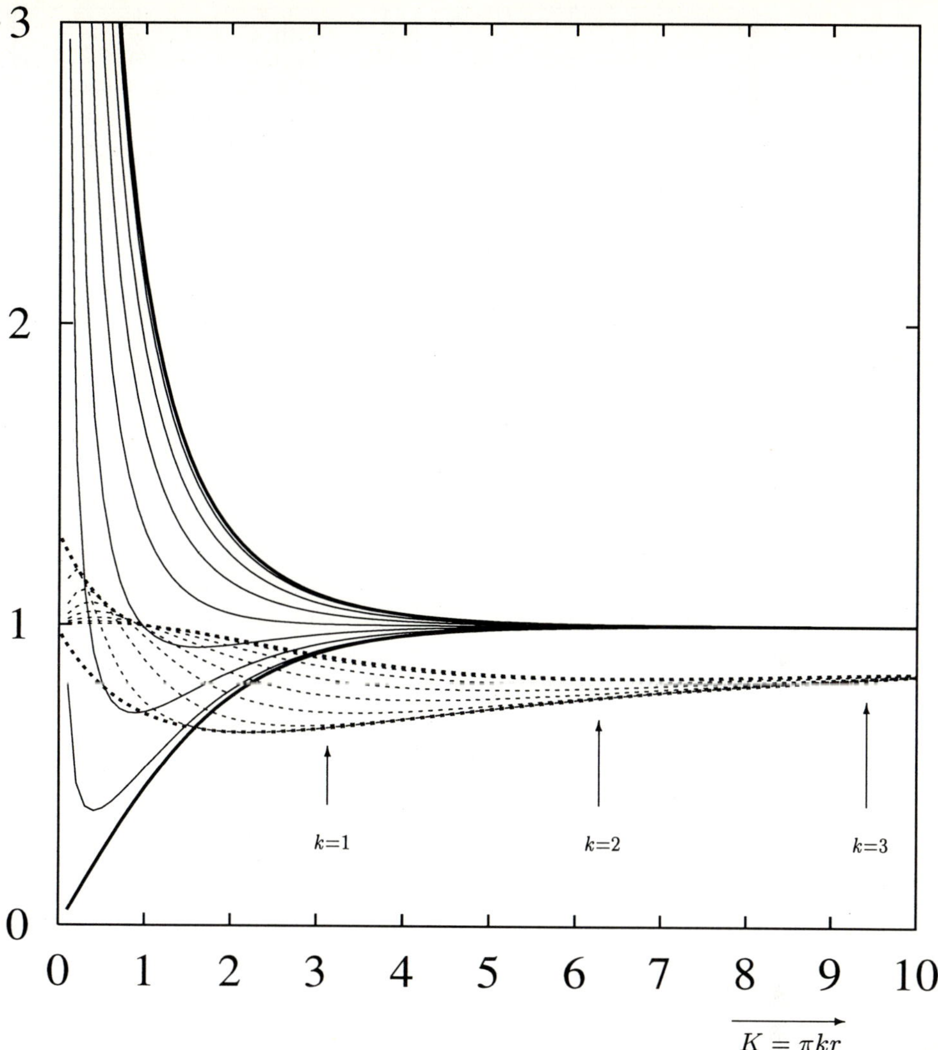

Figure 2 Visualizations of the eigenvalues for the preconditioned operator for $\sigma = 0$, $p = 8$. $K := \pi k r$ is displayed on the horizontal axis. The full lines correspond to the different values of $m = 1,...,p-1$ for the expression $\lambda_{k,m} g_k$ (see (2.8), (2.9)) and so for the preconditioner C_σ ($= C$ here). The dashed lines display the values of $\lambda_{k,m} g_{k,m}$, $m = 1,...,p-1$, when using the preconditioner (2.11). As an example, the arrows show the location of the eigenvalues for the preconditioned operators in the case $r = 1$. If r approaches zero, the eigenvalues move to the left, i.e. for (2.10), the condition number increases. If p is increasing the number and density of the curves (the eigenvalues) is affected, but not the upper and lower bound which are displayed as bold lines. For C_σ, these bounds coincide with the expressions $\lambda_{k,m} g_k$ for $m = 0$, $m = p$; for C_T the bounds have been calculated numerically. For $\sigma > 0$ we have to put $K := r\sqrt{\sigma B^2 + \pi^2 k^2}$. Even for this case, the figure shows the behaviour of the eigenvalues (under disregarding of a factor which is bounded between 1 and $\sqrt{2}$).

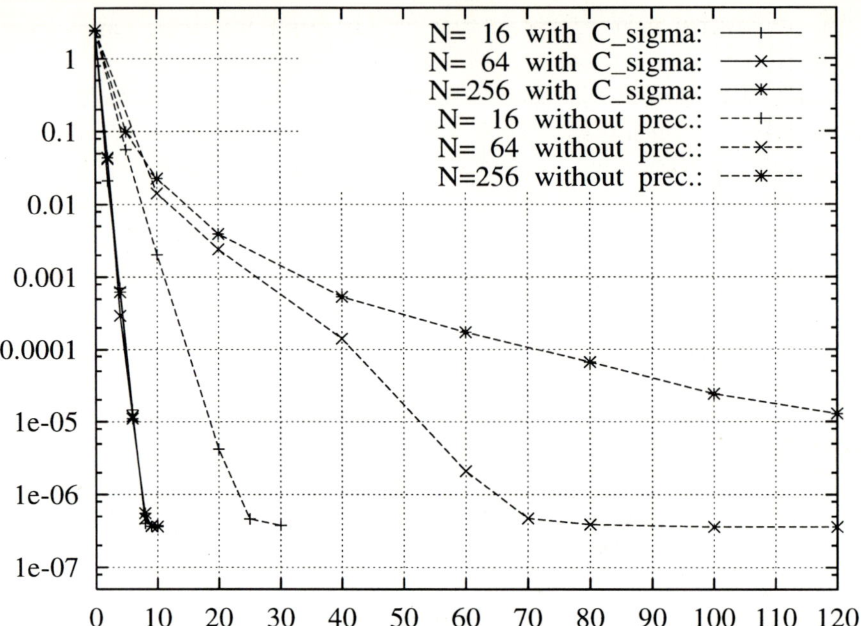

Figure 3 The error decay during the CGBI iteration. Dirichlet problem on $p=4$ spectral subdomains (with a highly oscillating CG starting vector to get a 'worst case' result), $r=1$. Full lines: using the preconditioner C_σ. Dashed lines: using no preconditioner. For C_σ the error decay rate does not depend on N.

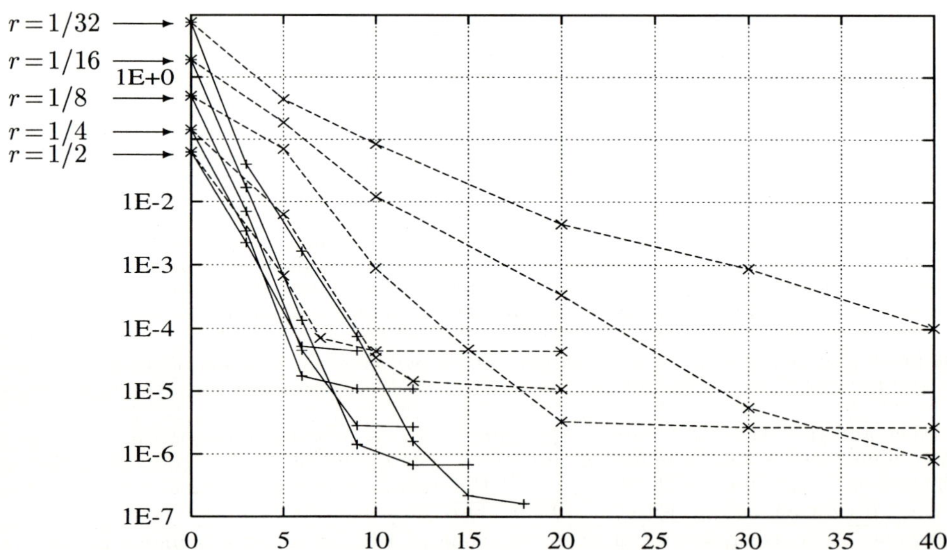

Figure 4 Error decay with preconditioner C_T (full lines) compared to C_σ (dashed lines) for $p=8$ spectral subdomains, $N=64$, for the Neumann problem; oscillating CG starting vector and subdomain aspect ratio $r = 1/2, 1/4, 1/8, 1/16, 1/32$. For C_T the error decay rate does not decrease.

3   On the advection-step in the spectral subdomains

As explained in introduction, the multi-domain multi-method Navier-Stokes solver makes use of three steps: 1) the transport-step, 2) the diffusion-step and 3) the projection-step. Concerning the transport-step, the two following questions must be answered:

1. Under which conditions is the characteristics method well adapted to solve the transport-step when spectral approximations are involved ?

2. Is the splitting transport-step / diffusion-step much more efficient than standard semi-implicit schemes ?

These two points are addressed in Sections 3.1 and 3.2 respectively.

3.1   Semi-Lagrangian approach for the transport-step

When assuming at time t_{n+1} a Q-order Backward Euler (BEQ) approximation of the material derivative of the velocity V, i.e.

$$D_t V(t_{n+1}) = \frac{\alpha_0 V^{n+1} + \alpha_1 \tilde{V}^n + \cdots \alpha_Q \tilde{V}^{n+1-Q}}{\tau} + 0\left(\tau^Q\right)$$

where τ is the time-step, α_q, $q = 0, .., Q$, a set of given coefficients and with :

$$\tilde{V}^{n+1-q} \approx V(\chi(x, t_{n+1}; t_{n+1-q}), t_{n+1-q})$$

where $\chi(x, t_{n+1}; t)$ solves the characteristics equation:

$$\frac{d\chi(x, t_{n+1}; t)}{dt} = V(\chi(x, t_{n+1}; t), t)$$

$$\chi(x, t_{n+1}; t_{n+1}) = x$$

then the transport-step consists, for each mesh-point x_k, in the calculation of the $\tilde{V}^{n+1-q}$, $q = 1, .., Q$.

The natural approach is then the characteristics method (see e.g. [1, 8, 9, 16, 5]), which needs (i) to locate all the characteristics feet and (ii) to interpolate the solution at these points.

However the standard characteristics method appears not suitable in the framework of spectral approximations because the spatial interpolants of high order (i) would be too much time consuming and (ii) might yield instabilities on Gauss-Lobatto grids.

Table 1   Stability results for the characteristics method

spatial interpol.	grid	stability cond.
linear	any	none
higher order	quasi-uniform	CFL numb. bounded
higher order	G.L.	always unstable
quadratic, modified[5] scheme	G.L.	$\tau = O(h^4_{max}) = O(h^2_{min})$

---

[5] The modified scheme uses local transformations onto an equidistant grid. Note that numerical tests seem to show that the condition $\tau = O(h^4_{max})$ may be too pessimistic.

Concerning the last point, the theoretical stability results obtained by S. Kräutle [10] on the 1D advection equation are given in Table 1. Note that quasi-uniform grids are defined such that $h_{min}/h_{max} \geq c > 0$, when $N \to \infty$ and so that Gauss-Lobatto-Chebyshev grids are not quasi-uniform, since $h_{max}/h_{min} \sim N$.

Since the characteristics method appears to be not adapted when spectral approximations are involved, an alternative method must be proposed to handle the transport-step in the "spectral sub-domains". We suggest the algorithm:

Determine the $\tilde{\boldsymbol{V}}^{n+1-q}$ by solving the Q auxiliary problems ($1 \leq q \leq Q$):

$$\left\{ \begin{array}{ll} \partial_t \boldsymbol{U} + \boldsymbol{V}.\nabla \boldsymbol{U} = 0 & t_{n+1-q} \leq t \leq t_{n+1} \\ \boldsymbol{U}(\boldsymbol{x}, t_{n+1-q}) = \boldsymbol{V}(\boldsymbol{x}, t_{n+1-q}). \end{array} \right.$$

Indeed, it is easy to show that:

$$\tilde{\boldsymbol{V}}^{n+1-q} = \boldsymbol{U}^{n+1},$$

where $\boldsymbol{U}^{n+1}$ is the numerical approximation of $\boldsymbol{U}(\boldsymbol{x}, t_{n+1})$. Such an algorithm is usually called the *semi-Lagrangian method*, which can be recasted in the so-called *operator - integration - factor splitting method* [14]. The basic idea is to transport the $\tilde{\boldsymbol{V}}^{n+1-q}$ at the mesh points. Let us remark that boundary conditions are only necessary for open flows, at the inlet part of the boundary, and that interpolation /extrapolation functions are now only needed in time (for $\boldsymbol{V}$).

In order to solve these auxiliary problems, an explicit time-scheme with large absolute stability region must be retained. The RK4 scheme appears well adapted. Interpolation /extrapolation functions are now only needed in time, for the required approximations of $\boldsymbol{V}$. Moreover, sub-time cycling is possible by using for the transport step a smaller time-step, τ/M, where M is a positive integer.

3.2   Comparisons of the semi-implicit and semi-Lagrangian methods

We want to compare :

- Semi-implicit (SI) schemes : diffusive terms are handled implicitly and convective terms explicitly (Adams-Bashforth extrapolation),

- Semi-Lagrangian (SL) schemes : based on the splitting transport-step /diffusion-step.

Accuracy and stability comparisons have been made for various problems. Hereafter we present results obtained for the 1D Burger's equation, the 2D regularized driven cavity problem and finally for a 3D driven flow in a cube. The full-length version of this study is given in [19].

1D viscous Burger's equation

Here we consider the following problem, which analytical solution is known [2]:

$$\partial_t u + u \partial_x u - 0.05\, \partial_{xx} u = 0 \qquad in\ ]0, \pi[,$$
$$u(0,t) = u(\pi,t) = 0,$$
$$u(x,0) = \sin(x).$$

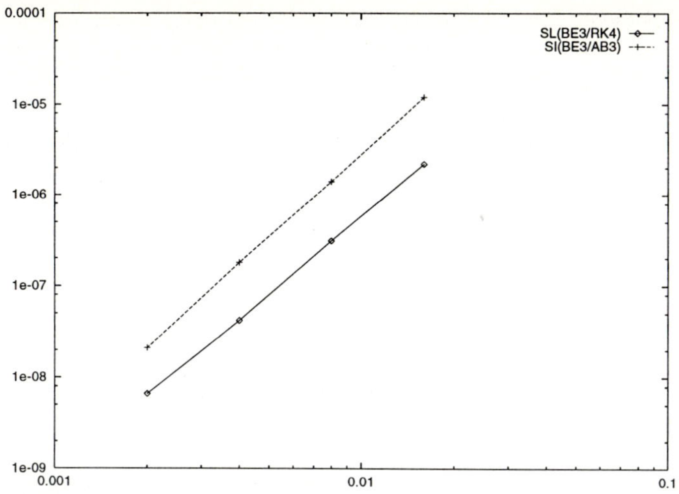

Figure 5 L^2-errors at time $t = 1$ vs the time step τ using the SL(BE3/RK4) ($\diamond$) and the SI(BE3/AB3) schemes (+).

The calculations have been made with Legendre-based spectral elements: $K = 3$ spectral elements with a polynomial approximation $N = 16$. Accuracy results are given in Fig. 5, for $Q = 3$. As may be observed, the accuracy of the semi-Lagrangian scheme is governed by the BEQ approximation. The global accuracy indeed equals $\max(Q,4)$, the RK4 time-integrator being of fourth order. Moreover, the SL scheme appears slightly more accurate than the SI one, probably due to the fact that there is for the SL scheme only one error leading term rather than two for the SI scheme (The BE3 approximation of the time-derivative and the AB3 extrapolation).

Regularized driven cavity problem

Comparisons have been made on a standard benchmark: the regularized driven cavity flow [6, 15]. The Reynolds number equals $Re = 1000$. The calculations have been made with a spectral element solver (see e.g. [13]), using $K = 100$ spectral elements and $N = 8$ for the polynomial approximation.
For the SI scheme, three different forms of the convective term have been considered: convective, conservative and skew-symmetric. Stability results are summarized in Table 2, where the critical time-steps obtained for the SL and SI schemes are presented.

Table 2 Critical time steps for the SL and SI schemes, with different treatments of the convection term (data are determined with a relative error up to 10%).

Re	SI(BE3/AB3)			SL(BE3/RK4)	
	convect.	conserv.	skew-symm.	$M = 1$	$M \neq 1$
$Re = 1000$	0.0042	0.0030	0.0036	0.015	0.16 (M=11)

Clearly, the SL scheme allows greater time-steps, especially if time sub-cycling is used.

Finally, tests have been made for a 3D flow in a cube with one homogeneous direction, using as boundary condition on one face a steady training velocity inducing a 3D unsteady flow. The calculations have been made using a Fourier-Chebyshev solver [17]. A visualization of this flow is given in Fig. 6.

Again the critical time-step is greater for the SL scheme. Nevertheless, due to the fact that a direct solver is used for solving, at each time-step, the generalized Stokes problem which results from the time-discretization, the time required to solve the transport step is here no longer negligible:

- SI(BE2/AB2), $\tau_c = 0.64\,10^{-2}$, CPU-time / time-step: $10s$
- SL(BE2/RK4), $\tau_c = 3.4\,10^{-2}$, CPU-time / time-step: $30s$ $(M = 1)$.

Consequently, in terms of CPU time the speed-up obtained by the SL scheme over the SI scheme is less convincing than in terms of critical time-steps, 1.77 and 5.31 respectively, which means that efforts are still needed to improve both the stability of the explicit scheme (see e.g. [7]) and the efficiency of the differentiation calculations.

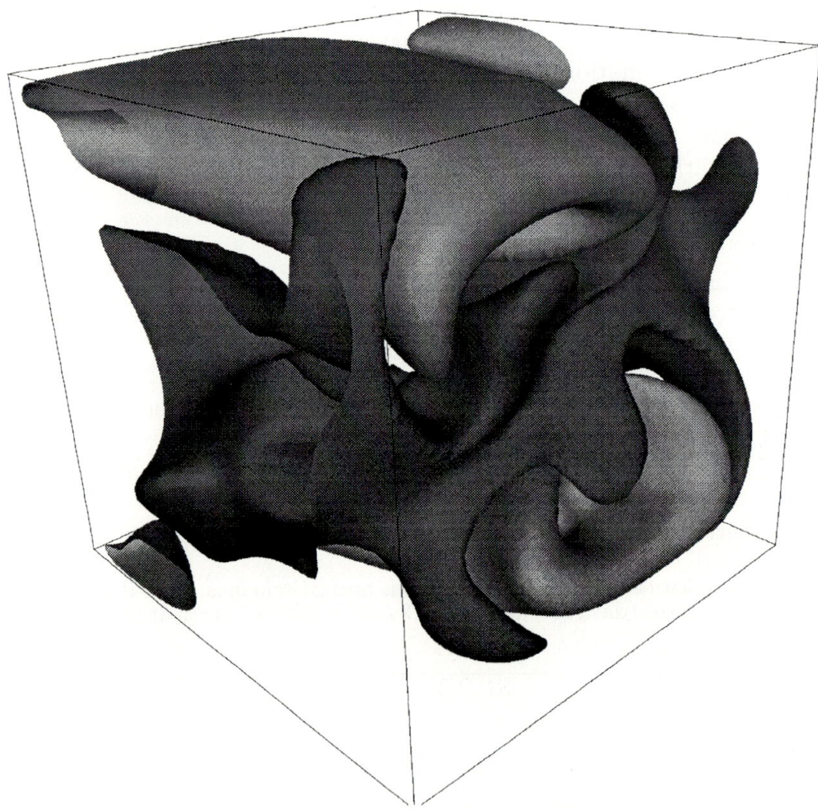

Figure 6 Isosurfaces $V_x = \pm 0.02$ of the x-component of the velocity at $t = 25$ (Chebyshev-Fourier *direct* Stokes solver, grid: 60^3)

4 Numerical results obtained with hybrid multi-domain couplings

In the following we present numerical results for finite element/finite difference - spectral Chebychev couplings including the computation of a Navier-Stokes flow.
First we consider the Poisson equation with Dirichlet and Neumann boundary conditions. In [4] we presented similar test cases, but for the FDM instead of the FEM.

4.1 The Poisson Equation with finite element and spectral Chebychev solvers

Test Configuration:

- 4 subdomains, $(N+1) \times (N+1)$ grid points on each subdomain with $N = 16, 32, 64, 128$.

- Poisson equation with the exact solution

$$u(x,y) = 56 \, K(x,y) \, (x^3 - Lx^2) \, y^2 \, (1-y)^2$$

on $\bar{\Omega} = [0,L] \times [0,1]$, $L = 4$, and homogeneous Dirichlet and Neumann boundary conditions, respectively. However, in order to solve a well posed "Neumann" problem, we assume a Dirichlet condition at the outlet.

- The CGBI is stopped when the residual δ_1 becomes stationary.

- Coupling FEM on the first subdomain $[0,1]^2$ with a Chebyshev spectral solver on the other subdomains, as shown in Fig. 7.

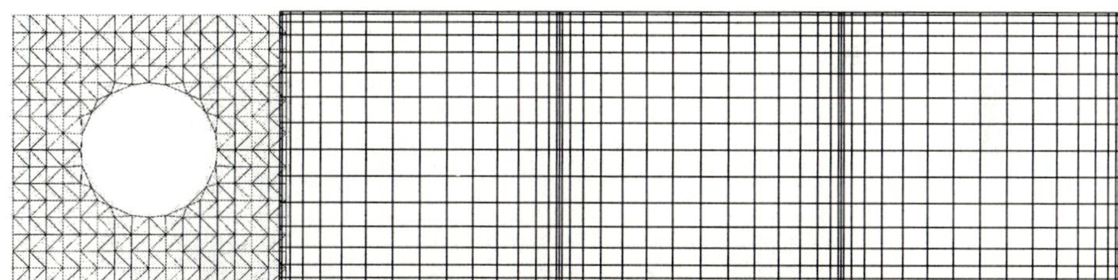

Figure 7 Mesh for the Dirichlet problem.

Fig. 8 displays the error decay for the Dirichlet and the Neumann problem. As expected, the FEM local solver determines the 2nd order convergence in N. However, the following test shows that, as in the case of FDM in [4], this effect does not destroy the spectral accuracy on subdomains far away from the first.

4.2 Error induced by a FEM-subdomain along the channel

Now we investigate the influence of the local FEM approximation on the Chebychev subdomains. We consider the following test configuration:

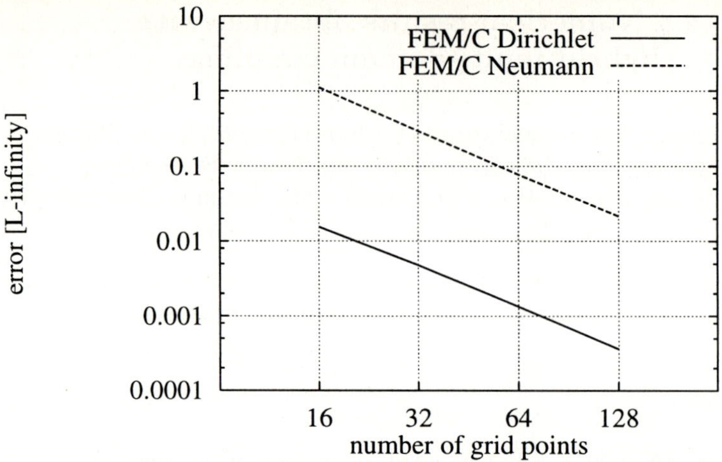

Figure 8 Error in the max norm vs N for the finite element-spectral Chebyshev solvers. For the Dirichlet problem we take $K = 0.25 - \sqrt{(0.5 - x)^2 + (0.5 - y)^2}$. For the Neumann problem we have chosen a FEM grid without an obstacle and $K = 1$.

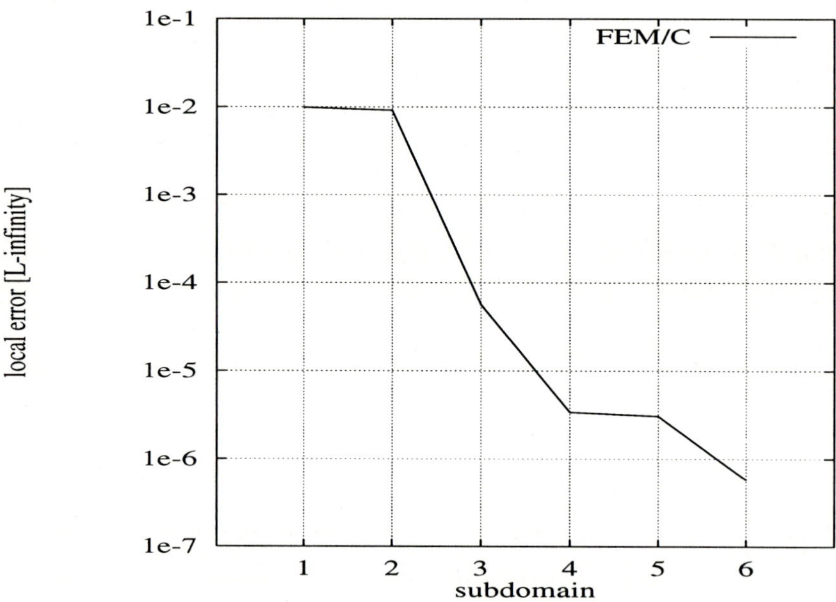

Figure 9 Error in the max-norm vs the subdomain index for the Poisson equation.

- 6 subdomains, 17×17 grid points each.

- Poisson equation with the same given exact solution like above with $L = 6$ and Dirichlet boundary conditions.

- Local Solvers: FEM on 1st subdomain, Chebyshev on the others.

Fig. 9 shows the decay of the L^∞-error along the channel for this test case. As predicted by theory, the error shows an exponential-like decay along the channel as it was already obtained with the FDM [4].

16

4.3 The Navier-Stokes equations with a finite difference - spectral Chebyshev coupling

In order to check the capabilities of the full Navier-Stokes solver, different steady and unsteady flows have been computed, using a finite difference approximation in the first subdomain and spectral Chebyshev in the others. Essentially, we have focused on jet or wake-like flows, but to preserve the simplicity of the geometry these flows have been modelled by using different velocity profiles at the inlet. Since in the present study we are essentially interested in the space approximation, the time scheme simply uses a first-order implicit backward Euler approximation and the transport step is achieved by a first-order characteristics method in all subdomains.

As an example, Fig. 10 shows the flow over a backward facing step. The height of the step is half of the channel width. The Reynolds number is 150 with respect to the maximum velocity and the step height. A Poiseuille flow is imposed at the inlet and homogeneous Neumann boundary conditions are used for the velocity at the outlet. The computational domain was decomposed into 12 square subdomains. In this example it is the C^1-singularity on the left boundary which justifies the use of the FDM in the first subdomain.

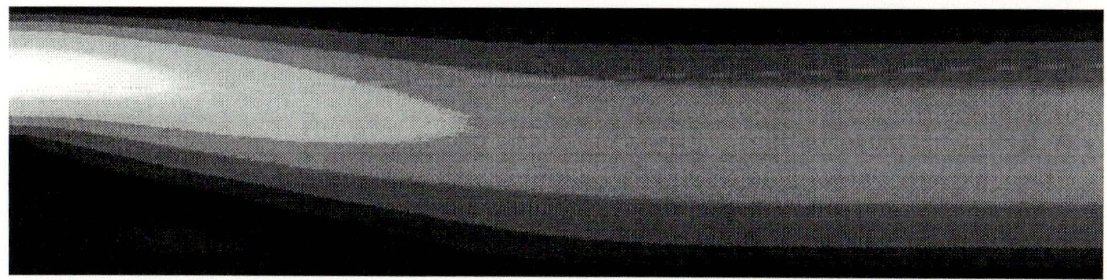

Figure 10 Visualization of the absolute value of the velocity for the Navier-Stokes flow over a backward facing step. Only the part $[0,4] \times [0,1]$ of the computational domain $[0,12] \times [0,1]$ is displayed. The recirculation zone covers about 1.6 subdomains.

References

[1] Benqué J.P., Ibler B., Keramsi A., Labadie G., *A new finite element method for Navier-Stokes equations coupled with a temperature equation*, in Proc.4th Int.Symp. on Finite Elements in flow Problems, Kawai T.(ed.), North-Holland, Amsterdam, 295-301, 1982.

[2] Benton E.R., Platzman G.W., *A table of solutions of the one-dimensional Burgers equation*, Quarterly of Applied math., 195-212, july 1972.

[3] Blazy, S., Borchers, W., Dralle, U.: *Parallelization methods for a characteristic's pressure correction scheme*, in: Flow simulation with high-performance computers II (Hrs. E.H. Hirschel), Notes on Numerical Fluid Dynamics, Vol. 38, 305-321, Vieweg, Braunschweig, Vieweg 1996

[4] W. Borchers, M.Y. Forestier, S. Kräutle, R. Pasquetti, R. Peyret, R. Rautmann, N. Roß, C. Sabbah, *A parallel hybrid highly accurate elliptic solver for viscous flow problems*, Numerical Flow Simulation I, NNFM 66 , Hirschel Ed., pp 3-24, 1998.

[5] Boukir K., Maday Y., Métivet B., *A high order characteristic method for the incompressible Navier-Stokes equations*, Comput.Methods Appl.Mech.Eng., **116**, 211-218, 1994.

[6] Bourcier M., François C., *Intégration numerique des équations de Navier-Stokes dans un domaine carré*, Rech. Aérospatiale, **131**, 23-33, 1969.

[7] Carpenter M.H., Kennedy C.A., *A fourth-order 2N-storage Runge-Kutta scheme*, NASA TM 109112, 1994.

[8] Douglas J., Russell T.F., *Numerical method for convection dominated diffusion problems based on combining the method of characteristics with finite element or finite difference procedures*, SIAM J.Numer.Anal., **19**(5), 871-885, 1982.

[9] Ewing R.E., Russell T.F., *Multistep Galerkin methods along characteristics for convection-diffusion problems*, in Advances in Computer Methods for Partial Differential Equations-IV, Vichnevetsky R. and Stepleman R.S.(eds.), IMACS, Rutgers University, New Brunswisk, New Jersey, 28-36, 1981.

[10] Kräutle S., *A parallel Navier-Stokes solver based on CGBI and the characteristics method*, Ph.D thesis, to be published.

[11] Lax, P.D.: *On the Stability of Difference Approximations to Solutions of Hyperbolic Equations With Variable Coefficients*, Comm. Pure Appl. Math., Vol.XIV, p.497-520, 1961

[12] Lions, J.L., Magenes, E.: *Non-homogeneous boundary value problems and applications I*, Springer Verlag, Berlin-Heidelberg-New York, 1982

[13] Maday Y., Patera A.T., *Spectral element methods for the Navier-Stokes equations*, In A.K.Noor (ed.), State-of-the-art surveys in computational machanics, ASME, New York, 71-143, 1988.

[14] Maday Y., Patera A.T., Ronquist E.M., *An operator-integration-factor splitting method for time-dependent problems: application to incompressible fluid flow*, J.of Scientific Computing, **5**(4), 263-292, 1990.

[15] Peyret R., Taylor T.D., *Computational methods for Fluid Flow*, Springer, New York, 1983.

[16] Pironneau O., *On the transport-diffusion algorithm and its applications to the Navier-Stokes equations*, Numer.Math., **38**, 309-332, 1982.

[17] Sabbah C., Pasquetti R., *A divergence-free multi-domain spectral solver of the Navier-Stokes equations in geometries of high aspect ratio*, J. of Comp. Phys.,**139**, 359-379, 1998.

[18] Triebel, H.: *Interpolation Theory, Function Spaces, Differential Operators*, North-Holland Publishing Company, Amsterdam New York Oxford, 1978

[19] Xu C.J., Pasquetti R., *On the efficiency of semi-implicit and semi-Lagrangian spectral methods for the calculation of incompressible flows*, Int. J. for Num. Methods in Fluids, to be published.

Self-Organizing Hybrid Cartesian Grid Generation and Solutions for Arbitrary Geometries

F. DEISTER[1], D. ROCHER[2], E.H. HIRSCHEL[1] and F. MONNOYER[2]

[1] Universität Stuttgart, I.A.G., Pfaffenwaldring 21, D–70550 Stuttgart, Germany

[2] Université de Valenciennes, L.M.E., Le Mont Houy – B.P. 311, F–59304 Valenciennes Cedex, France

Summary

An automatic adaptive hybrid Cartesian grid generator is presented together with solutions of flows around complicated two- and three-dimensional geometries. The primary computational grid is a Cartesian grid, which is generated based on an octree-data structure. A secondary grid may be added for resolving the boundary layer region of viscous flow around the solid body. It consists of quasi triangular-prismatic cells generated by marching the body surface triangulation outward. A modified unstructured flow solver is applied for the compressible Euler and Navier Stokes equations. For turbulent flows, the eddy viscosity is calculated using the Spalart-Allmaras turbulence model. The hybrid Cartesian grid is improved due to manifold adaptation features, which are optimized in order to avoid any user interaction. Therefore, the final grid depends only on the particular geometry and flow structure (self-organized grid).

1   Introduction

Grid generation is still one of the bottlenecks in application-oriented numerical fluid dynamics. The geometrical complexity of a realistic configuration makes lengthly and costly grid generation processes necessary, [1]. Hybrid Cartesian grid methods appear to have the potential to overcome this. In this paper a completely self-organizing hybrid Cartesian grid/solution system is presented, that works fully automatically. The CPU-time on a R10000 processor - for the initial grid generation around a complicated three-dimensional geometry - is reduced to a few minutes. Details of the hybrid Cartesian grid generation are provided in [2], [3], [4], [5], [6] and [7].

The grid/solution system distinguishes itself by the attribute of self-organization: starting from the surface triangulation of the body, the system generates the initial hybrid Cartesian grid by itself. After the flow solution is obtained, the system improves the hybrid grid accordingly. This procedure is repeated until all flow characteristics are sufficiently resolved. The cycle runs fully automatically without any user interventions: it depends only on the body geometry and the flow structure. In this sense, the grid is organized by the flow itself. In order to achieve this goal, the adaptation features are manifold.

This report is structured in the following way. Chapter 2 presents the generation of the hybrid Cartesian grid. The flow solvers are described briefly in Chapter 3. In Chapter 4 the self-organization of the hybrid grid is explained in detail. Euler and turbulent Navier-Stokes applications for two and three dimensions prove the validity of the concept, Chapter 5. A short conclusion together with a view to future work closes the report.

2 Generation of the Initial Hybrid Cartesian Grid

The surface of the geometry is described in form of a triangulation, which is adapted to the local curvature radius. It requires the attribute of water-tightness. For the adaptation of the triangulation the CAD-description of the body surface is required. The inviscid part of the flow field is resolved by the Cartesian grid. For viscous computations, it is not possible to resolve the boundary-layer and shear-layer regions with Cartesian cells efficiently. Therefore, the viscous regions of the flow field are discretized using the quasi triangular-prismatic grid.

2.1 Cartesian Grid Generation [2], [3]

The Cartesian grid bases on an octree-data structure providing a global data structure, which describes the connectivity between the Cartesian cells. The cells of the octree represent the Cartesian cells. Solid bodies merely blank out areas of the background Cartesian grid. For the case of a quasi-prismatic grid near the body surface, the outer surface of the quasi-prismatic grid represents the input triangulation for the Cartesian grid procedure. Thus the entire quasi-prismatic grid including the solid body is cut out of the Cartesian background grid rather than only the solid body, [4]. As a result arbitrarily shaped cut-cells arise around the solid body or the quasi-prismatic grid respectively. The shape of the cut-cells is as arbitrary as the number of cut-cells, in which a Cartesian cell is divided. It must be noted that the triangulated surface is cut out exactly. However, only the Cartesian cells or rather parts of them (cut-cells) inside the flow domain are taken into account for flow computations: they represent directly the control volumes for flow balances.

The size of the cut-cells may become arbitrarily small. In order to avoid numerical instabilities of the flow solver, these small cut-cells are merged in appropriate neighbour cells. The data-structure of the finished grid is face-based: every face inside the flow domain is related to both adjacent cells.

Two-dimensional grid data is extracted from a three-dimensional grid, [4]. The idea consists in first generating a three-dimensional grid and then in taking only the grid data of the cells located at the symmetry plane. Therefore the two-dimensional object needs a three-dimensional representation for the grid generation. It is obtained by connecting two object profiles with a simple triangulation. The first profile has to be located at the symmetry plane, whereas the second profile is shifted parallel into the flow field and is triangulated.

2.2 Quasi-Prismatic Grid Generation [4]

The quasi-prismatic grid for viscous flow simulations is generated by marching the body surface triangulation outward. The marching direction is given by the surface normals. They are calculated at each triangle vertex applying the iterative procedure of Karman [8] in a modified form, [4]. The height of the initial quasi-prismatic grid is simply predicted by the Blasius relation (laminar flow) and / or the $\frac{1}{7}$-power relation (turbulent flow) for the boundary layer thickness, [4]. The height must be locally reduced iteratively in order to prevent grid crossing in concave regions producing negative quasi-prismatic cell volumes. In addition, the height of the quasi-prismatic grid is to be reduced iteratively locally, if two quasi-prismatic cells intersect each other. This might be the case for two bodies located closely side by side.

After the computation of one prism for each surface triangle, each prism is subdivided in normal direction into a predefined number of quasi-prismatic cells. The data structure of the quasi-prismatic cells is face-based and is therefore identical to that of the Cartesian cells: the same unstructured flow solver is applied to the Cartesian and quasi-prismatic grids.

An appropriate resolution of the shear layers emanating from sharp trailing edges is important for accurate drag computations. In the presented approach the surface-triangular mesh of an airfoil or wing encloses part of the skeleton surface of the shear layer (imaginary Kutta panel, [4]), which leaves the sharp trailing edge. This extended triangle mesh is used for the quasi-prismatic grid generation.

3   The Flow Solvers

Two different flow solvers are applied for flow simulations. Recently, the French partner developed a **new Cartesian flow solver** (*NCFS*), whose internal data-structure is identical with the octree data-structure of the Cartesian grid, [5]: The cell connectivity, flux balance and gradient calculation base on the octree. It solves the compressible Euler equations with a cell-centered finite volume approach, which represents a second-order extension of Godunov's scheme (MUSCL-approach). After the piecewise linear reconstruction of the flow variables at the centers of the cell faces, the convective fluxes are discretized following the schemes of Roe, [9], and Steger-Warming, [10]. The space semi-discretized equations are integrated in time using the four-stage scheme of Tai, [11].

The flow solver of the German partner is a **modified unstructured flow solver** (*MUFS*) with a face-based data structure. The compressible Euler and Navier-Stokes equations are solved with a cell-centered finite volume solver, which is of Jameson type, [12]. In order to achieve numerical stability, a blend of second and fourth differences in the flow variables is added to the convective fluxes. The integration to steady state is done with a fully explicit five-stage hybrid time stepping scheme of Runge-Kutta type. Convergence to steady state is accelerated by using local time stepping, implicit residual smoothing and enthalpy damping (Euler computations). The discretization of the viscous fluxes is described in [4]. Turbulence is taken into account by calculation of the eddy viscosity, for which the turbulence modell of Spalart-Allmaras is applied, [13]. The flow solver takes benefit of the simplified geometry of the Cartesian cells. It is worth mentioning, that the solver keeps the attribute of full vectorization. This is achieved by sorting all cell faces.

4   Self-Organization of the Hybrid Cartesian Grid

The initial hybrid Cartesian grid is only of rough quality. Besides, the adaptation of only the volume grid does not suffice. The adjustment of the solid body discretization to the adjacent volume cells is as important. Several solution adaptation loops are applied until all flow characteristics are resolved accurately.

4.1   Adaptation of the Surface Triangular Mesh [6], [7]

A surface triangle is refined isotropically by splitting it into four smaller triangles. This is done by subdivision of the three triangle edges. Because hanging nodes are not allowed, the remaining triangles are split anisotropically into two sub-triangles, which hold a refined edge, [6]. If an anisotropic refined triangle is flagged for further refinement, firstly it must be coarsened and afterwards be refined isotropically in order to retain the regular triangulation. For the computation of new surface points bi-cubic B-splines are applied, which are generated from the CAD-description of the body surface. The hierarchical quadtree structure is applied for the triangle adaptation: Refinement or coarsening of a triangle is identical with adding or deleting a local branch in the quadtree.

If only the Cartesian grid is applied (inviscid computation), then the surface triangulation is adapted in order to achieve a comparable resolution of the surface triangles and the surrounding Cartesian cells. For an over-resolved body triangulation many cut-faces have to be calculated unnecessarily. If the body discretization is too coarse in comparison to the surrounding Cartesian cells, oscillations in the pressure distribution will occur. This happens, because the information of the local body curvature gets lost for the single cells: in fact the body appears as many lined up flat plates. A triangle is flagged for refinement, if its characteristic length L_{tri} is larger than the averaged edge length of the surrounding Cartesian cells $\bar{L}_{cart}$:

$$L_{tri} > \lambda_{refine} \cdot \bar{L}_{cart} \ . \tag{1}$$

The criterion for coarsening consists of:

$$L_{tri} < \lambda_{coarsen} \cdot \bar{L}_{cart} \ . \tag{2}$$

The characteristic length of a triangle is the square root of its area. The threshold values applied in this paper are set to $\lambda_{refine} = 1.4$ and $\lambda_{coarsen} = 0.5$.

For computations with the quasi-prismatic grid, prismatic cells are flagged for adaptation dependent on the flow solution. Because quasi-prismatic cells are refined or coarsened only in piles (Chapter 4.3), the underlying triangulation has to be adapted correspondingly. The refinement criterion of equation 1 is now applied in order to ensure a smooth transition between the quasi-prismatic and Cartesian grids. If the outer triangle is smaller than the surrounding Cartesian cell, then this time the triangle is not coarsened. Instead, the large Cartesian cell is flagged for refinement, if the condition holds:

$$L_{tri} < \lambda_{coarsen} \cdot L_{cart} \ , \tag{3}$$

with the cell length L_{cart} of the current Cartesian cell. Furthermore, the ratio between the characteristic triangle length and the height of the outmost quasi-prismatic cell of the pile, h, is restricted. If the ratio exceeds the threshold value $\lambda_{h,coarsen}$, then the triangle is flagged for refinement, for example $\lambda_{h,coarsen} = 4$:

$$\frac{L_{tri}}{h} > \lambda_{h,coarsen} \ . \tag{4}$$

4.2　Isotropic Adaptation of the Cartesian Grid [2], [3]

The isotropic refinement or coarsening of the Cartesian cells is identical with adding or deleting a local branch in the octree. This means for refinement, that a Cartesian cell is subdivided into eight smaller cells. Vise versa, eight octcells are merged together for cell coarsening.

The adaptation criterion applied in this paper is similiar to those of [14]. It consists of three parameters: divergence and curl of velocity and the strength of the numerical entropy. The divergence of velocity is used to find shock waves and the curl of velocity to locate shear layers. High values of entropy tend to indicate grid areas that are simply under-resolved. For each cell, weighted forms of the divergence, τ_d, and the curl of velocity, τ_c, and the strength of the numeric entropy, τ_e, are computed:

$$\tau_{d_i} = |\nabla \cdot \vec{v_i}| \, L_i^{\frac{3}{2}} \; , \quad \tau_{c_i} = |\nabla \times \vec{v_i}| \, L_i^{\frac{3}{2}} \; , \quad \tau_{e_i} = |\nabla p_i - a_i^{\,2} \nabla \rho_i| \, L_i^{\frac{3}{2}} \; . \tag{5}$$

L_i is the length-scale for each cell. The length-scale weight is used to find weaker features, which are in a coarser area of the grid. This allows the weaker features to be refined when the stronger features have been resolved, [14]. In order to set a threshold for refinement and coarsening, the standard deviation about zero is computed for each parameter of the criterion, with n the number of cells in the flow field:

$$\sigma_d = \sqrt{\frac{\sum_{i=1}^{n} \tau_{d_i}^{\,2}}{n}} \; , \quad \sigma_c = \sqrt{\frac{\sum_{i=1}^{n} \tau_{c_i}^{\,2}}{n}} \; , \quad \sigma_e = \sqrt{\frac{\sum_{i=1}^{n} \tau_{e_i}^{\,2}}{n}} \; . \tag{6}$$

A cell is flagged for refinement, if the following condition holds:

$$\tau_{d_i} > \lambda_{d_{refine}} \cdot \sigma_d \;\; OR \;\; \tau_{c_i} > \lambda_{c_{refine}} \cdot \sigma_c \;\; OR \;\; \tau_{e_i} > \lambda_{e_{refine}} \cdot \sigma_e \; . \tag{7}$$

On the other hand, a cell is flagged for coarsening, if the condition is satisfied:

$$\tau_{d_i} < \lambda_{d_{coarsen}} \cdot \sigma_d \;\; AND \;\; \tau_{c_i} < \lambda_{c_{coarsen}} \cdot \sigma_c \;\; AND \;\; \tau_{e_i} < \lambda_{e_{coarsen}} \cdot \sigma_e \; . \tag{8}$$

The threshold values are usually set to $\lambda_{x_{refine}} = 1.0$ and $\lambda_{x_{coarsen}} = 0.1$. But these values can be changed by the user to encourage or discourage adaptation around specific types of features, for example shock waves or shear layers.

4.3 Adaptation of the Quasi-Prismatic Grid [6], [7]

In contrast to the Cartesian cells, the quasi-prismatic cells are only refined or coarsened in piles and in directions parallel to the body surface. In normal direction the number of quasi-prismatic layers keeps constant, but the height of the quasi-prismatic grid is adapted locally to the computed boundary layer thickness. It is to emphasize, that the quasi-prismatic grid covers really only the viscous part of the flow domain.

Adaptation in Tangential Directions In tangential directions, the adaptation of the quasi-prismatic grid is reduced to that of the underlying triangle mesh because of the pile constraint. The advantage of this approach is the ability of geometric adaptation of the body surface using the CAD surface description without producing invalid cells near the body.

Due to the highly anisotropic nature of the flow in boundary layers and wakes, the parameters of the adaptation criterion have to be modified, [15]. For shock detection the difference of maximum and minimum pressure values at a vertex and its neighbours divided by the maximum pressure value is taken: τ_p. Because the quasi-prismatic cells are only adapted in tangential directions, the quotient is weighted with the tangential length scale L_t of the cell. In order to trace vortices, the same sensor τ_c is used as for the Cartesian cells, but now it is weighted with the normal length scale L_n of the cell. This is necessary in order to remove the highly stretched cells near the wall and wake center line from the computation of the sensor, because these cells hold very high values of boundary layer curl in comparison to the curl of vortices.

$$\tau_{p_i} = \frac{p_{max_i} - p_{min_i}}{p_{max_i}} L_{t_i}^{\frac{1}{2}} \; , \quad \tau_{c_i} = |\nabla \times \vec{v_i}| \, L_{n_i}^{\frac{3}{2}} \; . \tag{9}$$

Again, the standard deviation about zero is chosen as a threshold for refinement and coarsening. Like for Cartesian cells (equations 7 and 8), a quasi-prismatic cell is flagged for refinement, if one refinement parameter is satisfied. For coarsening both constraints must be fulfilled. A pile of quasi-prismatic cells is refined, if the number of flagged constituents, N_{refine}, is larger than a threshold value dependent on the number of cells in the pile, N_{pile}.

$$N_{refine} > \lambda_{refine} \cdot N_{pile} \; . \tag{10}$$

The same relation is applied for coarsening.

$$N_{coarsen} > \lambda_{coarsen} \cdot N_{pile} \; . \tag{11}$$

In this paper, the threshold ratios are set to $\lambda_{refine} = 0.3$ and $\lambda_{coarsen} = 0.9$.

Adaptation in Normal Direction   The local height of the quasi-prismatic grid is adapted to the computed boundary layer thickness. The viscous height is calculated following the approach of [16]: based on the Navier-Stokes solution, the viscous layer is investigated starting at the wall. Therefore, the decision is detected inside the viscous layer. For both, laminar and turbulent boundary layers, the height is evaluated using the diagnostic function F

$$F = Y^a \left[\frac{dU}{dY} \right]^b \; , \tag{12}$$

resulting in the boundary layer thickness $\delta = \varepsilon Y_{max}$. Here, Y_{max} is the wall distance, for which $F = F_{max}$. The constants for a turbulent boundary layer are evaluated from Coles velocity profiles and results in: $a = 1$, $b = 1$ and $\varepsilon = 1.936$. For laminar boundary layers, the constants are determined applying quasi- similiar solutions for compressible, laminar boundary layers including heat transfer effects: $a = 3.9$, $b = 1$ and $\varepsilon = 1.294$.

The second adaptation concerning the quasi-prismatic grid in normal direction consists in adjusting the height of the first cell above the body to a predefined value of the dimensionless wall distance $y+$. After changing the first layer height, the remaining cell heights are recalculated in order to achieve a smooth height distribution in normal direction.

Adaptation of the Kutta panel   The triangular grid in the near wake behind a sharp trailing edge (imaginary Kutta panel, [4]) is adapted to the skeleton surface of the shear layer, which leaves the sharp trailing edge. Figure 1 shows the impact of grid adaptation for the wake region behind the sharp trailing edge of an airfoil. Streamlines are represented by arrows. The quasi-prismatic grid is refined in tangential direction. Especially at the upper side of the airfoil, the quasi-prismatic height is increased in order to cover the entire boundary layer. Finally, the center line of the wake grid is adapted to the streamline emanating from the sharp trailing edge.

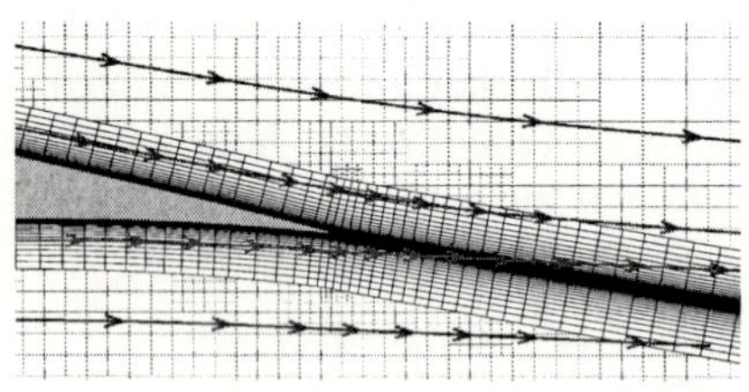

(a) Initial hybrid Cartesian grid.
(b) Adapted hybrid Cartesian grid.

Figure 1   Adaptation of the wake quasi-prismatic grid.

5   Sample Computations

It turned out that the **newly** developed Cartesian flow solver (*NCFS*), [5], is by far not as efficient as the **modified unstructured flow** solver (*MUFS*). On the one hand memory requirements are much higher, because for every cell all surrounding neighbours are stored explicitly. The identification of neighbouring cells using the octree is too time consuming during flow computation. And on the other hand the new solver cannot be vectorized completely for vector machines. With regard to quasi-prismatic grids, a second data-structure has to be implemented in the new flow solver, which differs from the octree data-structure completely. Problems concerning consistent discretization using different data structures may arise. In comparison, the unstructured flow solver does not distinguish between Cartesian and quasi-prismatic cells because of the face-based data-structure. Therefore, only the first two-dimensional sample for the Euler flow around the NACA 0012 airfoil is computed with the *NCFS*. The remaining samples are all obtained by applying the *MUFS*, [4]

Especially, the Euler flow around the Gulfstream Jet reveals the strength of the self-organizing system for the computation of highly complicated flows around complex geometries. The last flow computation shows the turbulent flow around the RAE-2822 airfoil (two dimensions) to evidence the manifold adaptations of the quasi-prismatic grid.

5.1   NACA 0012 Airfoil ($M_\infty = 0.85$, $\alpha = 1.0°$), *NCFS*

The inviscid flow around the NACA 0012 airfoil is computed applying the *NCFS*. The airfoil contour is discretized with 230 points. Figure 2(a) shows the initial Cartesian grid, which contains 1267 cells, of which 276 cells are intersected by the surface. After six adaptation steps, the grid is refined strongly in the vicinity of the two shocks and of the shear layer emanating from the sharp trailing edge. Finally, it consists of 14972 cells, whereby 322 cells are cut-cells (Figure 2(b)).

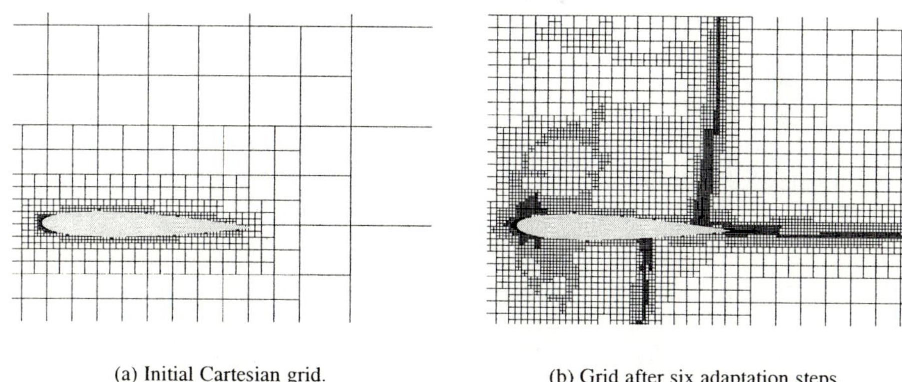

(a) Initial Cartesian grid.   (b) Grid after six adaptation steps.

Figure 2 Cartesian grid for NACA 0012 airfoil ($M_\infty = 0.85$, $\alpha = 1.0°$).

Both shocks are resolved well as the Mach number contours of Figure 3(a) show.

The pressure coefficient distribution is plotted in Figure 3(b) together with the computational data from [17]. Here, the present result is drawn as solid line, whereas the reference data is plotted as scatter points. Somehow they do not coincide very well. In comparison the pressure coefficient distribution calculated on the initial Cartesian grid is drawn as dotted line. Finally, the lift and drag coefficients result in $C_L = 0.3682$ and $C_D = 0.0716$

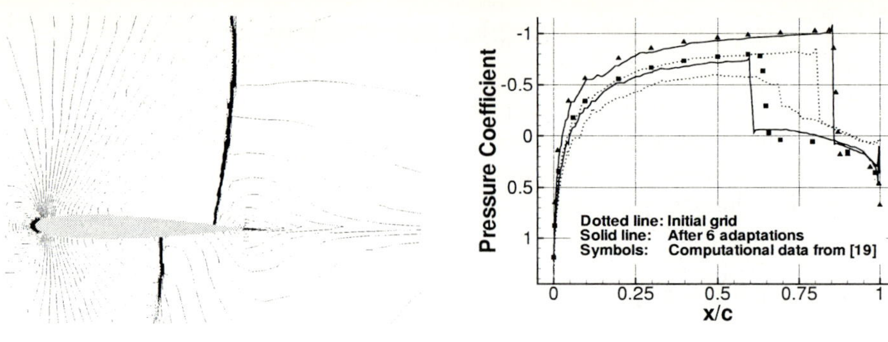

(a) Mach number contour.   (b) Pressure coefficient distribution.

Figure 3 NACA 0012 airfoil after six adaptation steps ($M_\infty = 0.85$, $\alpha = 1.0°$).

5.2 Generic Gulfstream Jet ($M_\infty = 0.85$, $\alpha = 3.0°$), *MUFS*

The Euler simulation of the flow around the Gulfstream Jet with the **modified unstructured flow solver** (*MUFS*) demonstrates the application of the self-organizing Cartesian grid approach to highly complex geometries. This configuration includes winglets and open nacelles. The computation is performed with free stream conditions at a Mach number of $M_\infty = 0.85$ and an angle of attack $\alpha = 3.0°$. For the generation of the initial Cartesian grid the Gulfstream Jet is discretized with 55249 surface triangles. Figure 4 shows the initial Cartesian grid, which contains 1191740 Cartesian cells. The number of surface intersected cells amounts to 88961 cells. The initial Cartesian grid is already refined automatically in the near wake region behind the sharp trailing edges and the nacelles. The CPU-time for the generation of the initial Cartesian grid takes 199 seconds on a workstation with a R10000 processor. This includes the bi-cubic B-spline generation also.

The initial Cartesian grid is adapted three times in order to resolve the various shocks and shear layers accurately. Finally, the adapted grid consists of 4982050 Cartesian cells, of which 322600 cells are cut by the airplane surface. The number of surface triangles increases up to 235448 triangles. For all adaptation stages, Table 1 states the number of entities.

Table 1 Adaptation history for Gulfstream Jet ($M_\infty = 0.85$, $\alpha = 3.0°$).

ADAPTATION STEP	NUMBER CUT-CELLS	NUMBER CELLS	NUMBER TRIANGLES	C_{lift}	C_{drag} $\times 10$	C_{mom} $\times 10^{-1}$
0	88961	1191740	55249	0.396	0.603	−0.113
1	132350	1372226	87683	0.414	0.507	−0.118
2	195982	2414673	139920	0.419	0.457	−0.119
3	322600	4982050	235448	0.419	0.436	−0.120

Figure 5 presents the highly adapted final Cartesian grid. Both, the Cartesian cells and the surface triangles are strongly refined in the region of the shocks and shear layers. The small picture in the upper left corner of Figure 5 magnifies the wing tip region: even the shock formation on the winglet is traced by the grid adaptation. Moreover, the Cartesian cells are refined strongly in the region of the exhaust jet.

Figure 4 Initial Cartesian grid for Gulfstream Jet ($M_\infty = 0.85$, $\alpha = 3.0°$).

Figure 5 Cartesian grid for Gulfstream Jet after three adaptation steps ($M_\infty = 0.85$, $\alpha = 3.0°$).

The corresponding Mach number contours are shown in Figure 6. Here, the typical rhombic pattern of the exhaust jet is due to the modelling as an underexpanded jet: the pressure at the exhaust is specified as 1.5 times the free stream pressure. In addition, the Mach number at the exhaust is taken as 1.2 and the inner energy as five times the free stream condition.

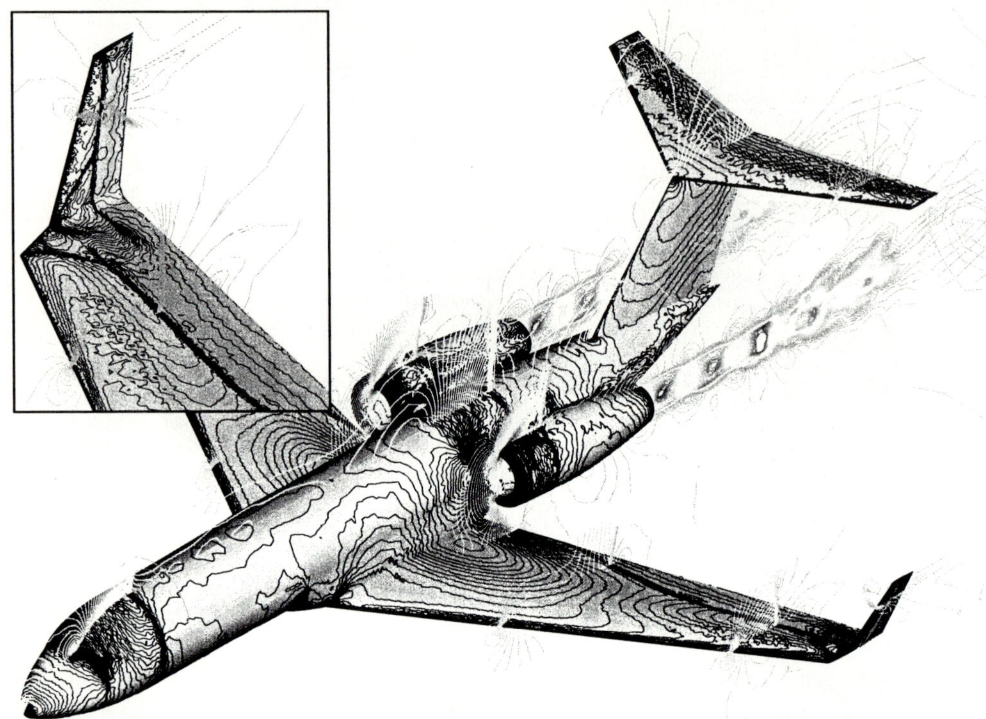

Figure 6 Mach number contours for Gulfstream Jet after three adaptation steps ($M_\infty = 0.85$, $\alpha = 3.0°$).

The nacelle region of the Gulfstream Jet is subject of Figure 7 in a magnified resolution. An oblique shock and a normal shock form on the upper nacelle surface, which join to become a single shock wave at some distance above the surface. Also a shock arises between the nacelle and the trailing edge of the wing. Because of the strong refinement of the Cartesian grid along all shocks, they are sharply resolved. Besides, at the edge of the exhaust jet and in the region of the jet expansion waves the Cartesian grid is refined. The location of the Mach disc is also visible in the Cartesian grid due to the strong adaptation.

Table 1 contains the aerodynamic coefficients for the Gulfstream Jet. Especially, the computed drag coefficient is reduced due to grid adaptation.

The shear layer at the edge of the exhaust jet is much stronger than the shear layers behind the wings. Therefore, the standard deviation of the curl adaptation sensor is dominated by the cells of the exhaust jet region. As a consequence the weaker shear layers behind the wings are not detected by the adaptation criterion. In order to avoid this, a separate adaptation domain with separate standard deviation is assigned to the region around the exhaust jet.

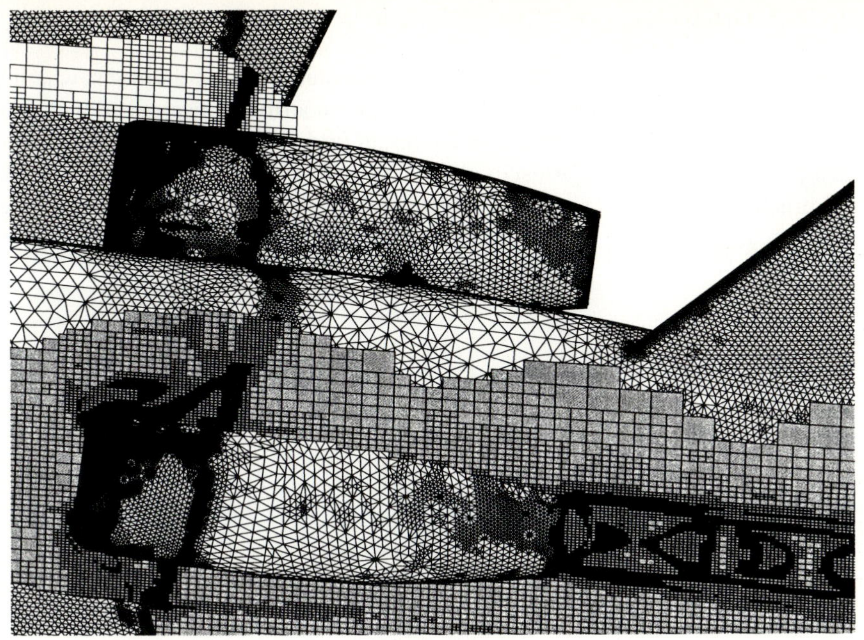

(a) Cartesian grid.

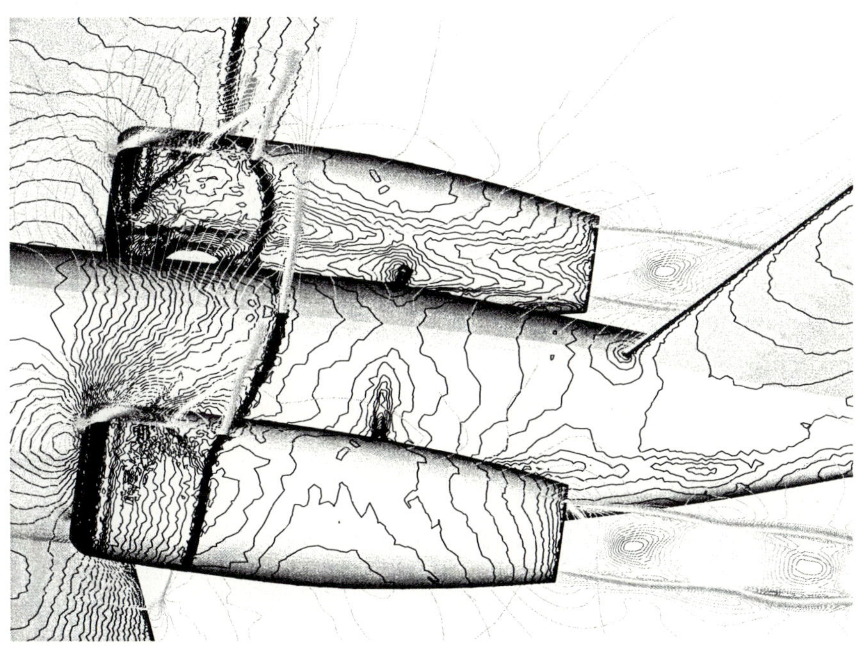

(b) Mach number contours.

Figure 7 Detail of Gulfstream Jet after three adaptation steps ($M_\infty = 0.85$, $\alpha = 3.0°$).

5.3 RAE-2822 Airfoil ($M_\infty = 0.73$, $\alpha = 2.79°$, $Re = 6.5 \times 10^6$), *MUFS*

The turbulent flow around the RAE-2822 airfoil is presented at flow conditions, which are identical to that in case 9 of AGARD-AR-138, [18]. The initial airfoil is discretized with 532 surface and 33 shear layer points. Figure 8(a) shows the initial grid for the entire airfoil, which contains 21350 quasi-prismatic and 8484 Cartesian cells. The quasi-prismatic grid consists of 35 layers. The initial direction of the Kutta panel is identical with the bi-sector direction of the trailing edge. The region of the shock at the upper side is magnified strongly in Figure 8(b), in which the initial Mach number contours are drawn.

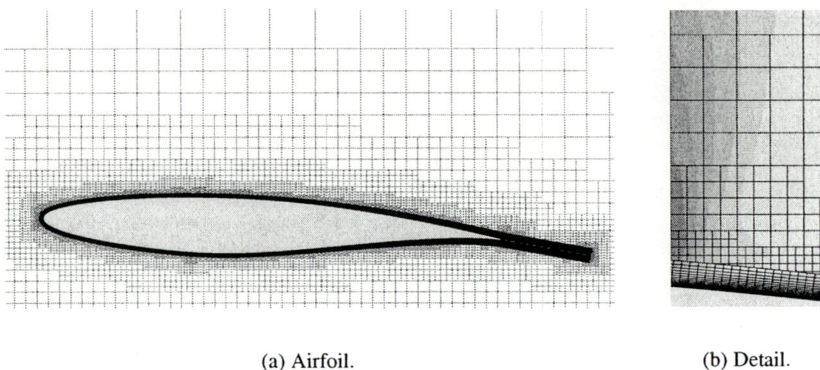

<table>
<tr><td>(a) Airfoil.</td><td>(b) Detail.</td></tr>
</table>

Figure 8 Initial hybrid Cartesian grid for RAE-2822 airfoil ($M_\infty = 0.73$, $\alpha = 2.79°$, $Re = 6.5 \times 10^6$).

After three adaptation steps, the quasi-prismatic and Cartesian grids are refined strongly in the shock and nose regions (1209 surface points, 52115 quasi-prismatic cells and 33854 Cartesian cells). Especially at the upper side of the airfoil, the quasi-prismatic grid height is increased in order to cover the entire boundary layer, Figure 9(a). Also the center line of the quasi-prismatic grid in the wake region is adjusted to the stream line emanating from the sharp trailing edge. The refinement of the Cartesian grid and the tangential refinement of the quasi-prismatic grid cause a sharply resolved shock. Figure 9(b) shows the grids and Mach number contours in the shock region.

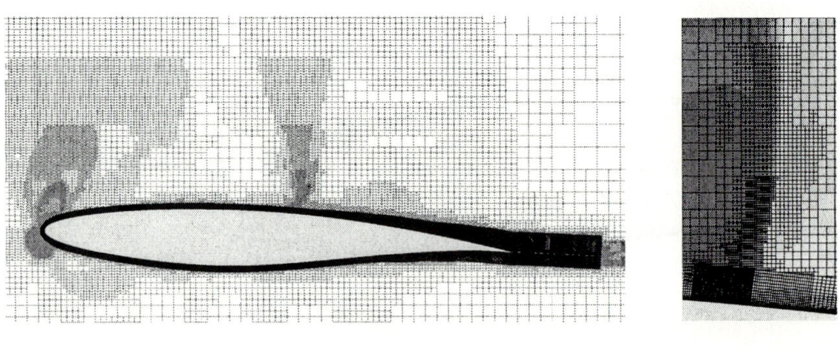

<table>
<tr><td>(a) Airfoil.</td><td>(b) Detail.</td></tr>
</table>

Figure 9 Hybrid Cartesian grid for RAE-2822 airfoil after three adaptation steps ($M_\infty = 0.73$, $\alpha = 2.79°$, $Re = 6.5 \times 10^6$).

Figure 10 shows the Mach number contours around the entire airfoil and in the shock region with a strongly magnified resolution.

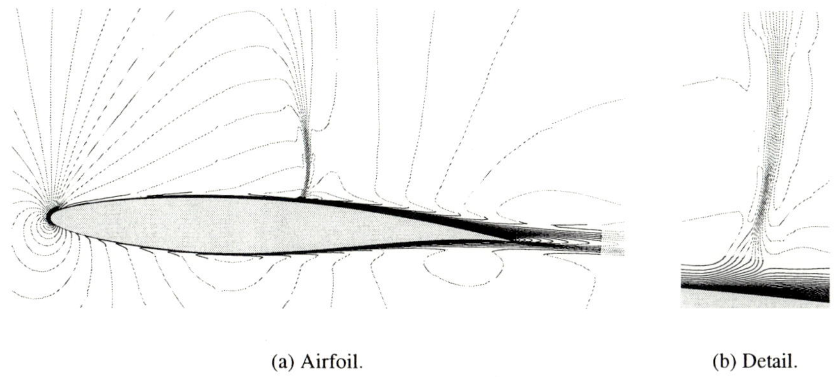

(a) Airfoil.   (b) Detail.

Figure 10 Mach number contours for RAE-2822 airfoil ($M_\infty = 0.73$, $\alpha = 2.79°$, $Re = 6.5 \times 10^6$).

Also the pressure coefficient distribution coincides well with the experimental results published in [18], Figure 11(a). The differences between the results obtained on the initial and the final adapted grid clarify the importance of adaptation. In Figure 11(b), the computed skin friction coeffient is compared with experimental values. The results of the initial grid are drawn with dotted lines.

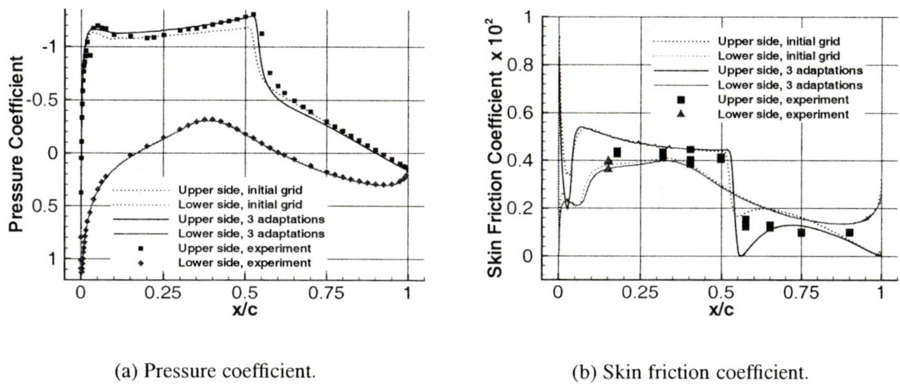

(a) Pressure coefficient.   (b) Skin friction coefficient.

Figure 11 Coefficient distributions for RAE-2822 airfoil ($M_\infty = 0.73$, $\alpha = 2.79°$, $Re = 6.5 \times 10^6$).

The impact of adjusting the height of the first quasi-prismatic cell layer around the airfoil is shown in Figure 12. After three adaptation steps, the target value for the dimensionless wall distance $y+$ equal to one is reached allmost over the entire airfoil. Only at the laminar nose, the shock region (because of a separation bubble) and near the trailing edge a $y+$ value equal to one could not be obtained.

The calculated aerodynamic coefficients become $C_L = 0.794\,(0.803)$, $C_D = 0.0164\,(0.0168)$, $C_M = -0.0874\,(-0.0990)$ and $C_F = 0.00545$. They are in good agreement with experimental results in [18], which are stated in brackets, except of the pitching moment.

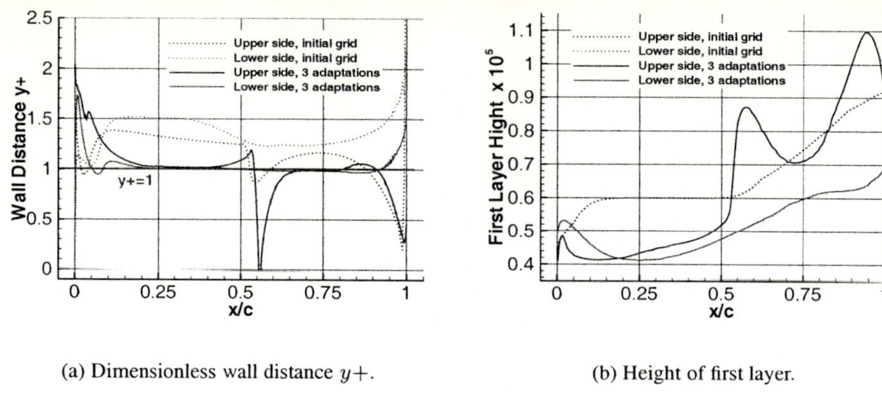

(a) Dimensionless wall distance $y+$.
(b) Height of first layer.

Figure 12 First layer height for RAE-2822 airfoil ($M_\infty = 0.73$, $\alpha = 2.79°$, $Re = 6.5 \times 10^6$).

6 Conclusion and Future Work

A hybrid Cartesian grid generator has been developed, which bases on the octree-data structure. For viscous simulations, a quasi-prismatic grid is added for the viscous layer around the body and in the near wake region. The whole grid is created automatically in only a few minutes (R10000 prozessor) for even complicated geometries. Flow results are obtained using a newly developed Cartesian and a modified unstructured flow solver. The grid generation and the flow solver are embedded in an adaptation cycle, which leads to a self-organized grid system: the final grid depends only on the particular geometry and the flow structure. The self-organized grid system runs fully automatically without any user interventions.

The grid adaptation features are manifold: the Cartesian grid is refined and coarsened isotropically. Furthermore, the quasi-prismatic grid is refined and coarsened in tangential directions. In normal direction, the quasi-prismatic cell heights are adapted to the local boundary layer height. In order to resolve the viscous sub-layer accurately, the height of the first quasi-prismatic cells around the body is adjusted to a predefined value of the dimensionless wall distance $y+$. Finally, the quasi-prismatic grid in the near wake region is adapted to the skeleton surface of the shear layer emanating from the sharp trailing edge.

The surface triangulation of the solid body is refined and coarsened during grid adaptation. Therewith, a comparable resolution between the triangles (surface discretization) and the surrounding Cartesian cells (volume discretization) is achieved. For the computation of the coordinates for new surface points, bi-cubic B-splines (CAD description of body surface) are applied.

Work is now underway for the improvement of the flow solver with regard to efficiency and convergence acceleration: multi-grid acceleration, implicit time integration scheme and parallelization of the flow solver. These improvements are the basis for three-dimensional viscous flow simulations around complete aircraft configurations. Afterwards, the extension to unsteady flows is scheduled, including moving and deforming bodies.

Acknowledgements

The authors would like to thank the CNRS and the DFG for grants under which the project "Self-Organizing Cartesian Grid-Generation System" is supported, embedded in the French- German project "Numerische Strömungssimulation - Simulation Numérique d'Ecoulements".

References

[1] E.H. HIRSCHEL AND W. SCHWARZ. "Mesh Generation for Aerospace CFD Applications". *Surveys on Mathematics for Industry*. Volume 4, 1995, pp. 249–265.

[2] F. DEISTER, D. ROCHER, E.H. HIRSCHEL AND F. MONNOYER. "Adaptively Refined Cartesian Grid Generation and Euler Flow Solutions for Arbitrary Geometries". *Numerical Flow Simulation 1, CNRS-DFG Collaborative Research Programme, Results 1996-1998* (E.H. Hirschel, ed.). Volume 66 of *Notes on Numerical Fluid Mechanics*, Vieweg Verlag, Braunschweig/Wiesbaden, 1998, pp. 25–49.

[3] F. DEISTER, D. ROCHER, E.H. HIRSCHEL AND F. MONNOYER. "Three-Dimensional Adaptively Refined Cartesian Grid Generation and Euler Flow Solutions for Arbitrary Geometries". *Computational Fluid Dynamics '98* (K.D. Papailiou, D. Tsahalis, J. Périaux, C. Hirsch and M. Pandolfi, eds.). Volume 1(1), John Wiley & Sons Ltd., Chichester - New York - Weinheim, 1998, pp. 96–101.

[4] F. DEISTER AND E.H. HIRSCHEL. "Adaptive Cartesian/Prism Grid Generation and Solutions for Arbitrary Geometries". AIAA-paper 99-0782, 1999.

[5] D. ROCHER, F. DEISTER, F. MONNOYER AND E.H. HIRSCHEL. "Flow Simulation on Adaptive Cartesian and Hybrid Prismatic-Cartesian Grids around Arbitrary Geometries". AIAA-paper 99-3313, 1999.

[6] F. DEISTER AND E.H. HIRSCHEL. "Self-Organizing Cartesian/Prismatic Grid/Solution System". Paper accepted for the ECCOMAS Conference, Barcelona, Spain, 2000.

[7] F. DEISTER AND E.H. HIRSCHEL. "Self-Organizing Hybrid Cartesian Grid/Solution System for Arbitrary Geometries". AIAA-paper 2000-4406, 2000.

[8] S.L. KARMAN, JR "Unstructured Cartesian/Prismatic Grid Generation for Complex Geometries". *Surface Modeling, Grid Generation and Related Issues in Computational Fluid Dynamics (CFD) Solutions*. NASA CP-3291, 1995, pp. 251–270.

[9] P.L. ROE. "Approximate Riemann Solvers, Parameter Vectors and Difference Schemes". *Journal of Computational Physics*. Volume 43, 1981, pp. 357–372.

[10] J.L. STEGER AND R.F. WARMING. "Flux Vector Splitting of the Inviscid Gasdynamic Equations with Application to Finite Difference Methods". *Journal of Computational Physics*. Volume 40, 1981, pp. 263–293.

[11] C.-H. TAI. "Acceleration Techniques for Explicit Euler Codes". PhD thesis, University of Michigan, 1990.

[12] A. JAMESON, T.J. BAKER AND N.P. WEATHERILL. "Calculation of Inviscid Transonic Flow over a Complete Aircraft". AIAA-paper 86-0103, 1986.

[13] P.R. SPALART AND S.R. ALLMARAS. "A One-Equation Turbulence Model for Aerodynamic Flows". *La Recherche Aérospatiale*. Volume 1, 1994, pp. 5–21.

[14] D.L. DEZEEUW. "A Quadtree-Based Adaptively-Refined Cartesian-Grid Algorithm for Solution of the Euler Equations". PhD thesis, University of Michigan, 1993.

[15] G.A. ASHFORD. "An Unstructured Grid Generation and Adaptive Solution Technique for High-Reynolds-Number Compressible Flows". PhD thesis, University of Michigan, 1996.

[16] H.W. STOCK AND W. HAASE. "A Feasibility Study of e^N Transition Prediction in Navier-Stokes Methods for Two Dimensional Airfoil Computations". IB 129-98/12, DLR, Institut für Entwurfsaerodynamik, Braunschweig, 1998.

[17] D.J. JONES. "Reference Test Cases and Contributors". *Test Cases for Inviscid Flow Field Methods*. AGARD-AR-211, 1985, pp. 5-1–5-24.

[18] P.H. COOK, M.A. MC DONALD AND M.C.P FIRMIN. "Aerofoil RAE 2822 - Pressure Distribution and Boundary Layer and Wake Measurements". *Experimental Data Base for Computer Program Assessment*. AGARD-AR-138, 1979, pp. A6-1–A6-77.

CD2D3D - a package to solve convection dominated problems employing ordering techniques

W. Hackbusch[†], R. Kriemann[†], S. Le Borne[‡], J.-F. Maitre[*]

[†]Max-Planck-Institute for Mathematics in the Sciences, Inselstr. 22-26,
04103 Leipzig, Germany

[‡]Christian-Albrechts-Universität zu Kiel, Mathematisches Seminar II,
Lehrstuhl für Praktische Mathematik, Olshausenstr. 40, 24098 Kiel, Germany

[*]UMR CNRS 5585, Ecole Centrale de Lyon, Dept. Mathematiques Informatique,
B.P. 163, 69131 Ecully Cedex, France

Summary

Recently ordering techniques have been studied in connection with their influence on the multi-grid solution of convection dominant problems [3], [6], [7]. Due to its potential for general applications, in this paper we will detail the characteristics of the software developed to perform numerical tests for the proposed ordering and solution schemes given in [3]. The purpose of this paper is twofold: Firstly, we show how ordering techniques can be incorporated in a complex software package developed for the solution of convection dominated partial differential equations, and secondly, we document the capabilities of the existing software and provide a general description of the employed data structures and interfaces so that users new to the package are able to easily experiment with the implemented ordering techniques or apply it to their own applications.

1 Introduction

The package CD2D3D, whose theoretical origins may be found in [3], solves an assortment of differential equations; the convection-diffusion equation

$$Lu := -\epsilon\Delta u + \mathbf{b} \cdot \nabla u + cu \quad = \quad f \quad \text{in } \Omega \subset \mathbf{R}^d, \tag{1.1}$$

$$u \quad = \quad \varphi \quad \text{on } \Gamma := \partial\Omega, \tag{1.2}$$

the Stokes equations with a convective term

$$-\epsilon\Delta\mathbf{u} + (\mathbf{b} \cdot \nabla)\mathbf{u} + \theta\mathbf{u} + \nabla p \quad = \quad \mathbf{f} \quad \text{in } \Omega \tag{1.3}$$

$$\nabla \cdot \mathbf{u} \quad = \quad 0 \quad \text{in } \Omega \tag{1.4}$$

$$\mathbf{u} \quad = \quad \varphi \quad \text{on } \Gamma = \partial\Omega, \tag{1.5}$$

as well as the Navier-Stokes equations

$$
\begin{aligned}
\mathbf{u}_t - \epsilon\Delta\mathbf{u} + (\mathbf{u}\cdot\nabla)\mathbf{u} + \nabla p &= \mathbf{f} &&\text{in } \Omega\times(0,T) && (1.6)\\
\nabla\cdot\mathbf{u} &= 0 &&\text{in } \Omega\times(0,T) && (1.7)\\
\mathbf{u}(x,t) &= \varphi(x,t) &&\text{on } \partial\Omega\times(0,T) && (1.8)\\
\mathbf{u}(x,0) &= \mathbf{u}_0(x) &&\text{in } \Omega. && (1.9)
\end{aligned}
$$

Here, Ω is a d-dimensional ($d = 2$ or 3) bounded domain while the parameter ϵ satisfies $0 < \epsilon \ll 1$ and $b : \mathbf{R}^d \to \mathbf{R}^d$ is a vector-valued function that characterises the convection which may differ in direction and magnitude over Ω.

CD2D3D was developed as a prototype to study the theoretical and practical aspects of numbering techniques for the multigrid solution of the above convection dominant problems. It was designed to handle (systems of) convection dominated elliptic partial differential equations on general two- or three-dimensional domains. The problem class addressed by this program should not be considered as the limit of the class of problems that could be successfully solved by the techniques embodied by this package. Conversely, we do not claim that every problem (formally) within this class can be solved using the existing code. Since the main focus of the authors was the analysis of ordering techniques, several such ordering techniques and corresponding iterative solvers have been implemented whereas the discretisation is restricted to continuous, piecewise linear elements on a triangulation. However, due to the modular structure of the program, extensions to different types of finite elements, refinement strategies, boundary conditions, etc., could be easily made and tested, with a variety of ordering techniques being already available. In this package time efficiency has been a secondary consideration to robustness, versatility and ease of maintenance.

The C++ source code is contained in a number of files which have been grouped into several subdirectories. These directories and the main functions of the files contained therein are summarized in table 1.

Table 1 Structure into directories.

subdirectory	subject
misc	miscellaneous (timer, type definitions, etc.)
net	grid management and refinement
discr	discretisation routines
matrix	algebraic matrix (-vector) operations
graph	numbering routines
solver	iterative solvers
visual	visualisation routines of grids, graphs, sparsity structures, etc.

The code was developed and tested under the *Solaris* and *Linux* operating systems using the *GNU-C*-compiler. For other operating systems or C-compilers some minor changes might be necessary (mostly regarding the configuration of the compilation-scripts).

The purpose of this paper is neither to provide a complete user's manual for the existing package nor to explain implementational tricks. In fact, its purpose is to illustrate how the ordering of variables interacts with other components of the overall solution process and its subsequential implementation. Given this, the reader will be introduced to the capabilities of the software developed by the authors so that an informed decision can be made by any individual considering its potential as a solver (in its complete or modular form), tailored for individual applications.

The remaining paper is organized as follows: The data structures of the initial triangulation and the methods for the subsequent grid refinement are documented in section 2. Section 3 deals with the solution of the given equation and comprises the (finite element) discretisation of the differential equations, the matrix storage format of the resulting systems of equations, reordering schemes for the unknowns and their interaction with the grid, the problem (matrix) and the linear solvers, the iterative solution of the linear systems by Krylov subspace methods or by the multigrid method with appropriate preconditioners or smoothing methods, respectively, and finally, the implementation of nonlinear solvers. In section 4, various possibilities to view the triangulations, matrix graphs, matrix sparsity patterns etc. are illustrated. Some elementary test problems are described in section 5.

2   Data structures and mesh generation

In this section, we define the data structures used to specify the domain Ω as a triangulation and the construction of a hierarchy of nets by an iterative refinement of the initial triangulation.

In the two-dimensional case, the initial triangulation has to be specified by the user in the form of its vertices (i.e. their coordinates), edges and triangles. The standard initial triangulation of the unit square in Figure 1 is, for example, described in the form displayed in Figure 2.

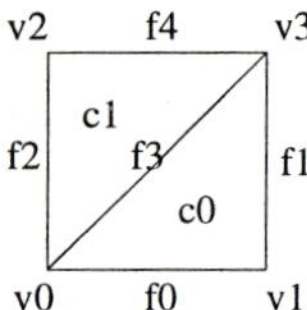

Figure 1   Initial triangulation

Here we first specify the dimension of the problem, i.e. 2 or 3, and then allow for different types of boundaries. Finally we enumerate the vertices, frontiers (i.e. edges) and cells (i.e. triangles). For the vertices we specify the coordinates and in the case of a boundary node the type of boundary, otherwise we declare it as interior.

For an edge we specify the two vertices that it connects and again its boundary type in the case of a boundary edge or interior otherwise (see, e.g., frontier f3). In the two-dimensional case all frontiers are of normal type, i.e. they are edges in contrast to, e.g., triangles or rectangles in the three-dimensional case. The cells of type 'triangle' are specified by three frontiers, ordered in a counterclockwise sense. The order of the two vertices specifying an edge must be such that the cell lying on the left side of this edge appears in the data file before the cell lying on the right side of the edge. E.g., in the above unit square example, the frontier f3 is specified by the vertices v3 and v0 in this order and, when interpreted as an edge from v3 to v0, then the cell c0 lies on its left side and c1 on its right side. Since c0 appears before c1 in the list of cells, the above condition is fulfilled. Furthermore we assign a refinement strategy. Here we choose a regular refinement where each triangle is refined into four congruent subtriangles by pairwise connecting the edge midpoints. Alternatively one could mark the triangle for different (possibly adaptive) refinement strategies.

In the three-dimensional case the user needs to specify the vertices, edges, frontiers (triangles in the case of tetrahedra) and cells (e.g. tetrahedra). Some representatives of these types are

36

```
//                              }                          frontier f1                  {
// dimension of the problem                                {                                nodes v3 v2
//                              node v2                         nodes v1 v3                  boundary = b0
dim = 2                         {                               boundary = b0                type = normal
                                    coord 0.0 1.0               type = normal            }
                                    boundary = b0           }
//                              }                                                        //
// boundary of our domain                                  frontier f2                  // the cells
//                              node v3                     {                            //
boundary b0 {}                  {                               nodes v2 v0              cell c0
                                    coord 1.0 1.0               boundary = b0            {
//                                  boundary = b0               type = normal                frontiers f0 f1 f3
// the vertices                 }                          }                                type = triangle
//                                                                                           refine = regular
node v0                         //                         frontier f3                  }
{                               // the frontiers            {
    coord 0.0 0.0               //                              nodes v3 v0              cell c1
    boundary = b0               frontier f0                     interior                {
}                               {                               type = normal                frontiers f3 f4 f2
                                    nodes v0 v1            }                                  type = triangle
node v1                             boundary = b0                                            refine = regular
{                                   type = normal          frontier f4                  }
    coord 1.0 0.0               }
    boundary = b0
```

Figure 2   Description of initial triangulation of unit square

```
node v0 {              edge e0 {              frontier f0 {            cell c0 {
    coord 0.0 0.0 0.0      nodes v0 v5            edges e0 e1 e2          frontiers f0 f1 f2 f3
    boundary = b1         boundary = b1          type = triangle         type = tetrahedron
}                     }                          boundary = b1           refine = bey
                                               }                        }
```

Figure 3   Examples for the description of the initial triangulation of the unit cube

given in Figure 3. Unlike the two dimensional case, the initial data specifying the nodes, edges and frontiers may be listed in an arbitrary order. The implemented refinement strategy *bey* for a tetrahedral grid in which a tetrahedron is divided into eight subtretrahedra of equal volume is described in [2].

The relevant classes that manage the grid and its refinement are in the *net* directory. More specifically, they are *TNetHierarchy* (a list of nets), *TNet* (a list of cells), *TCell* (an array of frontiers and an array of children cells) and *TFrontier* (two pointers to cells and an array of children frontiers)

The initial triangulation together with the refinement induces a natural ordering for the vertices which, on the level of the net, will not be changed again at a later time.

3   Equation solution

3.1   Discretization and sparse matrix storage

Let $\mathcal{T}_h$ denote a triangulation of Ω and let $\mathcal{M}^{\mathcal{T}_h}$ be the space of continuous, piecewise linear functions associated with $\mathcal{T}_h$. The parameter h characterises the size of the elements in the triangulation. The functions are represented using the standard nodal basis, i.e., a function can be specified giving its values at the vertices. Let $\mathcal{M}_\varphi^{\mathcal{T}_h}$ be the affine space of $\mathcal{M}^{\mathcal{T}_h}$ whose elements satisfy the Dirichlet boundary conditions at the vertices of $\mathcal{T}$ lying on $\Gamma = \partial\Omega$. The discrete equations that are solved are based on weak formulations of the problems and are given by:

Find $u \in \mathcal{M}_\varphi^{\mathcal{T}_h}$ such that

$$a(u,v) \quad = \quad f(v) \quad \text{for all } v \in \mathcal{M}_\varphi^{\mathcal{T}_h} \quad \text{with} \tag{3.10}$$

$$a(u,v) \quad := \quad \int_\Omega \epsilon \nabla u \nabla v + (b \cdot \nabla u + cu)v \; dx \tag{3.11}$$

in the case of the convection diffusion equation, and

Find $(\mathbf{u}, p) \in (\mathbf{M}_\varphi^{\mathcal{T}_h}, \mathcal{M}^{\mathcal{T}_H})$ such that

$$a(\mathbf{u}, \mathbf{v}) + b(p, \mathbf{v}) \quad = \quad f(\mathbf{v}) \tag{3.12}$$

$$b(q, \mathbf{u}) \quad = \quad 0 \qquad \text{for all } (\mathbf{v}, q) \in (\mathbf{M}_\varphi^{\mathcal{T}_h}, \mathcal{M}^{\mathcal{T}_H}) \tag{3.13}$$

in the case of the Stokes equations with a convective term. Here $\mathbf{M}_\varphi^{\mathcal{T}_h}$ is the product space $\mathbf{M}_\varphi^{\mathcal{T}_h} = (\mathcal{M}_\varphi^{\mathcal{T}_H})^d$ and $\mathcal{T}_h$ is the triangulation resulting from a refinement of $\mathcal{T}_H$. The (bi-)linear forms are given by

$$a(\mathbf{u},\mathbf{v}) \quad = \quad \int_\Omega \epsilon \nabla \mathbf{u} \cdot \nabla \mathbf{v} + (b \cdot \nabla \mathbf{u} + \theta\mathbf{u})\mathbf{v} \; dx,$$

$$b(p,\mathbf{u}) \quad = \quad -\int_\Omega p \cdot \operatorname{div} \mathbf{u} \; dx,$$

$$f(\mathbf{v}) \quad = \quad \int_\Omega f\mathbf{v} \; dx.$$

In the Navier-Stokes equations, we discretize the time derivative by a fractional-step-θ-scheme which is described in [5]. The resulting nonlinear set of equations can then be solved by a fixed point iteration or a fixed point defect correction method. These methods require the solution of linear subproblems which are Stokes equations with a convective term.

In either case the integral of the convective term $\int_\Omega b \cdot \nabla uv \; dx$ is replaced by an upwind scheme in order to ensure stability. To this end we employed a hybrid scheme that is due to Tabata [8](Chap III.3). In this scheme the convective part leads to at most two (three) off-diagonal matrix entries for a two- (three-) dimensional test problem. The subsequent ordering and solution techniques are not restricted to this discretisation method but applicable to a wide range of discretisation techniques possessing the following two properties:

- **(Relationship between the matrix sparsity structure and the triangulation)**: For finite elements with a nodal basis, the nodes in the triangulation correspond to the variables in the linear system, and the edges in the triangulation correspond to the nonzero entries in the resulting stiffness matrix (except for so-called accidental zeros).

- **(Relationship between the convection direction and the sizes of the matrix entries)**: Discretisation of the convective part with hybrid upwinding leads to a possibly large difference in magnitude between the entries $|a_{ij}|$ and $|a_{ji}|$, $(i \neq j)$. Typically, an entry $|a_{ij}|$ is large if the corresponding edge lies in the direction opposite to the convection direction, i.e., is nearly parallel with the convection, provided that the convection is dominant.

In the case of the Stokes equations, the resultant matrix has an inherent block structure due to the different types of unknowns (i.e. velocity components and pressure). Here the above properties can be observed for the first d diagonal blocks where d is the spacial problem dimension.

The assignment of degrees of freedom is implemented in the classes *TConvDiffDoFManager* and *TStokesDoFManager*. The classes *T2DConvDiffFE* and *T2DStokesFE* are used for (elementwise) assembling of the stiffness matrices and right-hand sides. The construction of a hierarchy of grids and corresponding discretizations as well as prolongation and restriction operators between the levels (e.g. for a subsequent multigrid solution) is realised by the classes *TConvDiffHierarchy* and *TStokesHierarchy*, respectively.

The global stiffness matrices are stored in a sparse block matrix format where only nonzero entries are stored in linear arrays. The corresponding classes are *TBlockMatrix* consisting of rows *TMatrixBlockRow* which in turn consist of matrix blocks *TMatrixBlock* which finally contain the actual matrix entries in a vector with entries of type *TMatrixData*. The diagonal entries are stored first followed by the remaining nonzero entries and the corresponding column indices. The stiffness matrix for a convection-diffusion problem, for example, results in a 1×1 block matrix structure, whereas the Stokes equations in d spacial dimensions result in a block matrix with $(d + 1) \times (d + 1)$ matrix blocks.

The classes *TVectorBlock* and *TBlockVector* model the corresponding (block) vectors. These matrix and vector classes are supplied with various algebraical methods such as matrix vector multiplication, multiplication of the upper (or lower or diagonal) part of the matrix with a vector, computation of an (incomplete) LU-decomposition of a matrix, etc., as well as some methods explicitly tailored to the orderings tested by this package. Regarding the reordering, these classes are equipped with methods that perform the necessary permutations (see also (3.14), (3.15)) based upon a permutation vector which is determined by one of the ordering routines described in the following section.

The relevant classes for the nonlinear iteration for the Navier-Stokes equations are *TFixedPointIteration* and *TFixedPointDefectCorrection* which are both derived from the base class *TNonLinearSolver*.

3.2   Ordering

In our implementation we have clearly separated the process of reordering a system of equations into the following two steps: Firstly, based upon a desired ordering strategy, a permutation vector is calculated, and in a second step this permutation is applied to the system of equations. Whereas for some of the ordering techniques there exist more efficient implementations in which the actual reordering is performed during the execution of the ordering algorithm, this separation into two steps provides the user of the software with the option to solely utilize the ordering algorithms to obtain a vector containing the permutation information, but except for that use his own software (i.e., for the grid and matrix generation, the actual reordering based upon the given permutation and the subsequent iterative solution). However, we point out that most orderings lead to improved convergence rates only in combination with certain iterative methods tailored to the particular ordering, i.e., combining an arbitrary ordering with an arbitrary iterative method is unlikely lead to success.

In the following we will provide implementational details for various ordering techniques; for further (theoretical) properties and an analysis of the efficiency of these techniques we refer to [3] and the references therein.

The common property of all ordering algorithms is that they operate on graphs which originate from the stiffness matrix. However, already the types of graphs differ for different orderings: The reverse Cuthill-McKee (RCM) algorithm works on an *undirected* graph (representing the matrix sparsity structure), the heuristic feedback vertex set algorithm (HFVS) requires a *directed* graph (representing the dominant matrix entries) while the planar algorithms for concentric cycles (CC) and a feedback vertex set (PFVS) additionally require coordinates for the vertices in the graph (which coincide with the corresponding vertex coordinates in the underlying triangulation) and assume the graph to fulfil the so-called *one-flow-direction condition*. An undirected graph can be constructed from just the stiffness matrix whereas a criterion to distinguish between 'large' and 'small' matrix entries is required for the construction of a directed matrix graph. The geometrical information can be extracted from the underlying triangulation if necessary.

Some of the ordering algorithms yield information in addition to the permutation vector that might be useful in a subsequent iterative method. E.g., ordering algorithms involving a feedback vertex set also give the size of the feedback vertex set (which could be used for iterative methods employing an (approximate) Schur complement w.r.t. the feedback vertex set vertices), and the ordering with concentric cycles yields information pertaining to the computed cycles (which could be used for block iterative methods).

The ordering techniques are implemented as methods of the classes *TBlockInfo* for the two-dimensional and *T3DPermBuilder* for the three-dimensional case. These classes and their methods will be illustrated in the following.

TBlockInfo - ordering techniques for two spatial dimensions

Fundamentally, there are three different algorithms implemented to compute a permutation: cyclic ordering, FVS ordering and RCM ordering. The relevant methods implementing these algorithms are

buildCYCPerm	– planar cyclic algorithm
buildFVSPerm	– planar/heuristic feedback vertex set algorithm
buildRCMPerm	– Reverse Cuthill-McKee algorithm.

The necessary data (matrix, geometrical information, etc.) for the methods are supplied via parameters or global variables. The parameters required for *buildCYCPerm* are described in Table 2. Here *OUT* implies that the content of the variable is not used prior to the method call and will be overwritten. The type *IN* on the other hand denotes a parameter whose content is used during the algorithm and will not be changed.

The vector *vecSize* consecutively contains the indices of the first and last vertices (w.r.t. the reorderd system) belonging to the computed cycles. Hence the difference of two such indices yields the size of a cycle, and the size of the vector is twice the number of cycles.

The corresponding list of parameters for the method *buildFVSPerm* is given in Table 3. The parameters *permute*, *arrVectors*, *mat*, *xi* and *criterion* are the same as in *buildCYCPerm*. Here, however, instead of the cycle information we now obtain the size of the computed feedback vertex set in *fvsSize*. Selections for either the planar and the heuristic feedback vertex set algorithm are performed via the parameter *fvsType*. The vector *arrVectors* is only used for the planar version.

Table 2 Parameters for *buildCYCPerm*

Parameter	Description	Type
permute	stores the computed permutation	OUT
vecSize	stores information about cycles (see below)	OUT
arrVectors	stores coordinates of the vertices	IN
mat	stiffness matrix	IN
xi, criterion	parameters for constructing graph of dominant entries	IN
ordering	specifies type of cycles (i.e., minimal or maximal)	IN

Table 3 Parameters for *buildFVSPerm*

Parameter	Description	Type
permute	stores the computed permutation	OUT
fvsSize	size of the computed feedback vertex set	OUT
arrVectors	stores coordinates of the vertices	IN
mat	stiffness matrix	IN
xi, criterion	parameters for constructing graph of dominant entries	IN
searchType	specifies the graph-traversal (i.e. depth first or breadth first search)	IN
fvsType	specifies FVS-algorithm (planar or heuristic)	IN

In the feedback vertex set ordering, the remaining acyclic graph after elimination of the feedback vertices is ordered by a downwind numbering. The *searchType* parameter determines whether this downwind ordering follows a depth- or breadth-first search.

The RCM algorithm requires only the stiffness matrix (in fact, it requires just the sparsity structure of the stiffness matrix) to compute a permutation vector, hence only the parameters *permute* and *mat* with their descriptions given in Table 3 are necessary.

For each ordering algorithm multigrid versions are available where permutation vectors are computed on all levels. Three different strategies have been implemented to compute orderings on the coarser levels which we refer to as projected, independent and nested multilevel ordering. The corresponding methods are denoted by *buildXXXProjPerm*, *buildXXXIndepPerm* and *buildXXXNestedPerm* where 'XXX' has to be substituted by 'CYC', 'FVS' or 'RCM', respectively. The parameters for these methods are analogous to those in the single grid case only that now you need to provide the neccessary information for all levels. In terms of the implementation, you have to supply arrays of the respective parameters, e.g. an array of permutation vectors, matrices, etc. One difference arises in providing the information about the vertex coordinates: In the multilevel case, the coordinates of the unknowns are not given by an array of coordinate vectors but are extracted from a given hierarchy of grids by employing methods of the class implementing the finite-element discretization. The parameters that differ from the single grid case are listed in Table 4.

T3DPermBuilder - ordering techniques for three spatial dimensions

The basic algorithms to compute a permutation in the three-dimensional case are again cyclic ordering, FVS ordering and RCM ordering which are realized in the methods *buildCYCPerm*, *buildFVSPerm* and *buildRCMPerm*, respectively. Since the cyclic ordering and the planar feed-

Table 4 Parameters for multilevel case

Parameter	Description	Type
fe	object implementing the finite element discretization	IN
listNets	a list (hierarchy) of grids	IN

back vertex set algorithm require the underlying graph to be planar, here we first decompose the three dimensional graph into subgraphs, so-called *flow surfaces*, which are planar and on which the above algorithms can then be executed. For the heuristic FVS algorithm one can choose between applying it to the separate flow surfaces or to the global graph. The RCM ordering is always done on the global matrix graph.

The necessary data (matrix, geometrical information, etc.) for the methods are again supplied via parameters or global variables. Since they differ from the two-dimensional case we will provide tables for the various methods illustrating the employed parameters.

The arguments for *buildCYCPerm* are listed in table 5.

Table 5 Parameters for *buildCYCPerm* and *buildFVSPerm*

Parameter	Description	Type
permute	stores the computed permutation	OUT
pNet	the underlying grid (triangulation)	IN
pEqn	describes the equation (e.g. convection direction)	IN
pMatrix	the stiffness-matrix	IN
xi, criterion	parameters for constructing graph of dominant entries	IN

The grid is used to supply the geometrical information of the unknowns. The construction of flow surfaces requires some information on the convection direction which is obtained from the parameter *pEqn*.

The type of feedback vertex set algorithm that will be performed depends on the choice of arguments that you provide for the method *buildFVSPerm*. If you supply a grid, the equation and the stiffness matrix then flow surfaces will be built to which the planar or the heuristic feedback vertex set algorithm can be applied. If only the matrix is given then the heuristic algorithm is applied globally. Additional to the parameters listed in Table 5 a parameter *fvsMethod* specifies the type of FVS algorithm to be employed (i.e. planar or heuristic).

Equal to the 2d-case, the only argument for the RCM-algorithm is the stiffness-matrix.

For these ordering algorithms there are again multigrid versions available which are implemented in the methods *buildMGCYCPerm*, *buildMGFVSPerm* and *buildMGRCMPerm*. The expected arguments are arrays of matrices and a hierarchy of grids.

Reordering

After the permutation (or a hierarchy of permutations in the multigrid case) is built, the indices in the matrix and the right-hand side have to be reordered. This is done by the *permute*-methods of a matrix and a vector. The only expected argument for this methods is the permutation vector.

If multigrid methods are to be applied then also the matrix-representation of the prolongations and restrictions have to be reordered. We have implemented the prolongations and restrictions as block matrices. The permute-method of these objects expects two arguments which specify permutations for the rows and the columns which correspond to the permutations on the fine and the coarse level, respectively.

3.3   Solving linear systems

Prior to the iterative solution of the sets of linear equations arising in the discretisation process, the unknowns can be reordered using one of the ordering techniques described in section 3.2. To this end, let $P \in \mathbf{R}^{n,n}$ be a permutation matrix, i.e., a bijective mapping $P : I \to I$ on the index set $I = \{1, \cdots, n\}$. For a system of equations $Kx = b$, the reordered system is given by

$$PKP^Ty = Pb. \tag{3.14}$$

The solutions of the original and the reordered systems are related by $x = P^Ty$.

In the case of the Stokes equations, only the velocity unknowns are reordered, i.e., given a permutation matrix $P \in \mathbf{R}^{n \times n}$ and the identity matrix $I \in \mathbf{R}^{m \times m}$, we reorder the system through

$$\begin{pmatrix} P & 0 \\ 0 & I \end{pmatrix} \begin{pmatrix} A & B^T \\ B & 0 \end{pmatrix} \begin{pmatrix} P & 0 \\ 0 & I \end{pmatrix}^T \begin{pmatrix} \tilde{x} \\ \tilde{y} \end{pmatrix} = \begin{pmatrix} P & 0 \\ 0 & I \end{pmatrix} \begin{pmatrix} f \\ g \end{pmatrix}. \tag{3.15}$$

In the case of the multigrid iteration, the stiffness matrices and right-hand sides as well as the prolongation and restriction operators are reordered on all levels.

To decouple the ordering from the solution methods, in practice we reorder the matrix and right-hand side vector and then apply the solver methods to the reordered systems. Given a vector that contains the desired permutation, the matrix and vector classes are equipped with the necessary reordering methods.

To solve the (reordered) sets of linear equations, the user is provided with the (global) multigrid method as well as the (conjugate gradient and) bicgstab method and a variety of smoothing and preconditioning methods, respectively. The implemented methods include the Richardson, Jacobi, (forward, backward and symmetric) Gauß-Seidel, SOR and the Kaczmarz iteration.

In the case of linear systems of equations arising from the discretisation of systems of partial differential equations such as the Stokes equations, the prolongations and restrictions are performed componentwise in a straight forward manner. As a smoother we implemented an iteration that is based on an incomplete block LU decomposition of the system matrix as suggested for the Stokes equations in [1], [4]. The appearing Schur complement is replaced by an approximation of its (pointwise) LU decomposition. For further details we refer to [3].

All solver classes are derived from a base solver class *TSolver* containing data members for the solver's name, the maximum number of iteration steps, a desired accuracy, a verbosity field and a mode in which the solver runs. The three possible modes are explained in Table 6.

The verbosity of the solver is bitfield encoded, its values and their meanings are summarised in Table 7.

Table 6 The modes of the iteration.

mode	description
0	The iteration continues until a desired absolute accuracy or the maximum number of steps is reached
1	the iteration continues until a desired relative improvement compared to the initial residual or the maximum number of steps is reached
2	the iteration continues until the maximum number of steps is reached independently of the development of the error

Table 7 The values of verbosity.

verbosity	description
0	no messages
1	print solver's name
2	print convergence history (reduction per step)
4	print average convergence rate at the end of iteration
8	print breakdown messages

3.4 Solving nonlinear systems

In the case of nonlinear systems like the Navier-Stokes equations, for each nonlinear step the convective term may have changed and require a new ordering with respect to the dominant matrix entries. Hence we monitor the change in the convection compared to the previous nonlinear step, and if this change is sufficiently large we adjust the ordering on all levels.

4 Graphics

We have implemented several visualisation methods for various structures, e.g. matrices (i.e. their sparsity patterns), grids, flows (convection directions) and graphs.

The output-format depends upon the spatial dimension of the object to be drawn. In the two dimensional case they are drawn using the *Postscript*-language. The output-format of all three dimensional objects can be chosen between the *Virtual Reality Modeling Language* (VRML) and *PovRay*. The latter format is not directly viewable and has to be *traced* by the *PovRay*-raytracer to produce images (see http://www.povray.org).

There exist visualisation-classes for each of the above mentioned objects, e.g. the class *TMatrixView* is capable of drawing matrices and *TNetView* is used to draw grids. All of these classes utilise a special printer-object which defines the output-format and has the ability to draw primitives (points, lines, circles etc.). These primitive methods are accessed to draw the actual objects.

Besides this rather complicated way of drawing there is also a class which encapsulates this process. The class *TDrawApp* has a collection of methods which take the given arguments, sets up the printer-object and calls the appropriate visualisation-objects for drawing. All these methods expect an argument *filename* which specifies the basename (without ending) of the file that is to be created. The ending depends upon the output-format: ".ps" for Postscript, ".wrl" for VRML and ".pov" for PovRay.

The drawing method for the grid can optionally be provided with a parameter *permutation* in which case the grid vertices are labelled in correspondence with this permutation.

When drawing the sparsity structure of a matrix, typically only the "strong" entries will be drawn, i.e., those that are specified by the parameter *xi* and the variable *criterion* of the global object *param*. These are identical to the parameters that were used to create the graphs in the ordering algorithms of section 3.2.

If the parameter *smaller* is set to *true*, then the "weak" matrix-entries will be drawn instead. With the parameter *pBlockStruct* one is able to define a partition of the indexset by storing the start-indices of each block. If m is the size of the array *pBlockStruct*, this leads to a $m \times m$-blockstructure of the matrix which will also be drawn. One application of this would be, if one wants to see the blockstructure which arises due to the decomposition of the grid into flow surfaces (see Figure 4).

Next to the graph of dominant entries several types of subgraphs can be visualised that are of special interest for a particular ordering. E.g., when constructing concentric cycles, a graph consisting of just these cycles can be viewed, and in the ordering with a feedback vertex set the graph from which such a feedback vertex set has been eliminated can be displayed. Again one can either have the original vertex names or the relabelled names displayed by providing a permutation vector.

To view the convection direction one has to supply an equation object and parameters (stepwidth) for a virtual grid. At the crosspoints of this grid, the equation is evaluated and the resultant convection direction is drawn.

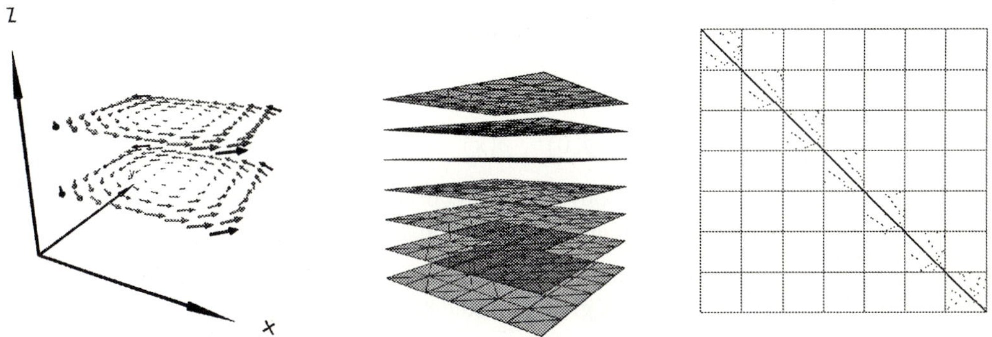

Figure 4   Convection (1st), flow surfaces (2nd) and reordered matrix (3rd)

In Figure 4 some examples for the three dimensional case are shown. The convection direction b, which is presented in the first picture, is defined by $b(x,y,z) = (0.5 - y, x - 0.5, 0)$. The constructed flow surfaces for this kind of convection are shown in the second picture. A permutation vector was computed on these flow surfaces using the planar cyclic ordering (see section 3.2) and then applied to the stiffness matrix. The result is shown in the third picture.

Some examples for the two dimensional case can be seen in Figure 6.

5   Test problems

In this section we briefly document the test problems that were solved with the CD2D3D package and which exercise most features of the package. The source code for the following test problems is included with the package:

- Test problem CONVDIFF2D: In this example, we solve the convection-diffusion(-reaction) equation (1.1), (1.2) on a two-dimensional domain.

- Test problem CONVDIFF3D: In this example, we solve the convection-diffusion(-reaction) equation (1.1) (1.2) on a three-dimensional domain.

- Test problem STOKES2D: In this example, we solve the Stokes equations with a convective term (1.3) (1.4), (1.5) on a two-dimensional domain.

- Test problem STOKES3D: In this example, we solve the Stokes equations with a convective term (1.3) (1.4), (1.5) on a three-dimensional domain.

- Test problem CAVITY2D: In this example, we solve the driven cavity problem on the unit square.

- Test problem LSE: In this example, we solve a linear system of equations $Ax = b$ where the matrix A and the vector b are read from a file.

In the case of the convection-diffusion equation we assign the boundary conditions and right-hand sides such that the exact solution is given by the function $u(\mathbf{x}) = \sum_{i=1}^{d} x_i^2$. In the case of the Stokes problem with a convective term, we pose the problem such that the exact solution is given by

$$u(\mathbf{x}) = (\sin x \sin y, \cos x \cos y)^T, \ p(\mathbf{x}) = 2 \cos x \sin y - 2 \sin 1(1 - \cos 1)$$

in the two-dimensional case and

$$u(\mathbf{x}) = (\sin x \sin y, \cos x \cos y, 0)^T, \ p(\mathbf{x}) = 2 \cos x \sin y - 2 \sin 1(1 - \cos 1)$$

in the three-dimensional case. The convection directions, boundary conditions and right-hand sides can be easily modified to create new test problems.

The user of these test problems has the choice to alter several parameters that control the solution process. The list of possible options is given in Table 8.

In the following, we will give an example for the CONVDIFF2D test problem. The program call

stokes -c circle -d -e 0.001 -g 7 2 -i grids/start1.tri -l 1 -m i -n 5 -o hfvs

Table 8 Options for solution process.

option	description
-c	name of convection direction (e.g. xline, curve, circle, fourcircles)
-d	draw
-e <value>	set epsilon
-g <number> <xi>	criterion and xi for graph definition
-h	to show this message
-i <filename>	start triangulation
-l <number>	minimal level (in multigrid iteration)
-m <sort>	ordering strategy in the case of several levels where <sort> can be 'p' (projected), 'n' (nested) or 'i' (independent).
-n <number>	number of levels
-o <order>	ordering technique where <order> can be chosen from n (none), c (concentric, minimal cycles), C (concentric, maximal cycles), pfvs (planar fvs), hfvs (heuristic fvs), ghfvs (global heuristic fvs) and r (rcm)
-r	use reaction term ("onereac")
-s <number>	number of iteration steps
-v <number>	verbosity level (w.r.t. ordering techniques)

```
main (...) ; chosen parameters :              time for net refinement : 0.020s
Start/Endlevel: 1 / 5                          time for discretization : 0.060s
Cycle(s) in MG: 1 — smoothing: 1/1            time for finding FVS : 0.050s
Number of Nets: 5                              Number of FVS nodes in net 0 : 0
Number of iterations: 100                      Number of FVS nodes in net 1 : 0
ordering: heuristic FVS                        Number of FVS nodes in net 2 : 0
search in graphs: dfs                          Number of FVS nodes in net 3 : 2
graph criterion / xi: E7 / 2 / edge-weight     Number of FVS nodes in net 4 : 6
Sort on Levels: independent                    building inverse permutations
epsilon: 0.001                                 time for permuting matrices and rhs's: 0.000s
convection direction: circle                   time to decompose on coarse grid (LU): 0.000s
reaction term: no
triangulation file: grids/start1.tri           backward Gauss-Seidel
                                               MG (F-BGS)::solve () : running (225 unknowns)
    net 0: 0                                    desired reduction in accuracy reached
    net 1: 1                                    average convergence rate after 6 steps = 0.202528
    net 2: 9                                    start accuracy = 6.562514e+00
    net 3: 49                                   reached accuracy = 4.528754e-04
    net 4: 225                                  time for multigrid: 0.040s
```

Figure 5 Output for CONVDIFF2D test problem

produces the output displayed in Figure 5 as well as files *fnet.ps*, *vecField.ps*, *graph.ps* and *matrix.ps* that are displayed in Figure 6 and show the refined grid, the vector field defined by the function b charactizing the convection in the convection-diffusion equation (1.1), the matrix graph (already reduced by the feedback vertex set vertices, here depicted for a coarser level), and the sparsity pattern (w.r.t. the 'large' matrix entries) of the reordered matrix.

In this example, the initial triangulation given in the file *grids/start1.tri* is refined four times which leads to the grid in Figure 6 (1st). Next the convection-diffusion equation with $\epsilon = 0.001$ and the convection *circle* defined by $b(x,y) = \begin{pmatrix} 0.5 - y \\ x - 0.5 \end{pmatrix}$ (see Figure 6, 2nd) is discretised. Based upon a criterion, here the so-called criterion $E7$ with parameter 2 (for various criteria to

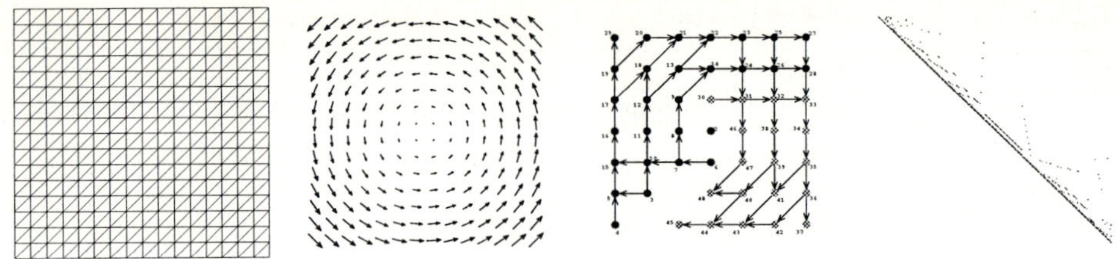

Figure 6 Triangulation (1st), convection (2nd), graph (3rd) and sparsity structure (4th)

define matrix graphs of dominant entries see [3]), a matrix graph of dominant entries is created for which a feedback vertex set is determined using the feedback vertex set algorithm for planar graphs. Such feedback vertex sets are determined independently on all levels. Next corresponding orderings (and their inverses) are determined and the corresponding permutations are executed. An exact LU decomposition is calculated on the coarsest level for the exact solution on this level in the multigrid method. Finally, the system of equations is solved using the multigrid V-cycle, with one presmoothing and one postsmoothing step of the backward Gauß-Seidel iteration. The results of the iteration which was run in mode 1 (see Table 6) are displayed according to the chosen level of verbosity (here 5, see Table 7).

Acknowledgement. The authors want to thank Thomas Probst for his valuable contribution to the development of the program.

References

[1] R. E. Bank, B. D. Welfert, and H. Yserentant. A class of iterative methods for solving saddle point problems. *Numerische Mathematik*, 56:645–666, 1990.

[2] J. Bey. Tetrahedral grid refinement. *Computing*, 55:355–378, 1995.

[3] Sabine Le Borne. Ordering techniques for two- and three-dimensional convection-dominated elliptic boundary value problems. *Computing*, (64):123–155, 2000.

[4] D. Braess and R. Sarazin. An efficient smoother for the Stokes problem. *Applied Numerical Mathematics*, 23:3–19, 1997.

[5] S. Müller, A. Prohl, R. Rannacher, and S. Turek. Implicit time-discretization of the nonstationary incompressible navier-stokes equations. In *Fast Solvers for Flow Problems*, volume 49 of *Notes on Numerical Fluid Mechanics*, pages 175 – 191. Vieweg, 1994.

[6] Thomas Probst. *Mehrgitterverfahren für Konvektionsdiffusionsgleichungen*. PhD thesis, University Kiel, 1999. dissertation.de, No. 100.

[7] H. Rentz-Reichert. *Robuste Mehrgitterverfahren zur Lösung der inkompressiblen Navier-Stokes Gleichung: Ein Vergleich*. PhD thesis, Institut für Computeranwendungen der Universität Stuttgart, 1996.

[8] H.G. Roos, M. Stynes, and L. Tobiska. *Numerical methods for singularly perturbed differential equations: convection diffusion and flow problems*, volume 24 of *Computational Mathematics*. Springer, Berlin, 1996.

Development of Navier-Stokes Solvers on Hybrid Grids

D. Hänel[1], A. Dervieux[2], O. Gloth[3], L.Fournier[4], S. Lanteri[5], R. Vilsmeier[6]

[1,3,6]: Uni-Duisburg, FB7/IVG, 47048 Duisburg, Germany
email: {hj454ha,oliver,hj000vi}@vug.uni-duisburg.de
[2,4,5]: INRIA, 2004, route des lucioles - 06902 Sophia-Antipolis, France
email: {Dervieux,Luc.Fournier,Lanteri}@Sophia.Inria.FR

Summary

The ongoing joint work of two research groups to exploit the advantages of hybrid grids for the numerical simulation in fluid dynamics is presented. After referring to some general issues of the present solution method, the development of an FAS multigrid method is outlined in more detail. Further on three alternate attempts for hybrid grid generation are presented.

1   Introduction

Considering the classical dispute weather structured or unstructured meshes are more promising for the use in computational fluid dynamics, both parties have strong points in their argumentations. Structured grids are usually build from quadrilateral elements which enable a high level of accuracy per grid point employed and nodes are numerable along grid lines. The latter property simplifies the development of efficient numerical methods on this grid type essentially and concerning efficiency, also hardware related aspects have to be mentioned, such as cash alignment and pipelining or memory requirements. Unstructured grid techniques by way of contrast, have their strong points on the geometrical flexibility and adaptivity. The obvious attempt to combine the advantages of both methods yield to hybrid techniques. The idea is generally to enable easy and automatic grid generation and adaptivity, while taking profit of the increased accuracy and efficiency of structured or semi-structured zones, wherever possible.

Many questions arise in the realization of the proposed aim. These include the formulation of efficient numerical methods on generalized grids as well as the generation and adaptation of the meshes itself. Beside efficient numerical methods, computational efficiency is important with special consideration of modern computer architectures, as distributed memory or virtual shared memory parallel. In the frame of this paper, the general computational approach is shortly presented, followed by a more detailed description of an explicit multigrid technique on hybrid grids and a some alternate methods for the generation of such grids.

2   Method of Solution

The principal method of solution employed here has previously been presented [1]. Therefore the description here is restricted to a short overview.

2.1   Governing Equations and Discretization

The general form of the unsteady conservation equations in integral form reads:

$$\int_V \vec{Q}\,\mathrm{d}A \;+\; \oint_{\delta V} \vec{H}\vec{n}\,\mathrm{d}S \;=\; 0 \; . \tag{2.1}$$

The equations are discretized via finite volume approach. The equation below describes the advance in time for a discrete Volume Vd:

$$\left.\frac{\Delta \vec{Q}}{\Delta t}\right|_{Vd} \;+\; \vec{Res}_{\Delta,Vd} \;=\; 0 \; . \tag{2.2}$$

Disregarding structured regions, the basic algorithms are unstructured and may not rely on a natural ordering of nodes and elements. Therefore an artificial ordering, a data structure, is required. Furthermore an element of the data structure will be called a molecule. To perform a time step for the equation given above, the discrete residual $\vec{Res}_{\Delta,Vd}$ has to be constructed upon such molecules:

$$\vec{Res}_{\Delta,Vd} = \frac{1}{V_{Vd}} \sum_{i=1}^{nr(Vd)} \vec{H}_{\tilde{i}(i)}\,\vec{n}_{\tilde{i}(i)}\,\Delta A_{\tilde{i}(i)} \quad , \tag{2.3}$$

$nr(Vd)$ is the number of molecules contributing to the residual, V_{Vd} is the corresponding volume. $\vec{H}_{\tilde{i}(i)}$ is the flux vector, $\vec{n}_{\tilde{i}(i)}$ the normal vector and $\Delta A_{\tilde{i}(i)}$ the area of the control interface supported by a molecule $\tilde{i}(i)$. The control interfaces have to enclose the volume completely.

The very general description of the finite volume approach opens up a large variety of possibilities how to define volumes and corresponding data structures. For the present studies it was found, that a node centered approach is useful.

2.2   Data structures employed

All methods are programmed in the object oriented language C++. It is important to note that, in the frame of object oriented programming, the data structures employed are less authoritarian than they usually are in sequential programs. The reason is, that additional structures required can easily be created and deleted within a running program. In sequential programs, especially with static memory as in Fortran, this is very difficult. Therefore algorithms are usually written around a once chosen data structure synonymous for a storage concept. Further on, due to the the possibility of virtual function calls (or also the template mechanism) the same task can be performed employing different data structures, as long as the data required are available. For the present node centered approach the previously introduced **Node-Node** structure is employed.

In contrast to an edge-based data structure, the **Node-Node** structures access edge and neighbour-node information as sets stored per node. Major advantage of this data structure is, that a partial residual evaluation for a set of nodes is simple, thus reducing interdependencies for parallel computations. Further on, the agglomeration process required for multigrid computations is relatively simple to formulate, since neighbouring nodes can be grouped recursively.

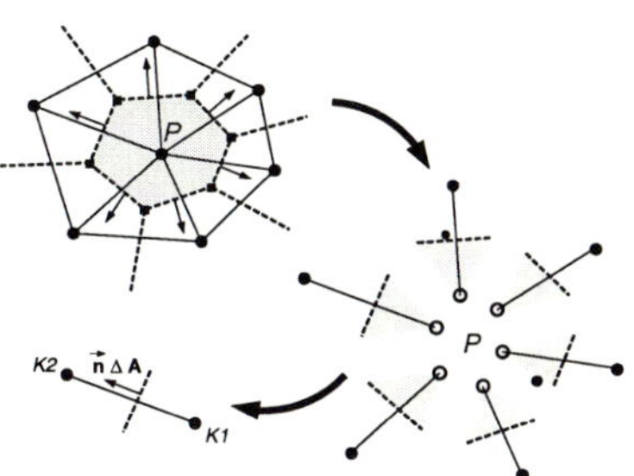

3 Development of a Hybrid Multi-Grid

The resolution of non linear PDEs can be done using time explicit as well as time implicit formulations. By saying explicit, it is meant that the variables on a new time level only depend on the state of the previous time level. Implicit schemes are generally based on solving a linear system of equations for every time step. They are known to be numerically efficient. Unfortunately, new physical problems or new discretizations require more implementation work. Consequently, explicit methods are suitable for an open development platform if their inefficiency for steady state problems can be overcome. Since a few decades, multigrid methods are used to accelerate the convergence properties of numerical methods. Due to the non linearity of explicit schemes, a Full Approximation Scheme (FAS) multigrid method (described in [2]) is useful. Explicit schemes provide an easy way of implementing different equations in the existing development platform for methods on unstructured grids. It seems to be natural to also find a sufficient way of generating mesh sequences for the FAS algorithm, without demanding too much user interaction. Please note, that user does not only stand for end users, but also for members of a research team working on methods for arbitrary physical problems employing a common basic software.

The volume agglomeration technique as presented in [3] is considered as a black box method for generating finite volume coarse levels. Some more recent works have proved the efficiency of this method in the field of computational fluid dynamics [4].

3.1 The FAS method

The FAS multigrid method is used to solve the following non linear equation:

$$\Phi(u) = S \quad ,$$

(3.4)

where u denotes the field of unknowns, Φ the non linear operator and S a source term. h and H, used as subscripts, refer to the fine and the coarse grid. As in every multigrid method we need inter-grid operators. I_h^H and $\tilde{I}_h^H$ are the operators to restrict the unknowns and the residual from the fine to the coarse grid. The prolongation of the correction from the coarse to the fine grid is written as I_H^h. The ideal two grid FAS algorithm can be described using four steps:

1. Pre-smoothing: If $u_h^{(k),0} = u_h^{(k)}$ is the initial value of the unknowns, we find a new estimate $u_h^{(k),1}$ of this field using several iterations of an explicit smoother (simply an explicit iteration scheme of the equation). $r_h^{(k),1} = S_h - \Phi_h(u_h^{(k),1})$ now represents the residual of this operation.

2. Solution of the coarse grid equation: The following equation in u_H is solved.

$$\Phi_H(u_H) = S_H = \Phi_H(I_h^H u_h^{(k),1}) - \tilde{I}_h^H r_h^{(k),1} \quad .$$

(3.5)

For the ideal two grid algorithm this has to be done until a sufficient level of convergence is reached.

3. Correction: Using the converged solution of [3.5], it is now possible to correct the variables on the fine grid.

$$u_h^{(k),2} = u_h^{(k),1} + I_H^h(u_H - I_h^H u_h^{(k),1}) \quad .$$

(3.6)

4. Post-smoothing: A new estimation $u_h^{(k+1)}$ of the field is found, using several iterations of the explicit smoother, with $u_h^{(k),2}$ as initial value.

3.2 Volume agglomeration technique

The agglomeration technique is used to build a set of coarse finite volume grids. A coarser level can be obtained by clustering neighbouring volumes. Each cluster of volumes will form a control volume of the coarsened mesh. Used recursively, this technique produces a sequence of grids with nested control volumes. Figure 1 shows the principle of this agglomeration algorithm applied to a mixed element sample mesh. The most used technique to agglomerate control volumes is a greedy-like algorithm. But for 3D problems, Mavriplis[4] proposes a priority list to initialize a pure greedy technique. In [5] a coloring edge-based method is found. These variants show an interest on the quality of the agglomerated cells. It can be observed that the regularity of the agglomerated levels influences the convergence rate.

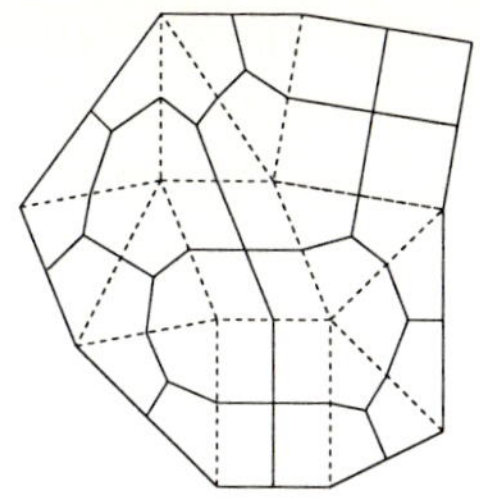 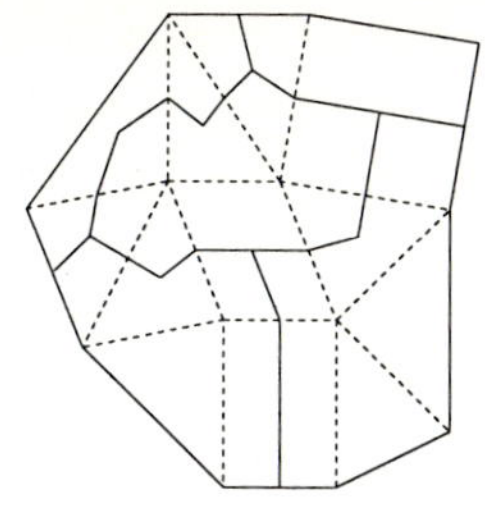

Figure 1 Elements and associated control volumes of fine grid (left) and agglomerated control volumes (right). Dotted lines: elements, solid lines: control interfaces.

Optimizing the shape of agglomerated cells

In the present work an optimization strategy based on the detection of so called 'horseshoe' cells is used. Every cell on every level of the grid sequence can be viewed as a closed polygon with a set of normal vectors. The idea of the optimization strategy is to construct an ellipse with a shape similar to the original cell (see figures 2 and 3). Once this shape has been constructed, the area ratio $\frac{A_{ellipse}}{A_{cell}}$ is a measure for the presence of horseshoes. To construct the replacing ellipse, values for r_{min} and r_{max} are needed. They are defined as follows:

$$r(\alpha) \quad = \sum_{i=1}^{N} \left| \begin{pmatrix} \cos\alpha \\ \sin\alpha \end{pmatrix} \cdot \vec{n}_i \right| \tag{3.7}$$

$$r_{min} \quad - \min(r(\alpha))\,, \alpha \in [0,\Pi]) \tag{3.8}$$

$$r_{max} \quad = \max(r(\alpha))\,, \alpha \in [0,\Pi]) \quad . \tag{3.9}$$

Where N is the polygon's number of segments and $\vec{n}_i$ is the normal vector of segment i. Please note that 'horseshoe' like cells appear to be bigger than they really are, when using this method to evaluate r_{max}. This gives a good tool to spot those cells and improve them.

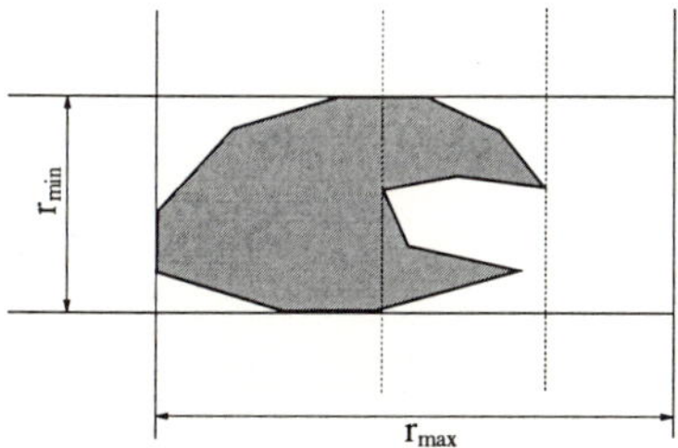 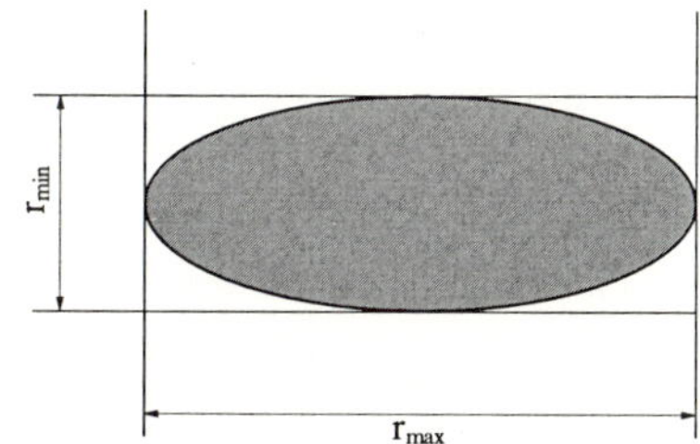

Figure 2 Cell to evaluate (left)

Figure 3 Replacing ellipse (right)

Besides this shape criterion other criteria can be employed. For example, the ratio $\frac{r_{max}}{r_{min}}$ can serve as an estimate for anisotropy of the agglomerate. Based on an already agglomerated level, the optimization is used to minimize a function, which consists of several evaluation criteria. Corrections are realized by swapping contributing fine grid cells between neighbouring agglomerates.

The following numerical experiment compares the convergence rate of an optimized and another not optimized agglomeration. As computational example serves a subsonic ($Ma = 0.5$) flow around a NACA0012 airfoil. The angle of attack is 2 degrees. Target discretization is a second order Roe-scheme and as driver a first order Roe-scheme is used. Figures 4 and 5 show a comparison of the coarsest level for the standard and optimized agglomeration. For this test case a small improvement in terms of convergence behavior can be observed, figure 6. For anisotropic regions the coarse grid quality will become a crucial requirement to obtain good convergence improvements (see also [6]). Another problem is introduced by quadrangular regions in hybrid grids. The natural orientation and node-numbering yields to sometimes unwanted effects (see also the NACA0012 test case later in this report). Very often the main axes of bilinear elements are turned by 45 degrees when going to a coarser grid level. This leads to rhombic elements, which are most unwanted, especially in anisotropic layers. Besides the above described shape optimization there is another approach to overcome this problem. In many cases bilinear regions could be described as structured subregions. This would offer the possibility to use efficient memory alignment, which is very important to obtain a good performance with modern CCNUMA machines. On the other hand it would offer a perfect agglomeration, since the structure and orientation are known in this case.

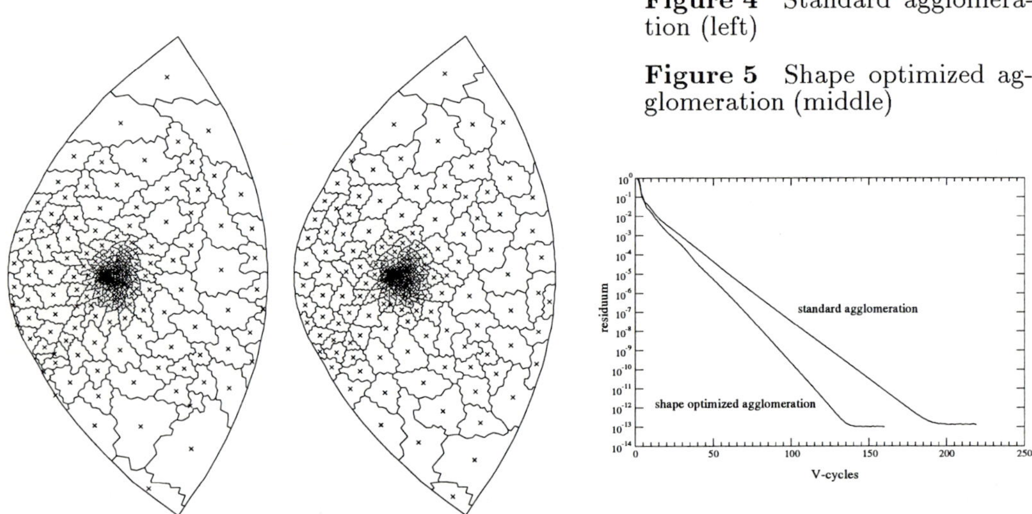

Figure 4 Standard agglomeration (left)

Figure 5 Shape optimized agglomeration (middle)

Figure 6 Convergence history

3.3   Implementation of an FAS multigrid

Spatial discretization using the defect correction technique

One major disadvantage of the above described coarsening technique is the difficulty to formulate higher order discretizations on the coarse levels. Typically, point coordinates would be needed for these. The agglomeration technique yields to sometimes irregular cells with point coordinates outside their domain. Furthermore experience shows that a second (or higher) order discretization on such an agglomerated grid is very critical in terms of stability for the explicit time stepping. Therefore it does not seem to be reasonable to apply a higher order approach on the coarse grids. To overcome this the following technique is used to have second order discretizations in space: The second order flux computation is performed on the fine grid using the MUSCL interpolation [7]. This technique is based on a reconstruction of the state on the cell interfaces, using linear extrapolation. The extrapolation is performed using the gradients of the variables, which are computed using Green's formula for field nodes and a least square approach for boundary nodes. The defect correction method [8] proposes to solve the equation [3.4] iteratively by building a convergent series, based on a problem which is simple to inverse. At each iteration the following problem has to be solved:

$$\Phi_{h,1}(u_h^{(n+1)}) = S_h^{(n)}$$
$$\text{with} \quad S_h^{(n)} = S_h + \Phi_{h,1}(u_h^{(n)}) - \Phi_{h,2}(u_h^{(n)}) \quad ,$$

(3.10)

where $\Phi_{h,1}(\ )$ and $\Phi_{h,2}(\ )$ respectively denote a first and a second order discretization on the fine level. Each system [3.10] is partially solved using one cycle of the FAS algorithm. When converged (i.e. $\Phi_{h,1}(u_h^{(n+1)}) = \Phi_{h,1}(u_h^{(n)})$) u_h is the solution of [3.4]. $\Phi_{h,2}(\ )$ is often called target discretization, whereas $\Phi_{h,1}(\ )$ is called driver discretization of the defect correction scheme.

Runge-Kutta for smoothing

The FAS algorithm allows to treat non linear equations. On every level the exact equation, including all boundary conditions, has to be modelled. In the presented FAS implementation an explicit Runge-Kutta is used for smoothing, as well as for solving the equation on the coarsest level. For linear multigrid methods a Jacobi or Gauss-Seidel relaxation is often used for smoothing. In [9] a study of this technique for a set of linear model equations can be found. Unfortunately, this requires a point implicit formulation of the discretization. To allow the method to be easily used with different discretizations, a fully explicit Runge-Kutta method has been chosen. The main problem when using Runge-Kutta as a smoother, is that the smoothing properties are dependent on the number of sub-steps, the coefficients used, and the CFL number. This is shown by a von Neumann analysis of the one dimensional convection equation [3.11]. It is discretized using an adjustable first order upwind scheme [3.12]. $\beta = 1$ represents a full upwind and $\beta = 0$ a central scheme. Time integration is done by a three step Runge Kutta scheme with the coefficients $\alpha_1 = \frac{1}{4}, \alpha_2 = \frac{1}{2}, \alpha_3 = 1$.

$$\frac{\partial \phi}{\partial t} + u \frac{\partial \phi}{\partial x} = 0 \quad ,$$

(3.11)

$$\frac{\partial \phi}{\partial t} + Res(\phi) = 0 \quad ,$$
$$Res(\phi) = \left(\beta \frac{\phi_i - \phi_{i-1}}{\Delta x} + (1 - \beta) \frac{\phi_{i+1} - \phi_{i-1}}{2\Delta x} \right) \quad . \tag{3.12}$$

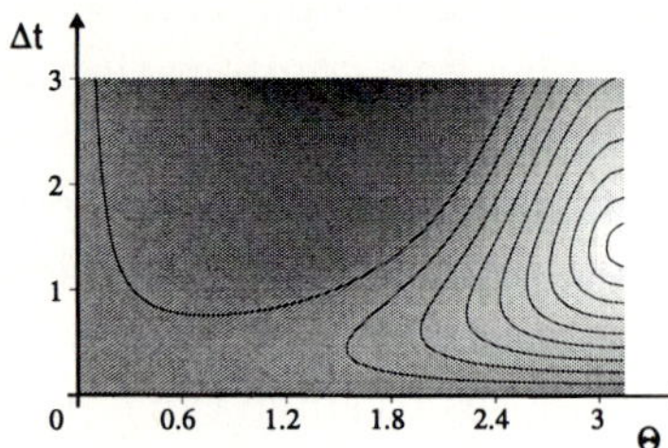
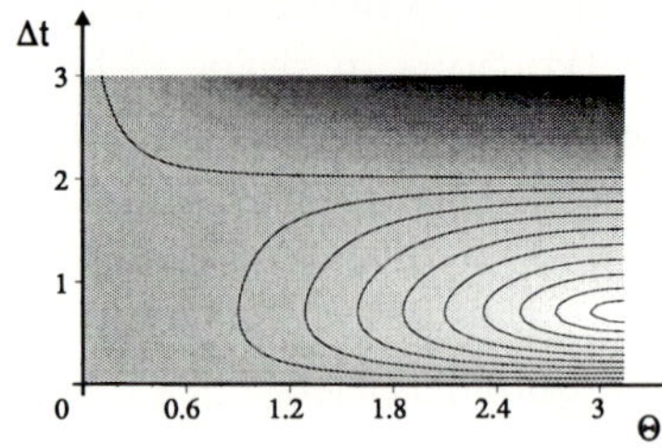

Figure 7
Amplification factor for $\beta = \frac{1}{2}$ (left)

Figure 8
Amplification factor for $\beta = 1$ (right)

The von Neumann analysis gives an amplification factor of this scheme as a function of the time step Δt and an angle $\Theta \in [0..\Pi]$. $\Theta = \Pi$ represents the shortest and $\Theta = 0$ the longest resolvable wavelength. As it is noted in [9], the smoother does not have to be convergent. It has to be efficient for high frequency disturbances. Unfortunately, within the scope of this work, it has not been possible to apply a theoretically unstable Runge Kutta scheme with good smoothing properties for high frequencies. Figures 7 and 8 show the amplification factor as a function of Δt and Θ. Dark color represents a high amplification. Isolines are drawn for amplification values of $(0.1, 0.2, 0.3, ..., 1.0)$. Please note that values > 1 indicate an unstable scheme. The plots show that the smoothing properties of a Runge-Kutta integration strongly depend on the time step width (CFL number). They also depend on the type of the discretization (half or full upwind). Unfortunately it is not possible to ensure a certain CFL number on a general mesh. Hopefully, future studies will lead to improved convergence rates.

Boundary conditions

The FAS multigrid algorithm needs a correct expression of boundary conditions on every grid level. As explained in 3.1 a partially converged solution is obtained on each grid. On every level the same set of equations is modelled. The only difference can be found in the position of the boundary nodes. Only on the finest mesh the nodes are on the boundary itself (half cells). This makes the implementation of strong (nodal) formulations for boundary conditions easy. For the agglomerated levels this is not the case anymore. The node position can no longer be assumed to be on the boundary. Figures 9 and 10 illustrate this.
Furthermore, geometric properties (e.g. surface normal vectors) are difficult, if not impossible, to reconstruct on the coarse levels. To still imply the boundary conditions on the coarsened levels, they are included in the source term of the defect correction scheme. The second order discretization $\Phi_{h,2}(u_h^{(n)})$ includes a nodal formulation of the boundary conditions (see equation [3.10]). This formulation works well in case of the compressible Euler equations. Here a nodal boundary

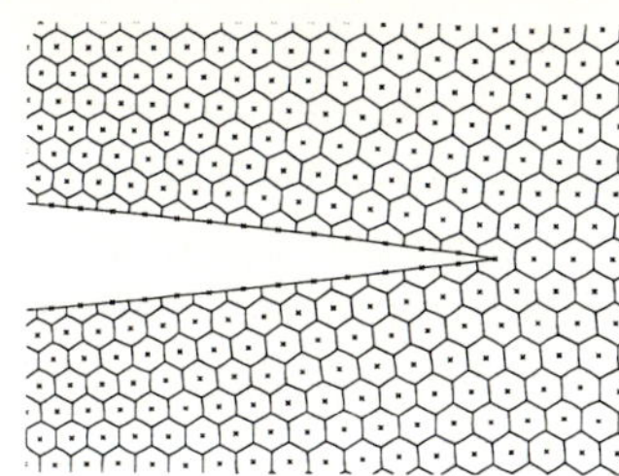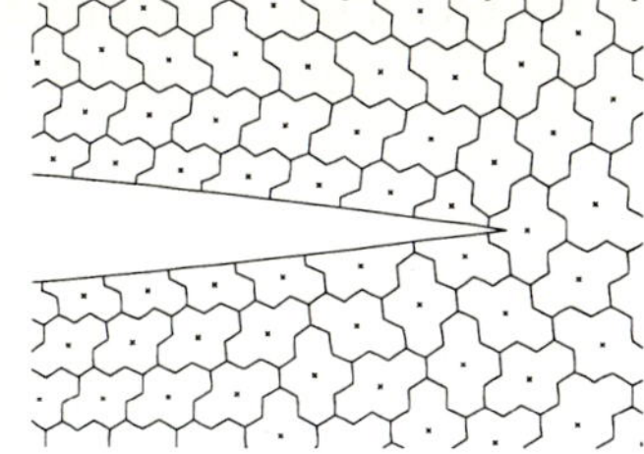

Figure 9 Fine grid control volumes (left)

Figure 10 1st agglomerated level (right)

condition forces the velocity vectors to be parallel to surfaces. For Navier Stokes computations, where a pure Dirichlet boundary condition is demanded (non slip condition on surfaces) it is unfortunately not working as well. The pure source term formulation of Dirichlet boundaries leads to strong stability problems. A so called ideal two grid cycle might lead to an unstable scheme in this case. Ideal cycle means that the iteration on the coarse grid has to converge for every V-cycle. In this case the point in time (iteration progress) where the boundary source time has been computed is far away from the actual state during the coarse grid iteration. This contradicts with the demand of a flexible algorithm. So the boundary condition problem is only partially solved by now. However, one explanation is that for the inviscid computations a weak formulation of the boundary conditions (pressure flux on surfaces) is always present. This is not the case for viscous computations. To compute a friction-caused flux on non-slip surfaces, velocity gradients are needed. Unfortunately, these gradients are very hard to compute on an agglomerated mesh. The boundary conditions are thus subject to further studies as well.

Inter-grid operators

Transfer operators are usually considered to be the weak part of the volume agglomeration technique. The restriction for the residual is simply a summation of all fine grid cells which contribute to one agglomerated cell. For the variable vector an averaging over all contributing cells is employed. Please note that this is done, using the control volume sizes as weighting factors. There are two different kinds of prolongation (interpolation) operators implemented. One is a canonical injection as in [3], meaning that the fine grid values are simply set to the corresponding coarse grid values. Another operator is an interpolation using a least squares approach. Furthermore there is the possibility to smooth the correction. Best results could be obtained with the simple canonical injection together with a smoothed correction.

3.4 NACA0012 test-case

As example for a hybrid multigrid computations serves the following inviscid computation of a NACA0012 airfoil in a channel. Please note that it is located slightly out of the channel centre. This test case is particularly hard to converge for an explicit scheme, due to the reflections on both ends of the channel. The inflow Mach number of this case was $Ma_\infty = 0.5$. Figure 3.4 shows the part of the mesh where the airfoil is located. The rest of the channel is filled with Cartesian elements. Figure 11 shows a zoom of the fine grid around the leading edge and a sequence of nested control volumes for the multigrid V-cycle. The whole mesh consists of approximately 55,000 nodes. A very good improvement for the convergence rate can be observed. See figure 3.4 for a comparison between the 5 level multigrid method and a simple explicit iteration. The horizontal axis of this plot has been scaled to represent the same CPU-time by means of work-units (1 work-unit: CPU-time for one explicit time step). In figures 3.4 and 3.4 iso-plots for the pressure and the Mach number are shown.

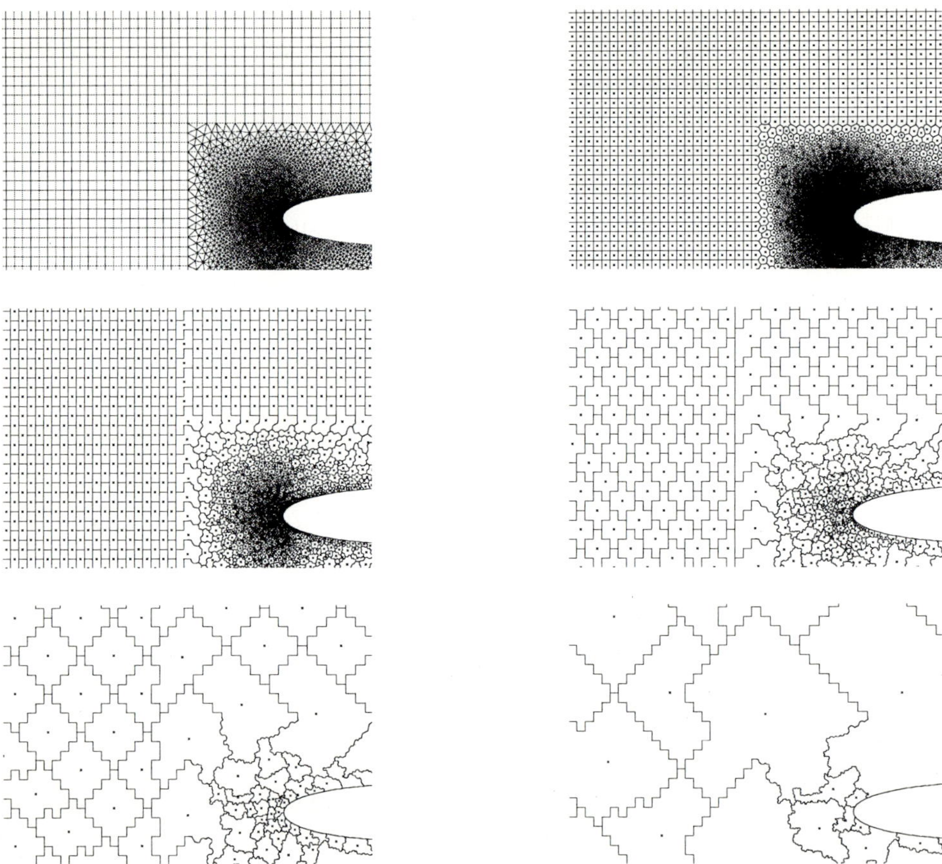

Figure 11 Sequence Initial fine mesh (top left) and corresponding control volumes (top right). Four figures below: Sequence of control volumes from first to 4th coarsening level

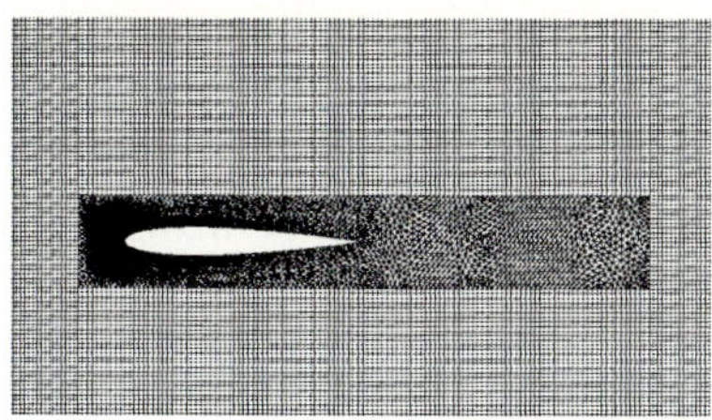

Figure 12 Computational mesh

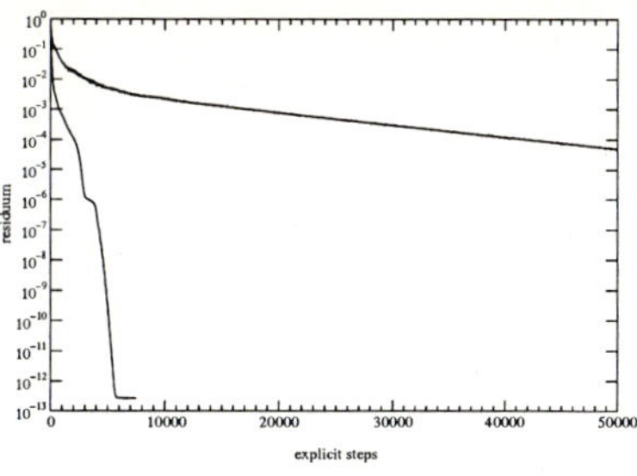

Figure 13 Convergence history in work-units. Explicit (slow) and FAS-method

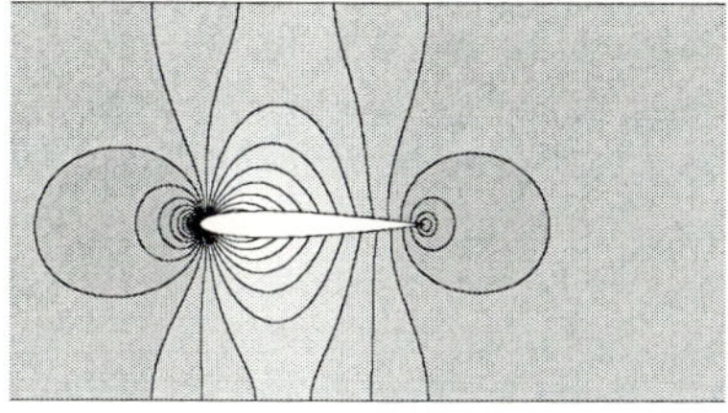

Figure 14 Pressure

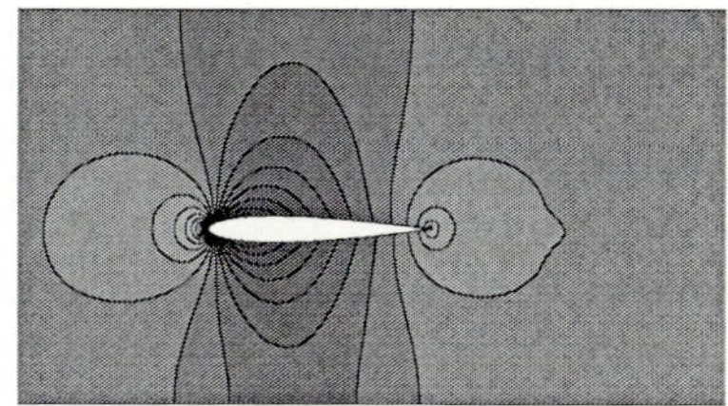

Figure 15 Mach number

4 Grid generation

The development of hybrid grid generators is an ambitious goal. The aim is a meshing tool, as flexible and automatic as a program to generate unstructured grids, but employing bilinear elements according to two major criteria:

- Since the use of simplex elements is critical in regions of strong anisotropy, bilinear elements can improve precision and robustness.

- Due to topological reasons, the use of bilinear elements yields to less computational work on Node-centered arrangements.

Major reason for the use of hybrid grids motivating this project is the first item. However it was found that the computational efficiency can also be heavily improved, taking profit of the natural ordering of structured blocks. Consequently, also the Cartesian orientation of mesh regions can be of profit if, in contrast to a general unstructured discretization, flux computations are transformed in the main axis of these. A corresponding mesh generation strategy, called here as "block out strategy" is explained in detail below.

According to the large amount of degrees of freedom, many approaches to hybrid grid generation are possible. The methods in development for the present project can be categorized in two major classes:

- Zonal approaches

- Methods based on element conversion

The present work is mainly focused on hybrid grid generation via element conversion. Zonal approaches, more classical and less challenging, have however been used as alternates, since these methods are relatively simple and require only independent meshing algorithms for structured and unstructured grid generation. Also, the block out strategy explained below can be considered as a member of the class of zonal methods.

4.1   General aspects on mesh generation

Before proceeding, some general aspects of grid generation are emphasized, mainly to justify the approaches chosen here.

Mesh generation as optimization process

From a very principal point of view, mesh generation can be seen as a mixed discrete analogue optimization process. The number of nodes and the connection of these to elements are discrete, while the exact coordinates of the nodes are analogue degrees of freedom.

Comparing the generation of unstructured and structured meshes, the main difference is, that in the structured case the topology of the meshes is given and thus the discrete degrees of freedom reduce to the choice of a number of nodes in each coordinate direction of a suitably transformed space. The optimization problem thus reduces predominantly to an analogue one for the position of the nodes used. In contrast, a complete optimization would, in the case of unstructured grids be computationally far too expensive due to a very high computational complexity. For practical use, reasonable approaches are thus required.

Conformal and non conformal meshes

Several methods for hybrid grid generation are proposed here. A first distinct differentiation is that of conformal or of non conformal mesh generation. Conformal grids are grids for which a clear and unique neighbourship relation exists between the elements. Such arrangements are, first of all more suitable for conservative discretizations. In non conformal mesh generation, sub-partitions of the mesh may overlap. The current and proposed studies relate to conformal meshes only.

4.2 Mesh generation via element conversion

This method has been previously presented and was published accordingly [10]. Most of the descriptions here are thus restricted to a short overview. However, until present a fully reliable system is not yet available, but the method is still very promising. On the other hand, this aim is considered very ambitious and a reliable method would solve most grid problems today encountered with unstructured grids.
The basis for the hybrid generation system is an unstructured field method. This method consists of a set of mesh modification tools applied in recurrence. These are:

- Insertion of additional nodes

- Deletion of nodes via edge-shrink operations

- Mesh reconnection via diagonal swap (5-node and full neighbour swap in 3D)

- Mesh smoothing in a minimization strategy based on circumcircles (circumspheres)

The extension of the method for the generation of hybrid grids includes additional element conversion tools. Several strategies are followed. First, bilinear cells can be created within triangular cells. The advantage of this strategy is, that the orientation of newly created bilinear elements is relatively simple, since it can rely on an edge (face) of the simplex, in which it is created.
A drawback of this method is, that it is not fully transferable to 3-D. Therefore an alternate method, the element conversion via edge creation/deletion is a matter of current and future studies. Splitting up a node of a triangle converts this to a quadrilateral. Similarly, a tetrahedron can, in a first stage be converted to a pyramid and in a second stage to a prism. Further details of this current development are outlined in [10].
A major challenge is the topological structure of regions with converted bilinear elements. These elements have been shown to be most useful, if aligned to anisotropic solution features, as boundary layers and it is required, that they satisfy quality criteria similar to the ones for purely structured grids. These are orthogonality and a smooth point distribution. Approaching the problem from unstructured mesh generation, these properties are however not easy to obtain. In a previous work on hybrid grids [1] the authors introduced a "crystallization process" for the growth of bilinear structured regions within an unstructured mesh. The idea of this approach is to build new bilinear elements (or convert simplex elements to bilinear elements) adjacent to already existing ones. This crystallization process was formulated in a way to enforce that new bilinear elements match the structure of existing parts of the mesh.

Concerning the creation of substructures within the mesh it has however been extremely interesting to observe self-ordering mechanisms for simplex mesh generation in 2-D. By this, the formation of fully regular regions is meant. These are regions within triangulations, containing at times up to some thousands of triangles in a structured ordering, although no ordering has been enforced.

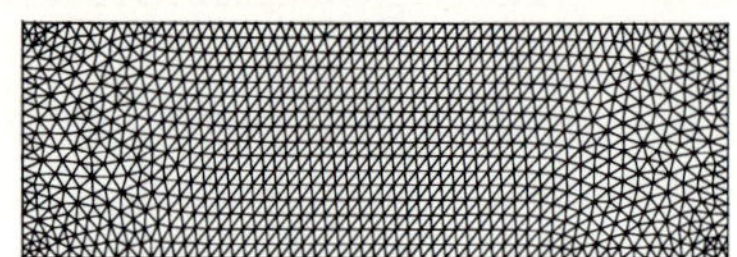

Figure 16: Example for self ordering in unstructured mesh generation.

The effect has not been much considered yet, since it does not play an essential role for unstructured meshes. Also, the mathematical background for this effect still bears unknowns. In the context of hybrid grids with corresponding structured sub-partitions, the effect might be of essential benefit. Unfortunately, self-ordering could not yet be obtained for unstructured quadrilateral grids.

4.3   Zonal methods for hybrid grid generation

As alternate to the element conversion methods above, the more classical zonal approaches are still interesting. In these methods, the computational domain is subdivided in several sub-domains, for which corresponding meshing techniques are employed. This method is relatively simple and some experiences were previously gained. The major drawback of the method is that the zones for which a certain type of mesh is required must be known in advance. This makes the meshing type less adaptive, specially if anisotropic adaptation is required. The method is however very useful for large scale aerodynamics, if boundary layers can be discretized on structured or semi-structured sub-grids, while the zones apart can be filled by unstructured meshes.

The question arises, how these methods can act as automatic generators Although major attention is drawn towards a fully automatic hybrid grid generation method, the zonal approach is still a topic to study, mainly due to two reasons:

- The advantages of structured and, even better, Cartesian mesh regions.

- Advances concerning the automatism of zonal mesh generation methods.

The corresponding approaches result in a meshing strategy, called the "block out" strategy here. This meshing strategy can somehow be assigned to the group of zonal methods.

4.4   Block out method

The idea of the "block out" method is relatively simple but effective: Consider an unstructured grid, consisting mainly of simplex elements. Large parts of the volume can be replaced by structured blocks of bilinear, Cartesian elements, while

the "empty spaces" between the blocks can be triangulated using the standard
unstructured grid generators. In 3-D it is however required, that the quadrilateral
surfaces of the blocks are covered by pyramids to provide triangulated faces to
attach the tetrahedra. The advantage of such meshes is a higher computational
efficiency without loss of flexibility. However, this strategy does not solve the
problem of discretizing highly anisotropic regions:

Topological advantages: According-
ing to the topology of bilinear elements the relation between the number of edges to the number of nodes
is lower. Since for the nodal schemes
employed here, flux contributions
are computed per edges, this relation
affects directly the efficiency of the
method.

Element type	Av. number of mol. per node
triangle (2D)	3
quadrilateral (2D)	2
tetrahedron (3D)	around 7
prism (3D)	5
reg. hexahedron (3D)	3

Alignment in storage: Much efforts are spent concerning a good alignment of
data for unstructured grids, also in a previous work of the authors [11]. It was
found, that structured grids offer a natural alignment of multiple data locality,
allowing a very efficient use of the cache memory of RISC-based machines. In the
unstructured case, the access sequences can be optimized, however this optimization does not allow the same performances.

Cartesian computations: The computation on Cartesian blocks can further on
be severely enhanced, if the coordinate system is locally transformed according
to the axis of the blocks. In this case, a Cartesian flux in a single direction is
sufficient in contrast to a scalar multiplication of the complete flux with a cell
interface normal vector. Performance winnings can be extrapolated upon the experiences with Cartesian grids from other projects and will depend on the flux
formulation and dimension of the problem.

Lower storage requirements: If the block is not stored in the sense of the
data structure employed for the unstructured part, the memory associated with
this data structure can be saved.

Accuracy wins in Cartesian regions: The discretization on Cartesian grids
bears advantages concerning the accuracy, at least for classical projection methods. These advantages can be translated in a lower node density, again yielding
to a higher efficiency.

The above mentioned advantages are very promising for large scale computations. However, in consequence, some extensions to the computational programs
have to be made, which are not yet implemented. To exploit the advantage of
lower storage requirements, corresponding iterators for the blocks, differing from

todays list-oriented iterators, have to be created. Further on, the advantage of restricted flux evaluation in the transformed Cartesian directions requires, that corresponding templates for reduced computations are provided for each flux.

Strategies for the construction of block out meshes

In comparison to classical block structured mesh generation, the method proposed here is relatively easy to implement. Since unstructured grids can always be used to fill the empty spaces between the blocks consistent conformal meshes are created and no special treatment of block interfaces is required. Also, the blocks do not fill the domain entirely and the placement is thus relatively free. According to this liberty, the solution to the problem is again a question of optimization, bearing a remarkable potential. At present, blocks are placed interactively, but automatic procedures are possible. In principal, two methods are promising:

Setting blocks: Within this method, suitable positions for the insertion of a block are sought. After finding the position, the corresponding part of the unstructured mesh overlapping the Cartesian block is deleted, including some additional free space, typically one additional neighbourship layer. The empty space between the newly created block and the remaining unstructured mesh is then re-triangulated, thus coupling the mesh. All further mesh optimization tools can be applied to the surrounding unstructured grids, thus merging the block in the triangulation. Figure 17 illustrates this strategy.

Reversed punching: The method of reversed punching follows the opposite strategy. Structured blocks can be placed on the computational domain, disregarding objects or other incompatibilities. In regions of overlap, the blocks are punched out, again some larger than the overlapping zone itself, and the empty spaces are triangulated accordingly, figure 18. As an example, the mesh for the NACA airfoil in the channel (figure 3.4) was created in this way.

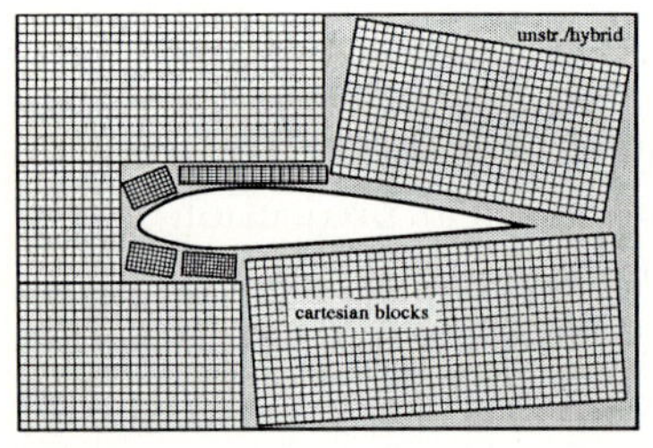

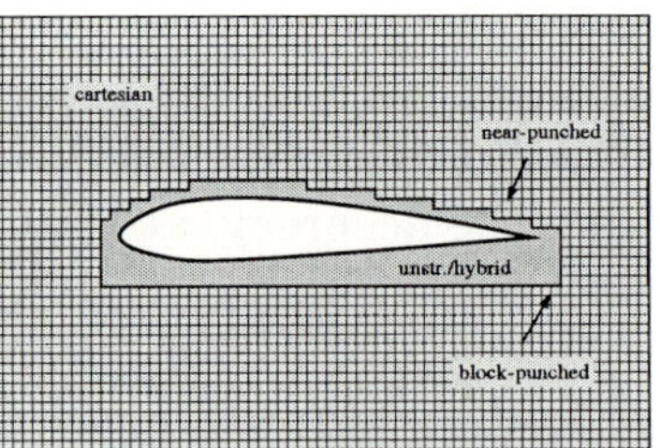

Figure 17 Non overlapping blocks and unstructured / hybrid regions (left)

Figure 18 Reversed punching method for grid generation (right)

As mentioned above, the method does not solve the problems encountered with highly stretched layers, where structured bilinear elements are preferred. To obtain layers of strong anisotropic resolution, the above methods have thus to be combined. Again an element conversion scheme can then be used to create bilinear layers, figure 19. Alternatively anisotropic clustering of overlapping blocks in these layers can be applied, figure 20. The latter option however spoils conformity and thus requires different algorithmic structures for the solution to treat overlapped partitions and non-conformal boundaries.

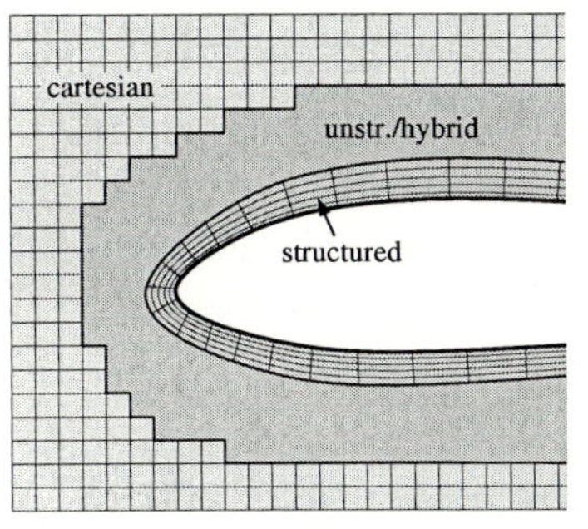

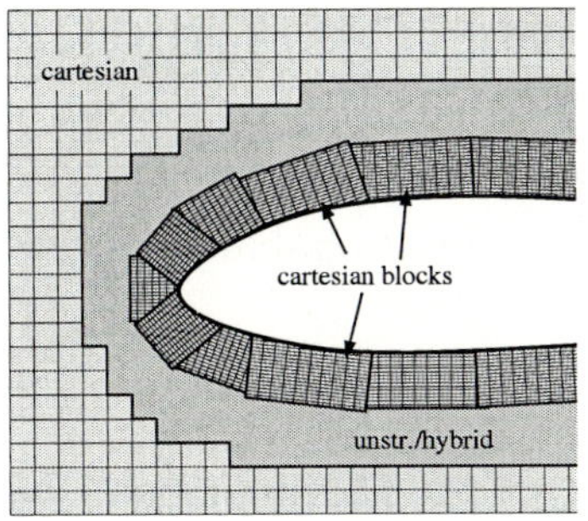

Figure 19
Stretched structured curvilinear sub-layer (left)

Figure 20   Non conformal overlapping set of boundary blocks for anisotropic layers (right)

As a very simple example the generation of a mesh for a 2D parabolic object in a triangular region is shown. A coarse Cartesian grid is first roughly adapted by a finer Cartesian block near the object. The object shape is then punched out up to a moderate distance from the object. In this test case, no triangulation of any empty space was performed, instead the punched out region was simply converted to triangular elements and points are moved towards the surface of the object, thus creating the surface nodes. In an appended optimization process, the body conformal boundary layer mesh region is created within the unstructured region. Figure 21 shows the complete mesh, figure 22 a detail.

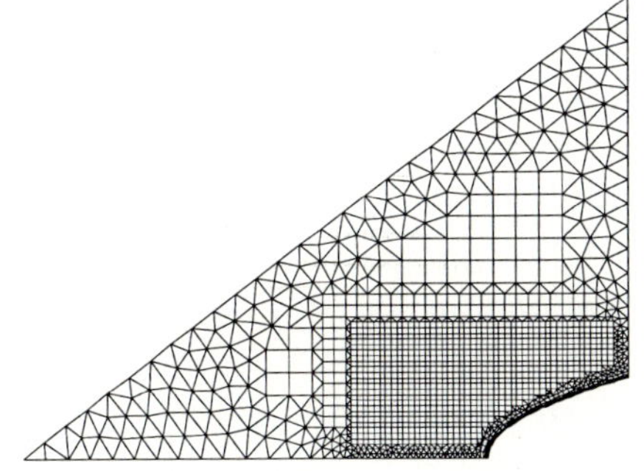

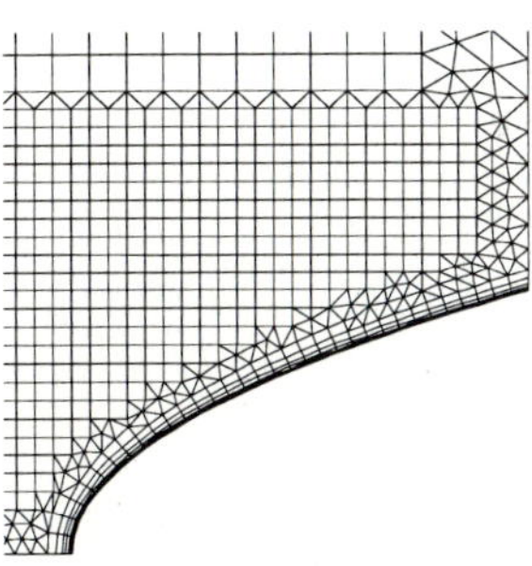

Figure 21
Sample hybrid mesh: Cartesian, unstructured, body conformal (left)

Figure 22
Sample hybrid mesh, detail (right)

Adaptation of meshes with Cartesian blocks

The use of Cartesian blocks is somehow contradictory to the idea of adaptation, since these require a very rigid geometry in relatively large parts of the domain. However, a structured block may always be logically converted into an unstructured part of the grid, if this is advantageous. Adaptivity can thus be introduced as follows:

1. The need for adaptivity can be evaluated as done for the rest of the computational domain.

2. If the limits of adaptation are exceeded, the block can be converted to a set of corresponding bilinear elements, further on considered as unstructured and thus free for all adaptation methods.

3. After adaptation is finished, new blocks can be created, if useful for the efficiency.

As an alternate, adaption methods related to the structures within the blocks are possible. However, due to the additional algorithmic efforts required, this strategy will not be followed in near future.

References

[1] D. Hänel, A. Dervieux, R. Vilsmeier, O. Gloth, C. Viozat, and L.Fournier. Development of navier-stokes solvers on hybrid grids. In E. H. Hirschel, editor, *Numerical Flow Simulation I, Notes on numerical Fluid Dynamics CNRS-DFG Collaborative Research Programme, Results 1996-1998*, pages 89–111. Vieweg Verlag, 1998.

[2] A. Brandt. Guide to multigrid development. In *Multigrid methods, in Lecture notes in Mathematics*, volume 960, pages 220–312. 1982.

[3] M. H. Lallemand, H. Steve, and A. Dervieux. Unstructured multigridding by volume agglomeration: current status. *Computers and Fluids*, 21:397–443, 1992.

[4] D. J. Mavriplis and V. Venkatakrishnan. A 3D agglomeration multigrid solver for the Reynolds-averaged Navier-Stokes equations on unstructured meshes. *Int. J. for Num. Meth. in Fluids*, 23:527–544, 1996.

[5] N. Okamoto, K. Nakahashi, and S. Obayashi. A coarse grid generation algorithm for agglomeration multigrid method on unstructure grids. In *Proceedings of 36th Aerospace Sciences Meeting and Exhibit*, volume 98-0615, AIAA, Reno, 1998.

[6] A. Dervieux, J. Francescatto, and G. Carré. Fast solver for unstructured finite volume methods. In *Proceedings of First International Symposium on Finite Volumes for Complex Applications*, 1996.

[7] B. Van Leer. Towards the ultimate conservative difference scheme V: a second-order sequel to Godunov's method. *J. of Comput. Phys.*, 32:361–370, 1979.

[8] J.A. Désidéri and P.W. Hemker. Convergence analysis of the defect-correction iteration for hyperbolic problems. *SIAM J. Sci. Comput.*, 16:88–118, 1995.

[9] P. Wesseling. *An introduction to multigrid methods*. John Wiley & Sons, 1991.

[10] R. Vilsmeier, O. Gloth, and D. Hänel. Generation of hybrid grids via element-conversion. In *Proceedings of the 6th International Conference on Numerical Grid Generation in Computational Field Simulation*, London, England, July 1998.

[11] R. Vilsmeier and D.Hänel. Central and upwind solutions of the conservation equations on 3-d unstructured meshes. In *Proceedings of the International Conference on Finite Elements in Fluids, Venezia*, 1995.

II.   CRYSTAL GROWTH AND MELTS

High Performance Computer Codes and their Application to Optimize Crystal Growth Processes, II

A. Degenhardt[3], P. Droll[2], M. El Ganaoui[4], L. Kadinski[1], M. Kurz[3], A. Lamazouade[4],
D. Morvan[4], I. Raspo[4], E. Serre[4],
P. Bontoux[4], F. Durst[1], G. Müller[3], M. Schäfer[2]

[1] Lehrstuhl für Strömungsmechanik, Universität Erlangen-Nürnberg
Cauerstr. 4, D-91058 Erlangen, Germany
[2] Fachgebiet Numerische Berechnungsverfahren im Maschinenbau,
Technische Universität Darmstadt
Petersenstr. 30, D-64287 Darmstadt, Germany
[3] Kristallabor am Institut für Werkstoffwissenschaften
Lehrstuhl Werkstoffe der Elektrotechnik, Universität Erlangen-Nürnberg
Martensstr. 7, D-91058 Erlangen, Germany
[4] Dpt. de Modélisation Numérique IRPHE - Réseau MFN CNRS
Université d´Aix-Marseille II
Technopôle de Chateau-Gombert,
38 Rue Frédéric Joliout Curie
F-13451 Marseille Cedex 20, France

Summary

The paper deals with the continuation of the development of high performance computer codes and their application to modelling of crystal growth processes started by the authors' research groups and reported in [1]. The mathematical model is based on the continuity equation and the conservation equations for momentum and heat transfer combined with mass transfer including chemical reactions. The thermal radiation analysis assumes a non-participating medium and semi-transparent walls. The radiation heat transfer is coupled with convection and conduction. The heat conduction includes thermal solid/fluid interactions between the gas and solid parts of the computational domain. The results of thermal calculations are used for the analysis of thermal stresses. The models are implemented in finite volume (both, block-structured and unstructured on non-orthogonal grids), and spectral and coupled finite volume/spectral numerical solution procedures.
The capabilities of the developed methods are demonstrated by computations of different practically relevant crystal growth processes. Calculations are carried out for Chemical Vapor Deposition of SiC and for Vertical Gradient-Freeze process both under practically used growth conditions.

1   Introduction

The objective of the collaborative work reported in this paper is the continuation of [1] developments and improvements of high performance computer codes and their application to the modelling, investigation, improvement and control of crystal growth processes as they are employed in practice.

An improvement of the growth processes and the relevant properties of the grown crystals and films of semiconductor materials can be achieved by a proper selection of the growth technique, improvements in the construction of the equipment and the proper selection of parameters used during the growth process. Together with economic impacts which favour larger crystals in length and diameter any improvements of industrial crystal growth process become increasingly time and money consuming. In this situation numerical simulations can effectively support the developments for improved crystal growth processes because they provide, in principle, an easy way to study the influence of changes of material properties, equipment geometry and growth parameters.

For a realistic modelling of crystal growth processes many physical aspects have to be taken into account: fluid and/or gas flow, heat transport by convection, heat conduction in the gas and solid parts, heat transport by radiation in transparent and semi-transparent media as well as multicomponent mass transport with homogeneous and heterogeneous chemical reactions, which may include different effects on the interface of the growing crystal like segregation effects, phase transformation and free boundary advective flows. For the prediction of defect formation, which is responsible for the crystal quality, the analysis of thermal stresses in the crystal is an important issue. For global simulations of crystal growth processes the above phenomena have to be considered in a coupled manner what makes such computations very challenging with respect to numerical techniques and computer resources.

In this work various codes involving different numerical techniques were employed in a complementary manner with the objective to allow reliable and efficient global simulations of practically relevant crystal growth processes. The numerical approaches comprise a block-structured finite-volume method (FASTEST), an unstructured finite-volume method (CrysVUN++), an enthalpy method and a pseudo-spectral method. Several code extensions, in particular for the pseudo-spectral approach, necessary to overcome limitations with respect to the applicability for real growth configurations are discussed. As exemplary practically important applications results for the CVD (Chemical Vapour Deposition) of SiC and the growth of GaAs by the VGF (Vertical Gradient Freeze) method are presented.

2   Mathematical formulation

The mathematical model used in the present paper is based on the solution of the coupled flow, heat transfer and mass transport equations including multicomponent diffusion and chemical reactions [1] and the references therein.

Continuity equation

$$\frac{\partial \rho}{\partial t} + \nabla \cdot (\rho \vec{v}) = 0 \tag{2.1}$$

with: ρ - the density and $\vec{v}$ - the velocity vector.

Momentum equations (Navier-Stokes):

$$\frac{\partial}{\partial t}(\rho \vec{v}) + \nabla \cdot (\rho \vec{v}\vec{v}) = -\nabla p + \vec{f} + 2\nabla \cdot (\eta \dot{S}) - \frac{2}{3}\nabla \cdot (\eta \nabla \cdot \vec{v}) \tag{2.2}$$

with: η - the dynamic viscosity coefficient, p - the dynamic (excess) pressure, $\vec{f}$ - the gravitational acceleration, $\dot{S}$ - the deformation rate tensor.

Energy equation:

$$c_p \frac{\partial}{\partial t}(\rho T) + \nabla \cdot (-\lambda \nabla T) + c_p \nabla \cdot (\rho T \vec{v}) - s = 0 \tag{2.3}$$

with: c_p - mass specific heat, λ - thermal conductivity, s - density of heat source.

Heat transfer by radiation for the transparent media with opaque or semi-transparent grey-diffuse radiating walls is described in the previous paper [1] and, in a very detailed way, in [2, 3].

Radiation propagation in nonscattering semitransparent gray media is described by the equation [4]:

$$\vec{s} \cdot \nabla I = \kappa I_B - \kappa I \tag{2.4}$$

with: $\vec{s}$ direction, I radiation intensity, κ absorption coefficient. The absorbtion of the radiation energy leads to an additional source term in the energy equation (2.3)

$$s_{rad} = \int_{4\pi} \left(I - 4\sigma T^4\right) d\Omega. \tag{2.5}$$

The adequate boundary conditions for diffusive reflection is given by

$$I(\vec{s}) = \epsilon \frac{1}{\pi}\sigma T^4 + (1 - \epsilon) \int_{2\pi} I(\vec{s}_o) d\Omega_o, \tag{2.6}$$

with: ϵ wall emissivity.

The radiation equation (2.6) may be simplified by many different ways (see e.g. [4]). In this paper so-called "P1 approximation" is used:

$$\nabla \cdot \left(\frac{1}{3\kappa}\nabla G\right) = -\nabla \cdot \vec{q} = \kappa(G - 4\pi I_b) = s_{rad}. \tag{2.7}$$

An adequate boundary condition for P1 approximation is derived from (2.7):

$$-\frac{2 - \epsilon}{\epsilon}\frac{2}{3\kappa}\mathbf{n}\nabla G + G = 4\sigma T^4 \tag{2.8}$$

with: $\mathbf{n}$ – unit vector normal to wall, G – total intensity integrated over all spatial angle and wavelengths, $I_b = \frac{1}{\pi}\sigma T^4$ – black body radiation intensity and s_{rad} is the source term in the energy equatiton (2.3).

Conservation of chemical species:

$$\frac{\partial}{\partial t}(\rho m_l) + \nabla \cdot (\rho \vec{v} m_l) = \nabla \cdot \left(\rho D_l \left(\nabla m_l + \alpha_l m_l \frac{\nabla T}{T}\right)\right) + R_l \tag{2.9}$$

with: m_l - mass fraction of species l, D_l - diffusion coefficient of species l, α_l - thermal diffusion coefficient of species l, R_l - rate of generation or consumption of chemical species per unit volume.

For gases the equation of state is valid:

$$\rho = \frac{P_0}{RT} \sum_{i=1}^{N} x_i M_i, \tag{2.10}$$

with: x_i - molar fraction, M_i - molar mass P_0 - constant operating (thermodynamic) pressure.

For the description of stress in solid bodies Cauchy's first and second law of motion are used. In cylindrical coordinates in the axisymmetrical, stationary case the equation may be written as ([5]):

$$\nabla \cdot \sigma + \vec{b} = 0 \tag{2.11}$$

with: σ – stress tensor, $\vec{b}$ – body force densities.

The stress–strain relationship for a thermoelastic anisotropic solid body is taken as described by Lambropoulos [6].

$$\sigma = C\epsilon - \alpha(T - T_{\mathrm{ref}}) \tag{2.12}$$

with: α – thermal expansion coefficient, T – temperature, T_{ref} – reference temperature for the relaxed body, ϵ – strain tensor, C – elasticity tensor.

The dependence of the components of C on the crystallographic orientation with respect to the cylindrical axis, i.e. the growth direction $\langle 1,1,1 \rangle$ or $\langle 1,0,0 \rangle$ and how to include anisotropic effects, as well as their correlation to the modulus of elasticity E and the Poisson's ratio ν, is described in more details in [6].

To close the set of equations the usual linear relation between displacements and strains are employed:

$$\epsilon = \frac{1}{2} \left(\nabla \vec{u} + \nabla \vec{u}^{\mathrm{T}} \right) \tag{2.13}$$

with: $\vec{u}$ displacement vector.

3   Numerical Methods

Spectral methods providing a high accuracy and a very good efficiency seem to be a method of choice for computing complex flow in the domains of moderate complexity. The coupling with the finite-volume method, featuring a good geometrical flexibility, allows the efficient computing of highly complex flows in global simulations. First investigations confirmed the expected good performance of the coupled code. The advancements concerning the improvement of the convergence behavior are shown in section 3.1.

The application of the coupled code to technical relevant problems is however still confined due to the restriction of the implemented spectral method:

- The computational domain has to be rectangular

- The material properties have to be constant.

To overcome this limitations, new developed algorithms were adopted and implemented:

- To extend the solver to moderately complex geometries a mapping technique has been used. The first results and the considered problems are discussed in section 3.2.

- With spectral methods, the set of linear equations is usually solved with an direct Helmholtz solver. This very efficient solver is however only suitable for calculations in rectangular domains and with constant material properties.

- The explicit treatment of the convective terms often used with pseudo-spectral methods leads to a time step size restriction which limits the stability of the algorithm. The advantages and limitations concerning efficiency, accuracy and time-step size are presented.

The extension of the two-dimensional code to three dimensions has been carried out for the Cartesian and the cylindrical coordinate system. With the cylindrical solver a coordinate transformation is applied, to avoid the problem at the axis. The transformation as well as some results are shown in section 3.5.

The pseudo-spectral method is based on a Chebyshev(-Fourier) approximation. The Chebyshev polynomials are evaluated at the Gauss-Lobatto points. The discretization of the time derivative uses a three-level-backward scheme. The velocity-pressure coupling is performed with a modified projection algorithm including a predictor for the pressure. For explicit treated convective terms, the solution reduces to a solution of Helmholtz and Poisson type equations.

The coupled code is based on the finite-volume code FASTEST, which employs a fully conservative FVM for the solution of the flow equations. FASTEST is also the base for the further developments of the modeling of CVD processes (section 4). The main features of the implemented method are summarized in [7] and embrace non-orthogonal boundary fitted block-structured numerical grids, a collocated (non-staggered) arrangement of dependent variables, for the coupled system of equations a pressure-correction approach of SIMPLE type, a nonlinear multigrid scheme for convergence acceleration and for the treatment of complex geometries the concept of block-structured grids is used, which forms also the base for the parallelization of the computations by grid partitioning. The solution method is formally second order accurate, since all approximations are performed in a central-difference manner. For discretization with respect to time a second-order fully implicit scheme is employed.

For global simulations of processes in crystal growth furnaces (section 5), the unstructured finite-volume code CrysVUN++ has been used [8]. The heat transfer and the fluid flow in the melt has been calculated applying the enthalpy method. This method is based on a global estimate of the flow and of the transfer in each phase (liquid an solid) using an averaging technique applied to the depending variables and physical properties. For the energy equation we use the enthalpy variable which accounts for the specific heat and the latent heat required for the phase change. For the momentum equations, the jump on the velocity between the liquid and solid phases is accounted by using a Darcy term derived from models of flow in porous media.

3.1 Coupled Finite-Volume - Pseudospectral Method

For the coupling of the SM and FVM we restrict ourselves to the case where the fluid parts are calculated fully with SM and the heat conduction in solid and fluid parts with the coupled SM-FVM formulation. For such cases, a coupling of the two methods within the energy equation is sufficient. At the solid-fluid interface Γ we have the continuity constraints:

$$T^{\Omega_s}\big|_\Gamma = T^{\Omega_l}\big|_\Gamma , \qquad (3.14)$$

$$\lambda_s \frac{\partial T^{\Omega_s}}{\partial x}\bigg|_\Gamma = \lambda_l \frac{\partial T^{\Omega_l}}{\partial x}\bigg|_\Gamma . \qquad (3.15)$$

The FVM code FASTEST uses the concept of block-structured grids. To preserve this structure also in the coupled SM-FVM code, an overlapping grid partitioning technique is used for the SM-FVM coupling. The overlapping is constructed by enlarging the finite-volume sub-domain into the spectral sub-domain. The boundary conditions of the spectral block at the interface are of Neumann type and the boundary conditions of the finite-volume block at the interface are of Dirichlet type [9].

The good efficiency and the satisfying accuracy of the coupled code has been demonstrated calculating a thermally driven cavity problem combined with a solid block in [9]. Further investigations of the convergence behavior of the coupled code showed a surprisingly low convergence order and accuracy (Fig. 1, dashed graphs) in comparison to a pure spectral code. The reason of this behavior was found in the pressure-correction scheme. The algorithm assumes homogeneous Neumann boundary conditions for the pressure at the walls. The improved pressure-correction scheme of HUGHES [10], et al. overcomes this

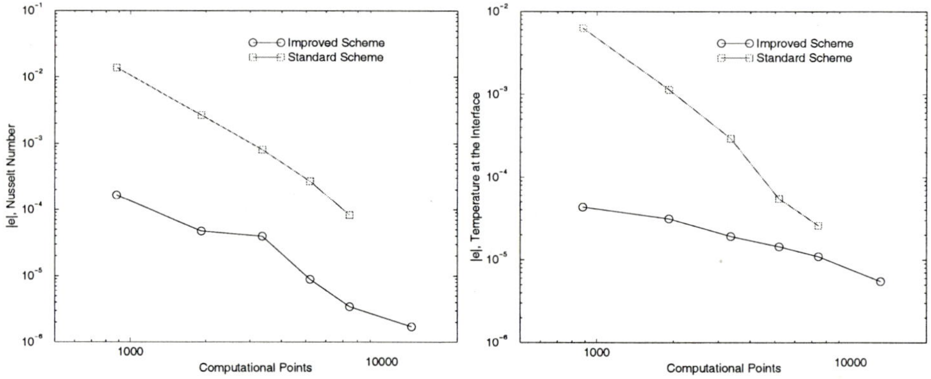

(a) Nusselt number Nu calculated at the interface wall Γ

(b) Temperature at the interface wall Γ

Figure 1 Relative error versus number of computational points for the Nusselt number and a temperature value at the interface between the solid (FV) and the fluid (SM) block for the standard and the improved pressure-correction scheme.

problem and leads to a distinct better accuracy and convergence (Fig. 1, solid graphs). Now, the accuracy for all variables is comparable with the accuracy of the pure spectral code. The relative error is computed with the solution of the coupled code (improved pressure-correction scheme) on the finest grid.

The coupled code allows efficient and very accurate simulations of fluid-solid problems. The application to technical relevant problems is however still limited due to the restriction to rectangular domains and constant material properties of the spectral (fluid) domain. The investigations in the next sections are the first steps to a more general formulation of the spectral method.

3.2 Domain Mapping

We focus here on the mapping in a two-dimensional square domain $[-1,1] \times [-1,1]$, defined due to the computational domain of the Chebyshev polynomials. The physical domain is described by four parametric equations:

$$\mathbf{x}_s(\xi), \ \mathbf{x}_n(\xi), \ -1 \leq \xi \leq 1, \tag{3.16}$$

$$\mathbf{x}_w(\eta), \ \mathbf{x}_e(\eta), \ -1 \leq \eta \leq 1. \tag{3.17}$$

The subscripts on $\mathbf{x}$ stand for *south, north, west, east* boundaries of the square domain $[-1,1] \times [-1,1]$. The spatial derivatives in the physical coordinates can be written as:

$$
\frac{\partial}{\partial x} = \xi_x \frac{\partial}{\partial \xi} + \eta_x \frac{\partial}{\partial \eta} \qquad \frac{\partial^2}{\partial x^2} = \xi_x^2 \frac{\partial^2}{\partial \xi^2} + \eta_x^2 \frac{\partial^2}{\partial \eta^2} + 2 \xi_x \eta_x \frac{\partial^2}{\partial \xi \partial \eta}
$$
$$\tag{3.18}$$
$$
\frac{\partial}{\partial y} = \xi_y \frac{\partial}{\partial \xi} + \eta_y \frac{\partial}{\partial \eta} \qquad \frac{\partial^2}{\partial y^2} = \xi_y^2 \frac{\partial^2}{\partial \xi^2} + \eta_y^2 \frac{\partial^2}{\partial \eta^2} + 2 \xi_y \eta_y \frac{\partial^2}{\partial \xi \partial \eta}.
$$

The partial differential equations may be transformed from physical space to computational space using equations 3.18. The term $\frac{\partial^2 \phi}{\partial \xi \partial \eta}$ occurring in the transformed equations is treated implicitly by an inner iteration process (for steady problems, the iteration in time can be regarded as outer iterations). The inner iteration process will be stopped, when the change of the $\frac{\partial^2 \phi}{\partial \xi \partial \eta}$ is smaller than a specified convergence criterion. As a consequence, the number of inner iterations increases when the term $\frac{\partial^2 \phi}{\partial \xi \partial \eta}$ gets dominant.

Neumann boundary conditions are realized due to a continuous update of the boundary values. The values are calculated in every inner iteration. Therefore, no extrapolation in time of the values is required. To investigate the properties of the implemented mapping technique, the flow in a inclined square cavity was calculated (Fig. 2) [11]. The effect of

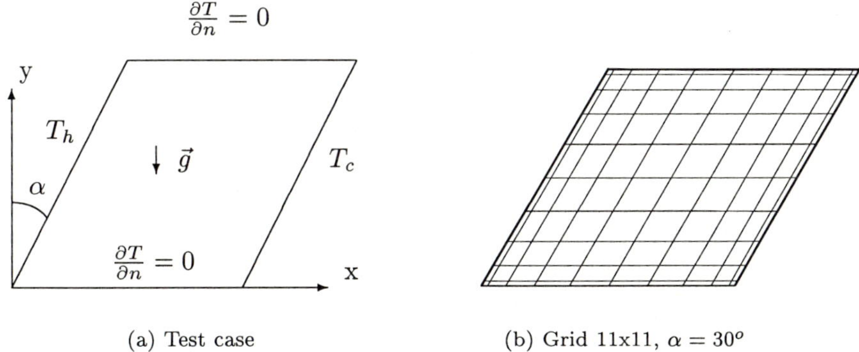

(a) Test case

(b) Grid 11x11, $\alpha = 30^o$

Figure 2 Test case with a non-orthogonal grid. $Ra = 10^5$, $Pr = 0.71$. The inclination angle α varies from 0° to 40°.

the inclination angle α has been studied.

The efficiency of the presented algorithm for moderate complex geometries decreases rapidly with an increase of the inclination angle due to the poor convergence behavior

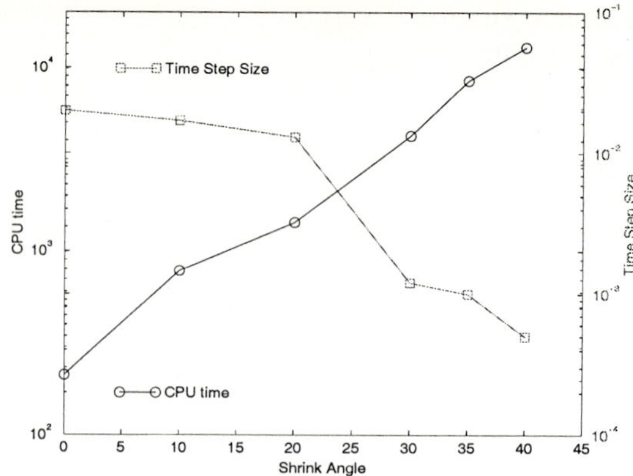

Figure 3 Domain Mapping. CPU time and time-step size versus inclination angle α for the calculation of fluid flow in an inclined square domain.

of the used iterative scheme (Fig. 3). Here, an iterative solver would allow an implicit discretization of the mixed terms and therefore feature a distinct better convergence.

The number of outer iterations/time steps to reach a steady state depends on the time-step size. With an increasing inclination angle, the time-step size had to be reduced to meet the stability criterion. The time-step size of the proposed scheme is limited due to the explicit treatment of the convective terms and due to the pressure-correction scheme. To overcome the first restriction, an implicit scheme is introduced.

3.3   Implicit Navier-Stokes Solver

A implicit formulation of the convective terms is strongly linked with the use of an iterative solver, because the fast direct solver looses its most important advantage: the possibility of calculating the eigenvalues and eigenvectors once in the preprocessing. Therefore the fast direct solver is not suitable for an implicit treatment of the convective terms. In the following, the formulation of an iterative solver is presented. All discretized equations (explicit and implicit) can be written in the general form:

$$\Gamma \left(A_{il}^2 \Phi_{ljk} + B_{jm}^2 \Phi_{imk} + C_{kn}^2 \Phi_{ijn} \right) \tag{3.19}$$
$$+ \quad \Lambda_1 A_{il}^1 \Phi_{ljk} + \Lambda_2 B_{jm}^1 \Phi_{imk} + \Lambda_3 C_{kn}^1 \Phi_{ijn} + \sigma \, \Phi_{ijk} = F_{ijk},$$

with matrices A^1, B^1, C^1 composed of the derivative operators for the first spatial derivatives and A^2, B^2, C^2 composed of the derivative operators for the second spatial derivatives. This six matrices include also the type of the boundary conditions (Dirichlet or Neumann). The (non-constant) coefficient Γ describes the diffusion coefficient and Λ_j, e.g., the velocity of the convective term. F_{ijk} can be noted as $f_{ijk}^\Phi + F_{ijk}^{BC}$, where f_{ijk}^Φ includes the source terms and F_{ijk}^{BC} describes the values of the boundary conditions. σ is a scalar and may be time dependent.

This set of equations derived from discrete Helmholtz equations can be solved efficiently using Krylov subspace iterative methods with a preconditioning methodology derived from fast direct methods. The basic principle behind fast direct solvers is to apply an inexpensive transformation to break down a problem into a number of lower-dimensional but independent problems. Fast direct methods are standard tools for solving the Poisson equation on regular domains with Dirichlet, Neumann, or periodic boundary conditions.

In the following the various components of the iterative solver [12] are described, where, for simplicity, we restrict ourselves to three-dimensional problems with Chebyshev polynomial ansatz functions in all directions.

Iterative Solver using a Fast Direct Solver as a Preconditioner

To solve equation (3.19), shortly denoted by $A\Phi = F$, with preconditioned CG-methods, one can write

$$QA\Phi = QF \tag{3.20}$$

with Q as a preconditioning matrix.

The CG method does not require the coefficient matrix A and the preconditioning matrix Q in matrix form; only the result of the matrix-vector products $Y = AX$ and $Y = QX$ are needed [13]. To minimize the number of operations, the matrix-vector products are evaluated using equation (3.19). In the case of the product $Y = AX$, the matrix-vector product alters to three matrix-matrix products.

To evaluate the product $Y = QX$ for the preconditioning the fast direct solver is used. Instead of calculating the matrix-vector product, the equation $Q^{-1}Y = X$ has to be solved. The inverse matrix of the preconditioning matrix Q is derived from A. The simplest approximation of Q^{-1} we obtain, if we neglect the convective part, $\Gamma_j = 0$ and use a mean value of the diffusion coefficient $\overline{\Gamma}$. This means, that only the diffusive terms are used for the preconditioning. As a consequence, such a preconditioning will be efficient only for diffusion dominated problems.

To solve the preconditioned set of equations, we use a restarted version of the GMRES method, [14], a very robust method for non-symmetric systems. Although the RGM-RES algorithm is often less efficient than the Bi-CGStab(l), proposed by SLEIJPEN and FOKKEMA, [15], it is less sensitive to badly preconditioned matrices caused by dominant convective type terms, which are not yet included in the preconditioning algorithm.

In order to demonstrate the implicit spectral method, we applied the proposed method to the well known buoyancy driven cavity problem [11]. The Rayleigh number was chosen to 10^5 and the Prandtl number to 0.71. For the considered reference value, the relative error of the Nusselt number, the solution with the implicit method is of higher accuracy (Table 1). However this effect vanishes with an increasing number of computational points and an increasing time-step size. Because of the severe time step restriction of

Table 1 Relative error of the Nusselt number calculated with the explicit and the implicit formulation for different time-step sizes $\Delta t_1 < \Delta t_2$. The explicit algorithm is unstable for Δt_2.

Grid	explicit, Δt_1	implicit, Δt_1	implicit, Δt_2
21x21	3.0E-4	4.1E-6	1.4E-4
41x41	7.1E-7	2.6E-7	1.5E-6

the explicit formulation, the implicit method gains from the possibility to cope with larger time steps and reaches the steady state on the 21×21 grid for a 4 times larger time step size i.e. (with 4 times less time steps). With the same time-step size as the explicit method the CPU time is approximately 2 times higher, but with a larger time step the implicit method outperforms the explicit one. But nevertheless, the efficiency

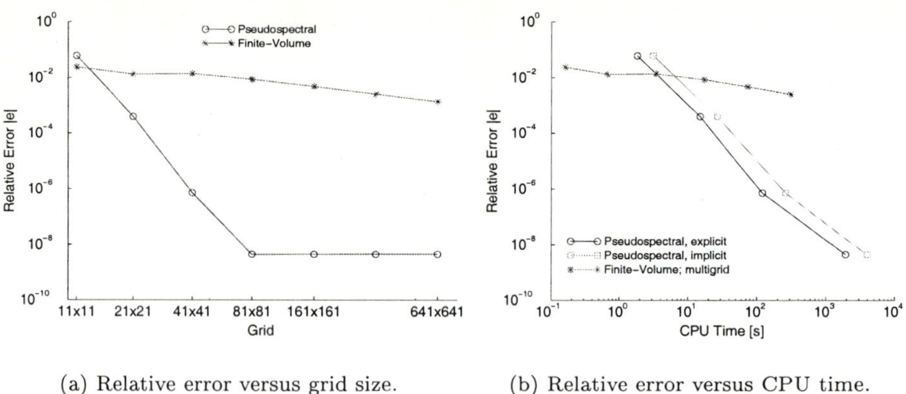

(a) Relative error versus grid size.   (b) Relative error versus CPU time.

Figure 4 Relative error of the Nusselt number. The calculations were done with an explicit and an implicit spectral code and with a finite-volume code (FASTEST) using a full multigrid algorithm for convergence acceleration. The explicit and implicit calculations were done with the same time step size.

is still distinct higher than the finite-volume code FASTEST using the full multigrid algorithm for convergence acceleration (Fig. 4), if one requires a high accuracy. Therefore we can conclude, that the efficiency of the implicit formulation of the spectral method is comparable with an explicit one. The accuracy is tendentious better. The time step size of the implicit algorithm is dominated by the pressure correction scheme. This leads to small time step sizes on fine grids. In such cases, no benefit can be gained using an implicit formulation. But on the other hand, also no disadvantages occur. Due to the use of an iterative solver, the algorithm is also well suited for problems with moderately complex geometries by the domain mapping technique described in section 3.2.

3.4   Non-constant Material Properties

The developed and implemented algorithms have been used to calculate flow with temperature dependent material properties. To verify the implementation and to demonstrate the high accuracy of the spectral code, a thermally driven cavity problem was selected. Similar to the well known cavity benchmark problem with constant material properties, a square with a side wall length L encloses a fluid. The temperature of the left and right walls are T_c and T_h, respectively. The horizontal walls are insulated.

The temperature dependence of the viscosity, density and heat conduction is modeled by functions usually used for the gas flow in CVD simulations with non-constant material properties. The Rayleigh number is close to 10^5. The problem is governed by the Low-Mach number equations and was solved by a finite-volume and a spectral code to compare the results.

In Fig. 5 the relative error of the mean temperature gradient at the hot wall is presented. The error is calculated with the spectral result on a 161x161 grid. It can be seen, that the spectral code already on the first grid obtains an accurate solution and outperfoms the finite-volume code concerning the accuracy and efficiency. The numerical algorithm has been applied to calculate the flow in a CVD reactor. Due to the large temperature differences, the material properties are non-constant. The three-dimensional

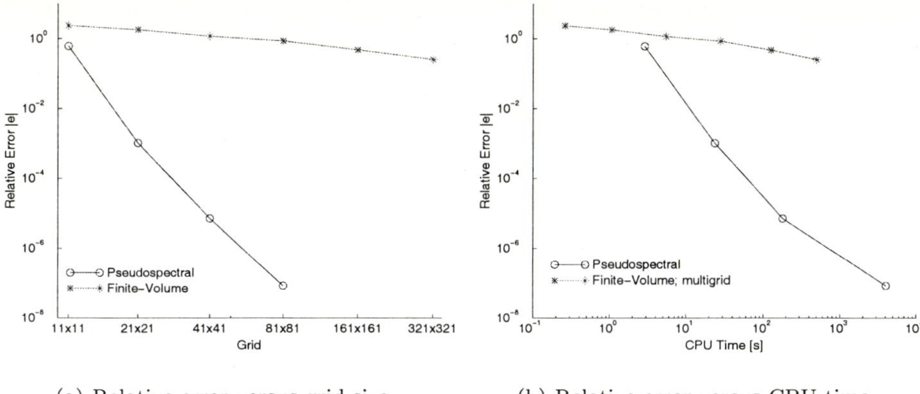

(a) Relative error versus grid size.   (b) Relative error versus CPU time.

Figure 5 Relative error of the mean temperature gradient at the hot wall versus grid size and CPU time. The calculations were done with an implicit spectral code and with a finite-volume code (FASTEST) using a full multigrid algorithm for convergence acceleration.

example presented here, is deduced from the CVD reactors presented in chapter 4. In Fig. 6 the spectral solution for the three-dimensional reactor is shown.

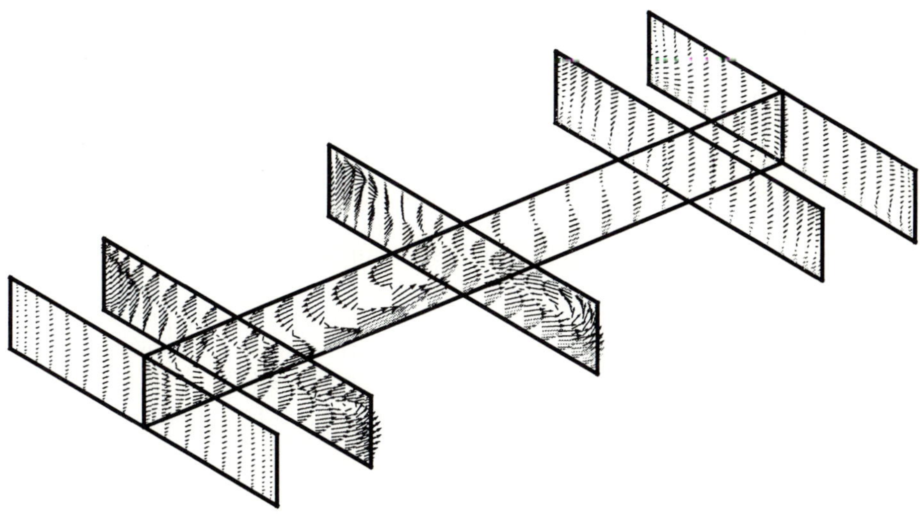

Figure 6 Velocity vectors in the reaction zone of a CVD reactor, three-dimensional calculation.

3.5 Three-Dimensional Cylindrical Solver

The flow of the melt in the crucible induced due to bouancy forces and rotating boundaries is complex and usually time-dependent. The flow structures can be generated on small length scales and their boundary layers can become exceedingly thin when the rotation rate is increased. Therefore, it seems appropriate to develop solutions based on

79

the highly accurate spectral method which uses expansions in Chebyshev polynomials
to approximate the solution in the non-azimuthal directions. Here, we present a direct
pseudospectral algorithm for in rotating cylindrical cavities. This represents an extension of earlier studies on developing such algorithms [10] and on physical investigation
of the problem (see for more details [16]). The aspect ratio is given by $A = \Delta R/2h$. The
Reynolds number is defined by $Re = \Omega R^2/\nu$, where R is the radius of the cavity, Ω the
speed of the rotating disk and ν the kinematic viscosity.

When using cylindrical coordinates to calculate the solution of the 3D Navier Stoces
equations in a cylinder, the complexity is further increased by the singular behavior
of the coefficients when the radial coordinate tends to zero. Moreover, as there is no
natural boundary condition at the axis ($r = 0$), a change of dependent variables (velocity,
pressure) has been chosen to enforce a boundary condition at the axis. The chosen change
of dependent variables $\Psi = (u,v,w,p)$ is as follows :

$$\Psi = r^{-1}\widetilde{\Psi}. \tag{3.21}$$

This variable change yields the conditions $\widetilde{\Psi} = 0$ at $r = 0$. This transformation yields
new operators and the 2D equations to be solved for each Fourier wave (after time
discretization) are no longer the real Helmholtz or Laplace equations. Furthermore, the
new operators to be diagonalized can have complex eigenvalues and involve the use of
Complex Fast Fourier Transformation. The spatial and temporal accuracy of the method
is tested on analytical solutions (Fig. 7). The error E_ψ, is evaluated from the discrete L^2
error of each flow variable Ψ and computed at all collocation points. We notice (Fig. 7a)
that the spectral accuracy is obtained for the three velocity components, with an error
of about 10^{-12} as soon as the resolution becomes sufficient (i.e. $N = M = K = 48$). We
can observe that for the pressure the error is small (about 10^{-10}), but higher than that
for the velocity . The error E'_Ψ in terms of the time step is given n Fig. 7b, where the
temporal error is shown to be in $O(\Delta t^2)$ for each dependent variable. To obtain insights
on the slip velocity, we define $E'_{\Psi bound}$ as the values of E'_Ψ at boundary points. $E'_{\Psi bound}$
is displayed in Fig. 7c and our method exhibits a temporal behavior of $O(\Delta t^3)$, retaining
the same behavior as the previous study [10].

To demonstrate the efficiency of our method two complex physical problem are considered in cylindrical rotor-stator cavities. For $A = 2$, three-dimensional instability patterns
are shown in the Bödewadt layer close to the stationary disk (Fig. 8). The characteristic
parameters of these structures (wavelength, frequency angle..) are shown to be consistent
with type I and type II instabilities in rotating flows. These results are consistent with
earlier experimental investigations [17] and theoretical studies [18].

For smaller aspect ratios a bubbles-type vortex breakdown occurs on the axis (Fig. 9).
The phenomenon of vortex breakdown, which is observed in a variety of flow situations
involving concentrated vortices, is characterized by an abrupt transition, usually leading
to the appearance of an on-axis stagnation point and region of reversed flow. Attempts to
numerically simulate vortex breakdown require the specification of hypothesized boundary conditions both upstream and downstream of the breakdown. Then, the flow inside
an enclosed rotor-stator cavity is a particularly attractive candidate for numerical studies
because the boundary conditions are not required. An axial flow is generated along the
axis by the Ekman suction induced by the rotating end wall. One to three steady and
unsteady vortices are numerically obtained and a 3D behavior of the flow in the near
axis region is exhibited. Bubbles structure and transition to a time-dependent flow are
in good agreement with the experimental results [19].

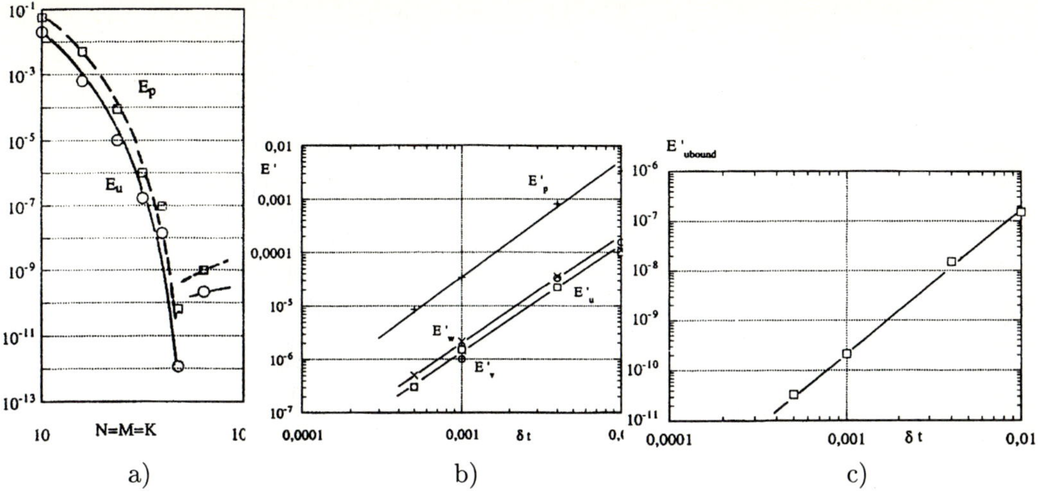

a)   b)   c)

Figure 7 Display of the errors for $Re = 2500$, $A = 2$: (a) E_Ψ for u and p, versus the grid $N = M = K$ for a stationary analytical solution, (b) E'_Ψ for u , v , w and p versus the time step δt, $N = M = K = 32$, (c) $E'_{\Psi bound}$ versus the time step δt, $N = M = K = 32$.

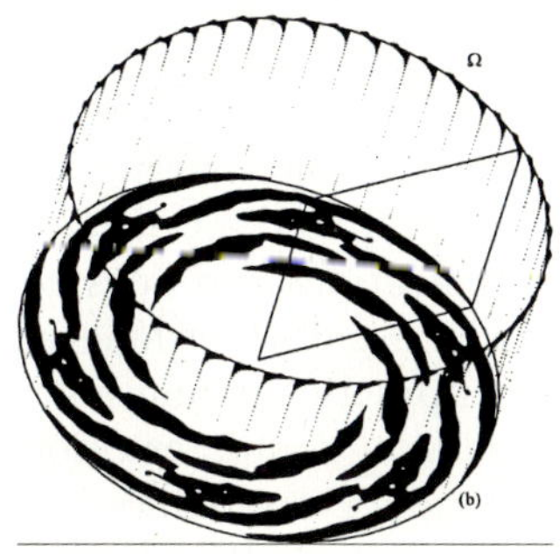

Figure 8 Three-dimensional instability of the Bödewadt layer in a cylindrical cavity, A=2, Re=30000. ISO-surface of axial velocity.

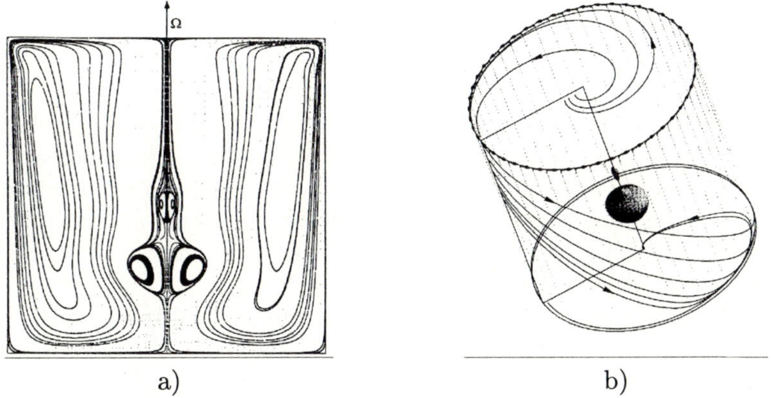

a)   b)

Figure 9 Vortex breakdown phenomenon with two bubbles. A=0.5 and Re=1850. (a) Display of particle paths. (b) ISO-surface of axial velocity.

4 CVD Simulations

4.1 CVD of SiC

Chemical vapor deposition (CVD) of SiC epitaxial layers is important technique to manufacture electronic devices for various applications, the most promising among them are devices for high-power, high-frequency and high-temperature applications.

Optimal CVD processes have to make possible reproducible growth of epitaxial layers with uniform distribution of properties - thickness, doping concentrations, etc. To achieve this goal a reliable control of transport of Si- and C-containing species to the growing SiC epitaxial layer and Si/C ratio over the wafer surface is of decisive importance.

The goal of the present work is modelling of flow dynamics, heat and mass transfer during CVD of SiC in the vertical reactor [20] to understand the impact of flow regimes on the growth mechanism.

Model of the SiC CVD Process

To develop a consistent model of CVD of SiC in the vertical reactor it is necessary to describe as accurately as possible the pathways of chemical decomposition of the species introduced into the reactor - silane and propane. The most advanced model of homogeneous chemical processes during CVD of SiC was proposed and tested in [21]. A set of many gas phase reactions was proposed and kinetical data were collected and summarized in this publication. But, in the proposed set of chemical reactions the major part of the species exists in very low concentrations and does not influence transport of Si- and C-atoms to the growing surface. The use of the whole set of the proposed homogeneous chemical reactions makes the model practically unusable for relevant CVD processes because of extremely high requirements for computer resources.

Sensitivity analysis shows that a significantly reduced set of the homogeneous chemical reactions provides a reasonably accurate results for major species concentrations in the reactor volume and transport of Si and C to the growing SiC epitaxial layer. Selection of the most relevant S- and C-species is supported also by extensive thermodynamical consideration of the Si-C-H system performed in [22].

In the present modelling work a set of the 7 major species important for accurate prediction of growth rate are taken into account:
SiH_4, SiH_2, Si, C_3H_8, CH_3, CH_4, C_2H_5, H and H_2. [23]

A consistent set of homogeneous chemical reactions used in the calculations is presented in the table 2. The Arrhenius parameters of the direct reactions are taken from [21], and data for the reverse reactions are calculated using the equilibrium constants. The described set of homogeneous chemical reactions represents an example of chemical system with widely varying reaction rates and, therefore, the "'stiff chemistry"' algorithms are needed to calculate the chemistry.

Two methods are used for the solution of stiff sets of chemical reactions in combination with the mass transport equations for the chemical species in the flow domain [24] and the references inside. In the first method the artificial time step is calculated for each species in the gas mixture for each point separately based on the local reaction rate. Although this approach in general improves the convergence behaviour, the total number of iterations may still be high in complex chemical problems since locally very small time steps may be required. The second method treats each computational as an ideally mixed reactor, isolated from neighbor cells except for the transport terms which are calculated in a previous iteration. The approach looks for the solution of the equations, which are discretized in the spatial dimensions, and solves the balance equations for the gas mass fractions using Newton solver for the simultaneous solution of the mass fractions of all species in a computational cell.

To summarize, these two methods for stiff chemistry in laminar reactive flows offers considerable reduction in calculation time but only in a limited number of problem.

Modelling results: Distribution of chemical species and growth rate

The velocity vectors and isotherms in the reactor are shown in Fig. 10 and 11. The results of the computations have been confirmed experimentally by the visual control of the flow pattern over the susceptor which is possible in the EMCORE system used for CVD of SiC due to existence of an irradiating layer in the gas above the substrate holder [25].

The typical distribution of SiH_4, SiH_2, Si, C_3H_8, CH_3, CH_4 are shown in Fig. 12, 13, 14, 15, 16 and 17, for other calculated growth regimes the results are comparable (with different total values). In the upper part of the reactor the concentration of silane and propane is decreased by diffusion and only in the hot region above wafer and holder decomposition takes place.

In this region a maximum of the products of decomposition (SiH_2, Si, CH_3 and CH_4) is formed, depending on their occurrence as intermediate product and the diffusive transport of these species to the growing surface.

The calculations on the grid with two blocks and with 2842 CV on a workstation SUN ULTRA2 were obtained in approximately 1 hour. The calculated growth rates are listed in Table 3 and plotted in comparison to experimental values in Fig. 18 and 19. Although the absolute values of the calculated growth rate are between a factor of 3 and 4 to high, the dependence of the growth rate from the silane and propane flow is comparable. At constant propane inflow the growth rate is increasing linear with the silane inflow until the input ratio silane to propane is equal one. For higher input ratio the slope seems

Table 2 Gas phase reactions and Arrhenius parameters in moles, cubic centimeters, seconds, Kelvin and kcal/mole for the reaction rate constant $k = A \cdot T^\beta \cdot \exp\left(-\frac{\Delta E}{RT}\right)$.

	gas phase reactions	A_0	β	ΔE	Ref.
H1	$2\,H + H_2 \rightleftharpoons 2\,H_2$	$9.2 \cdot 10^{16}$	-0.6	0.0	[21]
R1	reverse reaction to H1	$2.493 \cdot 10^{17}$	-0.568	104.131	(calc.)
H2	$C_3H_8 \rightleftharpoons C_2H_5 + CH_3$	$1.698 \cdot 10^{16}$	0.0	84.84	[21]
R2	reverse reaction to H2	$2.39 \cdot 10^6$	1.777	-3.948	(calc.)
H3	$C_2H_5 + H \rightleftharpoons 2\,CH_3$	$1.0 \cdot 10^{14}$	0.0	0.0	[21]
R3	reverse reaction to H3	$4.809 \cdot 10^{10}$	0.632	9.572	(calc.)
H4	$CH_4 + H \rightleftharpoons CH_3 + H_2$	$2.2 \cdot 10^4$	3.0	8.75	[21]
R4	reverse reaction to H4	$1.276 \cdot 10^2$	3.205	7.337	(calc.)
H5	$SiH_4 \rightleftharpoons SiH_2 + H_2$	$6.671 \cdot 10^{29}$	-4.795	63.45	[21]
R5	reverse reaction to H5	$1.14 \cdot 10^{23}$	-3.397	5.467	(calc.)
H6	$SiH_2 \rightleftharpoons Si + H_2$	$1.06 \cdot 10^{14}$	-0.88	45.0	[21]
R6	reverse reaction to H6	$5.045 \cdot 10^{11}$	-0.198	1.532	(calc.)

Table 3 Process data used for the calculations and resulting growth rates

No.	Pressure P	Silane flow	Propane flow	Growth rate
1	50 Torr	4 sccm	6.6 sccm	4.6 μm/h
2	50 Torr	9.9 sccm	6.6 sccm	11.5 μm/h
3	50 Torr	17 sccm	6.6 sccm	20.0 μm/h
4	50 Torr	17 sccm	1.5 sccm	5.5 μm/h
5	50 Torr	17 sccm	3.3 sccm	12.2 μm/h

to be decreased. Varying the propane inflow at high Si/C-ratio also results in a linear dependency of the growth rate. Below a Si/C input ratio of approx. 1.7 (experimental) respectively 1 (calc.) the growth rate is nearly constant.

The difference in absolute values and shifting of the saturation of the growth rate may be explained by the assuming of the formation of Si-droplets by exceeding the saturated vapor pressure of Si over solid or liquid silicon [26]. By this the flux of Si to the surface and with that the growth rate would be reduced as long as the carbon transport is not the limiting factor. Due to the complexity of models of formation and transport of small solid or liquid droplets and clusters in a gas phase, a simple extension of the used mathematical model is not possible. This is a subject of the forthcoming studies.

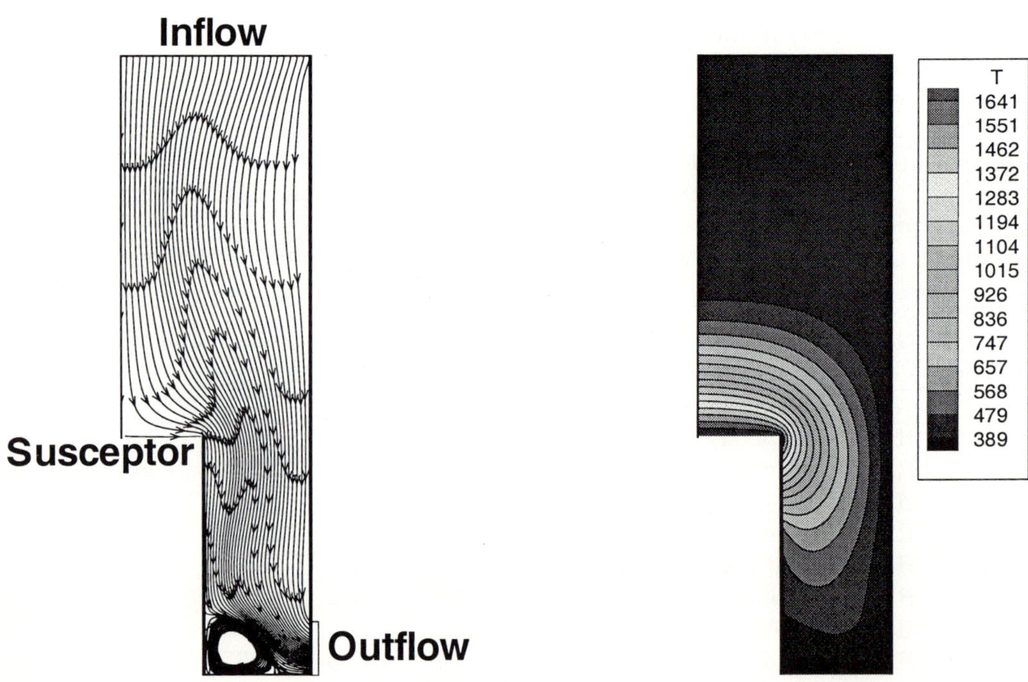

Figure 10 Flow in the reactor   **Figure 11** Temperature distribution

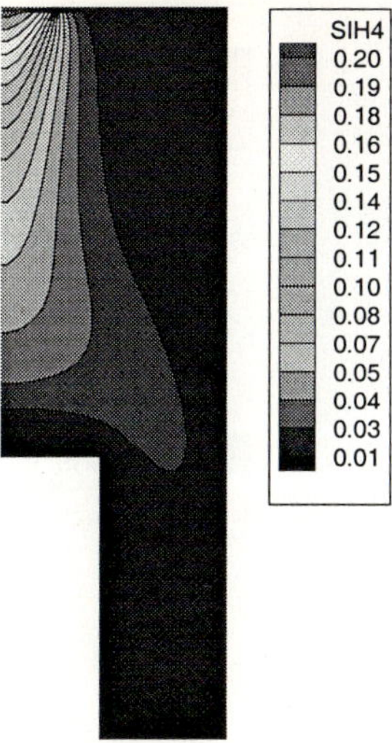

Figure 12 Molar fractions of SiH$_4$

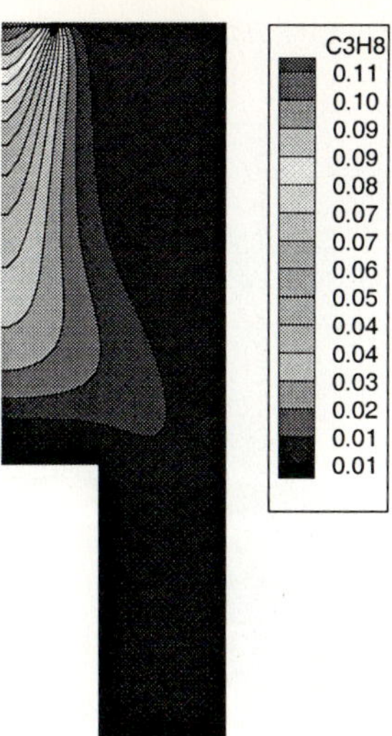

Figure 13 Molar fractions of C$_3$H$_8$

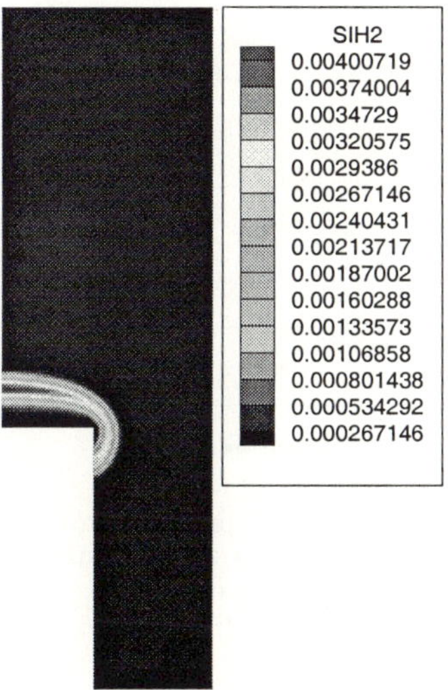

Figure 14 Molar fractions of SiH$_2$

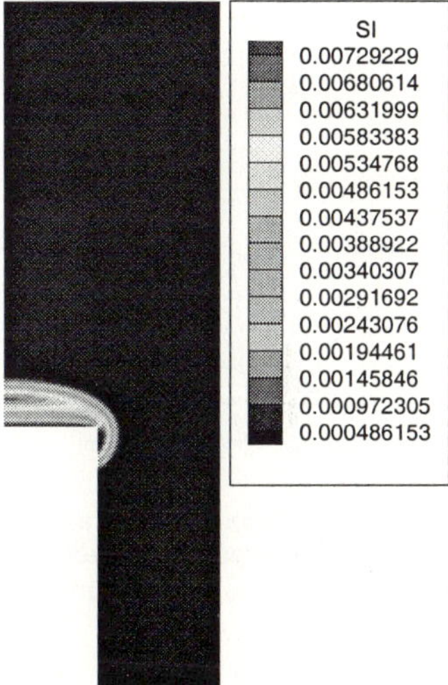

Figure 15 Molar fractions of Si

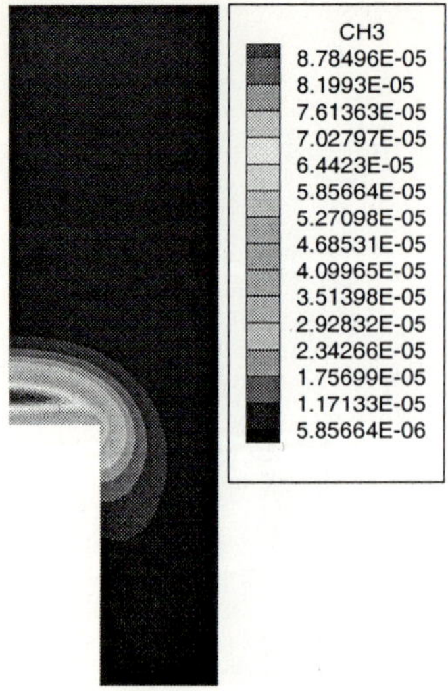

Figure 16   Molar fractions of CH_3

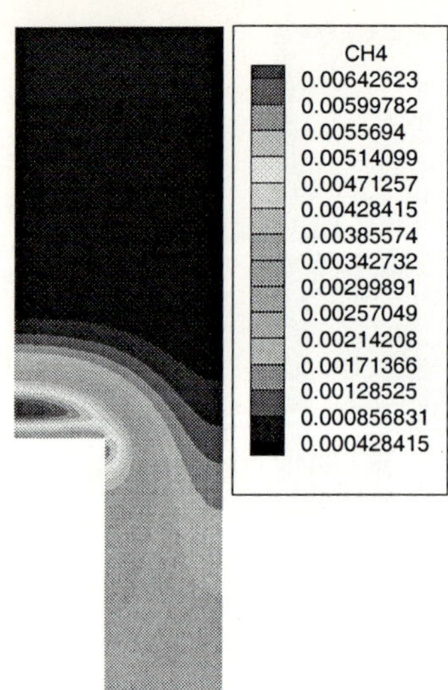

Figure 17   Molar fractions of CH_4

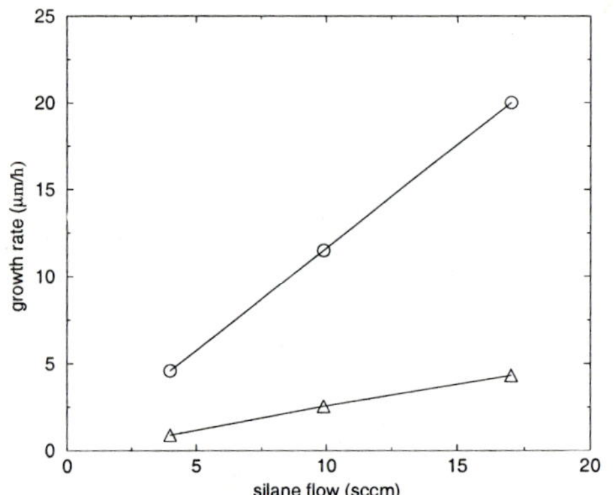

Figure 18   Influence of silane flow on experimental measured ($\triangle$) and calculated ($\circ$) growth rates ($F_{C_3H_8}$=6.6 sccm)

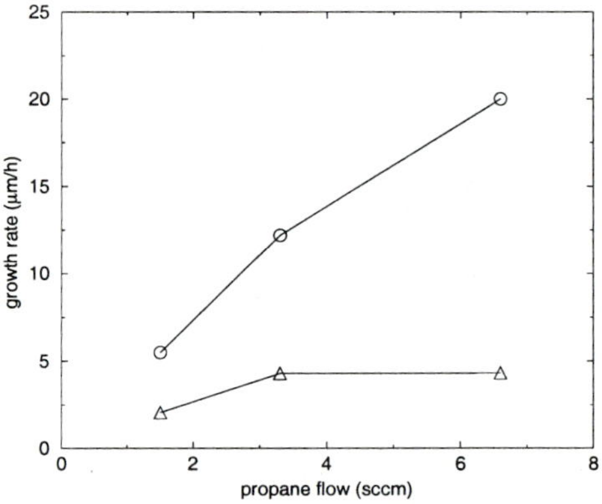

Figure 19   Influence of propane flow on experimental measured ($\triangle$) and calculated ($\circ$) growth rates (F_{SiH_4}=17 sccm)

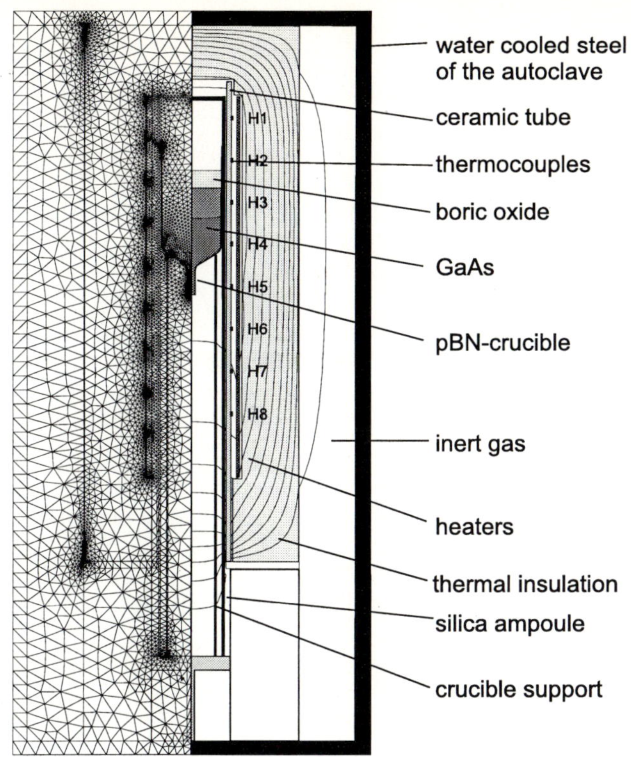

Figure 20 The right side shows Sketch of a Vertical Gradient Freeze (VGF) or Bridgman (VB) furnace for the growth of GaAs crystals with 3" diameter. The furnace consists of 8 heaters H1 – H8. In the experiment the power of each heater is controlled by a thermocouple beside the heater on the ceramic pipe. On the left hand side, the numerical grid for the numerical analysis is shown.

5 Crystal Growth by the Vertical Gradient Freeze Method

Figure 20 shows sketch of a furnace for the growth of 3" GaAs single crystals by the Vertical Gradient Freeze (VGF) method. The raw GaAs material is piled up in the conal and the cylindrical part. This assembly is covered by a boric oxide layer to avoid loss of arsenic, which is undesired because of a decrease of the crystals homogeneity and because of its toxicity.

In the beginning, the furnace is heated up in order to melt the raw material. Then in a distinct thermal process the temperature field with certain gradients in the solid and the liquid is shifted into axial direction. This is performed by heater controlling. The movement of the melting temperature through the liquid leads to a directional growth of the crystal. The quality of the material strongly depends on the growth conditions — i. e. on the homogeneity of the temperature gradient on the crystal and on the resulting thermal stress. These conditions can be optimized numerically with `CrysVUn++` [8].

5.1 Influence of Anisotropic Heat Conduction in the Crucible

In the VGF–growth of GaAs the typical applied material for the crucible is pyrolytic Boron Nitride (pBN). Due to its production procedure (CVD on a fibre), it has thermal heat conductivities that depend on the direction. Thus the heat conductivity λ in equation (2.3) becomes a tensor. In the concrete example of pBN, the conductivities differ by more than a factor of 30: $\lambda_{||}^{\mathrm{pBN}} = 62.7\,\mathrm{W/mK}$ and $\lambda_{\perp}^{\mathrm{pBN}} = 2.0\,\mathrm{W/mK}$. The conductivities

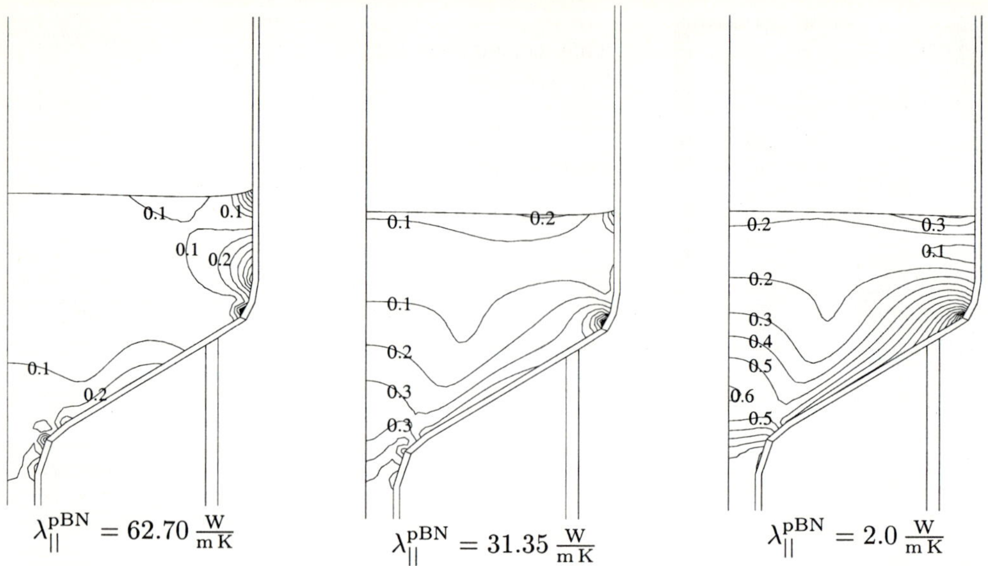

Figure 21   Isolines of the v. Mises stress resulting for three different values of $\lambda_{\parallel}^{\mathrm{pBN}}$.

of GaAs are , $\lambda_{\mathrm{solid}}^{\mathrm{GaAs}} = 7.1\,\mathrm{W/mK}$ and $\lambda_{\mathrm{liquid}}^{\mathrm{GaAs}} = 17.8\,\mathrm{W/mK}$.

We consider a certain stage of the growth, where the solid liquid interface is in the bulk crystal near the cone. In [27] the heating power was optimized in order to achieve a planar solid liquid interface and only low stresses near the solidification front. In order to demonstrate the influence of the anisotropic heat conductivity of the crucible material on the shape of the solid-liquid interface and the thermal stresses computations with different vertical heat conductivities of the boric oxide were carried out. The corresponding stress distributions are shown in figure 21.

We recognize that the shape of the interface changes from concave to convex with decreasing heat conductivity $\lambda_{\parallel}^{\mathrm{pBN}}$. Of course, these changes result from a modification of the temperature profile, also leading to a modification of the thermal stress. The solidification takes place for stress values smaller than 0.2 MPa in the case of the optimized process with the real material parameters while this value is greater than 0.2 MPa. The results show that the anisotropic conduction has a significant influence on the thermal stress and thus on the quality of the growing crystal. If one wants to optimize the growth process, this effect must be taken into account.

5.2   Study of the Effects of the Thermal Convection on the GaAs-Si Solidification

A global simulation of a crystal growth experiment has been performed with CrysVUN++. The resulting temperature distribution in the crucible serves as a realistic thermal boundary condition for the enthalpy method [28, 29, 30] in order to study the influence of thermal convection on the temperature distribution in the melt and of the shape of crystal/melt interface [31]. In a second step the dopant distribution during the process is investigated [32].

We use here an axisymmetric formulation of the enthalpy code. The conservation equations are made dimensionless according to the thermal diffusive time. It leads to the dimensionless Rayleigh number $Ra = g\beta\Delta T L^3/\nu\kappa$ and Prandtl number $Pr = \nu/\kappa$ with thermal diffusivity κ and viscosity ν. The temperature is made dimensionless with the temperature difference between the top and the bottom of the vertical wall $\Delta T = 50.555K$. The length scale is made dimensionless with the size of the vertical wall $\Delta y = 8.4cm$. The GaAs-Si phase diagram gives a partition ratio of $k = 0.9$. That means that Silicon segragates less than GaAs in the solid phase.

The results concern the boundary conditions given in Figs. 22, 23 and 24 for the growth time $t = 47\,h$ after the seeding. The profiles correspond to dimensional temperature with zero value for the solidification temperature. For this thermal field the physical properties for GaAs correspond to a thermal Rayleigh number $Ra = 6 \times 10^5$ and a Prandtl number $Pr = 6.6 \times 10^{-2}$. The solutal diffusivity parameter becomes $D = 1.8 \times 10^{-3}$.

The mesh used is 100×200 in horizontal and vertical directions respectively, the time step is $\delta t = 10^{-4}$ and the simulation is done for 20000 times steps First a conductive solution of the thermal field is obtained using the nonlinear thermal boundary conditions indicated in Fig. 25 giving the position of the solid-liquid interface. This solution is used as input to compute the velocity field and the transport of dopant. It should be remarked that in the considered configuration only the thermal field has an influence on convection, the solutal convection is neglectable.

Concerning the flow pattern, which is maintained during the simulation after a transitory regime (Fig. 26), a convective cell centered near the wall develops and occupies all the fluid phase (Fig. 27). This convective motion induces an increase of the Si-dopant concentration in the liquid part nearby the vertical wall. Figure 28 gives 4 stages of the concentration fields. In the beginning the Si-dopant is concentrated near the boundaries of the fluid phase (interface, wall and the top). The convective motion provoques a solutal transfer from the interface to the top of the ampoule. At successive time steps the Si-dopant concentration driven by the convective motion increases near the wall of the ampoule and decreases near the axis of the ampoule.

The Si-dopant fraction is represented by the profile in Fig. 29 at the interface level as a function of time and radial position. From t_1 to t_4, due to the segregation phenomena, the Si-dopant decreases linearly with the distance to the axis. When the convection develops another behaviour is observed: the mass fraction of the Si-dopant decreases until the half of the ampoule when the convection effect is more important than the segregation so that the concentration has a maximum value near the point $x = 0.4$ and it decreases near the the wall. The concentration gradient of Si-dopant is concentrated near the melt/solid interface and increases with the convective motion. Structural cells of dopant surrounding the convective solution develop in the bulk. In the course of time a progressive degeneration of Si-dopant at the interface near the axis and an enrichment towards the ampoule are observed. This observation is related to the evolution of the shape of the interface plotted on Fig. 30 showing the interface position along the axis. This allows to interpret the insight of segragation on the Silicon distribution indicated in Fig. 29. We see an enrichment of dopant in the liquid part which is in competition with the solutal transport which tends to degenerate the dopant distribution nearby the wall. At the time t_{10} we observe that the Si-dopant concentration has a maximum level of saturation near the wall of the ampoule ($x = 0$) and decreases near the axis ($x = 0.48$). At this time we have an approximate ratio of 2 between the concentrations in the two zones. The equilibrium is near the time t_8.

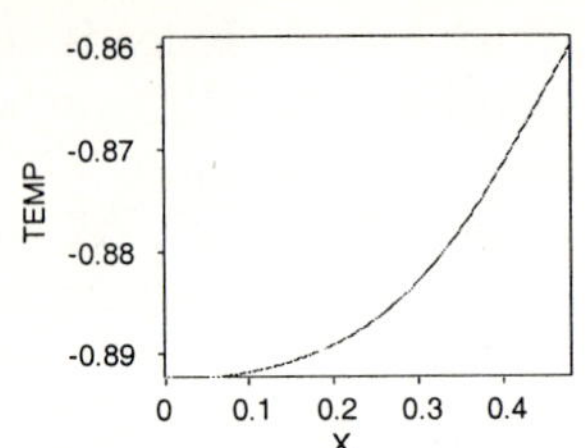

Figure 22 South boundary thermal condition

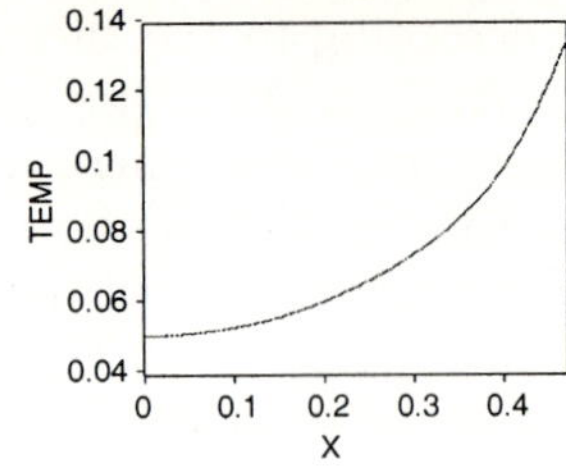

Figure 23 North boundary thermal condition

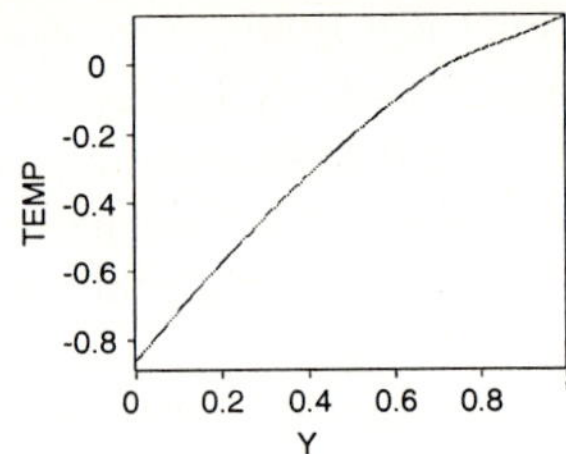

Figure 24 East boundary thermal condition

Figure 25 Initial thermal field in the crucible

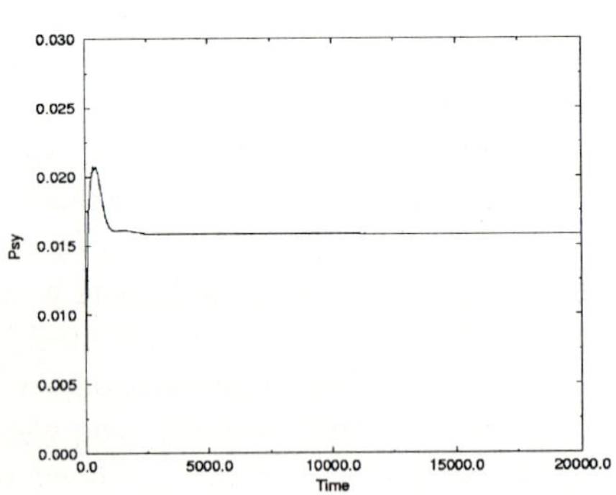

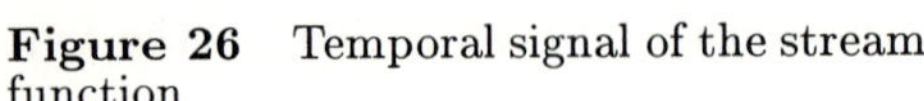

Figure 26 Temporal signal of the stream function

Figure 27 Streamlines and solid fraction isovalues calculated at one stage of the simulation.

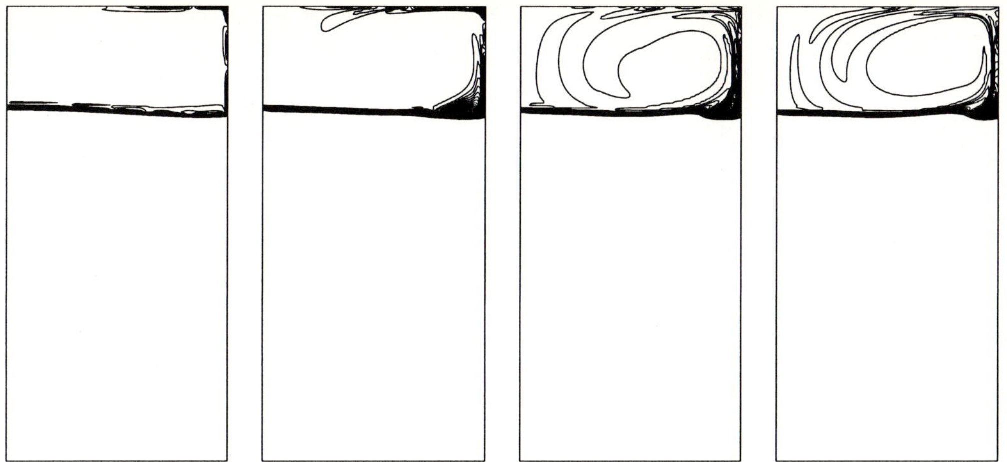

Figure 28   Solute concentration isovalues calculated at four stages of the simulation.

6   Growth of Semitransparent Crystals

Many materials that are industrially grown as single crystals are semitransparent in some part of the optical spectrum, i.e. the absorption length is of similar order of magnitude as the dimensions of the grown crystals for those materials and parts of the spectrum. This is the case for materials used for optical components as well as for semiconductors at lower temperatures, found when cooling them down after single crystal growth from the melt. This semitransparent behaviour is very important for modelling the thermal field in a growth furnace. Often the heat transport through the crystal happens mostly by radiation. Neglecting the semi-transparency completely would therefore lead to much too high temperature differences in the crystal.

To get an approximate value for the thermal flux through the crystal, especially when the absorption length is noticeably smaller than the dimensions of the crystal, it is sufficient to modify the thermal conductivity to model the thermal and radiative flux through the material. This approximation is known in the literature as Diffusion Approximation (see [4], p.485 ff). While giving reasonably good results for the overall temperature field in the furnace and the needed heater powers, this approximation does not describe the temperature field close to the border of the crystal in a layer of about one absorption length thickness. Temperature differences in this boundary layer can dominate the thermo-elastic stress in the whole crystal, as we will show later. Therefore, if predictions on the occurring thermo-elastic stress are to be made, it is necessary to use a better approximation.

One possible choice for such an approximation is the well-known P_1-Approximation as outlined in section 2. It describes the thermal behaviour in the boundary layer in good approximation, especially for the case of a not too large absorption length (see [4]) and it has the advantage of not leading to long computing times as, for instance, when using a P_N- or Discrete-Ordinate-Approximation. Furthermore it is comparably simple to implement, especially if curvilinear coordinates have to be used.

The following example demonstrates the importance of the boundary effects mentioned above. Fig. 31 shows the numerical setup used. It is the interior of a realistic VGF-furnace

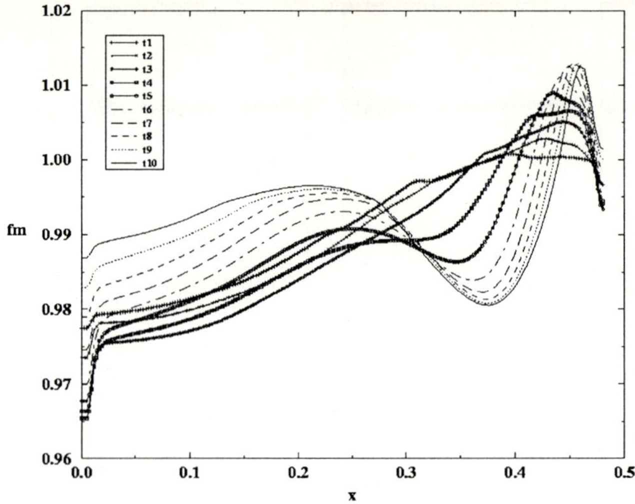

Figure 29 Solute distribution in the crystal calculated at ten stages of the simulation.

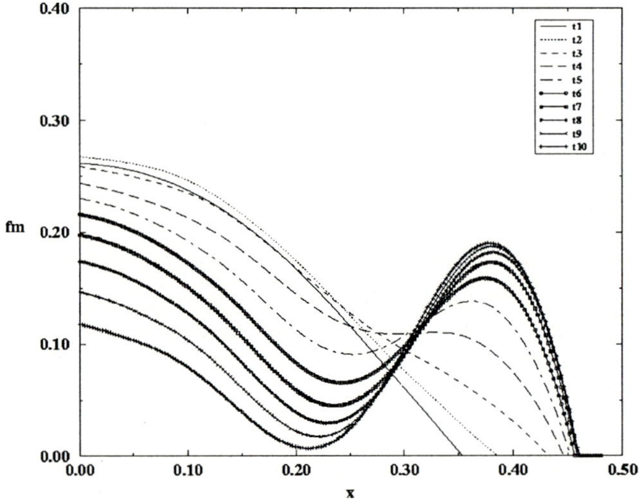

Figure 30 Interface position calculated at ten stages of the simulation.

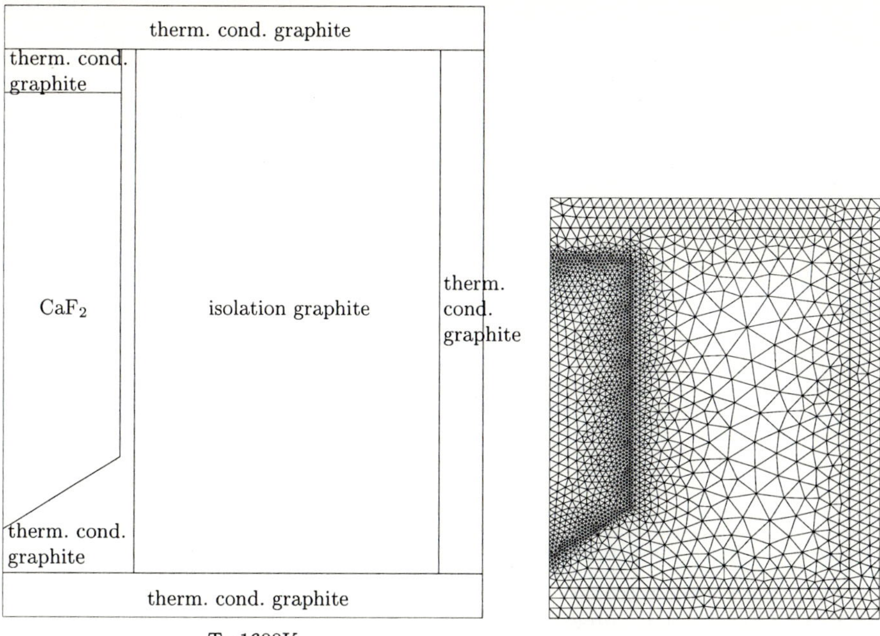

Figure 31   Problem configuration (left) and numerical grid (right). The inner diameter of the crucible is 8 cm, the hight of the crystal is 30 cm including a 4 cm cone. The crucible is made of a typical graphite material with an assumed thermal conductivity decreasing from 55 W/Km at 1000K to 30 W/Km at 2000K. Its walls are 1 cm thick. It is standing in a box, and covered with a lid of the same material, both 3 cm thick. It is surrounded by a 21 cm layer of insulation material with an assumed thermal conductivity increasing from 0.3 W/Km at 1000K to 0.6 W/Km at 2000K.

for CaF$_2$. The grid used is shown in fig. 31

The following three cases are considered: (1) a constant thermal conductivity of CaF$_2$, which was adapted to model thermal and radiative heat flux at 1680 K (near melting temperature), (2) a thermal conductivity proportional to T^3 according to the Diffusion Approximation as mentioned above and (3) a normal thermal conductivity only describing the conductive heat flux and P$_1$-Approximation for the radiative heat flux.

The corresponding axial temperature profiles and the thermo-elastic stresses are given in Figs. 32 and 33, respectively. It can be seen that the small deviation of the P$_1$-curve near the bottom and top of the crystal dominates the stress all over the crystal. Note that differences in temperatures between calculations with constant thermal conductivity and according to the Diffusion Approximation have only very little effect on the thermo-elastic stress.

The above example shows, that an adequate model is needed for calculations of thermo-elastic stresses in semi-transparent materials. The P$_1$-Approximation implemented describes the relevant boundary effects with reasonable accuracy.

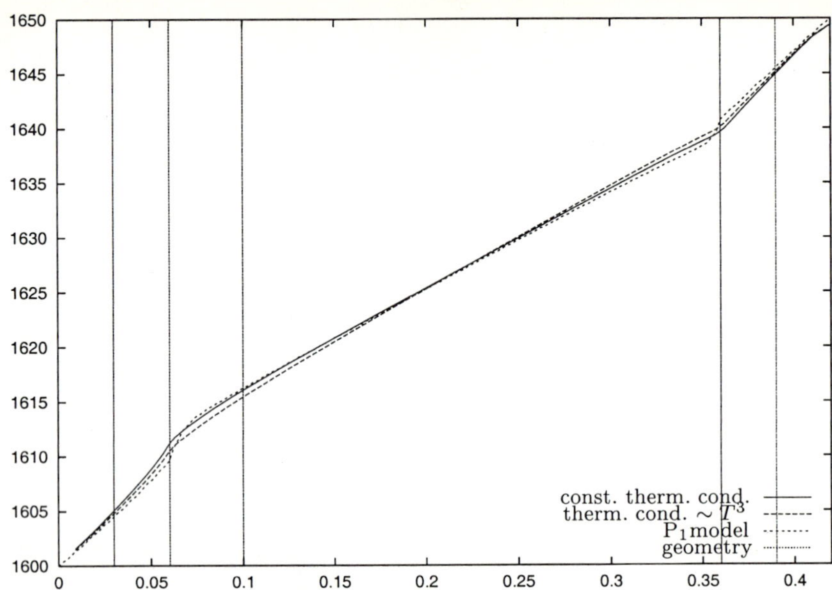

Figure 32 Axial temperature profile for constant thermal conductivity, thermal conductivity proportional T^3 (Diffusion Approximation) and for P_1-Approximation. The vertical lines mark the box walls (0-3 cm, 39-42 cm), the crucible (3-6 cm), the conic (6-10 cm) and cylindric (10-36 cm) parts of the crystal and the lid (36-39 cm)

7   Conclusions

Spectral methods seem to be the method of choice for computing complex flow in moderate complex domains. Due to the global nature, the methods are usually applied only to simple geometries. The development of a coupled spectral/ finite-volume code using a robust implicit iterative solver for the spectral blocks seem to be a solution to overcome this bottleneck.

The advantages of the implicit iterative solver are also important for simulations with non-constant material properties. Parts of the coupled code have been already successfully applied to calculate the flow in a CVD reactor, where the viscosity, the heat conduction and the density are strongly temperature dependent.

The efficiency of the spectral method was also preserved in the coupled method, which is especially necessary for the extension of the coupled code to three-dimensions. Future work will concern the further increase of the applicability of the coupled code to technical relevant problems. Therefore, the main working areas will be the mapping and domain decomposition of the spectral blocks.

Modeling of CVD of SiC in a rotating disc reactor is performed with the goal to study the deposition mechanism. It is shown that the modeling without taking into account the droplet formation gives overpredicted deposition rate. The further physical model extensions are the aims of the forthcoming studies. It was shown that such kinds of chemical systems may be successfully calculated by the developed "stiff chemistry" algorithms.

The thermal model of VGF-furnaces for growing GaAs or CaF_2 bulk crystals was improved in order to optimize the growth conditions. It has been shown, that the new

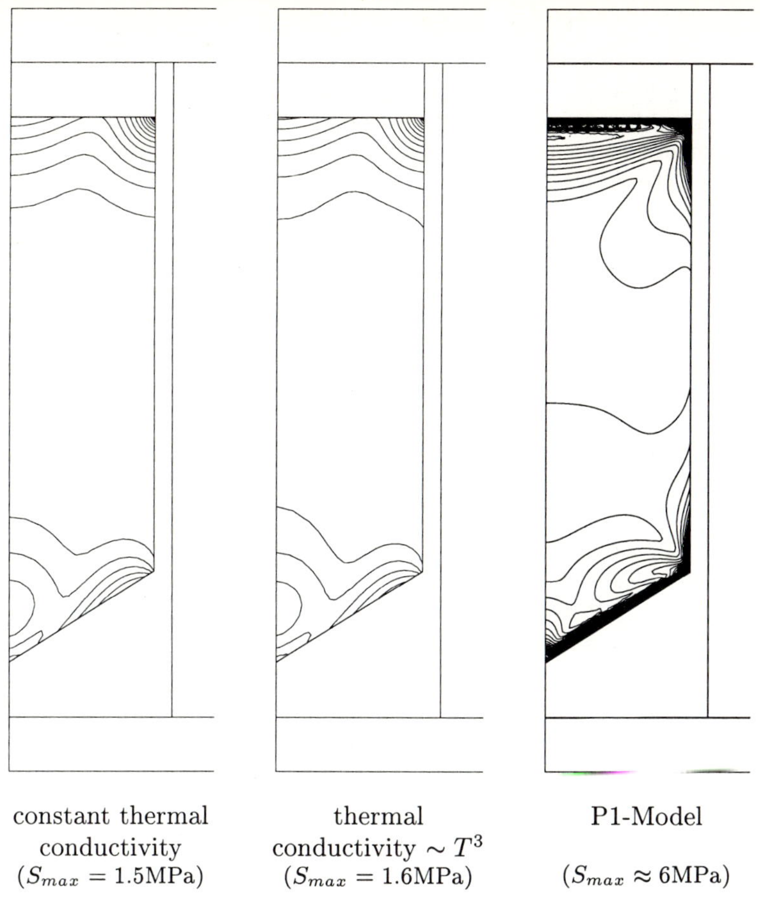

Figure 33 Thermo-elastic von Mises stress S in the crystal for constant thermal conductivity, thermal conductivity proportional T^3 (Diffusion Approximation) and for P_1-Approximation. Isoline are drawn with a spacing of 0.1 MPa.

physical features, anisotropic heat conduction and semi-transparency, have great influence on the thermal stresses and so to the dislocations in the crystal. Also they seem to cause only minor changes to the global temperature profile in the crystal, anisotropic heat conduction and semi-transparency may have great influence on the local thermal regime in the near of the phase boundary, which is the crucial point for crystals quality.

8   Acknowledgments

Financial support was provided by DFG and CNRS within the French–German research Program *Numerical Flow Simulation II*. One of the authors is grateful to Y. Egorov (LSTM) for his help with the calculations.

References

[1] P.Droll, M.El Ganaoui, L.Kadinski, M.Kurz, A.Lamazouade, O.Louchart, D.Morvan, M.Naamoune, A.Pusztai, I.Raspo, P.Bontoux, F.Durst, G.Müller, J.Ouazzani, and M.Schäfer. High Performance Computer Codes and their Application to Optimize Crystal Growth Processes. In *Numerical Flow Simulation I*, volume 66 of *Notes on Numerical Fluid Mechanics*, pages 115–143. Vieweg Verlag, 1998.

[2] F. Durst, L. Kadinski, and M. Schäfer. A multigrid solver for fluid flow and mass transfer coupled with grey-body surface radiation for the numerical simulation of CVD processes. *J. Crystal Growth*, 146:202–208, 1995.

[3] L. Kadinski. *Mathematische Modellierung und numerische Simulation von CVD-Prozessen in der Halbleitertechnik.* PhD thesis, Friedrich-Alexander-Universit"at zu Erlangen, 1996.

[4] Modest. *Radiative Heat Transfer.* McGraw-Hill, 1993.

[5] Timoschenko and Goodier. Theory of elasticity. *McGraw–Hill Book Company, Inc.*

[6] D. Lambropoulos J. The isotropic assumption during czochralski growth of single semiconductors crystals. *Journal of Crystal Growth*, 84:349–358, 1987.

[7] F. Durst and M. Schäfer. A parallel blockstructured multigrid method for the prediction of incompressible flows. *Int. J. for Num. Meth. in Fluids*, 22:549–565, 1996.

[8] M. Kurz. Development of CrysVUN++, a software system for numerical modelling and control of industrial crystal growth processes. *Ph. D. thesis, university of Erlangen*, 1998.

[9] P. Droll, M. Schäfer, O. Louchart, and P. Bontoux. Coupling of a finite-volume method with a pseudospectral method. *ECCOMAS 98, Proceedings*, 1 (2):1240–1245, 1998.

[10] S. Hugues and A. Randriamampianina. An improved projection scheme applied to pseudospectral methods for the incompressible Navier-Stokes equations. *Int. J. for Num. Meth. in Fluids*, 1997.

[11] M. Hortmann and M. Perić. Finite volume multigrid prediction of laminar natural convection: Bench-mark solutions. *Int. J. for Num. Meth. in Fluids*, 11:189–207, 1990.

[12] P. Droll and M. Schäfer. An implicit pseudospectral method for the solution of the incompressible Navier-Stokes equations. *Special Issue of CFD Journal*, 9, to appear in 2000.

[13] C. D. Dimitropoulos, B. J. Edwards, K.-S. Chae, and A. N. Beris. Efficient pseudospectral flow simulations in moderately complex geometries. *J. of Comp. Physics*, 144:517–549, 1998.

[14] Y. Saad and M.H. Schultz. GMRES: A generalized minimal residual algorithm for solving nonsymmetric linear systems. *SIAM J. Sci. Stat. Comput.*, 7:856–869, 1986.

[15] G.L.G. Sleijpen and D.R. Fokkema. BiCGSTAB(L) for linear matrices involving unsymmetric matrices with complex spectrum. *ETNA*, 1:11, 1993.

[16] E. Serre and J. P. Pulicani. A 3d pseudospectral method for rotating flows in a cylinder. *Int. J. of Computers and Fluids*, Paper in print.

[17] O. Savas. Stability of bödewadt flow. *J. of Fluid Mech.*, 183:77–94, 1987.

[18] H.P. Greenspan. *The theory of rotating fluids.* Cambridge University Press, 1972.

[19] M.P. Escudier. Observations of the flow produced in a cylindrical container by a rotating endwall. *Exp. in Fluids.*, 2:189–196, 1984.

[20] R. Rupp, P. Lanig, J. Völkl, , and D. Stephani. First results on silicon carbide vapour phase epitaxy growth in a new type of vertical low pressure chemical vapor deposition reactor. *J. Crystal Growth*, 31:41–146, 1995.

[21] M. D. Allendorf and R. J. Kee. A model of silicon carbide chemical vapor deposition. *J. Electrochem. Soc.*, 138:841–852, 1991.

[22] M. D. Allendorf. Equilibrium predictions of the role of organosilicon compounds in the chemical vapor deposition of silicon carbide. *J. Electrochem. Soc.*, 140:747–753, 1993.

[23] P. Kaufmann and Yu. N. Makarov. Modelling of flow, heat transfer, mass transport during CVD of SiC in a vertical reactor. Technical report, LSTM, University Erlangen-Nürnberg, 1994. unpublished.

[24] T. Weber. *Die numerische Simulation reaktiver Strömungen als Basis zukünftiger Reaktormodellierungen.* PhD thesis, Friedrich-Alexander-Universität zu Erlangen, 2000.

[25] R. Rupp, P. Lanig, J. Völkl, and D. Stephani. Mater. *Res. Soc. Symp. Proc*, 423:253ff, 1996.

[26] A.N. Vorobev, A.E. Komissarov, A.S. Segal, Yu.N. Makarov, S.Yu. Karpov, A.I. Zhmakin, and R. Rupp. Modeling analysis of gas phase nucleation in silicon carbide chemical vapor deposition. *Mat. Sci & Eng, B*, 61–62:176–178, 1999.

[27] M. Kurz and G. Müller. Control of thermal conditions during crystal growth by inverse modelling. *J. Crystal Growth.*

[28] El Ganaoui M. and Bontoux P. An homogeneisation method for solid-liquid phase change during directional solidification. *ASME H.T.D.,* Numerical and Experimental methods in Heat Transfer, 361:453–469, 1998.

[29] M. El Ganaoui. Modélisation de la convection instationnaire en presence d'un front de solidification déformable. application à la croissance cristalline. *Phd Thesis*, 1997.

[30] D. Morvan, M. El Ganaoui, and P. Bontoux. Numerical simulation of 2d crystal growth problems in a vertical bridgman-stockbarger furnace, latent heat effects and crystal-melt in interface morphology. *Int. J. Heat Mass Transfer*, 42:573–579, 1999.

[31] El Ganaoui M., Bontoux P., and et Morvan D. Capture d'un front de solidification en interaction avec un bain fondu instationnaire. *C. R. Acad. Sciences Paris*, t. 327(Srie II b):41–48, 1999.

[32] Lamazouade A., El Ganaoui M., Morvan D, and et Bontoux P. Simulation numérique par une approche porosité enthalpie de la convection thermique et solutale dans une ampoule de bridgman. *Int. J. of Thermal Sciences (Revue Générale de Thermique)*, 8(N 38), 1999.

Spline Volume Tracking for Interfacial Flows

I. GINZBURG[1], G. WITTUM[2], and S. ZALESKI[3]

[1] Universität Kaiserslautern, ITWM, Erwin-Schrödinger-Straße 49, D-67663 Kaiserslautern, Germany

[2] Universität Heidelberg, IWR, INF 368, D-69120 Heidelberg, Germany

[3] Université Pierre et Marie Curie, Laboratoire de Modélisation en Mécanique, 4 place Jussieu, 75005 Paris, France

Summary

In this paper we present a 2D Volume of Fluid two phase flow model with surface tension. The model is based on an implicit in time discretization of Navier-Stokes equations on unstructured grids, using staggered finite volumes. A cubic spline interface interpolant which preserves the volume fraction distribution is introduced. Interface adaptive and/or interface aligned deformable grids are reconstructed at each time step with help of either a linear or spline interface approximation. Anomalous currents around bubbles are significantly reduced with help of cubic spline interpolants. The simulations of buoyant bubbles are compared with known theoretical, numerical and experimental results.

1 Introduction

There are different approaches in the direct simulation of interfacial flows, which can be classified into two main directions, *front tracking* and *front capturing*.

In front tracking methods usually marker particles are used to explicitly represent the interface. For a given distribution of these markes particles, the surface tension force is calculated via an interpolation curve and used to calculate the momentum balance. Afterwards the interface is advected and the marker points are redistributed, if necessary. Among the implicit front capturing methods there are VOF and Level sets. For a more detailed introduction in both methods, the problems associated therewith etc., we refer the reader to Rider & Kothe [15], Scardovelli & Zaleski [17] or Ginzburg & Wittum [8] and the references cited therein.

In our approach the curvature is calculated via cubic spline interpolants, the governing equations are discretized on interface grids (adaptively refined and/or interface aligned) while the advection of the interface is done on a regular mesh.

The paper is organised as follows. In the next section the basic equations are presented, followed by a description of the interface representation and grids in Section 3. Section 4 presents an overview of the solution procedure, followed by some numerical results in Section 5.

2 Governing Equations

Assume that two immiscible phases, liquid and gas, occupy a 2D rectangular domain Ω. The velocity field $\vec{u} = (u_x, u_y)$ and pressure p of each phase obey the incompressible Navier-Stokes equation

$$\rho \frac{\partial \vec{u}}{\partial t} + \rho \nabla \cdot (\vec{u} \otimes \vec{u}) = -\nabla p + \rho \vec{g} + \vec{F}_s + \nabla \cdot (2\mu \mathbf{D}), \quad \mathbf{D} = \frac{1}{2}(\nabla \vec{u} + \nabla^t \vec{u}) , \tag{1}$$

$$\nabla \cdot \vec{u} = 0 . \tag{2}$$

In bulk, $\rho = \{\rho_l, \rho_g\}$ is the density and $\mu = \{\mu_l, \mu_g\}$ is the viscosity. For immiscible phases they are constant along a particle path, therefore the following relations hold

$$\frac{\partial \rho}{\partial t} + \vec{u} \nabla \rho = 0 ; \quad \frac{\partial \mu}{\partial t} + \vec{u} \nabla \mu = 0 . \tag{3}$$

The surface tension force acts on the interface S between the fluids. When the surface tension coefficient σ is constant, the surface tension force per unit area $\vec{F}_s$ is defined as

$$\vec{F}_s = \sigma \kappa \vec{n} \delta_S . \tag{4}$$

Here $\kappa(\vec{r}) = -(\nabla \cdot \vec{n})|_{\vec{r}}$ is a local curvature, taken positive if the center of curvature is in the gas phase; $\vec{n}(\vec{r})$ is the unit normal to S at $\vec{r}$, directed from the continuous phase into the gas phase. δ_S is concentrated on the interface S. A precise numerical computation of the curvature is difficult unless the interface is represented by a sufficiently smooth curve, e.g. twice continuously differentiable. Lafaurie et al. [11] have shown, that $\vec{F}_s$ can be included into eq. (1) in the momentum conserving form

$$\vec{F}_s = -\nabla \cdot \mathbf{T} , \qquad \mathbf{T} = \sigma (\mathbf{I} - \vec{n} \otimes \vec{n}) \delta_S . \tag{5}$$

The velocity is assumed to be continuous across the interface: $[\vec{u}]_S = 0$. Here and below, $[\psi]_S = \psi_l - \psi_g$ denotes the jump across the interface S. Momentum conservation supplies the additional interface conjunction condition

$$[2\mu \mathbf{D} \cdot \vec{n} - p\vec{n}]_S = \vec{F}_s . \tag{6}$$

At the boundary of the physical domain Ω, we impose either noslip or freeslip boundary conditions.

3 Interface reconstruction and advection

In VOF methods, the distribution of the gas phase is represented by its volume fraction C_k for each cell k. PLIC ([22], [12], [15]) reconstructs the interface by linear slope segments perpendicular to a given normal, which is given by

$$\vec{n} \approx \nabla C . \tag{7}$$

Finite difference approximations of this relation are only first order accurate.

3.1 Cubic Splines

Cubic splines [6], [7], [13], going through a *fixed position of marker points* (x_k,y_k), are smooth enough to ensure the accurate numerical computations of the curvature in (4). The third order polynomial parametric function $(x(s),y(s))$, $x(s) \in C^2$, $y(s) \in C^2$, has the following form in the interval between (x_{k-1},y_{k-1}) and (x_k,y_k), $k \geq 1$

$$\psi(s) = \frac{M_{k-1}^{\psi}}{6L_{k-1}}(s_k - s)^3 + \frac{M_k^{\psi}}{6L_{k-1}}(s - s_{k-1})^3$$

$$+(\frac{\psi_k}{L_{k-1}} - \frac{M_k^{\psi}L_{k-1}}{6})(s - s_{k-1}) + (\frac{\psi_{k-1}}{L_{k-1}} - \frac{M_{k-1}^{\psi}L_{k-1}}{6})(s_k - s) . \tag{8}$$

Here $\psi(s)$ denotes $x(s)$ or $y(s)$; s_k is a value of the parameter s at point (x_k,y_k): $s_k = \sum_{j=1}^{k}((x_j - x_{j-1})^2 + (y_j - y_{j-1})^2)^{\frac{1}{2}}$ and $L_{k-1} = s_k - s_{k-1}$. The continuity of the first derivatives at the points (x_k,y_k) supplies the necessary linear equations to determine the second derivatives M_k^x and M_k^y.

In VOF models, the points lying on the interface are unknown *a priori*. Our basic idea is to construct a cubic spline interpolant (8) which exactly cuts the phase distribution. Assume that the spline intersects the boundaries of the regular $h \times h$ interfacial cell k with boundary $\partial\Sigma$ at points $\mathbf{C} = (x_{k-1},y_{k-1})$ and $\mathbf{D} = (x_k,y_k)$ (see Fig. 1). Suppose that we know *which* edges should be cut by the interface but we do not know *where* the interface cuts them. Then one of the coordinates of each cross point is fixed. Let ξ_k denote the unknown coordinate. Following [13], we use Stokes theorem to express the volume V_k of the cell k cut by the cubic spline as

$$V_k = \oint_{\partial\Sigma} x dy . \tag{9}$$

This is schematically shown by the sequence of points $\mathbf{C},\mathbf{D},\mathbf{E},\mathbf{F},\mathbf{C}$ in Figure 1.

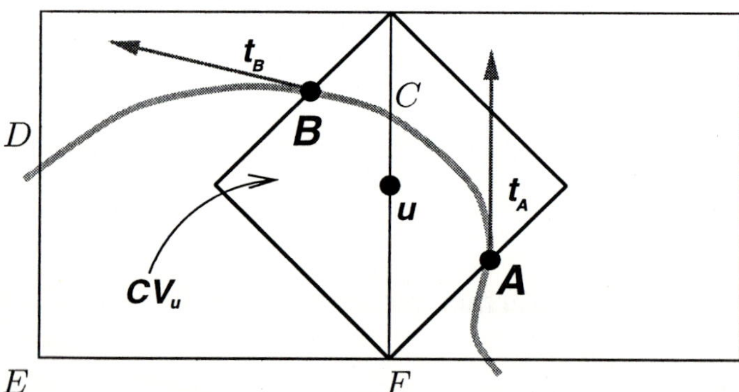

Figure 1   Sketch for spline interface.

Our purpose is to satisfy the volume conservation relations

$$V_k = C_k h^2 , k = 1..N . \tag{10}$$

When the function $\int_{s_{k-1}}^{s_k} x(s)y'(s)ds$ is determined from (8), V_k is written in terms of the unknowns $\{ M_{k-1}^x, M_k^x, M_{k-1}^y, M_k^y, \xi_{k-1}, \xi_k \}$. Together with the conditions of the continuity of the first derivatives at the points (x_k,y_k), relations (10) supply $3N$ equations to find $3N$ unknowns, $\{(M_k^x,M_k^y,\xi_k), k = 0,...,N-1\}$, in case of self-connected interfaces considered in this paper. Otherwise, additional boundary conditions should be imposed for the derivatives of the cubic spline at the boundary points (see [14]). When solving this nonlinear system with the Newton method, the Jacobian can be computed analytically.

We have to describe now how we define which edges of the interfacial cells should be cut by the interface. First, we construct the discontinuous linear interface with PLIC. Then a preliminary cubic spline, referred to as *spline A*, is drawn through the centers of the linear-slope segments.

When phase distribution is not continuous, *spline A* happens to intersect the non-interfacial cell. When the interfacial cell has more than two interfacial neighbours, the neighbour cells intersected by *spline A* are fixed by going along the PLIC interface.

The cross points (x_k^0,y_k^0) of *spline A* with the cell edges determine those which are assumed above to be cut by the interface. A mass violating cubic spline, referred to as *spline B*, goes through these points and provides an initial guess $(M_k^{x,0},M_k^{y,0},\xi_k^0)$ in order to solve the nonlinear system for cubic *spline C* which preserves phase distribution.

Two main problems should be emphasized. First, the solution of nonlinear problems can be not unique and a special, problem dependent criterion, should be used to fix one of the solutions. The solution is obviously not unique, for instance, in case of axisymmetric 2D bubbles considered below. Second, the solution can go outside the cell along the given edge. In this case, it does not keep the phase volume in the given cell but spreads it between two (or more) neighbouring cells. When this situation is encountered, while using fully convergent Newton method [14], we accept the solution lying inside the interfacial cells that minimizes the residual of the nonlinear equations. If the second derivatives (M_k^x,M_k^y) are set equal to zero, the solution of the nonlinear system (10) corresponds to a *continuous linear interface* reconstruction which maintains the phase distribution. However, the same solvability problems still remain.

3.2 Advection

The volume fraction is advanced on regular meshes via solving

$$\frac{\partial C}{\partial t} + \vec{u} \cdot \nabla C = 0 \,. \tag{11}$$

for incompressible fluids. This is done with a flux based approach similarly to the SURFER code (see [12], [9], [17]).

As an alternative approach, a Lagrangian moving of the PLIC interface as described in the appendix of [8] is also used. The test example is shown in Fig. 11.

3.3 Computational Grids

The problem is solved using three types of grids: regular, adaptively refined and interface aligned.

The regular grid is used to propagate the interface as sketched above.

As an indicator for the refinement we use the norm of the normal $\vec{n}$ as given by (7). Fig. 2 (a),(b) show adaptively refined grids.

The idea of the aligned grid is to separate the phases in order to calculate the pressure and density jumps accordingly. The aligned grids are constructed either by the use of the PLIC interface or by the use of cubic spline interpolants.

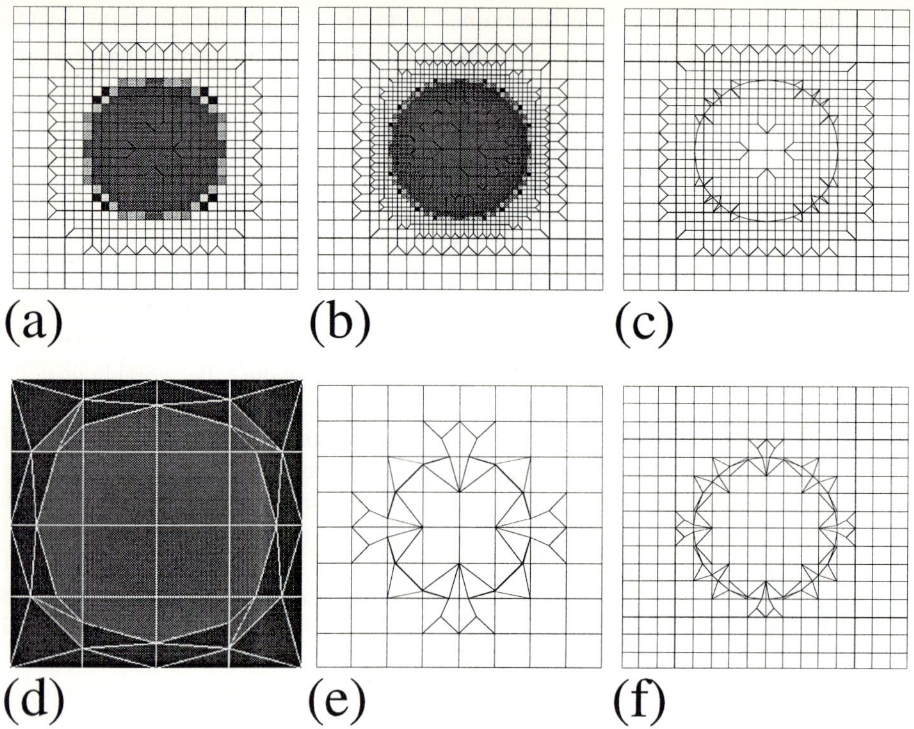

Figure 2 Phase distribution corresponds to circle $R = 0.25$ except of case (d) where it is done for a square. One (a) and two (b) levels adaptively refined grids. (c): *spline B* aligned grid, constructed over one adaptive level. (d), (e), (f): PLIC aligned grids.

Spline aligned grids contain less "bad" elements than PLIC aligned, since the mesh is intersected continuously by the spline, contrary to the PLIC segments (see [8]).

4   Solution Procedure

The equations are discretized in space using staggered finite volumes (see [8]). The pressure is addressed to the center of mass of each control volume while *both* components of the velocity are stored at the edges (see Fig. 3).

The pressure is assumed to be elementwise constant. The density and viscosity are averaged through their volume fraction on the regular grid (on the aligned grid they correspond to their phase, of course). Velocity approximations are calculated using nonconforming rotated Q elements, [5].

The surface tension force is calculated through

$$\int_{CV_u} \vec{F}_s = - \oint_{\partial CV_u} \mathbf{T} \cdot \vec{n}_u ds \ . \tag{12}$$

δ_S can be approximated as $|\nabla C|$, [11]. Finite element approximations are used to compute $\mathbf{T}$ from the nodal values.

If the interface is described by a spline going through the points $\mathbf{A}$ and $\mathbf{B}$ (see Fig. 1), the intersections of the spline with the edges of the control volume, the surface tension force can be integrated using

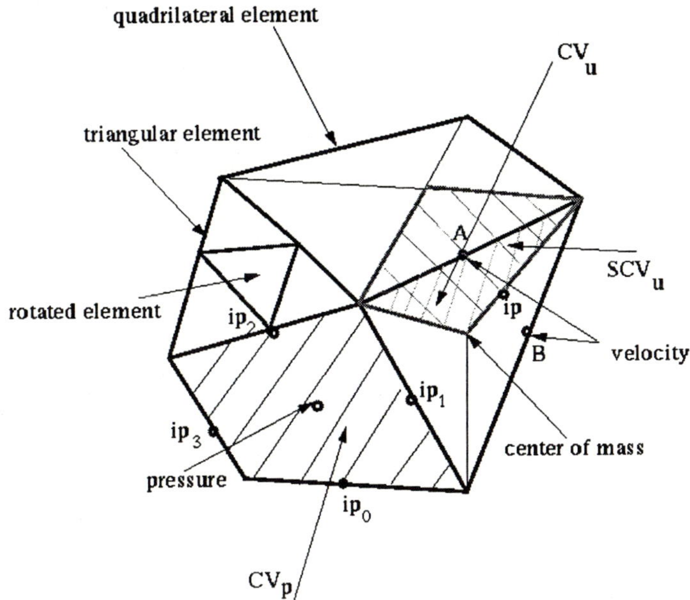

Figure 3 Sketch for staggered nonuniform grid.

$$\int_{CV_u} \vec{F}_s = \oint_A^B \sigma \kappa \vec{n} ds = \sigma(\vec{t}_B - \vec{t}_A); \ \vec{t} = \frac{(x'(s), y'(s))}{(x'^2(s) + y'^2(s))^{\frac{1}{2}}} \ , \tag{13}$$

see [13].

The time discretization is done implicitly with finite differences.

The nonlinear problem resulting from the convective part is solved with the Newton method. A multigrid method with BiCGstab acceleration ([1], [18]) is used to solve the linear saddle-point problem for the unknown pressures and velocities.

Alternatively, the convective term is linearised by

$$\mathbf{V}^{t+\Delta t} = \mathbf{V}^t + \frac{\partial \mathbf{V}^t}{\partial \vec{u}} \frac{\partial \vec{u}}{\partial t} \Delta t \ ; \qquad \frac{\partial \vec{u}}{\partial t} \approx \frac{\vec{u}^{t+\Delta t} - \vec{u}^t}{\Delta t} \qquad \mathbf{V} = \vec{u} \otimes \vec{u} \ . \tag{14}$$

More details about the discretization and solvers etc. can be found in [8].

5 Examples

In this section we show the results of some calculations. First, the Laplace law will be considered as standard test for surface tension approximation. Afterwards some experiments of rising bubbles will be presented.

5.1 Laplace law

For a stationary spherical bubble of radius R we consider the pressure balance at the interface

$$p_{in} - p_{out} = \frac{\sigma}{R} \,.$$
(15)

In order to compare the accuracy of the curvature representation the amplitude of the spurious currents is also calculated for increasing resolution.

Table 1 Results for static bubble.

	Amplitude of the anomalous currents			
	$\vec{F}_s$: rel. (12), (5)	$\vec{F}_s$: rel. (13)		
h	exact circle	exact circle	*Spline B*	*Spline C*
1/16	0.0413	8.67×10^{-9}	1.00×10^{-2}	1.64×10^{-3}
1/32	0.0293	9.22×10^{-9}	6.00×10^{-3}	1.89×10^{-4}
1/64	0.0289	1.23×10^{-8}	1.76×10^{-3}	4.89×10^{-5}
	Laplace law: $(\bar{p}_{in} - \bar{p}_{out})\frac{R}{\sigma} = 1$			
1/16	0.936	1.0	1.00282	0.999905
1/32	0.969	1.0	1.00087	1.000010
1/64	0.988	1.0	1.00019	1.000000

Both fluids have the same densities and viscosities equal to one. The surface tension force $\vec{F}_s$ with $\sigma = 1$ is applied in both forms: conservative (12), (5) and exact (13). The amplitude of the anomalous currents, $\max(\|\ \vec{u}\ \|)$, and measurements of the Laplace law are given in Table 1. The behaviour of the currents on the aligned, adaptively refined or regular grids is similar when the relations (12), (5) are used: $\max(\|\ \vec{u}\ \|)$ is proportional to σ/μ_l and it varies weakly with increasing of space resolution. This follows the same lines as the results that Laufarie et al. [11], Popinet & Zaleski [13] obtain with finite difference approximations of (12) on uniform grids. On the other hand, since the pressure gradient across the interface is correctly described on the aligned grid, where phase separation inside a control volume is done by the boundary between two elements, the anomalous currents almost vanish when (13) is used with the exact curvature and exact values of the normal vector $\vec{n}$. Consequently, an exact pressure distribution (15) is obtained in this case (see second column in Table 1).

Secondly, we construct the cubic splines B and C, corresponding to the volume fraction distribution for a circle $R = 0.2$. In order to fix *spline C*, we iterate the solution of the nonlinear problem until the curvature at the nose becomes equal to one at the neighbouring point. $\vec{F}_s$ is implemented then in the form (13) on the spline aligned grids. Table 1 shows that the amplitude of the spurious currents is reduced significantly, especially in case of the mass conserving *spline C*. In both cases, it decreases with the spatial resolution. These results confirm the ability of cubic spline aligned grids to represent the surface tension accurately.

5.2 Rising bubbles

The conditions of the calculations are the following: The dimension of the box is $1cm \times 2cm$; the initial bubble radius is $R = 1/6$ cm; the start position of its center is at $(0.5cm, 0.75cm)$; $g = 980$ cm/s^2; the density of the liquid $\rho_l = 1.0 g/cm^3$. Bubble velocity U is computed from the change in the position of the bubble nose using the central differencing in time.

The relative loss $M^{rel.}(t)$ of the total mass Ma of the gas phase is defined as

$$M^{rel.}(t) = \frac{Ma(t = 0) - Ma(t)}{Ma(t = 0)} .$$

(16)

The convection term is mainly discretized with (14). We have not found any noticeable difference between these solutions and implicit Euler discretization, at least for the medium range Reynolds numbers under consideration. Surface tension force is implemented in the exact form (13) on the spline aligned grids. Otherwise, we use (12) with (5) except in the third experiment, where spline interpolants are also used on not aligned grids.

First Experiment

Buoyant two dimensional bubbles with three different Eötvös numbers: $E_0 = 1$, $E_0 = 10$ and $E_0 = 104$ are modeled by Unverdi & Tryggvason [21] with a front-tracking method. They use a finite difference MAC-type discretization on a 65×129 regular grid. Four experiments with different M are considered for each E_0. The Morton number M decreases with the fluid viscosity μ_l, whereas the bubble viscosity μ_g and the density ratio are kept constant; so μ_l/μ_g decreases going from the calculations with higher to smaller M.

Here, we calculate the bubbles with the same E_0, M, μ_l/μ_g and $\rho_l/\rho_g = 40$ as in [21]; freeslip boundary conditions are used. The results are shown in Fig. 4 and Fig. 5. All calculations, except those of the skirted bubbles, are done on *spline B* aligned computational grids, constructed over 32×64 regular mesh. Skirted bubbles displayed in Fig. 5(e) are obtained on regular 64×128 regular grid while spline technique has not yet been implemented to deal with the reconnection. The time step $\Delta t = 5 \times 10^{-5} s$ except for Fig. 4(b) where it is doubled. Relative mass loss (16) is given at $t = 0.06 s$. When the moving is done on the same mesh and with the same time step, $| M^{rel.} |$ increases with Re (cf. $M^{rel.}$ in cases (c) and (d)).

At steady state, bubbles shapes and streamlines are found in good qualitative agreement with the results displayed in Fig.3-4 in [21]. Consider first small bubbles or the bubbles with high surface tension when $E_0 < 40$. Following the classification in [3], they are found in a spherical or ellipsoidal regime (see Fig. 4). For $E_0 = 1$, the top of the bubble becomes slightly flatter than the back and the wake appears due to stronger deformation when M decreases. This agrees with the steady state axisymmetric shapes drawn by Ryskin & Leal ([16], Fig.2). Oblate ellipsoids obtained for $E = 10$ are in close agreement with [21]; when M decreases, the separation occurs at the rim of the ellipsoid and "egg-shaped" closed wake appears behind it.

At $E_0 > 40$, the bubbles are found in spherical cap regime. When Re increases, the bubbles likewise change from oblate ellipsoidal to the oblate-ellipsoidal cap shapes, and then to the spherical-cap shapes (see photos presented by Bhaga &Weber [2], Fig. 2-3). Large bubbles in high M systems at Re of order 10 to 50 can develop thin annular films of dispersed fluid, usually referred to as "skirt". Skirted bubbles are studied by Hnat & Buckmaster [10]; in their experiments, the values $We/Re > 2.32$ imply the appearance of a skirt for $Re > 9$. Although this criterion is not satisfied by the 2D bubble displayed in Fig. 5(e), the skirt partially envelopes the wake similar with the idealized skirted spherical cap bubble drawn in Fig. 2 in [10]. The wake

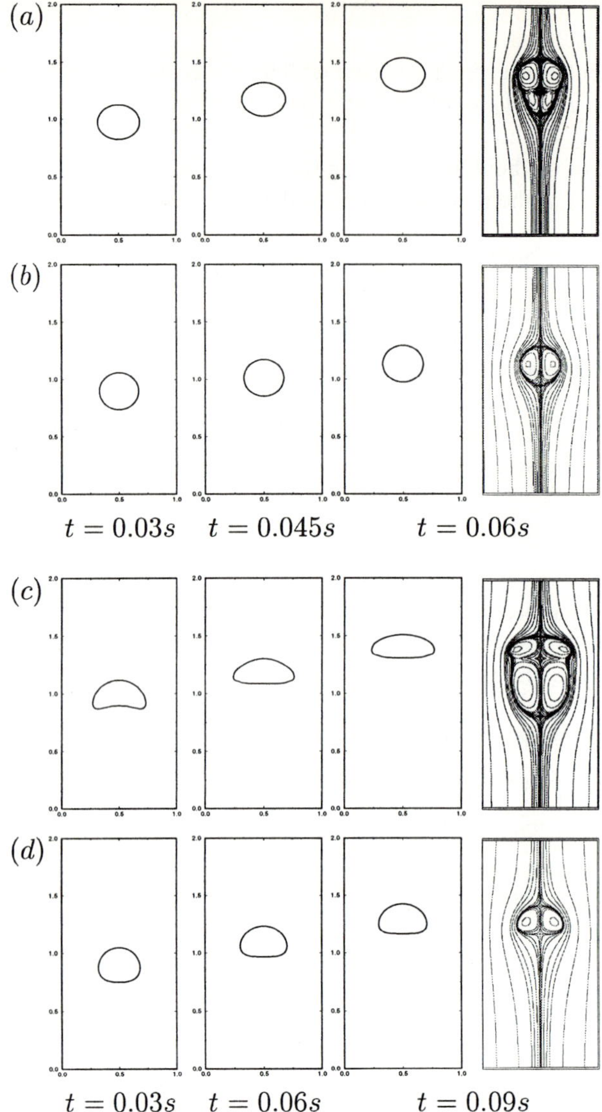

Figure 4 Rising bubbles with $E_0 = 1$ (a, b) and $E_0 = 10$ (c, d). Further details are given in the table following Fig. 5.

structures agree with the schematic wake diagram for skirted bubbles (see [3], Fig.8.5): internal circulation consists of two vortexes while toroidal vortex develops behind the bubble.

Second experiment

Large gas bubbles are modeled with the axisymmetric Level Set method in the recent work of Sussman & Smereka [20]. Their dimensionless parameters, calculated with $L = R$, coincide with the experimental parameters of *bubble A* and *bubble C* in Table I of Hnat & Buckmaster [10]:

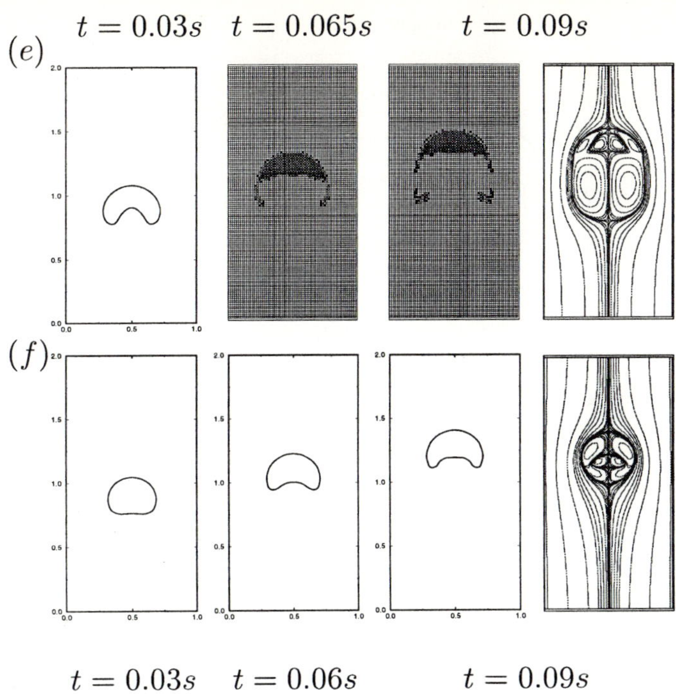

Figure 5 Rising bubbles with $E_0 = 104$. Data for this and the previous pictures of Fig. 4:

	μ_l/μ_g	E_0	M	Re_0	Re	We	$M^{rel.}$
(a)	88	1	10^{-7}	56.23	42	0.58	2.0×10^{-4}
(b)	493	1	10^{-4}	10	4.46	0.2	5.2×10^{-5}
(c)	88	10	10^{-4}	56.23	21.55	1.47	5.3×10^{-5}
(d)	493	10	10^{-1}	10	3.5	1.18	2.8×10^{-5}
(e)	85	104	10^{-1}	57.9	22.5	15.6	-1.3×10^{-5}
(f)	479	104	10^{2}	10.3	3.58	12.53	2.2×10^{-5}

$Re = 9.8$, $Fr = 0.76$, $We = 7.6$ and $Re = 24.4$, $Fr = 0.88$ and $We = 27.2$, accordingly. In both cases, $\mu_g/\mu_l = 0.0085$ and $\rho_g/\rho_l = 0.0011$. Both bubbles have the same Morton number ($M = 0.065$) and equal surface tension coefficient σ. Their volumes are different and correspond to $E_0 = 39.3$ and $E_0 = 123.1$, respectively.

We consider here two bubbles of the same radius $R = 1/6cm$ but with different σ. The dimensionless parameters coincide with the data given above; $\Delta t = 5 \times 10^{-5}s$. The evolution of the bubbles is displayed in Fig. 7 and Fig. 8. The form of the wakes outside the bubbles agrees with the experimental data (see [10], Fig. 1). In agreement with Bhaga & Weber [2], in intermediate time the lower edge becomes sharper when Re increases. The bubbles shapes agree quite well with the evolutions presented in Fig. 4 and Fig. 6 in [20]. Similar to the results of Sussman & Smereka, *bubble C* develops the skirts and then continues to rise with nearly the constant speed at the nose as expected from the experimental measurements (see [2],[3]). Rise velocity of *bubble C* is shown in Fig. 6.

The skirts break off and travel behind the bubble. This behaviour is similar to the motion of the large bubble displayed in Fig. 5(e) which has close E_0 and M values.

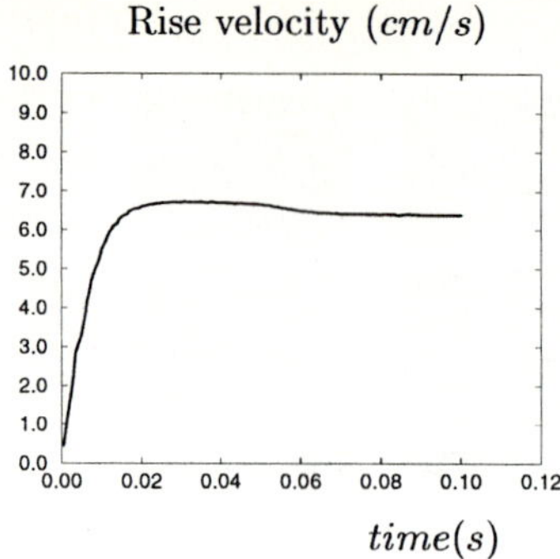

Figure 6 Rise velocity of *bubble C*.

Since our experiment is in two dimensions, the steady state nose velocity U is less than the expected value U^{3d} in 3D case: $U^{3d} = (FrgR)^{\frac{1}{2}}$. Collins [4] derives the velocity of a two-dimensional spherical capped bubble rising along the axis of a channel of finite width $2b$ as

$$U^{2d} = [\frac{gb}{6\pi}(3 - \tanh^2 \alpha) \tanh \alpha]^{\frac{1}{2}} \, , \alpha = \frac{\pi \bar{c}}{2b} \, . \tag{17}$$

Here $2\bar{c}$ is the length of the body. Collins shows that until $\bar{c}/b$ exceeds 0.4, $\bar{c}$ is equal to the radius of curvature at the front stagnation point $\bar{a}$. In order to compare U with the predictions (17), noslip boundary conditions are used. The radius of curvature $\bar{a}$ is estimated by fitting a circle through interface points closest to the nose. We obtain: $U \approx 6.37 cm/s$, $\bar{a} = 0.298 cm$, $U^{2d} \approx 6.849 cm/s$ for *bubble A* and $U \approx 6.39 cm/s, \bar{a} = 0.33 cm, U^{2d} \approx 6.958 cm/s$ for *bubble C*. We believe that the discrepancy with the prediction (17) may be related to the unaccuracy in curvature estimation. Moreover, noslip boundary condition on the top of the box can slow the flow. Besides that, the steady shapes of the obtained bubbles still more resemble the oblate-ellipsoidal cap than the spherical cap.

When the moving is done on the same advection mesh, bubble shapes obtained on the regular and adaptively refined grids practically coincide (see Fig. 7, Fig. 8). They are in close agreement with the shapes obtained on the aligned grid in the case of *bubble A* (cf. middle and bottom rows in Fig. 7). In the case of *bubble C*, some difference appears between the shapes when the skirts develop (see middle row in Fig. 8). Indeed, we usually obtain that the indentation at the rear is a bit less developed on the aligned grids. We relate this to the difference in treating of interfacial cells while discretizing on regular and aligned grids. Moreover, the skirts look thicker and mass is preserved better on the aligned grid since its corresponding advection mesh (underlying regular mesh) is coarser than in case of not aligned grids but equal time steps are used in all computations.

Third Experiment

Bubbles that rise in viscous liquids have been studied by Bhaga & Weber [2]. They have found that for $M > 4 \times 10^{-3}$ and $Re < 110$, a closed toroidal wake develops behind the bubble. For $Re > 110$ the wake appears to be open and unsteady. In their work, Bhaga & Weber show the steady shapes and the streamlines around rising bubbles in four situations with decreasing M (see Fig. 19 in [2]: it corresponds to cases (a), (d), (f), (g) in their Table 3). The Morton number decreases with the viscosity of the liquid, from $M = 848$ to the critical value $M_{cr} = 4.63 \times 10^{-3}$, which is relatively high since very viscous liquids are used. In this section, we take

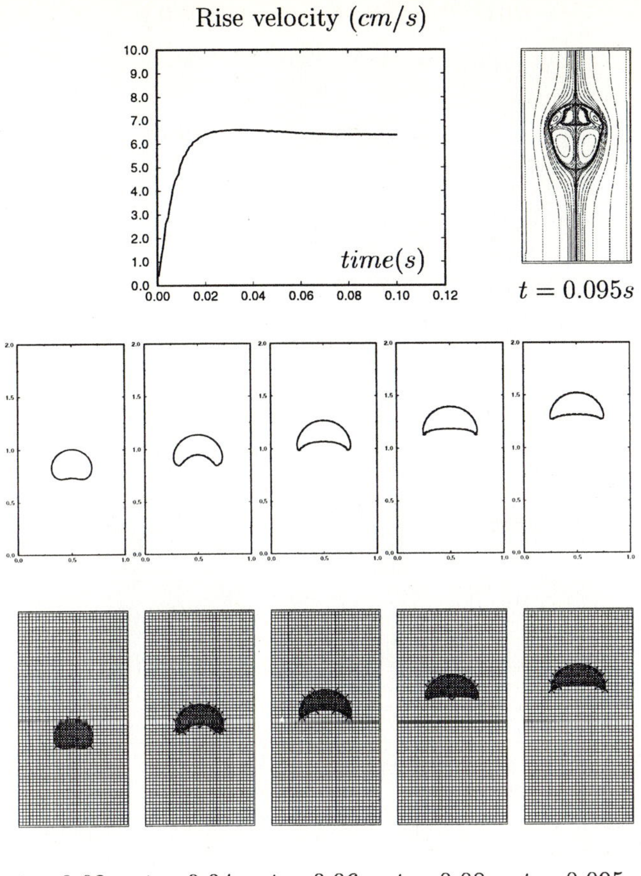

Figure 7 Evolution of *bubble A*: $\mu_g/\mu_l = 0.0085$, $\rho_g/\rho_l = 0.0011$, $E_0 = 39.3$, $M = 0.065$, $Re_0 = 31.1$, $Re = 11$, $We = 4.9$. **Bottom row:** Aligned computational grids in course of the motion. **Middle row:** Cubic splines correspond to computations on regular 64×128 mesh (dashed line) and adaptively refined grid (solid line), which is constructed over 32×64 regular mesh. At $t = 0.06s$, $M^{rel.}$ on the aligned, adaptively refined and uniform grids is -9.45×10^{-6}, 2.32×10^{-5} and 2.73×10^{-5}, accordingly. **Top row:** Rise velocity is plotted in time. Streamlines are shown at steady state.

$\rho_l = 1.314 g/cm^3$ and $\rho_l/\rho_g = 1090.5$, corresponding to an *air* bubble ($\mu_g = 1.78 \times 10^{-4}$) in liquids [2]. We show in Fig. 9(a) and (b) the evolution of air bubbles in liquids used by Bhaga & Weber for two Morton numbers, $M = 848$ ($\mu_l/\mu_g = 1.5 \times 10^5$) and $M = 5.51$ ($\mu_l/\mu_g = 4.2 \times 10^4$). The solutions for air bubbles at $M = 0.103$ ($\mu_l/\mu_g = 1.5 \times 10^4$) and $M = 4.63 \times 10^{-3}$ ($\mu_l/\mu_g = 7184$), are obtained on the aligned grids up to the moment when the skirts break off. The computations on the aligned grids are performed with the $\vec{F}_s$ in the form (13). When $\vec{F}_s$ is computed in the conservative form ((12), (5)), the parasite currents quickly grow because of the small bubble viscosity and destroy the interface, whether the grid is aligned to the interface or not. By increasing the bubble viscosity, we reach the stable solutions in these cases (see Fig. 9(c), (d) and the corresponding viscosity jumps in Table 2).

As the alternative approach, we discretize $\vec{F}_s$ on the cubic spline interpolants also in case of *not aligned grids*. For this purpose, we use the relations (13) for each pair of points defining the intersections of the cubic *spline B* with the boundary of the control volume. Then, stable

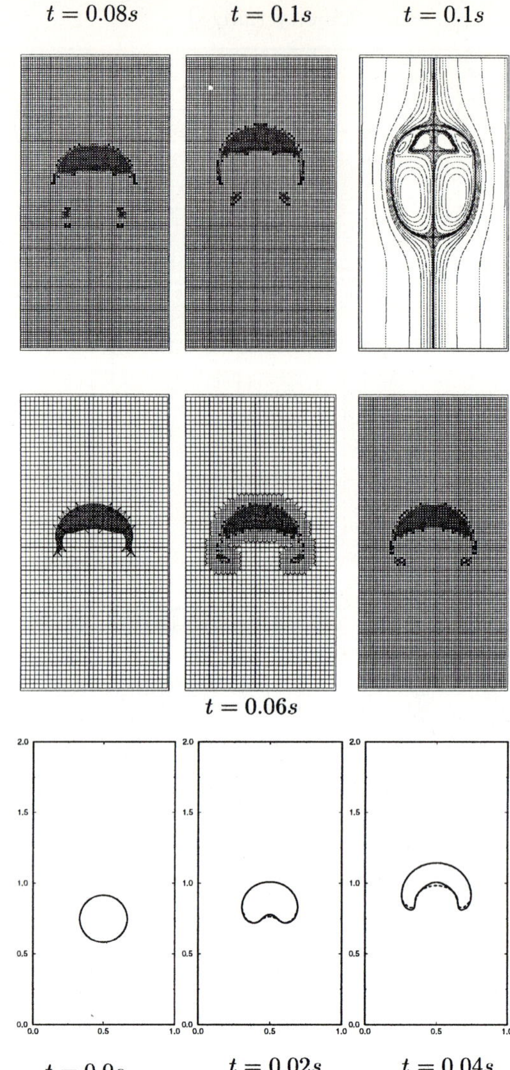

Figure 8 Evolution of *bubble C*: $\mu_g/\mu_l = 0.0085$, $\rho_g/\rho_l = 0.0011$, $E_0 = 123.1$, $M = 0.065$, $Re_0 = 73.2$, $Re = 25.9$, $We = 15.4$. **Bottom row:** Cubic splines correspond to computations on the regular 64×128 mesh (dotted line), the adaptively refined grid (solid line) and the aligned grid (dashed line). **Middle row:** Phase distribution is displayed on the aligned, adaptively refined and regular mesh. The corresponding values of $M^{rel.}$ are -1.82×10^{-5}, -6.73×10^{-5} and -6.56×10^{-5}. **Top row:** Further motion of the skirted bubble is done on the adaptive grid.

Table 2 Data for the calculations of Fig. 9

	μ_l/μ_g	E_0	M	Re_0	Re	We	$\Delta t(s)^-$
(a)	1.5×10^5	116	848	6.55	1.69	7.75	1×10^{-4}
(b)	4.2×10^4	116	5.51	23.1	7.93	13.72	5×10^{-5}
(c)	250	116	0.103	62.4	22.09	14.54	5×10^{-5}
(d)	100	115	4.63×10^{-3}	134.6	47.75	14.46	2.5×10^{-5}

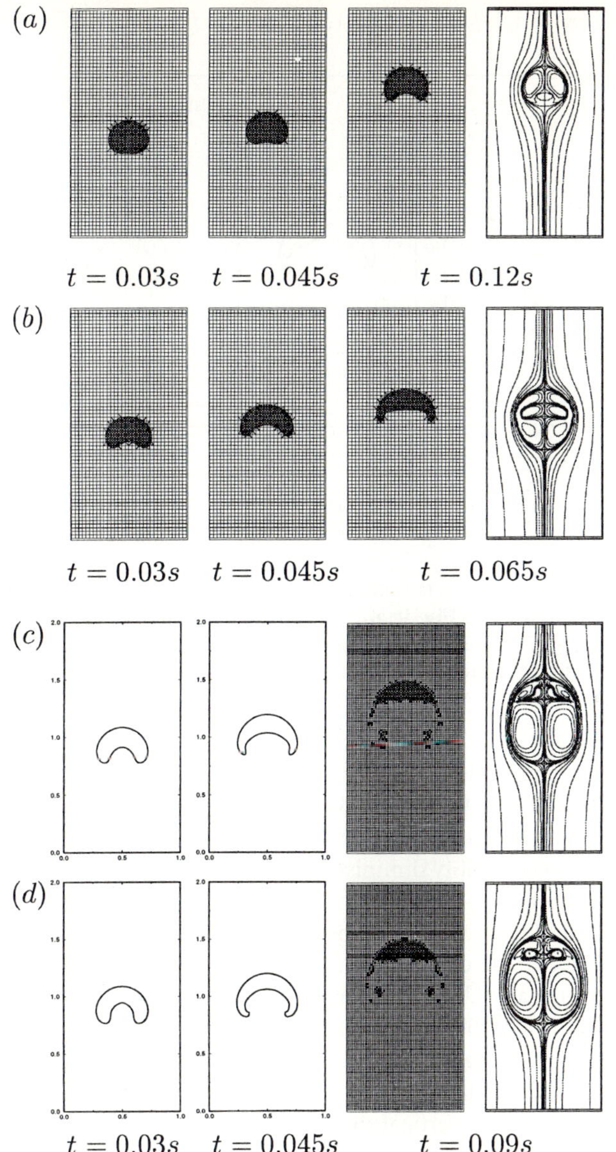

Figure 9 Eötvös and Morton numbers correspond to the experiment of Bhaga & Weber. The evolution of the bubbles is modeled with the rotated discretization. (a): aligned grid; (b), (c) and (d): adaptively refined grids. Data given in Table 2.

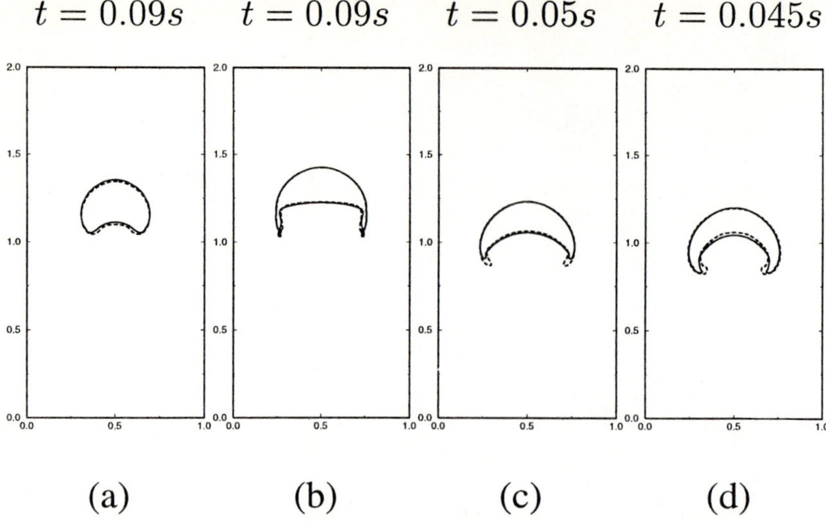

Figure 10 Bubble shapes are obtained with the rotated (solid line) and MAC (dashed line) discretizations. (a), (b), (c), (d): E and M correspond to Fig. 9; μ_l/μ_g correspond to air bubble (1.5×10^5, 4.2×10^4, 1.5×10^4 and 7184, accordingly). Rotated discretization: 32×64 aligned mesh in case (a), adaptively refined grids otherwise. MAC discretization is done on regular grids: 32×64 in case (a), 64×128 mesh in other cases.

solutions for air bubbles are obtained for all considered M as in cases of the aligned grids. This is demonstrated in Fig. 9(b) and Fig. 10. Until now, no special account of the interface position is included while discretizing the pressure gradients on not aligned grids. In spite of this, computing with the spline interpolants considerably diminishes the anomalous currents on them.

The bubble shapes and the behaviour of the surrounding liquid qualitatively agree with the experimental results. The bubbles take oblate ellipsoidal cap shapes. The spherical cap form is not yet reached in case (d), unlike in the experiment of Bhaga & Weber, since the attained Reynolds numbers are smaller than in real $3D$ experiments. A strong indentation at the rear during the motion results in the appearance of closed toroidal wakes. Bhaga & Weber fit the frontal surface of the bubble and compare the boundary of the wake with the boundary of an ellipsoid in the case of an oblate ellipsoidal cap. We find that the wake grows continuously when going from smaller to higher Re. The wake is greater in vertical direction than the frontal ellipsoid, except for the smallest Re. In this way, the study of the wakes provides results similar to the experimental measurements. The bubbles with $M = 5.51$, $M = 0.103$ and $M = 4.63 \times 10^{-3}$ are skirted. This is not so in the physical experiment but agrees with the previous numerical computations of large bubbles (see Fig. 5(e) and Fig. 8) as well as with the results obtained with our MAC-type description discussed below.

For comparison, the MAC-type central finite-difference (f.d.) approximation of (1)-(2) on a regular grid was implemented. Explicit as well as implicit time discretizations can be used to treat both convection and diffusion terms. A linear saddle-point problem is solved for all unknown pressures and velocities, similar as for the current discretization. Surface tension is implemented with a f.d. approximation of (5), following [11], [9]. It should be underlined that this

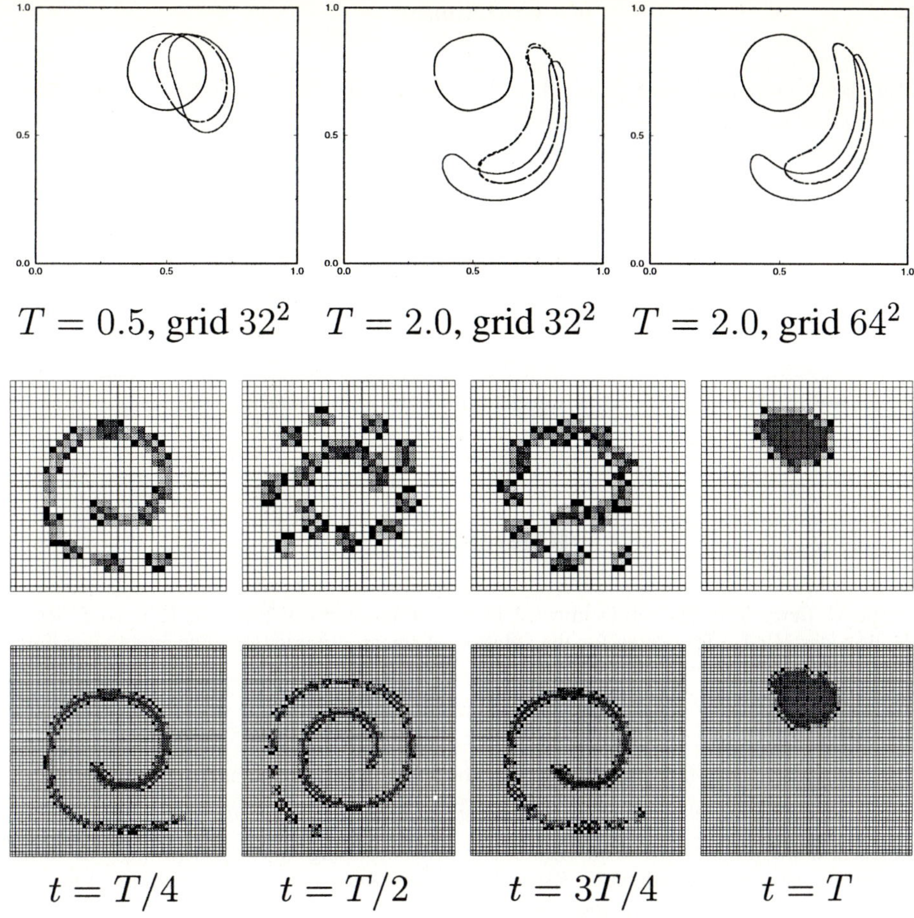

$T = 0.5$, grid 32^2 $T = 2.0$, grid 32^2 $T = 2.0$, grid 64^2

$t = T/4$ $t = T/2$ $t = 3T/4$ $t = T$

Figure 11 Results for a circular body, placed in the time-reversed single vortex-flow field. Advection scheme is Lagrangian moving based on first order PLIC method. **Top row:** bubble shapes are plotted with *spline A* interpolant. $t = T/4$ (dashed), $t = T/2$ (dot-dashed), $t = 3T/4$ (long-dashed), $t = T$ (solid). **Middle row:** $T = 8.0$, 32^2 grid. **Bottom row:** $T = 8.0$, 64^2 grid.

discretization remains stable also for small bubble viscosities in all experiments discussed in this section. The interface shapes, plotted with the cubic spline interpolant, are compared in Fig. 10. In case (a) of the smallest Re and the greatest viscosity ratio, small difference in bubble shapes is mainly related to the difference in the discretizations on the regular and aligned grids. The skirts agree quite well in the next three experiments where we mainly attribute the difference to the different treating of the surface tension force. This is confirmed by the computations in cases shown in Fig. 9(c) and (d), when the relations (5) are used in both discretization schemes. Finally, we do not think that the advection schemes are responsible for skirt formation in our model since preliminary computations with use of Unsplit Advection Scheme [15] provide very close results. Remarkably, the 2D Level Set method [19] and Axisymmetric Level Set method [20] also exhibit a tendency to form skirts.

6 Conclusion

A model for two dimensional two-phase flow was presented. The Navier-Stokes equations are discretized on unstructured grids. An accurate computation of the surface tension was possible via the use of cubic splines in combination with the staggered finite volumes. The anomalous currents were reduced significantly.

Acknowledgements

We would like to acknowledge support from the SFB412 and the CNRS-DFG French-German research programme "Numerische Strömungssimulation - Simulation Numérique d'Ecoulements".

References

[1] R. Barrett, M. Berry, T. F. Chan, J. Demmel, J. Donato, J. Dongarra, V. Eijkhout, R. Pozo, C. Romine, and H. Van der Vorst. Templates for the solution of linear systems: Building blocks for iterative methods. *SIAM*, 1994.

[2] D. Bhaga and M. E. Weber. Bubbles in viscous liquids: shapes, wakes and velocities. *J. Fluid Mech.*, 105:61, 1992.

[3] R. Clift, J. R. Grace, and M. E. Weber. *Bubbles, Drops, and Particles.* Academic Press, 1978.

[4] R. Collins. A simple model of the plane gas bubble in a finite liquid. *J.Fluid Mech.*, 22:763, 1965.

[5] M. Crouzeix and P.-A. Raviart. Conforming and nonconforming finite element methods for solving the stationary stokes equations i. *Revue Française d'Automatique, Informatique et Recherche Opérationnelle. Sér. Rouge Anal. Numér.*, R-3:33, 1973.

[6] B. J. Daly. A technique for including surface tension effects in hydrodynamic calculations. *J. Comput. Phys.*, 4:97, 1969.

[7] D. E. Fyfe, E.S. Oran, and M. J. Fritts. Surface tension and viscosity with lagrangian hydrodynamics on a triangular mesh. *J. Comput. Phys.*, 76:394, 1988.

[8] I. Ginzburg and G. Wittum. Two-phase flows on unstructured grids with spline volume tracking. *Preprint IWR*, 1999.

[9] D. Gueyffier, J. Lie, A. Nadim, R. Scardovelli, and S. Zaleski. Volume of fluid interface tracking with smoothed surface stress methods for three-dimensional flows. *J. Comput. Phys.*, 152:423, 1999.

[10] J. G. Hnat and J. D. Buckmaster. Spherical cap bubbles and skirt formation. *Phys. Fluids*, 19:182, 1976.

[11] B. Lafaurie, C. Nardone, R. Scardovelli, and S. Zaleski. Modeling merging and fragmentation in multiphase flows with surfer. *J. Comput. Phys.*, 113:134, 1994.

[12] J. Li. Calcul d'interface affine par morceaux. *C.R.Acad.Sci.Paris*, 320, série IIb:391, 1995.

[13] S. Popinet and S.Zaleski. A front-tracking algorithm for accurate representation of surface tension. *Int. J. Numer. Meth. Fluids*, 30:775, 1999.

[14] W. H. Press, S. A. Teukolsky, W. T. Wetterling, and B. P. Flannnery. *Numerical Recipes in C.* Cambridge, 1992.

[15] W. J. Rider and D. B. Kothe. Reconstructing volume tracking. *J. Comput. Phys.*, 141:112, 1998.

[16] G. Ryskin and L. G. Leal. Numerical solution of free-boundary problems in fluid mechanics. part 2. buoyancy-driven motion of a gas bubble through a quiescent liquid. *J. Fluid Mech.*, 148:19, 1984.

[17] R. Scardovelli and S. Zaleski. Direct numerical simulation of free-surface and interfacial flow. *Annu. Rev. Fluid Mech.*, 31:567, 1999.

[18] G. L. Sleijpen, H. A. van der Vorst, and D. R. Fokkema. Bicgstab(l) and other hybrid bi-cg methods. *Numerical Algorithms*, 7:75, 1994.

[19] M. Sussmann and S. Osher P. Smereka. A level set approach for computing solutions to incompressible two-phase flow. *J. Comput. Phys.*, 114:146, 1994.

[20] M. Sussmann and P. Smereka. Axisymmetric free boundary problems. *J. Fluid Mech.*, 341:269, 1997.

[21] S. H. Unverdi and G. Tryggvason. A front-tracking method for viscous, incompressible, multi-fluid flows. *J. Comput. Phys.*, 100:25, 1992.

[22] D. L. Youngs. Time-dependent multi-material flow with large fluid distortion. *(In Numerical Methods for Fluid Dynamics)*, 1986.

Modeling of Free Surfaces in Casting Processes

J. Neises[1], Y. Delannoy[2], M. Medale[3], G. Laschet[1], M. Stemmler[1], G. Fontaine[2]
[1] ACCESS e.V./ Intzestr. 5, D-52072 Aachen, Germany
[2] EPM / MADYLAM CNRS UPR 9033 BP 95, F-38402 St Martin d'Heres Cedex, France
[3] IUSTI Technopôle de Château-Gombert 5, rue Enrico Fermi, F-13453 Marseille Cedex 13, France

Summary

Casting still is a challenge for numerical simulation and free surfaces play an important role in casting processes. There is the moving solid-liquid interface during solidification and the moving liquid-gas interface during mold filling. Thus, the simulation of fluid flow with free surfaces is a key part of a tool simulating casting processes.

Three approaches modeling fluid flow applied to casting are presented. There are a Finite Element (FE) using an ALE method treating free surfaces, a Finite Volume (FV) method applying a volume-tracking approach and a hybrid FV method based on stabilized FE approximations using bubble functions, which applies a phase field method to model free surfaces occurring in mold filling.

These methods will be compared on benchmark problems designed for casting processes:

- Melting of pure substances driven by natural convection - AMETH.

- Filling of a plate using a tall sprue - MCWASP Benchmark 1995.

First results obtained using the presented approaches are shown in this paper.

1 Introduction

Though casting is an ancient art, it is still a challenge for numerical simulation. A simulation tool necessarily has to cover the wide range of physical phenomena, which are involved. These phenomena range from fluid flow and free surface dynamics, which govern the mold filling process and convection of the melt, to heat transfer and phase-change processes while the part solidifies. Mold filling affects the subsequent casting process and the final quality of the casted part. Moreover, the solidification pattern is severely influenced by the flow field. An accurate modeling of phase change is also a very important task in simulating casting processes [2, 16]. Within this framework, the modeling of fluid-flow and free surfaces is the key building block of simulating casting processes, as a reliable prediction of the final product's properties depends on the correct simulation of mold filling and phase change. Casting simulation as an industrial tool demands methods with accuracy, fidelity and robustness. Physical meaning in relation to the computational resources becomes an important topic. Thus, efficiency is a major topic of the used methods [11, 7].

116

Three alternate approaches are selected to simulate fluid flow applied to casting processes. The tools, which are developed at the three research groups ACCESS, IUSTI and MADYLAM, are used to model free surface flows coupled to phase change phenomena, predicting properties of solidified bodies. Moreover, the selected methods represent the main different approaches of numerical simulation in this field.

In chapter 2 an FE based ALE method is presented. Due to the remeshing, which is used in ALE methods, interfaces can be tracked accurately and interfacial phenomena can be integrated [6].

Alternatively FV methods are often applied. Nowadays unstructured FV methods are under development [5, 8]. In the third chapter such a method is introduced.

In the fourth chapter a hybrid FV approach based on a stabilized FE method [1, 12, 14] is described. Bubble functions are used to derive a stabilized FE method, which is reordered to an FV-like method using virtual finite volumes [8].

Finally, numerical results achieved using these methods are presented in chapter 5. Two benchmarks were selected to test free surface flows applied to casting processes. The AMETH benchmark [2] tests the coupling of fluid flow and melting of a pure substance. The second one is a well known test of mold filling presented at the MCWASP VII[11].

2   A Parallel ALE Finite Element Method for Moving Interface Problems

The parallel FE method developed at IUSTI is based on the Arbitrary-Lagrangian-Eulerian (ALE) technique to track interfaces as solidification fronts or free surfaces [14, 15]. The ALE method uses a fixed grid, which is locally remeshed regarding the moving interface [6]. This way the interface is represented accurately and interfacial phenomena, e.g. latent heat release or marangoni-convection, can be discretized by surface integrals. The method covers 2D and axisymmetric problems.

Incompressible flow is modeled using a projection method, which was implemented by a fractional step approach. At first each time step the heat transfer problem is solved using the known velocity. Then the fluid flow is calculated in two steps. At first an intermediate velocity is determined solving the momentum equation. Then in a projection step pressure and the divergence free velocity are calculated. The corresponding FE discretization leads to:

Momentum equation (calculating an intermediate velocity):

$$\int_\Omega \left[\left(\rho \frac{\vec{u}^{k+1} - \vec{u}^k}{\Delta t} + \left(\vec{u}^{k+1} \cdot \nabla \right) \vec{u}^{k+1} \right) \delta\vec{u} + \rho\nu \left(\nabla\vec{u}^{k+1} : \nabla\delta \right) \right] dV =$$
$$\int_\Omega \delta\vec{u} \left[\vec{f}_V^{k+1} - \nabla \left(2p^{k+1} - p^k \right) \right] dV + \int_{\delta\Omega} \delta\vec{u} \left(\sigma_V^{k+1} \cdot \vec{n} \right) dS,$$

Continuity equation (projection of velocity and solving for pressure):

$$\int_\Omega \left[\nabla \left(p^{k+1} - p^k \right) \nabla\delta p \right] dV = \frac{\rho}{\Delta t} \int_\Omega \delta p \left(\nabla \cdot \vec{u}^{k+1} \right) dV.$$

Dirichlet boundary conditions are imposed to both equations at each step.

The program has been parallelized using the PETSc [10] library of Argonne National Laboratory. The PETSc library provides scalable massive parallel iterative solvers. This way an efficient parallelization is obtained without the necessity of dealing with message passing. Since PETSc uses libraries as MPI, BLAS and LAPACK the code remains portable.

3 An Unstructured FV Method Using Object Oriented Programming

The second selected method, which is developed at MADYLAM, is based on an unstructured FVM method in 2D [5] using object oriented programming (OOP) techniques. The application of OOP promises a greater flexibility than classical programming approaches. The elements are chosen as finite volumes leading to a cell-vertex approach. A collocative system is selected to store the unknowns.

Different upwind schemes have been investigated. A first order (UDS1) and a second order (UDS2) scheme have been implemented as well as an ENO scheme [9, 15]. In (UDS 1), the value of the advected unknown at the center of the face is taken upwind, whereas the mass flux is centered. Using a second order Scheme (UDS 2), the value of the unknown at the center of the face is taken upwind, plus a component relative to the rate of the upwind gradient multiplied by the distance vector from the center of the upwind element to the center of the face. The Second Order ENO Scheme (ENO 2) has a similar structure as UDS2, except that the gradient is an anisotropic reconstruction of the gradient in order to reduce oscillations [9]. The different advection schemes were tested, using a simple benchmark: the transport of a passive scalar (e.g. temperature) for a given velocity.

A uniform flow diagonal to a quadrilateral uniform mesh with 20*20 elements was assumed as test case. The initial distribution includes a jump across the mesh in flow direction. The boundary conditions fix the temperature at the left and lower boundary. An exact scheme should preserve the initial state. Any observed disturbance will be due to the applied upwind scheme.

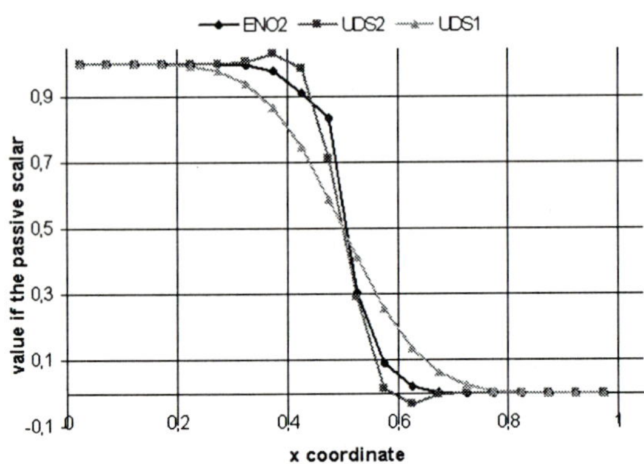

Figure 1 Results of the advection test at the second diagonal achieved with the upwind schemes UDS1, UDS2 and ENO2.

Due to the 1st order discretization, using UDS1 severe artificial diffusion can be observed. The reconstruction of the solution is more accurate using UDS2. Less diffusion is introduced, but with the price of overshoots and undershoots. Applying Quasi-ENO the oscillations are diminished (Fig. 1), but with about twice the computational time as for UDS 2. Nevertheless a fairly good reconstruction of the function without oscillations is obtained by the ENO2 scheme. Thus, among the three schemes the ENO scheme is the most accurate one.

To model incompressible flow in casting processes, SIMPLE based methods [15] are selected to resolve the pressure state. Later a front-tracking method (PLIC[4]) will be implemented in this unstructured FV method. For testing purposes a front-tracking algorithm has been designed as user routines of FLUENT UNS4.2 at first. This method has been used for the AMETH benchmark [2].

4   A Hybrid Finite-Volume Method Stabilized by Bubble Functions

The integrated casting simulation tool CASTS of ACCESS is based on a hybrid FE/FV formulation. The geometrical and mathematical features of the FE method are combined with the computational simplicity of the FV method using an FE enmeshment. The hybrid FE/FV method works on virtual finite volumes without any geometrical description of the cells [8]. Fluid flow is modeled by a stabilized equal order approximation [1].

Bubble functions were introduced to the FE method for Stokes-flow by the MINI element [1, 12]. The MINI element (Fig. 2) fulfills the inf-sup condition resulting in a stable discretization. By static condensation the degrees of freedom can be reduced to the same number as for equal-order discretizations.

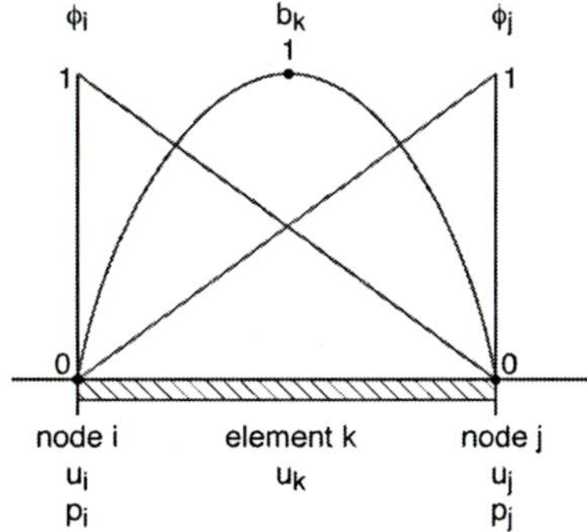

Figure 2   MINI element in 1D using first order velocity plus bubble function and first order pressure approximations

The transient linearized incompressible Navier-Stokes equation is the starting point of the method. Applying the condensation procedure to the momentum equation, the weak equation for the bubble mode is considered componentwise:

$$\sum_j \int \rho \left(\vec{u}_t^I\right)_j dt \int_{V_k} b_k \Phi_j dV + \int \rho \left(\vec{u}_t^I\right)_k dt \int_{V_k} b_k b_k dV + \tag{4.1}$$

$$\sum_j \int \rho \nu \vec{u}_j^I dt \int_{V_k} \nabla \Phi_j \nabla b_k dV + \int \rho \nu \vec{u}_k^I dt \int_{V_k} \nabla b_k \nabla b_k dV +$$

$$\sum_j \int \rho \vec{v} \vec{u}_j^I dt \int_{V_k} \nabla \Phi_j b_k dV + \int \rho \vec{v} \vec{u}_k^I dt \int_{V_k} \nabla b_k b_k dV +$$

$$\sum_j \int p_j dt \int_{V_k} \left(\nabla \Phi_j b_k\right)^I dV = \sum_j \int \rho \vec{g} \vec{u}_j^I dt \int_{V_k} \Phi_j b_k dV, I = 1,2,3.$$

According to the FE/FV framework the thermophysical properties and the linearized velocity $\vec{v}$ are approximated by average values. The solution of equation (4.1) for the bubble mode is performed approximately. As the reconstruction of the bubble mode may be dropped [12], the change of the bubble during a time-step (4.2) is neglected:

$$\int \rho \left(\vec{u_t^I}\right)_k dt \int_{V_k} b_k b_k dV = 0. \tag{4.2}$$

This way, the static condensation is reduced to a simple task. Equation (4.2) is solved for the bubble mode. Then, the discretized incompressible Navier-Stokes equation can be stabilized. Considering the upwind stabilization for the momentum equation, a conventional upwind scheme is advantageous compared to stabilization by classical bubble functions [1]. For stabilizing the continuity equation a selective approach is selected limiting the bubble entry by a pressure stabilizing Petrov-Galerkin method. The condensed mean bubble mode is substituted into the continuity equation. By introducing the mean bubble mode, the following stabilizing enhancement term (4.3) is derived:

$$\sum_I \sum_j \int \frac{1}{\nu} \left(\vec{u_t}\right)^I dt \sum_k \frac{\int_{V_k} \Phi_j b_k dV}{\int_{V_k} \nabla b_k \nabla b_k dV} \int_{V_k} \left(b_k \nabla \phi_i\right)^I dV + \tag{4.3}$$

$$\sum_I \sum_j \int \frac{\vec{v}}{\nu} \vec{u}^I dt \sum_k \frac{\int_{V_k} \nabla \Phi_j b_k dV}{\int_{V_k} \nabla b_k \nabla b_k dV} \int_{V_k} \left(b_k \nabla \phi_i\right)^I dV +$$

$$\sum_I \sum_j \int \frac{1}{\rho \nu} p_j dt \sum_k \frac{\int_{V_k} \left(\nabla \Phi_j b_k\right)^I dV}{\int_{V_k} \nabla b_k \nabla b_k dV} \int_{V_k} \left(b_k \nabla \phi_i\right)^I dV -$$

$$\sum_I \sum_j \int \frac{1}{\nu} \vec{g_t^I} dt \sum_k \frac{\int_{V_k} \Phi_j b_k dV}{\int_{V_k} \nabla b_k \nabla b_k dV} \int_{V_k} \left(b_k \nabla \phi_i\right)^I dV = 0.$$

Due to the limited stabilizing capabilities of the classical bubble functions [1] a restricting switch is required for the pressure stabilization. The terms resulting from bubble function stabilization occur in similar form, if a Pressure Stabilizing Petrov Galerkin (PSPG) method [24] is applied. The PSPG method is a stabilization method using first order shape functions. It is constructed analogously to a SUPG. Thus, PSPG consists of the addition of a parameterized divergence of the momentum residual to the continuity equation (4.4):

$$\nabla \cdot \vec{u} \frac{\tau}{\rho} \nabla \cdot (momentum\,residual) = 0. \tag{4.4}$$

As for other stabilization methods of Petrov-Galerkin type, PSPG depends on the proper choice of τ. Tezduyar et al. [14] suggest to adopt the same choice as for SUPG:

$$\tau_{PSPG} = \tau_{SUPG} = \frac{h}{|\vec{u}|} \left[\frac{1}{C^2} + 1 + \frac{9}{Re^2}\right]^{-1}, \tag{4.5}$$

where $C = \frac{|\vec{u}| \Delta t}{h}$ and $Re = \frac{|\vec{u}| h}{2\nu}$ are respectively the element Courant and Reynolds numbers. Within the hybrid framework these parameters are evaluated at the links between two adjacent nodes interpreting the elemental width h as the distance between the nodes. The stabilization parameter is always less than Δt. In the limit of a vanishing time-step, the introduced stabilization reduces to zero.

120

Now, the continuity equation is stabilized either by bubble functions or by PSPG. A switch from bubble functions to PSPG ensures this limitation and the performance of the method. The bubble function terms are selected only if their absolute values are less than the absolute values of the PSPG terms. Thus, for increasing Reynolds or Courant number the PSPG terms are used predominantly.

Concerning free surface flows in 3D, sophisticated ALE (see chapter 2) or front reconstruction techniques, such as PLIC [4], lead to geometrical calculations on the finite elements or the finite volumes. Thus, both methods are contradictory to the philosophy of the hybrid FE/FV approach, which works on virtual finite volumes. Therefore an enhancement, which was inspired by level set or phase field methods to the VoF method was developed [8].

5   Application to Casting Processes

The AMETH problem [2] deals with melting of a pure substance driven by natural convection. Consider a 2D square cavity of equal height and width initially filled with a solid material at freezing temperature T_F. The simulation starts raising the temperature at one vertical wall above freezing temperature. The other vertical wall is maintained at the initial temperature. The horizontal walls are adiabatic. No-slip conditions are imposed at all walls. After an initial transient, where heat conduction is dominating, thermal convection develops in the liquid phase. Due to the heat transport, the primarily vertical solidification front is distorted. The material melts faster at the top. With increasing Raleigh number the influence of this instability rises. The less stable the problem is, the earlier the distortion occurs (Fig. 3).

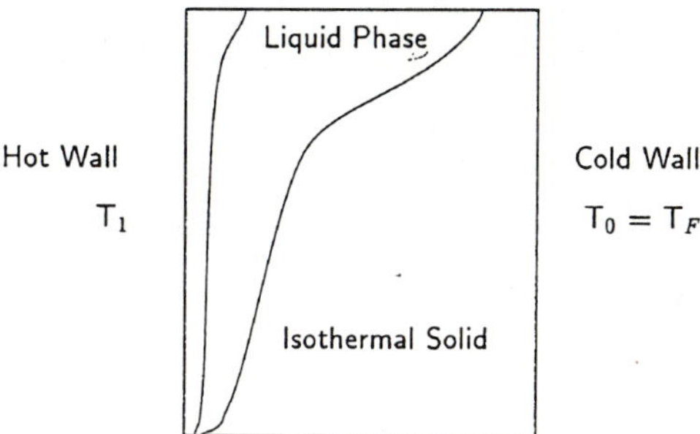

Figure 3   This figure shows the problem schematically. A square cavity is given. The top and bottom walls are adiabatic. The left wall is heated, the right wall is held at the initial temperature. The substance melts and convection starts and deformes the planar front.

The thermo-physical properties of the material are assumed to be constant. Several Prandtl, Stefan and Raleigh numbers are tested within four test cases. In the first two test cases, having low Prandtl and Stefan number, tin was selected as material to be molten.

$$c_p = 200 \frac{J}{kgK}, \rho = 7500 \frac{kg}{m^3}, \lambda = 60 \frac{W}{mK}, \alpha = 4*10^{-5} \frac{m^2}{s}, \beta = \frac{8}{3}*10^{-4} \frac{1}{K}. \tag{5.6}$$

The high Prandtl number test cases correspond to the melting of paraffin (octadecane) with the following properties:

$$c_p = 1250\frac{J}{kgK}, \rho = 800\frac{kg}{m^3}, \lambda = 0.2\frac{W}{mK}, \alpha = 2*10^{-7}\frac{m^2}{s}, \beta = 0.002\frac{1}{K}. \qquad (5.7)$$

Several tests were performed by using the three presented methods. The two test cases for paraffin, with Prandtl number 50 and Stefan number 0.1, were used to test the abilities of the planned front-tracking method of MADYLAM, which will be used within the unstructured FV method. The test case with Raleigh number 10^8 was selected. This test case is the most unstable one. Thus, any problem, which might be inherent of the approach should be detected. A one-domain approach applying an enthalpy formulation for the latent heat release was used. Phase change was modeled based on the source term method by Voller et al. [16]. This method was implemented by user subroutines in FLUENT UNS 4.2. Front-tracking lead to local refinement and coarsening of the FV mesh. The initial mesh consists of 3300 triangles. The front-tracking lead to oscillations of the front in the lower part of the cavity. After 3000s a too large distortion of the originally planar melting front is observed (Fig. 4). Thus, a sophisticated refinement is required to ensure a smooth evolution of the front.

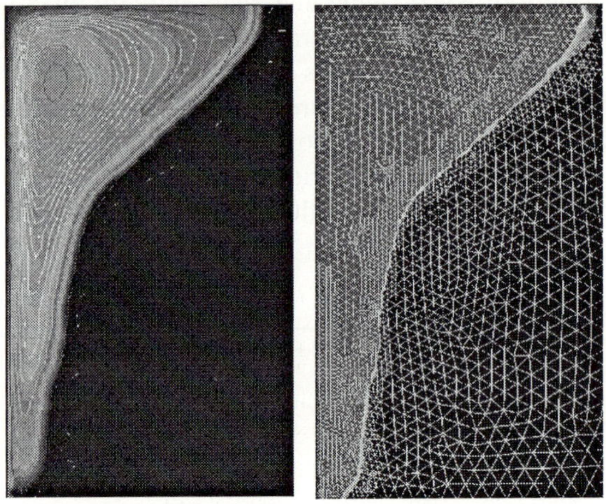

Figure 4 AMETH test case 4 simulated using FLUENT UNS 4.2 and a user supplied front-tracking technique at 3000s. The left figure shows the temperature distribution andi streamlines. The right part shows the adapted mesh.

The high Raleigh number test case for tin was performed using the ALE method of IUSTI. Also a one domain approach using the enthalpy approach and treating the latent heat contribution as a source term has been used. Furthermore, for this test case the ALE method was used to investigate the influence of the boundary conditions on the solution. Three different boundary conditions have been used:

- no-slip condition,
- perfect slip condition,
- Marangoni convection.

The tests reveal an important impact of the selected boundary conditions on the temperature distribution and the position of the solidification front (Fig. 5). Imposing a no-slip condition clearly is the most restricting choice. Thus the melting front is the less advanced during the simulation. Perfect slip leads to a more advanced front. Regarding Marangoni convection results in the most advanced interface.

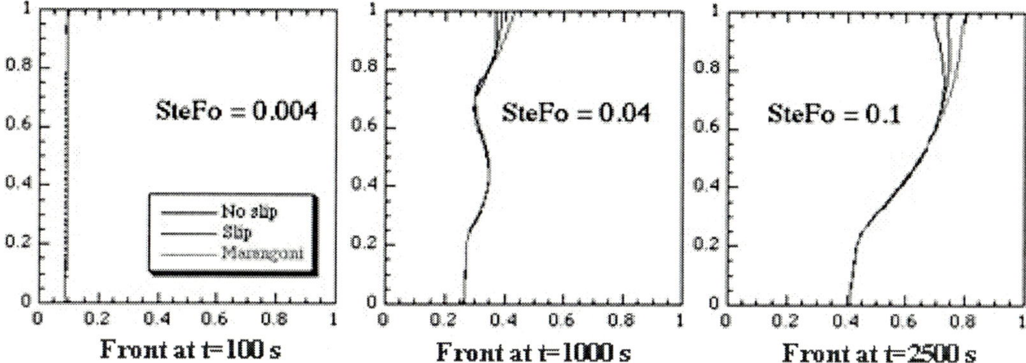

Figure 5 Isotherms on the AMETH test case 2 using different boundary conditions at the solidification front with the ALE method of IUSTI. The more sophisticated the boundary condition at the melting front, the further this front evolves.

The same test case was simulated using the hybrid FE/FV method. An apparent heat capacity approach has been applied to this heat transfer problem [3]. But, the apparent heat capacity is less suitable for pure substances, where no mushy region exists. Indeed, the discretization of the apparent heat capacity used with the hybrid method requires a temperature i interval of at least 1K within which the latent heat is released. Thus, this method is not suitable for this test case. The apparent heat capacity method leads to a much faster melting, than expected (Fig. 6). Therefore for such tests a more accurate method, the discontinuous integration method [3], is under investigation to model phase change of pure substances.

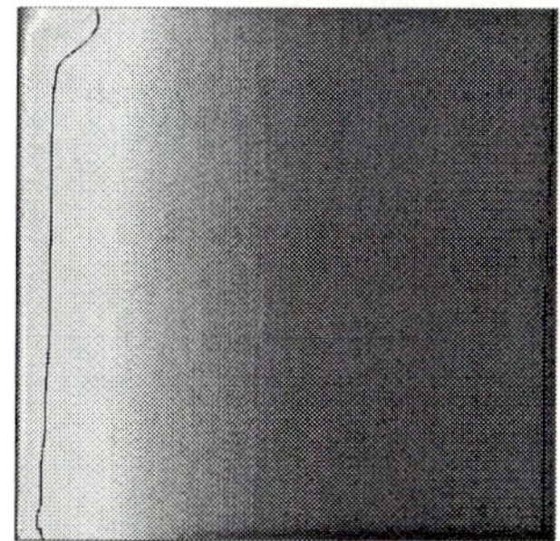

Figure 6 This figure shows the result of a test convection driven melting using the hybrid approach with an apparent heat capacity method. The material melts much too fast. Thus, the front is already distorted by convection after 20s.

The tests cases reveal large variations between the different methods and boundary conditions.

These tests still are challenging for coupled simulation of fluid flow and heat transfer due to the difficulties in modeling phase change of pure substances. The influence of the boundary conditions at the surface has been demonstrated clearly. Thus, front-tracking methods should outperform one-domain approaches. The latter at least need some sub-grid resolution to model the phenomena at the solidification front.

At the 7th conference on Modeling of Casting, Welding and Advanced Solidification Processes the results of a benchmark problem of mold filling were presented [11]. The overall geometry of the casting is a simple plate, with a bottom gated running system (Fig. 7). A tall sprue was chosen generating turbulence in the runner and the gate. This is a challenging test of any computational tool simulating mold filling. The benchmark was accompanied by a series of experiments to evaluate the simulations. This benchmark was simulated using the developed stabilized hybrid FE/FV method.

The mold was discretized by a mesh of 10464 hexahedrons and 7908 nodes. No slip boundary conditions were imposed. The pouring basin is kept full during filling. Within the mold vacuum is assumed. The same configuration was used to test an earlier approach [8]. Falling down the sprue now the melt forms a thinning profile. In opposite to the earlier results, now the front remains compact, due to the better resolution of the flow field. The front reaches the bottom of the sprue after 0.26s, which is about 10 per cent later than in experiments. High velocities are observed when the melt leaves the sprue and enters the runner. The melt fills the whole cross-section of the runner, due to smearing of the front. The melt forms the characteristic mushroom kind front entering the plate. Then, it spreads aside and swashes against the left and right wall rising there. This pattern is also observed in the experiments. Then the front flattens and the melt rises vertically with an almost planar front. Using the stabilized equal order method, the plate is completely filled after 2.65 seconds (Fig. 7). This is a deviation of about 20 per cent in comparison to the experimental filling time of 2.24s. Nevertheless this result is achieved without any calibration of parameters and is within the range of the results achieved by the participants of the benchmark test and constitutes an improvement compared to the earlier results of about 20s [8]. Thus, the former artificial compressibility method is clearly outperformed by the stabilized equal order approach.

Since this benchmark is a challenging problem and several simplifying assumptions have been made (e.g. laminar flow), it will be a remaining problem for checking our further developments. The implementation of the stabilized method revealed, that the coupling of the flow field and the mass transport and boundary conditions still are open questions. The coupling depends severely on time stepping, due to the sequential calculation of the flow field and the mass tranport. Additionally, due to the no slip boundary condition the boundary layer cannot be resolved, if coarse meshes are used. Thus, sub-mesh technologies as multi-scale methods are under investigation as further development for the FE/FV approach.

6   Conclusions

Three approaches to model free surface flows applied to casting processes have been presented. These methods range from a pure FE method using ALE techniques and a hybrid satbilized FE based finite volume approach to a finite volume method.

The unstructured FV method is under development yet. Especially the implementation of a pressure correction scheme and a front-tracking method are on the roadmap. Thus, more and improved results should be obtained in the near future.

The hybrid FE/FV approach suffers from the linearization (and thus smearing) within each

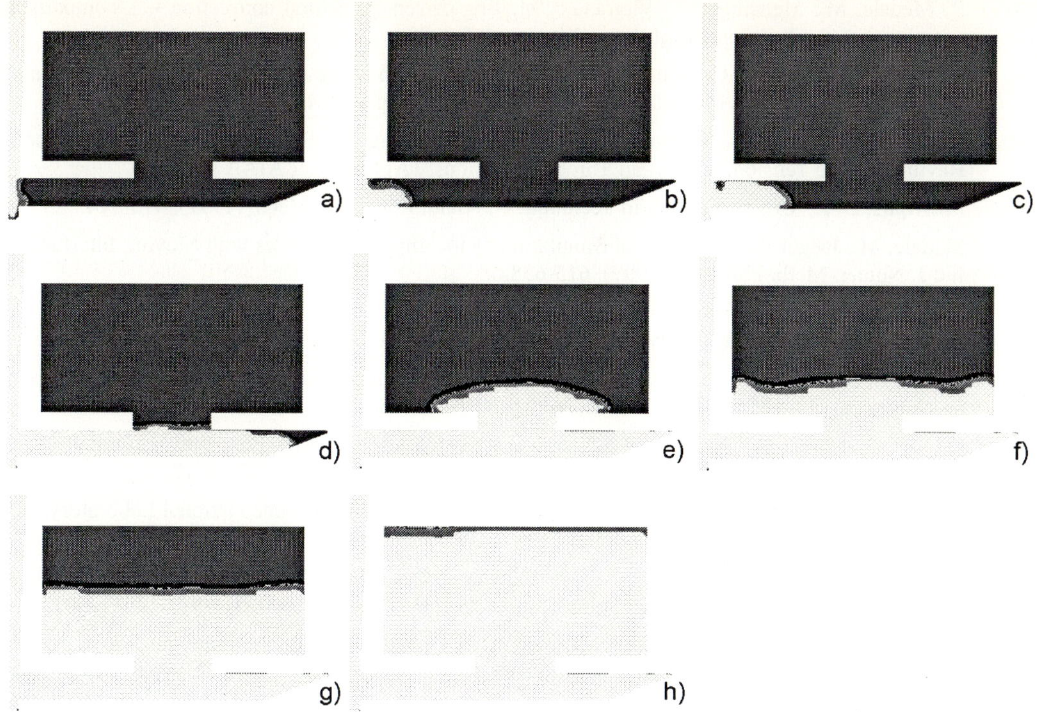

Figure 7 Simulation of the benchmark problem. The melt leaves the sprue (a) and enters the runner (b) filling the complete cross-section. The transition region remains compact (c). The typical mushroom shape of the front can be observed (d), when the plate is entered by the melt. The melt flows aside and the front flattens (e). When the melt reaches the sides of the plate it rises and splashes back (f). An almost flat front is formed and the plate is filled vertically (g). The plate is filled at 2.7 seconds (h).

time-step. Therefore, some steps as the implementation of a Newton-Raphson method for time integration are under development. Nevertheless, the MCWASP results show the significant progress of the FE/FV method by the introduction of stabilized methods.

Today the pure FE method is the most accurate method among the three presented approaches. The adaptive mesh refinement gives an accurate representation of interfaces. Thus, special boundary conditions like Marangoni-convection can be implemented to obtain more realistic results. The current method is restricted to quasi 2D problems. The enhancement to 3D is rather complicated, but is under investigation.

References

[1]   Brezzi, F., Franca, L.P., Hughes, T. J. R., Russo, A.,: ”Stabilization Techniques and Subgrid Scale Capturing”, The State of the Art in Numerical Analysis, based on the proceedings of the Conference on the State of the Art in Numerical Analysis, York, England, April 1996, IMA Conference Series, Vol. 63 (I.S. Duff and G.A. Watson, eds.), Oxford University Press, 391-406.
http://www-math.cudenver.edu/lfranca/research/papers/.

[2]   Bertrand, O, Binet, B, Combeau, H., Couturier, S., Delannoy, Y., Gobin, D., Lacroix, M., LeQuéré,

P., Medale, M., Mencinger, J., Vieira,G.: "Melting driven by natural convection - A comparison exercice:first results". Journal of thermal sciences, Feb.98.

[3] Idelsohn, S., Storti, M., Crivelli, L.: "Numerical methods in phase-change problems", Archives of Computational Methods in Engineering - State of the art reviews, Vol. 1 (1994), 49-74.

[4] Kothe, D. B., Rider, W.J., Mosso, S.J., Brock, J. S., Hochstein, J. I.: "Volume Tracking of Interfaces Having Surface Tension in Two and Three Dimensions", AIAA Paper AIAA 96-0659.

[5] Mavriplis, D. J: "Unstructured Grid Techniques", Annu. Rev. Fluid Mech. (1997), 473-514.

[6] Medale, M., Jaeger, M.: "Numerical Simulation of Incompressible Flows with Moving Interfaces", Int. J. Numer. Meth. Fluids 24 (1997), 615-638.

[7] Medale, M.: "Parallel finite Element Model for Coupled Fluid Flow and Heat Transfer Problems: Application to Melting driven by Convection", USNCCM99, Book of Abstracts (1999), 516.

[8] Neises, J., Steinbach, I., Delannoy, Y.: "Modeling of Free Surfaces in Casting Processes", in: Notes on Numerical Fluid Mechanics, Vol. 66, Vieweg (1998), 168-186.

[9] Ollivier-Gooch, C.F.: "Quasi-ENO Schemes for Unstructured Meshes Based on Unlimited Data-Dependent Least-Squares Reconstruction", Journal Comp. Phys. 133 (1997), 6-17.

[10] Portable, Extensible Toolkit for Scientific Computation (PETSc) Argonne National Laboratory, http://www.arc.unm.edu/Workshop/libraries/pmatlib/petsc.html.

[11] Sirrell, B., Holliday, M., Campbell, J.: "The Benchmarktest 1995", Proc. 7th Int. Conf. on Modeling of Casting, Welding and Advanced Solidification Processes, TMS (1995), 915-932.

[12] A. Soulaimani, M. Fortin, Y. Ouellet, G. Dhatt, F. Bertrand: "Simple Continuous Pressure Elements for Two- and Three Dimensional Incompressible Flows", Comput. Meth. Appl. Mech. Eng. 62 (1987), 47-69.

[13] Swaminathan, C. R., Voller, V. R.: "Streamline Upwind Scheme for Control-Volume Finite Elements, Part I Formulations", Numerical Heat Transfer, Part B, vol. 22 (1992), 95-107.

[14] Tezduyar, T. E., Mittal, S., Ray, S. E., Shih, R.: "Incompressible flow computations with stabilized bilinear and linear equal-order-interpolation velocity-pressure elements", Comput. Meth. Appl. Mech. Eng. 95 (1992), 221-242.

[15] Thomadakis, M., Lechziner, M.: "A Pressure-Correction Method for the solution of Incompressible Viscous Flow on unstructured Grids", Int. J. Num. Meth. Fluids 22 (1996), 581-601.

[16] Voller, V., Cross, M., Walton, P.: "Assessment of weak solution numerical techniques for solving Stefan Problems", in: Numerical Methods in Thermal Problems, Lewis, R. W., Morgan, K. (eds.), Pineridge Press, Swansea U.K. (1979).

Parallel Computation of the Saturation Process in a Nonlinear Dynamo Model

Egbert Zienicke[1], Hélène Politano[2], Annick Pouquet[2]

[1] Universität Ilmenau, Fakultät Maschinenbau, Postfach 100565,
D-98684 Ilmenau, Germany

[2] CNRS, UMR 6529, Observatoire de la Côte d'Azur, B.P. 4229,
F-06304 Nice Cedex 4, France

Summary

In this paper the backreaction of a growing magnetic field in a nonlinear dynamo on the flow is investigated. The hypothesis that the magnetic field by the action of the Lorentz force supresses Lagrangian chaos of the flow is checked by direct numerical simulations of the MHD equations. As a measure of the level of chaos the Lyapunov exponent of a set of 128×128 trajectories of fluid particles is computed in the linear growth phase of the dynamo and in the saturated phase of the dynamo when the magnetic field has reached its final strength. The numerical code, based on a pseudospectral algorithm, is developed for parallel computation on a multiprocessor system (Cray–T3E). The trajectories for the computation of the Lyapunov exponent are advanced in a timestep parallel to the timestep of the MHD-solver. Magnetic Reynolds numbers up to 240 and scale separations between the wavelength of the hydrodynamical forcing and the scale of the computational domain up to four are reached. For the runs where the kinetic Reyold number is high enough that the hydrodynamical bifurcation sequence to a more chaotic flow already has taken place, the mean value of the Lyapunov exponent is noticeable diminished in the saturated phase compared to the growth phase of the dynamo.

1 Introduction

Magnetic fields are practically ubiquitious in the Universe: planets, stars, the interstellar medium and the galaxy as a whole have magnetic fields. If one does not assume that these magnetic fields are primordial — what may be the case for the galactic magnetic field — there is a mechanism needed which is able to explain how magnetic fields of a body arise and how they are maintained.

The common feature of all these bodies having a magnetic field is the presence of electrically conducting fluid: plasma in the atmospheres of stars and in the interstellar medium, and matter in a metallic state in the core of planets (for example the Earth or Jupiter). A continous transfer of mechanical energy from the convection inside the fluid into magnetic field energy by induction effects sustains the magnetic field which is observed.

A quantity of electrically conducting fluid with a given flow $\mathbf{u}(\mathbf{x},t)$ inside of a bounded region is a dynamo, if an arbitrary small magnetic field perturbation $\delta\mathbf{B}$ is growing exponentially, i. e. if the state $\mathbf{B} \equiv 0$ is unstable against small perturbations. This is the statement of the *linear* or *kinematic dynamo problem.* The term 'linear' stems from the fact that the induction equation, giving the law for the time evolution of the magnetic field, is linear in **B** for a given flow:

$$\frac{\partial \mathbf{B}}{\partial t} = \nabla \times (\mathbf{u} \times \mathbf{B}) + \eta \nabla^2 \mathbf{B}. \tag{1.1}$$

Here $\eta = 1/\mu_0 \sigma$ is the magnetic diffusivity.

The linear dynamo problem has been investigated extensively in the past, theoretically as well as numerically on model systems. Two main mechanisms, which have both already an extended theory behind, are often quoted: (1) the *turbulent dynamo* based on *helical turbulent fluctuations* of the underlying three-dimensional flow and (2) the *fast dynamo* based on the stretch-twist-fold mechanism on small scales provided by *chaotic motion* of the underlying flow.

The theory of the mean field dynamo or turbulent dynamo starts by splitting the velocity and the magnetic field into a mean part and a fluctuating part: $\mathbf{B} = \overline{\mathbf{B}} + \mathbf{B}'$, $\mathbf{u} = \overline{\mathbf{u}} + \mathbf{u}'$, where the bar denotes an ensemble average. In the case of isotropic turbulence of the velocity field the analysis leads to the following time evolution for the mean magnetic field (see [1], [2])

$$\frac{\partial \overline{\mathbf{B}}}{\partial t} = \nabla \times (\overline{\mathbf{u}} \times \overline{\mathbf{B}}) + \alpha \nabla \times \overline{\mathbf{B}} + (\beta + \eta)\triangle\overline{\mathbf{B}}. \tag{1.2}$$

The coefficient α, which is in general a pseudo tensor, allows for amplification of the magnetic field. Because of the choice of the letter α in early influential work the amplification of magnetic field by turbulent fluctuations has become known as α -effect. The other coefficient β describes a turbulent diffusion. α and β are functions of the fluctuations of the given velocity field:

$$\alpha = -\frac{1}{3}\tau\overline{\mathbf{u}' \cdot (\nabla \times \mathbf{u}')}, \qquad \beta = \frac{1}{3}\tau\overline{\mathbf{u}'^2}, \tag{1.3}$$

where τ is the correlation time. α is mainly determined by an ensemble average over the helicity $\mathbf{u}' \cdot (\nabla \times \mathbf{u}')$ of the fluctuations of the velocity field. The helicity is a measure for the presence of helical motions and for the non-axisymmetry of the fluctuating flow. Helicity of the flow arises in a natural way in the motion of a fluid in a rotating frame by the Coriolis force. Important for the action of the α-effect is a spatial *scale separation* between the small scale of helicity fluctuations and the large scale of the developing mean magnetic field.

If equation (1.3) is transformed into non-dimensional units by using a typical velocity scale U and a typical length scale L of the flow, the magnetic diffusivity is replaced by the factor $1/R^M$ in the diffusion term. R^M is the magnetic Reynolds number $R^M = LU/\eta = LU\mu_0\sigma$. A high magnetic Reynolds number means that the diffusion of the magnetic field is weak. In the limit $R^M \to \infty$ (the limit of infinite conductivity) diffusion totally vanishes and the magnetic field is transported by the flow as if the magnetic field lines were attached to the fluid particles. This is often also called the approximation of "frozen magnetic field", because the magnetic field lines behave as if they were frozen into the fluid. For astronomical bodies the magnetical Reynolds number normally is high (core of Earth: 10^3, core of Jupiter: 10^6, solar convection zone: 10^8), more because of the large extension L of astronomical objects than because of high conductivity σ.

A dynamo is called fast, if the growth rate of magnetic energy γ as a function of the magnetic Reynolds number remains greater than zero in the limit $R^M \to \infty$:

$$\lim_{R^M \to \infty} \gamma(Re^M) > 0. \tag{1.4}$$

Fast dynamo action is based on the stretch-twist-fold mechanism, which can be visualized in a simple model: Consider a flux tube of magnetic field shaped as a torus. The flow of the electrically conducting fluid first stretches the tube to twice its size, then it is twisted in the middle so that the torus has the shape of the number eight. In the last step it is folded by the flow in such a way that the two loops ly above each other and that the magnetic field lines of the two loops show in the same direction. In the whole cycle the magnetic field has doubled its strength. If the cycle is repeated several times, the magnetic field grows exponentially. This cycle is of course a simple model, but one can imagine many types of flows, where a stretch-twist-fold mechanism can be realized (see [3]). Especially, when a flow is *chaotic* in the sense as defined in the theory of nonlinear dynamical systems, stretching, twisting and folding takes place down to arbitrary small scales (see [4],[5]). This is the reason why Lagrangian chaos of the underlying flow is of such importance for fast dynamo action.

The linear phase of a dynamo lasts as long as the magnetic field is so small that the Lorentz force $\mathbf{j} \times \mathbf{B}$ is not able to change the velocity field. But, since the magnetic field is growing exponentially in the linear phase, one has only to wait long enough that a realizable backreaction on the velocity field will take place. This is the beginning of the *nonlinear phase* where a saturation of the magnetic field will take place in the end by means of the action of the Lorentz force on the velocity field and by means of Ohmic dissipation. To describe the saturation one has to solve the full magnetohydrodynamical equations — i.e. the Navier-Stokes equation with Lorentz force *and* the induction equation — in a regime were both the velocity field and the magnetic field are not negligible in the nonlinearities of these equations:

$$\frac{\partial \mathbf{u}}{\partial t} + (\mathbf{u} \cdot \nabla)\mathbf{u} \;=\; -\frac{1}{\rho}\nabla p + \nu \triangle \mathbf{u} + \frac{1}{\rho}\mathbf{j} \times \mathbf{B} + \mathbf{f}, \tag{1.5}$$

$$\frac{\partial \mathbf{B}}{\partial t} + (\mathbf{u} \cdot \nabla)\mathbf{B} \;=\; (\mathbf{B} \cdot \nabla)\mathbf{u} + \eta \triangle \mathbf{B}. \tag{1.6}$$

Because of the Maxwell equation $\mathbf{j} = \nabla \times \mathbf{B}/\mu$, the Lorentz force also is a nonlinearity in the Navier-Stokes equation, additionally to the inertia term $(\mathbf{u} \cdot \nabla)\mathbf{u}$.

To extend the linear dynamo theory into the nonlinear regime one early began to construct heuristic phenomenological models to introduce a saturation effect. For the turbulent dynamo theory these models are called α-quenching, because the amplification factor α is supposed to decrease with growing magnetic field strength. Often a dependence of the form

$$\alpha = \frac{\alpha_0}{1 + \alpha_1 |\overline{\mathbf{B}}|^n} \tag{1.7}$$

is used, see for example [6], [7] [8]. Concerning the strength of the quenching the discussion is not yet settled: Some authors ([9], [10], [11]) propose a different behaviour of α with a much stronger quenching for high magnetic Reynolds number. A saturation at a much lower level as equipartition of magnetic and kinetic energy would follow with the consequence that the dynamo effect could not explain equipartition of magnetic and kinetic energy as observed in astronomy.

There are some attempts to find models to understand better the back reaction of the magnetic field on the velocity field and to find mechanisms for the process of saturation. Using closures of turbulence in [12], [13] it is found, that saturation can occur in the presence of a large scale magnetic field through helical Alfvén waves. The large scale field originates from an inverse cascade of magnetic helicity, which is due to the invariance of the total helicity $H^M = < \mathbf{B} \cdot \nabla \times \mathbf{B} >$. Models [12], [14] and computations in the incompressible [15], [16] and compressible cases - both subsonic [17] and supersonic [18] - indicate linear growth of magnetic helicity, also in the nonlinear dynamic regime.

An important role in saturation is played by a large scale magnetic field. A strong homogenous field (largest possible scale) is known to suppress turbulent fluctuations parallel to the magnetic field lines. This is of importance, both for the turbulent as well as the fast dynamo. Anti-dynamo theorems suggest, that an inherent three-dimensional motion of the flow is necessary for dynamo action, whereas a strong magnetic field has the tendency to force the flow into a two-dimensional structure. Helical fluctuations (necessary for the turbulent dynamo) as well as stretch-twist-fold operations at small scales (as necessary for the fast dynamo) are only possible in three dimensions. Nevertheless, the magnetic field developing in a dynamo process normally does not have the simple structure of a homogenous field. Often a strong filamentary intermittent structure of the growing magnetic field is observed in astrophysical flows (for example in the sun athmosphere) as well as in numerical simulation [19], [20], [21]. If the magnetic field is localized in strong flux tubes the suppression of small scales of flow also would be localized, whereas in the rest of the volume the flow would be only weakly influenced and still be able to produce magnetic field on small scales. The strength of the suppression of magnetic field production then would be strongly influenced by the spatial structure of the magnetic field (see [22]).

Before going into the details of spatial structures in this work we just concentrate on prooving the effect as it should be stated for the fast dynamo: *whether a growing magnetic field is able to suppress Lagrangian chaos of the flow on small scales.* This interesting question has been raised in [23] and was investigated numerically on a dynamo model with strong confinements on the development of the nonlinear terms of the flow: (i) The advection term responsible for turbulence in an incompressible fluid is totally neglected, allowing for a simple decomposition of the linear/nonlinear phase of the dynamo, (ii) The Lorentz force is averaged in z-direction to let the flow retain a two-dimensional (but time dependent) structure. With these assumptions, a clear diminution of chaos — as diagnosed by a two-dimensional map of finite-time Lyapunov exponents — is obtained. As the nonlinearity, which is supposed to suppress Lagrangian chaos is present in these calculations (although in reduced form), we find this result encouraging. On the other hand, the neglection of the advection term cannot be without influence on the development of the velocity field, especially on the development of small scales promoting turbulence and also Lagrangian chaos. We therefore investigate the same problem on a different dynamo model, solving the full MHD-equations in three dimensions and dropping all simplifying assumptions.

In the next section we introduce the dynamo model, which will be investigated. Section 3 gives some detail about the numerical implementation of the three-dimensional, parallel MHD-solver and how Lyapunov exponents are calculated in the code to measure the level of Lagrangian chaos. Section 4 is devoted to the presentation of the numerical results.

2 The Model

Although our numerical investigation mainly is aimed at the fast dynamo, we nevertheless have chosen a model system which shows both, chaos of the underlying flow as well as helical flow with the possibility of scale separation. The conducting fluid is situated in a box with sides of length L. We apply periodic boundary conditions in all three space directions. The MHD-equations are computed in Alphénic units, this means the magnetic field has also the dimension of a velocity by the transformation $\mathbf{b} = \mathbf{B}/\sqrt{\rho\mu}$. The equations then read

$$\frac{\partial \mathbf{u}}{\partial t} + (\mathbf{u} \cdot \nabla)\mathbf{u} = -\nabla P + \nu\Delta\mathbf{u} + (\mathbf{b} \cdot \nabla)\mathbf{b} + \mathbf{f}_{ABC}, \qquad (2.8)$$

$$\frac{\partial \mathbf{b}}{\partial t} + (\mathbf{u} \cdot \nabla)\mathbf{b} \;=\; (\mathbf{b} \cdot \nabla)\mathbf{u} + \eta \triangle \mathbf{b}, \qquad (2.9)$$

where the pressure field $P = p/\rho + \mathbf{b}^2/2$ is the sum of the static and the magnetic pressure. Additionally, the magnetic field and the velocity field fulfill $\nabla \cdot \mathbf{b} = 0$ and $\nabla \cdot \mathbf{u} = 0$, the latter because we assume the fluid to be incompressible. The velocity field is forced by the ABC-forcing:

$$\mathbf{f}_{ABC} = \nu k_0^2 \mathbf{u}_{ABC} = \nu k_0^2 \begin{pmatrix} A \sin k_0 z + C \cos k_0 y \\ B \sin k_0 x + A \cos k_0 z \\ C \sin k_0 y + B \cos k_0 x \end{pmatrix}. \qquad (2.10)$$

The ABC-flow $\mathbf{u}_{ABC}$ consists of three orthogonal Beltrami waves with amplitudes A, B and C. The wavenumber k_0 allows to introduce a scale separation between the forcing and the largest lengthscale, which is the length of the box. ABC-flows are exact solutions of the Euler equation and also of the Navier-Stokes equation, if the above forcing $\mathbf{f}_{ABC}$ is applied. Because of the Beltrami property $\nabla \times \mathbf{u}_{ABC} \propto \mathbf{u}_{ABC}$ they have strong helicity. This is the reason why they early were studied in the context of the α-effect ([24]).

If all three amplitudes are non-zero the flow is non-integrable (in the sense of the theory of dynamical systems) and shows a mixture of regular islands and chaotic regions in a Poincaré-section (see [25], [26], [27]). Because of the existence of Lagrangian chaos in the flow the ABC-dynamo is a candidate for fast dynamo action. This question was investigated in some numerical studies in the framework of the linear dynamo theory: ([28], [29], [21], and [30]). The results are positive, but the highest magnetic Reynolds number reached are about 10^3, so that the answer is not entirely decisive.

ABC-flows without magnetic field were investigated in [31]. For the case $A = B = C = 1$ the ABC-flow is a stable solution of the Navier-Stokes equation up to a Reynolds number about 13. At a critical Reynolds number of 13.04 a bifurcation to a timedependent solution takes place. Further bifurcations in a regime still at low Reynolds number are observed. The flow which evolves under ABC-forcing becomes more and more turbulent for higher Reynolds numbers.

The nonlinear dynamo with ABC-forcing has already been investigated under different questions: In [19] among other topics scaling effects by varying the parameter k_0 between 1, 2 and 4 are studied for the nonlinear dynamo. It is found that the level of saturation increases significantly by increasing k_0. [32] and [33] investigate the bifurcation structure of the nonlinear dynamo for low kinetic and magnetic Reynolds numbers (up to $Re^V = Re^M = 20$) and classify the symmetries of the solutions they find.

3 Numerical Implementation

3.1 The MHD-solver

The MHD-solver was developed by one of us (E. Zienicke) as a parallel code on the Cray-T3E. It is of spectral type using expansions into exponential functions for the space dependence of $\mathbf{u}(\mathbf{x},t)$ and $\mathbf{b}(\mathbf{x},t)$:

$$\mathbf{u} = \sum_{\mathbf{k}} \mathbf{u}_{\mathbf{k}} e^{i\mathbf{k}\cdot\mathbf{x}}, \qquad \mathbf{b} = \sum_{\mathbf{k}} \mathbf{b}_{\mathbf{k}} e^{i\mathbf{k}\cdot\mathbf{x}}, \qquad (3.11)$$

where $\mathbf{u}_{\mathbf{k}}$ and $\mathbf{b}_{\mathbf{k}}$ are the time-dependent expansion coefficients giving the amplitudes for the modes $\mathbf{k} = (2\pi/L)(n_1,n_2,n_3)^T$ in Fourier space (n_1, n_2 and n_3 denote integer numbers). The partial differential equations (2.8) and (2.9) are in this way represented by an infinite number of ordinary differential equations for the Fourier coefficients:

$$\dot{\mathbf{u}}_{\mathbf{k}} = \mathbf{w}_{\mathbf{k}} - ik P_{\mathbf{k}} - \nu k^2 \mathbf{u}_{\mathbf{k}} + \mathbf{f}_{\mathbf{k}}^{ABC}, \tag{3.12}$$

$$\dot{\mathbf{b}}_{\mathbf{k}} = \mathbf{c}_{\mathbf{k}} - \eta k^2 \mathbf{b}_{\mathbf{k}}. \tag{3.13}$$

$\mathbf{w}_{\mathbf{k}}$ and $\mathbf{c}_{\mathbf{k}}$ denote the Fourier coefficients of the nonlinearities of the Navier-Stokes equation and of the induction equation, respectively:

$$w_j = b_i \frac{\partial b_j}{\partial x_i} - u_i \frac{\partial u_j}{\partial x_i} = \frac{\partial}{\partial x_i}(b_i b_j - u_i u_j), \tag{3.14}$$

$$c_j = b_i \frac{\partial u_j}{\partial x_i} - u_i \frac{\partial b_j}{\partial x_i} = \frac{\partial}{\partial x_i}(b_i u_j - u_i b_j). \tag{3.15}$$

The second form of the nonlinearity, often called conservative form, is obtained using $\nabla \cdot \mathbf{u} = \nabla \cdot \mathbf{b} = 0$.

The pressure in the Navier-Stokes equation (3.12) can be eliminated in Fourier space. Taking the divergence of the Navier-Stokes equation one gets for the pressure $P_{\mathbf{k}} = -i\mathbf{k} \cdot \mathbf{w}_{\mathbf{k}}/k^2$, which can be inserted back again into (3.12) with the result:

$$\dot{v}_{\mathbf{k}}^{(j)} = \left(\delta_{ij} - \frac{k_i k_j}{k^2} \right) w_{\mathbf{k}}^{(i)} - \nu k^2 v_{\mathbf{k}}^{(j)} + f_{\mathbf{k}}^{ABC,(j)}. \tag{3.16}$$

Equations (3.16) and (3.13) are integrated using an Adams-Bashforth/Crank-Nicolson timestep, i.e. an explicit timestep of second order for the nonlinear terms and the forcing and an implicit timestep of second order for the diffusion terms.

The computation of the nonlinearities in each timestep would need $\mathcal{O}(N^6)$ operations if they would be computed directly in Fourier space. This operation count can be reduced significantly by using a pseudo spectral method (see [34] and [35]). The nonlinearity is first transformed into physical space by a fast Fourier tranformation needing $\mathcal{O}(N^3 \log(3N))$ operations. There it can be computed pointwise by $\mathcal{O}(N^3)$ operations. Finally the result is transformed back by a second FFT into the Fourier space again. The leading order in the expression for the total number of operations for each timestep is then determined by the Fast Fourier transformation.

The crucial step in parallelizing a code is to find the best distribution of the data onto the processors allocated for the computation in order to avoid time consuming communication between processors. For the MHD-solver this concerns mainly the distribution of the 3d-arrays for the velocity and the magnetic field as well in Fourier space as in physical space.

The pseudo spectral algorithm can be parallelized rather easily. The timestep itself and the computation of the nonlinearities can be done by each processor on its own data. Communication between different processors only is necessary in the Fast Fourier transformations. But for this task optimized parallel library routines are available on the Cray. To use the parallelized FFT's in a Fortran code the second and third dimensions of the 3d-arrays are distributed on a two-dimensional array of processors. In our implementation the total number of processors has to be a power of 2 as is also the case for the spatial resolution N in the three space dimensions. Where communication was necessary, especially in input and output routines, the *shmem*-subroutines of the Cray-T3E were used. With 64 processors the code needs the following computation times for one timestep: 0.11s for a resolution of 64^3 gridpoints, 0.68s for resolution 128^3 and 6.5s for resolution 256^3.

3.2 Computation of Lyapunov exponents

As a measure of stretching we compute the largest Lyapunov exponent of the flow in the growth phase and in the saturated phase of the dynamo. The Lyaponow exponent measures the exponential growth in time of the distance between two initially nearby fluid particles. Denoting by $\mathbf{x}_0$ and $\mathbf{x}_0 + \delta \mathbf{x}$ the initial positions of the fluid particles at time t_0 the Lyapunov exponent is defined as

$$\lambda(\mathbf{x}_0) = \lim_{t \to \infty} \frac{1}{t} \log \left(\frac{|\mathbf{x}(t; \mathbf{x}_0, t_0) - \mathbf{x}(t; \mathbf{x}_0 + \delta \mathbf{x}, t_0)|}{|\delta \mathbf{x}|} \right). \tag{3.17}$$

If the flow is exponentially stretching, the Lyapunov exponent is greater than zero, else it is equal to zero.

To compute the Lyapunov exponent for a starting point $\mathbf{x}_0$ one has to integrate the trajectories $\mathbf{x}(t; \mathbf{x}_0, t_0)$ and $\mathbf{x}(t; \mathbf{x}_0 + \delta \mathbf{x}, t_0)$ of the corresponding fluid particles by integrating the three-dimensional ordinary differential equation

$$\dot{\mathbf{x}}(t; \mathbf{x}_0, t_0) = \mathbf{u}(\mathbf{x}(t; \mathbf{x}_0, t_0), t). \tag{3.18}$$

To get an overwiev of the Lyapunov exponents in the whole phase space we start two neighbouring trajectories at 128×128 equally spaced starting points $\mathbf{x}_0^{(i,j)}$ in an intersection plane $x = const$ of the computational domain at a chosen time. All these trajectories are followed for a finite time and the finite time Lyapunov exponent is computed numerically (for more details see [4] and [5]). To be mathematically correct one would have to integrate the trajectories an infinite time, however at least in the growth phase of the dynamo we are limited from above by the time when the magnetic field begins to saturate and the kinematic dynamo phase ends. We compare with Lyapunov exponents computed for a time interval of the same length in the saturated phase.

Numerically the computation of the Lyapunov exponents is included in the MHD-code integrating velocity field and magnetic field in time. Parallel to the timestep of the MHD-solver all 128×128 trajectories are advanced in a timestep of the differential equation (3.18) using the actual velocity field to calculate the right hand side.

The FFT delivers the values of the velocity in physical space only on a cartesian grid of points, while the trajectories reach arbitrary points in the computational domain. One could get, on the other hand, an exact value according to the truncation of the Fourier expansion for the velocity by computing

$$\mathbf{u}(\mathbf{x}) = \sum_{\mathbf{k}} \mathbf{u}_{\mathbf{k}} e^{i \mathbf{k} \cdot \mathbf{x}} \tag{3.19}$$

directly, but this needs $\mathcal{O}(N^3)$ operations for 128×128 trajectories in each timestep. This is too much, because evaluating the velocity in any way needs communication between processors (see below). So one is forced to use an interpolation, either in physical space or in Fourier space. We decided rather to use a trigonometric interpolation in Fourier space up to a wavenumber k_m than quadratic or cubic interpolation in physical space for two reasons: First the amplitude of the modes is decreasing rapidly for the higher shells as can be seen from the energy spectra. The second reason has to do with the parallelization of the Lyapunov part of the code. We distribute the trajectories over the available processors. Each processor has to compute the right hand side for his own trajectories. But the trajectory can be in a part of the computational domain, which is stored on another processor if interpolation in physical space is used. This would cause an amount of communication that would make it impossible to run the code. But if one uses a trigonometric interpolation up to a wavenumber k_M (sufficiently low) the corresponding Fourier coefficients are stored only on a few processors (ideally four) and can be broadcasted to all other processors for the computation of the right hand side.

All in all the computation of the Lyapunov exponents is not so well suited for parallelization as the spectral algorithm of the MHD-solver. Nevertheless, with the constraints described above it is feasible. Using all modes up to shell $k_M = 15$ we got the following computation time for one timestep with computation of Lyapunov exponent for 128×128 trajectories: 21.5s with 16 processors at resolution 64^3 ($\approx$ factor 53 compared to MHD-solver running alone) and 6.5s with 64 processors at resolution 128^3 ($\approx$ factor 9 compared to MHD-solver running alone).

4 Numerical Results

Before we begin with the presentation of the results we first define the lengthscale and the timescale, in which we express our data. The length of the computational domain is chosen as $L = 2\pi$. Therefore k_0 is an integer number and equal to the scale separation $L/l_0 = L/(2\pi/k_0) = k_0$, where l_0 denotes the wavelength of the forcing. We define the kinetic and the magnetic Reynolds numbers corresponding to the length $L^* = l_0/2\pi$ (the length scale of the forcing divided by 2π) and to the velocity $U^* = \sqrt{(A^2 + B^2 + C^2)/3}$ (by this choice one has $U^* = 1$ for $A = B = C = 1$):

$$R^V = \frac{\sqrt{(A^2 + B^2 + C^2)/3}}{k_0\nu}, \qquad R^M = \frac{\sqrt{(A^2 + B^2 + C^2)/3}}{k_0\eta}. \qquad (4.20)$$

This is the same definition as was used in [19]. All times in the following are measured in units of the turnover time

$$\tau_{NL} = \frac{L^*}{U^*} = \frac{l_0/2\pi}{U} = \frac{1}{k_0\sqrt{(A^2 + B^2 + C^2)/3}}. \qquad (4.21)$$

In this time scaling the value of the Lyapunov exponent for pure ABC-flow will scale by a factor of k_0 for different scale separations because of the factor $1/t$ in its definition (3.17). As a first check of the Lyapunov part of the code and to have a comparison for the following calculations in the growth phase and in the saturation phase of the nonlinear dynamo we computed the Lyapunov exponents for the pure ABC_{k_0}-flow for $k_0 = 1,2$ and 4. We started with 128×128 equally distributed initial conditions in the plane $x = \pi/2$. In figures 3 and 5 one can see the result for the ABC_1-flow after 80 time units and for the ABC_2-flow after 160 time units as grey scale image: bright regions correspond to small Lyapunov exponents and dark regions to high Lyapunov exponents. There are small bands of chaos visible belonging to a chain of hyperbolic fixpoints. Figure 3 can be compared with figure 10 of [27]. For $A = B = C$ the chaos is rather weak taking the mean value $\lambda\,(\text{ABC}_{k_0})$, because the larger part of phase space corresponds to integrable respective nearly integrable flow (KAM-tori). Convergence of the mean Lyapunov exponent is reached for a time interval greater than 800 (change less than 4%). For $\Delta t = 160$ the mean values for the LE are 0.0243 for $k_0 = 1$, 0.0536 for $k_0 = 2$ and 0.1251 for $k_0 = 4$, which confirms roughly our statement above: the ratio should be 1:2:4 and actually is 1:2.21:5.15. With the exception of run 1 the flow is different from the ABC-flow in the growth phase and the saturated phase of the dynamo, and the convergence is better than for the pure ABC-flow.

We calculated Luapunov exponents for four different runs with scale separations $k_0 = 1,2$ and 4 and different kinetic and magnetic Reynolds numbers. For the amplitudes A, B, and C of the forcing we restricted ourselves to the case of equal amplitudes $A = B = C = 1$, which is the best investigated up to now (see [21], [19]). The runs are listed in table 1.

The procedure how the magnetic seed is introduced is different for run 1 compared to the other runs. The reason is that the kinetic Reynolds number for run 1 is still in a regime where the

Table 1 List of runs. The first four columns give the scale separation, the Reynolds numbers and the resolution for the runs. $\chi = E^M/E^V$ gives the ratio of magnetic to kinetic energy in the saturated phase of the nonlinear dynamo. In the last four columns are listed the following data: the time intervall for the computation of the Lyapunov exponent, the mean value of the Lyapunov exponent for 128×128 trajectories in the growth phase and the saturated phase, and the ratio of mean value of Lypunov exponents between the saturated phase and the growth phase.

run	k_0	R^V	R^M	N	χ	Δt	$\overline{\lambda_g}$	$\overline{\lambda_s}$	$\overline{\lambda_s}/\overline{\lambda_g}$
1	1	12	12	16	0.03	320	0.019	0.053	2.74
2	1	60	240	64	0.12	80	0.115	0.073	0.64
3	2	60	240	128	1.0	160	0.189	0.090	0.48
4	4	12	12	64	1.4	160	0.437	0.256	0.59

ABC_1-flow is a stable solution of the pure hydrodynamical equation. Therefore run 1 is started with the initial condition $\mathbf{u} = \mathbf{u}_{111}^1 + \delta\mathbf{u}$ for the velocity and $\mathbf{b} = \delta\mathbf{b}$ for the magnetic field, where $\mathbf{u}_{111}^1$ is the ABC-flow for $k_0 = 1$ and $\delta\mathbf{u}$ and $\delta\mathbf{b}$ are small perturbations with an energy about 10^{-9} in the first six shells of k-vectors. The instability of the ABC-flow in this run arises only because of the presence of the magnetic field perturbation.

For the three other runs we demonstrate the procedure of setting the magnetic seed on run 2 (see figure 1). The ABC-flow now is hydrodynamically unstable. To separate between the purely hydrodynamic instability and the dynamo effect we first let the ABC-flow destabilize in a pure hydrodynamic run with inititial condition $\mathbf{u} = \mathbf{u}_{111}^1 + \delta\mathbf{u}$ for the velocity and zero magnetic field $\mathbf{b} \equiv 0$. The perturbation grows exponentially until the ABC-flow (shell 1 in figure 1) breaks down (before time unit 50 in figure 1) to a time dependent state with all modes excited. This ABC-forced state is the flow field were now the magnetic seed $\delta\mathbf{b}$ is introduced (time unit 150 in figure 1) and the growth phase of the dynamo begins. When the magnetic field strength is high enough the nonlinear phase begins and one can see, that the velocity field is influenced by the magnetic field (time unit 350 to 450 in figure 1). Then the nonlinear dynamo saturates, i.e. the magnetic field and the velocity field reach their final state.

We now turn to a more detailed description of the results for the individual runs. As already mentioned in the growth phase for run 1 the flow is very near to the ABC-flow. This expresses itself in an only slightly higher value of the mean Lyapunov exponent than for pure ABC-flow. The finite time Lyapunov exponent is computed for the time intervals [0, 320] in the growth phase and [2920, 3240] in the saturated phase. The level of saturation measured by $\chi = E^M/E^V$ is about 3% for this run. In this run the dynamo bifurcation *enhances* the level of chaos from the growth phase to the saturation phase. In figure 2 (left hand side) one can see, that besides a peak of very low value of the Lyapunov exponent there is a long tail of higher Lyapunov exponent values in the saturated phase. This result can be explained as follows: The ABC-flow is a stationary flow and the flow for the saturated dynamo is periodic (i.e. the bifurcation to the magnetic state is a Hopf bifurcation). A time dependent flow of course generates more chaotic trajectories than a stationary flow. Run 1 thus can not be considered as generic concerning the question of suppression of chaos by a growing magnetic field out of the following reasons: (i) Reynolds number is low, (ii) the flow is stationary and not in the turbulent regime, i.e. the peculiarities of the dynamics in the transitional regime influence the result, (iii) the magnetic field still is weak, so the effect is rather due to pure hydrodynamical effects (that appear because the ABC-flow is destabilized by the magnetic field) than the action of the Lorentz force.

Keeping the scale separation but increasing the kinetic and magnetic Reynolds numbers to

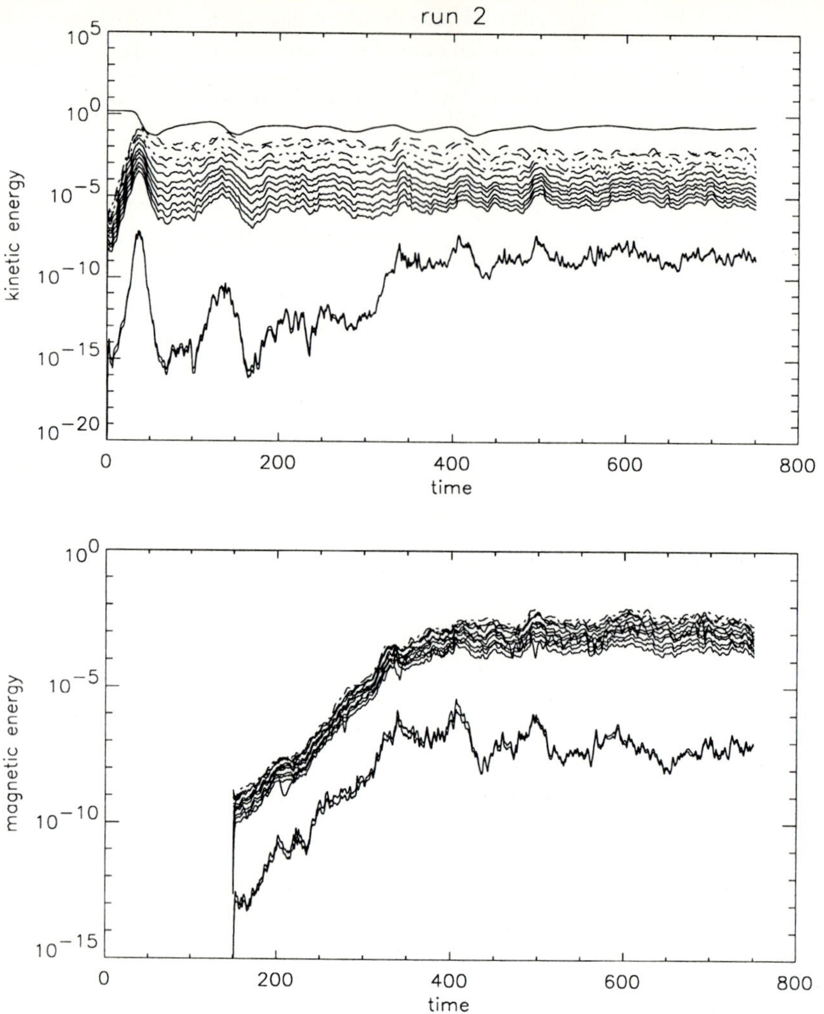

Figure 1 Time evolution of kinetic and magnetic energies in different shells of k-vectors for run 2. The solid line with the highest kinetic energy corresponds to the ABC_1 flow. It is hydrodynamically unstable and destabilizes without the influence of magetic field. At time 150 the magnetic seed field is introduced. It grows exponentially in the kinematic phase of the dynamo. At time 350 the nonlinear saturation of magnetic field begins. The two lowest curves in both plots correspond to shells 31 and 32.

$R^V = 60$ and $R^M = 240$ run 2 was performed. The distribution of Lyapunov exponents is now represented in figure 2 and figure 3 in two different ways. In figure 2 the histogram of Lyapunov exponents now shows a Gaussian distribution with a mean of 0.115 in the growth phase. That the flow in the growth phase has a much higher level of Lagrangian chaos than the pure ABC-flow is visualized in figure 3, where the greyscale images of the ABC-flow and the flow in the growth phase are opposed to each other. In the saturated phase, were the magnetic field energy now is 12% of the kinetic energy of the flow, chaos now clearly is diminished: the corresponding histogram in figure 2 is shifted to the left, and in figure 3 the corresponding grey scale image

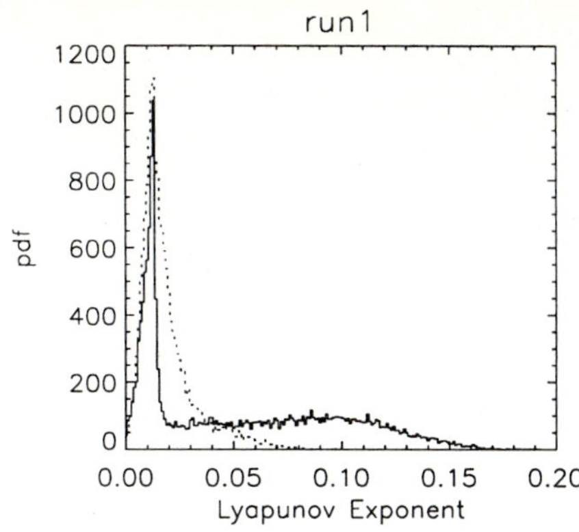
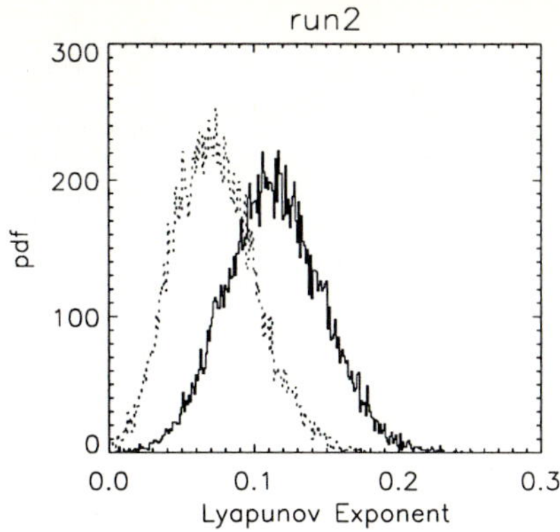

Figure 2 Histograms showing the number of trajectories with a Lyapunov exponent inside given intervalls of length $\Delta\lambda$ for run 1 and run 2. The width of the intervalls for both plots is given by $\Delta\lambda = 0.001$; the total number of trajectories is 128^2. The solid line is plotted for the growth phase and the dashed line for the saturated phase of the dynamo.

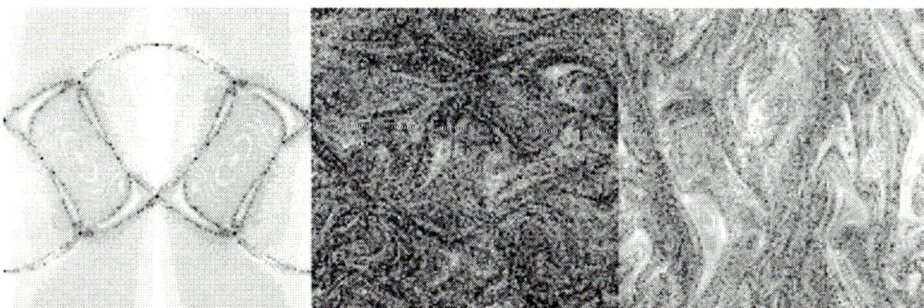

Figure 3 Grey scale images of finite–time Lyapunov exponents ranging from 0 (white) to 0.25 (black). The first image on the left side is computed for the ABC_1–flow and shown as comparison to the ABC-forced flow in the growth phase of the dynamo of run 2 (center). The image on the right side finally is computed for the saturated phase of run 2.

is visibly brighter than in the growth phase. The mean value of the Lyapunov exponent in the saturated phase is 0.0731, which is 64% of the growth phase.

In run 3 we kept the values of the Reynolds numbers but switched to scale separation $k_0 = 2$. A doubling of scale separation also demands a doubling in resolution in each space direction (from 64^3 in run 2 to 128^3 in this run). The level of saturation now is considerably increased, we reach equipartition of kinetic and magnetic energy: $\chi \approx 1.0$. Here also the strength of chaos in the flow of the growth phase of the dynamo is considerably higher than in the ABC_2-flow as becomes clear from figure 4. Now the backreaction of the magnetic field by the Lorentz-force reaches its full strength, which shows itself in the data for the Lyapunov exponent: see the histogram in figure 5 (left hand side) and the pixel graphic in figure 4. The mean value of the Lyapunov exponents in the growth and saturated phases are 0.1892 respective 0.0901 showing a diminution to 48% in the saturated phase compared to the growth phase.

In run 4 the scale separation is increased to $k_0 = 4$, but the kinetic and magnetic Reynolds

Figure 4 Grey scale images of finite–time Lyapunov exponents ranging from 0 (white) to 0.35 (black). Again the ABC flow (now for scale separation $k_0 = 2$, left image), the flow of the growth phase (center) and of the saturation phase (right) are shown, now for run 3.

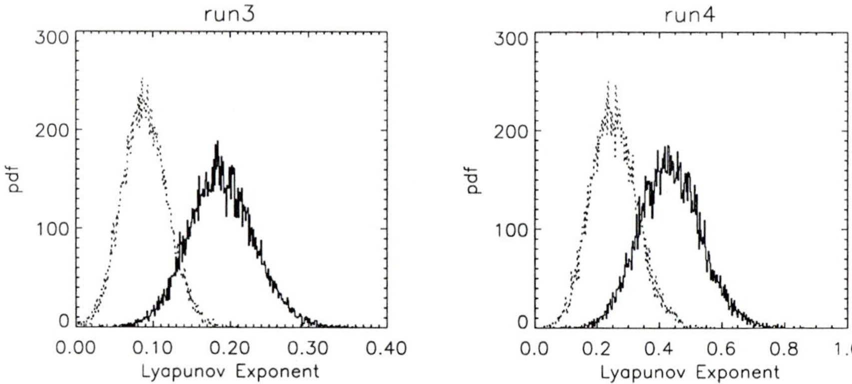

Figure 5 Histograms showing the number of trajectories with a Lyapunov exponent inside given intervals of length $\Delta\lambda$ for run 3 and run 4. The width of the intervalls is given by $\Delta\lambda = 0.001$ for run 3 and by $\Delta\lambda = 0.0025$ for run 4. The solid line represents the growth phase and the dashed line represents the saturated phase of the dynamo.

numbers are reduced to 12. Nevertheless, the ABC_4-flow already is hydrodynamically unstable and a more chaotic state is reached without magnetic field, before the seed field is introduced. A high value of the ratio of $\chi = E^M/E^V \approx 1.4$ is found. Also in this run we find a diminition of chaos from the growth phase to the saturated phase as can be seen from the histogram in figure 5 (right hand side) and table 1: the mean value of the LE in the saturated phase is 59% of that in the growth phase.

5   Conclusions

The investigation of the saturation process of a nonlinear dynamo is numerically challenging insofar as it is necessary to integrate the full MHD-equations in three-dimensional space. Besides the inertia term of the Navier-Stokes equation also the Lorentz force and the interaction between the velocity field and the magnetic field add nonlinear terms in the MHD-equation. All these nonlinearities must be included to check the hypothesis whether the backreaction of a growing magnetic field suppresses Lagrangian chaos of the underlying flow, which is the motor to amplify the magnetic field at small scales in the fast dynamo. The magnetic Reynolds numbers reached in our calculations are still small compared with the magnetic Reynolds numbers of astronomical objects and thus we could not deal with the fast dynamo properly (fast dynamo means the limit $R^M \to \infty$). Nevertheless, the tendency from our results seems to be clear: Provided that the kinetic Reynolds number is high enough, the mean Lyapunov exponent for the flow in the exponential growth phase of the dynamo is higher than in the saturated phase when the magnetic field has reached its full strength. This supports the hypothesis that the Lorentz force acts in a way that chaos of the underlying flow is suppressed by the growing magnetic field.

Besides the Lyapunov exponent there are other interesting measures for the flow and the magnetic field that can be investigated to compare for the growth phase and the saturated phase. The important quantity for the α-effect dynamo is the kinetic helicity of the flow. For $k_0 = 4$ the ABC-forced dynamo also develops a large scale magnetic field. The magnetic energy in shells 1 and 2 becomes larger than all other shells in run 4 (see [36]). This happens after the exponential growth of small scales. The total helicity shows significant changes at the times were the energies in shells 1 and 2 are growing.

Another very important question is the spatial structure of the flow as well as the magnetic field. A crude measure for the intermittency of the magnetic field can be defined by the ratio b_{max}/b_{rms}, where b_{max} is the maximum value of the magnetic field on all collocation points in the computational domain. b_{max}/b_{rms} changes by a factor between 1/3 and 1/4 between growth and saturated phase in run 3 (see [36]).

The question of spatial structure probably is closely connected with the question at which level $\chi = E^M/E^V$ the magnetic field saturates for the dynamo (see [22]). As already mentioned in the introduction some authors express the opinion that the suppression of the dynamo mechanisms becomes so strong for very large Reynolds number that equipartition of magnetic and kinetic energy could not be reached. If the magnetic field is strong only in small parts of space (filamentary structure) there would remain large regions of space were chaos and helical turbulence of the flow could exist to create a sufficient amount of magnetic field to reach equipartition. The space would be divided in two types of regions, one type where the magnetic field is generated and the other where the magnetic field is transported and stored by the flow.

These issues should be further investigated in the future. Numerically, it would be desirable to reach as high magnetic Reynolds number as possible to find out, wether there is an enhancement of the suppresive effects on magnetic field generation by an already existing magnetic field for large magnetic Reynolds numbers.

Acknowledgements: Computations were done on T3E/IDRIS (Orsay). We are pleased to acknowledge financial support from CNRS–1202–MFGA & EEC–ERBCHRXCT930410.

References

[1] H.K. Moffatt, Magnetic Field Generation in Electrically Conducting Fluids, Cambridge University Press, Cambridge 1978

[2] F. Krause and K.-H. Rädler, Mean Field Magnetohydrodynamics and Dynamo Theory, Akademie Verlag, Berlin 1980

[3] S. Childress and A.D. Gilbert, Stretch, Twist, Fold: The Fast Dynamo, Springer, Berlin 1995

[4] J. Guckenheimer and P. Holmes, Nonlinear Oscillatons, Dynamical Systems and Bifurcations of Vector Fields, Springer, New York 1983

[5] E. Ott, Chaos in Dynamical Sustems, Cambridge University Press, Cambridge 1993

[6] P.H. Roberts and A.M. Sowards, Dynamo Theory, Ann. Rev. Fluid Mech. **24** (1992) 459-512

[7] S.A. Jepps, Numerical Models of Hydromagnetic Dynamos, Journ. Fluid Mech. **67** (1975) 625 - 646

[8] N.I. Kleeorin and A.A. Ruzmaikin, Mean Field Dynamo with Cubic Non-Linearity, Astron. Nachr. **305** (1984) 265 - 275

[9] S.I. Vainshtein and F. Cattaneo, Nonlinear Restrictions on Dynamo Action, Astrophys. Journ. **393** (1992) 165 - 171

[10] A.V. Gruzinow and P.H. Diamond, Self Consistent Theory of Mean-Field Electrodynamics, Phys. Rev. Lett. **72** (1994) 1651 - 1653

[11] F. Cattaneo and D.W. Hughes, On the Nonlinear Saturation of the Turbulent α - Effect, Phys. Rev. E **54** (1996) R 4532

[12] A. Pouquet, U. Frisch, J. Léorat, Strong MHD helical turbulence and the nonlinear dynamo effect, Journ. Fluid. Mech. **77** (1976) 321-354

[13] A. Bhattacharjee and Y. Yuan, Astrophys. Journ. **449** (1995) 739B

[14] U. Frisch, A. Pouquet, J. Leorat, A. Mazure, Possibility of an inverse cascade of magnetic helicity in magnetohydrodynamic turbulence, Journ. Fluid Mech. **68** (1975) 769-778

[15] A. Pouquet and G.S. Patterson, Numerical simulation of helical magnetohydrodynamic turbulence, Journ. Fluid Mech. **85** (1978) 305-323

[16] M. Meneguzzi, U. Frisch, A. Pouquet, Helical and nonhelical turbulent dynamos, Phys. Rev. Let. **47** (1981) 1060-1064

[17] R. Horiuchi and T. Sato, Phys. Fluids **31** (1988) 1142

[18] D. Balsara and A. Pouquet, The formation of large-scale structures in supersonic magnetohydrodynamic flows, Phys. Plasmas **6** (1999) 89-99

[19] B. Galanti, P.L. Sulem, A. Pouquet, Linear and Non-linear Dynamos Associated with ABC Flows, Geophys. Astrophys. Fluid Dyn. **66** (1992) 183 - 208

[20] A. Brandenburg et al., Magnetic structures in a dynamo simulation, Journ. Fluid Mech. **306** (1996) 325-352

[21] D.J. Galloway and U. Frisch, Dynamo Action in a Family of Flows with Chaotic Streamlines, Geophys. Astrophys. Fluid Dyn. **36** (1986) 53 - 83

[22] E. Blackman, Overcoming the Backreaction on Turbulent Motions in the Presence of Magnetic Fields, Phys. Rev. Lett. **77** (1996) 2694 - 2697

[23] F. Cattaneo, K. Hughes, E. Kim, Suppression of Chaos in a Simplified Nonlinear Dynamo Model, Phys. Rev. Lett. **76** (1996) 2057 - 2060

[24] S. Childress, New Solutions of the Kinematic Dynamo Problem, Journ. Math. Phys. **11** (1970) 3063 - 3076

[25] V.I. Arnold, Sur la topologie des écoulements stationnaires des fluides parfaits, C. R. Acad. Sci. Paris **261** (1965) 17 - 20

[26] M. Hénon, Sur la topologie des lignes de courant dans un cas particulier, C. R. Acad. Sci. Paris A **262** (1966) 312 - 314

[27] T. Dombre et al., Chaotic Streamlines in the ABC Flows, Journ. Fluid Mech. **167** (1986) 353 - 391

[28] V.I. Arnold and E.I. Korkina, The Growth of a Magnetic Field in a Threedimensional Steady Incompressible Flow, Vest. Mosk. Un. Ta. Ser. 1, Matem. Mekh., no. 3, (1983) 43 - 46

[29] D.J. Galloway and U. Frisch, A Numerical Investigation of Magnetic Field Generation in a Flow with Chaotic Streamlines, Geophys. Astrophys. Fluid Dyn. **29** (1984) 13 - 18

[30] Y.-T. Lau and J.M. Finn, Fast Dynamos with Finite Resistivity in Steady Flows with Stagnation Points, Phys. Fluids B **5** (1993) 365 - 375

[31] O. Podvigina and A. Pouquet, On the Non-linear Stability of the 1:1:1 ABC Flow, Physica D **75** (1994) 471 - 508

[32] F. Feudel, N. Seehafer, O. Schmidtmann, Bifurcation Phenomena of the Magnetofluid Equations, Math. and Comp. in Sim. **40** (1996) 235 - 245

[33] F. Feudel, N. Seehafer, B. Galanti, S. Rüdiger, Symmetry-Breaking Bifurcations for the Magetohydrodynamic Equations with Helical Forcing, Phys. Rev. E **54** (1996) 2589-2596

[34] D. Gottlieb and S. Orzag, Numerical Analysis of Spectral Methods: Theory and Applications, SIAM-CBMS, Philadelphia 1977

[35] C. Canuto, M. Hussaini, A. Quarteroni, T. Zang, Spectral Methods in Fluid Dynamics, Springer, Berlin 1988

[36] E. Zienicke, H. Politano, A. Pouquet, Variable Intensity of Lagrangian Chaos in the Nonlinear Dynamo Problem, Phys. Rev. Lett. **81** (1998) 4640 - 4643

III.   FLOWS OF REACTING GASES

Investigation of the Flow Characteristics Occurring in Flame Stabilization Processes

M. Buffat*, J. Yan, L. Duchamp de Lageneste*, T. Rung, O. Guerriau*, F. Thiele,

Hermann-Föttinger-Institut für Strömungsmachanik
Technische Universität Berlin, 10623 Berlin, Germany

* Laboratoire de Mécanique des Fluides et d' Acoustique
Ecole Centrale de Lyon (ECL) - 36, avenue Guy de Collongue
BP163 - 69131 Ecully Cedex, France

Summary

This paper presents numerical simulations of sub-critical flows inside a combustion chamber using both Large Eddy Simulation (LES) and Reynolds-Averaged Navier-Stokes equations (RANS). An experimental setup, "ORACLES", which is a simplified combustion chamber, is selected to study the flow physics. Simulations are performed on unstructured grids using LES and RANS, and on structured grids applying RANS with various turbulence models. To study the influence of the combustion processes, the diffusion flame inside a combustion chamber with a bluff body is investigated by RANS approaches on structured grids applying various turbulence models. Both results are compared with experimental data.

1   Introduction

Flame stabilization within combustion processes occurring in a wide range of practical systems of industrial or domestic relevance, such as aero- and car-engine combustion chambers and industrial burners, is a major issue in the framework of combustion in general. The understanding and analysis of the associated flow is central for the contribution to the progress of combustion technology. Numerical flow simulations on high-performance massively parallel computers are now available for the computation of turbulent flows with increasing complexity as those encountered in most engineering systems. This promotes the description of the characteristic motion in stabilization processes. The elaboration of a consistent route towards the modelling of reactive flows in general and stabilization processes in particular indisputably requires a solid basis for the prediction of the complex structures encountered in the associated turbulent flow: the interaction of turbulent structures and mean flow on one hand, and the transient evolution on the other hand. For instance, this can be achieved by means of sophisticated statistical turbulence models incorporating relevant elements to handle turbulence anisotropy. The assessment can be approached by comparison with full or large-scale simulations.

The stabilization processes encountered in both premixed and non-premixed flames are, at this time, not well understood, but are of great interest for a wide range of industrial applications. The principle aim of the investigations proposed is directed to an improved knowledge of the stabilization processes, and to an implementation of the findings into an appropriate computational model. These objectives can be achieved by accomplishing a close link between different numerical simulations and experimental studies. The computational modelling of such complex flows is an extremely challenging task with regard to both the accuracy of the applied physical models and the numerical approximation. Although the available techniques to account for combustion are still based on modelling rather than on scale determination, a better understanding of the mixing phenomena within combustion chambers can be approached by LES.

It is known that simple statistical turbulence and combustion models are a major contributor to the misrepresentation of the stabilization processes. This reflects the need for further effort to improve their respective predictive accuracy. A significant progress has been accomplished with respect to the refinement of various low- and high-Reynolds number eddy-viscosity based turbulence models, including realizable, local, nonlinear and RNG variants. With preliminary efforts devoted to the improvement and application of low-Re high-order turbulence models for external aerodynamic flows, further refinement is under investigation for complex internal flows. Here, the preliminary effort will be devoted to the application of low-Re models and Reynolds stress turbulence models within the RANS approach.

Subsequently, various parameters will be investigated in a collaborative computational study in order to improve the understanding of the physical phenomena occurring in the stabilization processes. The parametric study will concentrate on the geometry used for the stabilization of the flame and the turbulence intensity distribution in the flow region. Further emphasis of this investigation is placed on the supplementary refinement of presently used turbulence models in order to account for the influence of combustion in greater detail. One of the objectives of the investigation presented is to study the performance of different combustion models in the prediction of turbulent combustion. Since the probability density function (PDF) is an important concept in turbulent combustion modelling and widely used, the presumed PDF methods will serve as the basic approach. The flamelet concept is also used for its simplicity. All these combustion models will be applied to investigate the influence of combustion models on the stabilization processes in flames.

The configurations under investigation are a simplified combustion chamber (ORACLES) without chemical reaction to study the flow physics and a diffusion flame inside a combustion chamber. Simulations will be performed on unstructured grids using LES and RANS. In addition, RANS calculations are also preformed on structured grids using advanced turbulence models. To study the influence of the combustion processes, the diffusion flame will be investigated by the RANS approach on structured grids applying various turbulence models. The computed results are compared with experimental data.

2 Turbulence Modelling

Since turbulence plays an important role in the mixing process of the various components involved in the overall system, the quality of a turbulence model is of major importance in the computation of reacting flows. The most prominent and simple representative model is the isotropic-viscosity model based on linear Boussineq relations. Despite its simplicity and numerical advantages, the eddy-viscosity concept is afflicted by a variety of deficits. Various theoretical and computational studies have demonstrated the inability of the linear eddy-viscosity approach to mimic the important processes associated with the complex interaction mechanisms between different types of stress and strain fields. In the past years, intensified research has been undertaken to remedy the shortcomings of the linear eddy-viscosity model, while retaining important numerical advantages of the concept. A typical feature of these attempts is their foundation on the constitutive relation linking the Reynolds stresses to nonlinear expressions of strain-rate and vorticity components, which effectively gives rise to an anisotropic eddy-viscosity coefficient. The investigations of complex turbulent flows based on nonlinear eddy-viscosity modelling have shown encouraging results [2, 5, 11]. The Explicit Algebraic Stress Models (EASM) [15] are good alternatives and viable approximations at computational costs comparable to eddy–viscosity models. However, they were found to produce inappropriate unrealizable results, like negative turbulent kinetic energy components even in simple shear flows. A rigorous analysis of the realizability of nonlinear stress-strain relationships for Reynolds-stress closures can be found in [16], where a new realizable quadratic eddy-viscosity model has been developed, which has shown encouraging predictive capabilities for a category of flows of practical interest.

When using EASM in low-Re form, wall distances or normal unit vectors to the wall appear in damping functions or wall-reflection parts of the pressure-strain term, which are difficult to define when the flow geometry is complex. Thus, a wall-parameter-free (or local) low-Re EASM is of great benefit to the prediction of complex wall-bounded turbulent flows. Based on the analysis of the near-wall behaviour of the Reynolds-stress anisotropy, a local low-Reynolds number modification of EASM is reported by Guo et al. [9].

Within the code NATUR [12], a set of different Algebraic Stress Models (ASM) have been incorporated, e.g. the Launder, Reece and Rodi (LRR) model [10] and the Speziale, Sarkar and Gatsi (SSG) model [17]. Recently, a two-layer model proposed by Chen and Patel [4] has been implemented in the code NATUR, which combines the standard $k - \epsilon$ model or a 2nd order (LRR) model with a simpler one-equation model to resolve the flow near a solid wall. The assessment of these different models has been achieved through their application to a wider category of incompressible flows, involving strong interaction of streamline curvature and turbulence.

3 Large Eddy Simulation

A major effort has been devoted to the investigation and development of LES on un-structured, finite-element-based meshes for the flow simulation inside the combustion chamber. The LES code NATURLES [7] solves the three-dimensional Navier-Stokes equations for compressible turbulent reacting flows. A mixed Finite-Volume (FV)/Finite-Element (FE) method [6] is used for the spatial approximation of the Navier-Stokes equations on unstructured meshes. The hyperbolic parts of the equations are treated employing the FV approach, while the diffusive and source terms are handled using FE. This method was originally developed in a RANS code NATUR at ECL to compute turbulent flows using a standard $k - \epsilon$ model and an implicit scheme to improve the numerical efficiency.

The subgrid scale models implemented in NATURLES have been validated by comparative testing on the case of a temporal mixing layer (see Figs. 1 – 2). As seen here, the classical Smagorinsky model shows too much diffusion. Therefore, it is reasonable to choose the dynamical version of the Smagorinsky model as the basic model for the flow simulations, which will be applied to the computation of the ORACLES configuration.

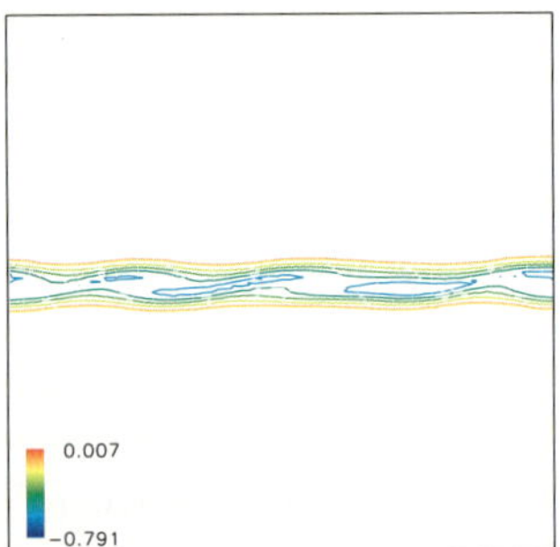
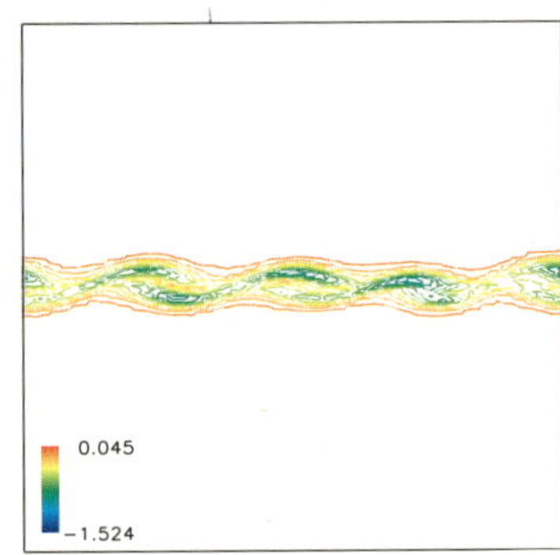

Figure 1 Comparative testing of the classical Smagorinsky model (left) and the dynamical Smagorinsky model (right). Vorticity iso-lines obtained at t = 40 s.

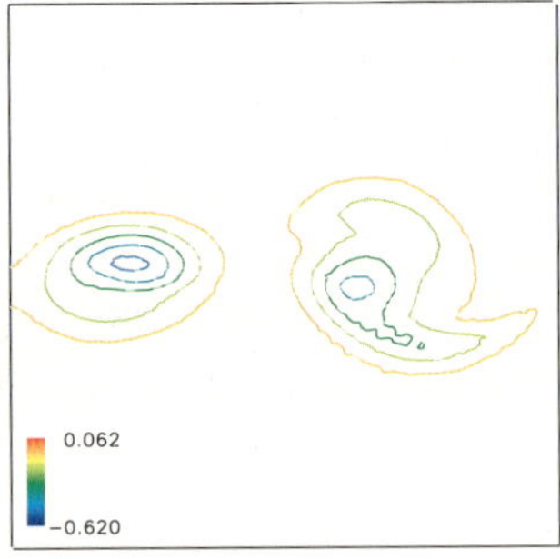
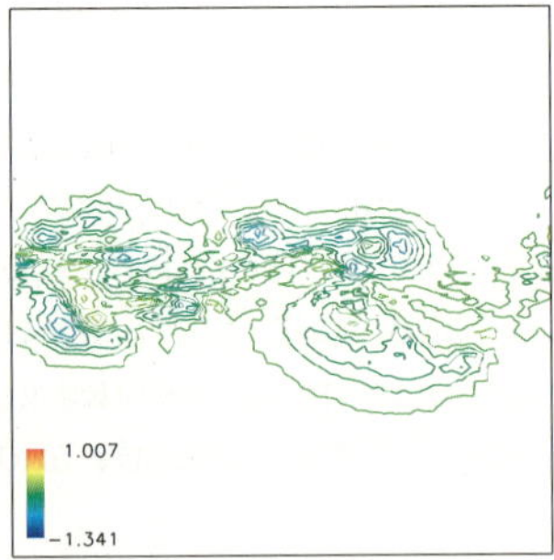

Figure 2 Comparative testing of the classical Smagorinsky model (left) and the dynamical Smagorinsky model (right). Vorticity iso-lines obtained at t = 80 s.

4 Combustion Modelling

Compared to non-reacting systems, the governing equations for reacting flows are completed by the following equations:

species continuity equation:

$$\frac{\partial \overline{\rho} \widetilde{Y_n}}{\partial t} + \nabla \cdot (\overline{\rho} \widetilde{v} \widetilde{Y_n}) = \nabla \cdot (\frac{\mu}{Sc} + \frac{\mu_t}{Sc_t}) \nabla \widetilde{Y_n} + \widetilde{\dot{\omega}_n} \, , \tag{1}$$

energy equation:

$$\frac{\partial \overline{\rho} \widetilde{h}}{\partial t} + \nabla \cdot (\overline{\rho} \widetilde{v} \widetilde{h}) = \frac{\partial \widetilde{p}}{\partial t} + \widetilde{v} \cdot \nabla \widetilde{p} + \sigma_{ij} \frac{\partial \widetilde{u_i}}{\partial x_j} + \overline{\rho} \epsilon + \nabla \cdot (\frac{\mu}{Pr} + \frac{\mu_t}{Pr_t}) \nabla \widetilde{h} \, , \tag{2}$$

where $\widetilde{h}$ represents the enthalpy and its temperature dependence:

$$h_n = h_{n,ref} + \int_{T_{ref}}^{T} c_{p_n} dT \, , \tag{3}$$

thermal equation of state

$$\overline{p} = \overline{\rho} R \overline{T} = \overline{\rho} R \sum_{n=1}^{N} \frac{\widetilde{Y_n}}{M_n} \overline{T} \, . \tag{4}$$

In Eq. (2), ϵ is the turbulence dissipation rate, the Newton stress tensor σ_{ij} is calculated from:

$$\sigma_{ij} = \mu \left(\frac{\partial \widetilde{u_i}}{\partial x_j} + \frac{\partial \widetilde{u_j}}{\partial x_i} \right) - \frac{2}{3} \mu \frac{\partial \widetilde{u_k}}{\partial x_k} \delta_{ij} \, . \tag{5}$$

Since the chemical source terms $\widetilde{\dot{\omega}_n}$ are highly nonlinear and strongly dependent on temperature and species concentrations, a moment closure approach is inadequate. Therefore, the modelling of the chemical source terms in the calculation of turbulent reacting flows is still a challenging problem. In this context, methods analyzing the evolution of the PDF are of particular interest because they offer an elegant treatment of chemical reactions.

For diffusion combustion, the presumed β-PDF is widely used. The presumed β-PDF is a function with two parameters, mixture fraction $\widetilde{Z}$ and its variance $\widetilde{Z''^2}$. For these two variables, two transport equations are solved:

$$\frac{\partial \overline{\rho} \widetilde{Z}}{\partial t} + \nabla \cdot (\overline{\rho} \widetilde{v} \widetilde{Z}) = \nabla \cdot (\frac{\mu}{Sc} + \frac{\mu_t}{Sc_t}) \nabla \widetilde{Z} \, , \tag{6}$$

$$\frac{\partial \overline{\rho} \widetilde{Z''^2}}{\partial t} + \nabla \cdot (\overline{\rho} \widetilde{v} \widetilde{Z''^2}) = \nabla \cdot (\frac{\mu}{Sc} + \frac{\mu_t}{Sc_t}) \nabla \widetilde{Z''^2} + 2 \frac{\mu_t}{Sc_t} (\nabla \widetilde{Z})^2 - C_z \overline{\rho} \frac{\epsilon}{k} \widetilde{Z''^2} \, , \tag{7}$$

where $C_z \sim 2$ is a model constant.

4.1 Fast Chemistry Model Using Presumed PDF Approach

A practical idealized approach in non-premixed combustion systems is based on the assumption that the chemistry is infinitely fast. Its first principal feature is the neglect of all intermediate reactions, so that pure fuel and pure oxidant will react to form the products at the moment they are in contact. Its second main feature is the assumption that the effective diffusivity coefficient of all species is the same. With these assumptions, the quantity can be described by differential equations that do not contain chemical source terms. In this way, the main problem in turbulent combustion is circumvented. The mean species concentrations can be computed through PDF and the relations between species concentration and mixture fraction:

$$\tilde{Y}_n = \int_0^1 Y_n(Z) P(Z) dZ \,. \tag{8}$$

4.2 Flamelet Model for Turbulent Diffusion Flame

The flamelet concept views a turbulent diffusion flame as an ensemble of laminar diffusion flamelets. The main advantage of the flamelet concept is the fact that the chemical time and length scales need not to be resolved in a multi-dimensional CFD code [13, 18]. This concept has recently gained popularity for its simplicity. The steady flamelet concept can be described as:

$$-\rho \frac{\chi}{2} \frac{\partial^2 Y_n}{\partial Z^2} = \dot{\omega}_n \,, \tag{9}$$

where $\chi = 2D \frac{\partial Z}{\partial x_j} \frac{\partial Z}{\partial x_j}$ is the scalar dissipation rate. For the turbulent diffusion flame, this formulation can be written as:

$$-\frac{1}{2}\bar{\rho} \int_0^1 \tilde{\chi}_Z \frac{d^2 Y_n}{dZ^2} P(Z) dZ = \widetilde{\dot{\omega}_n} \tag{10}$$

with the conditional mean dissipation rate $\tilde{\chi}_Z$ prescribed as

$$\tilde{\chi}_Z = C_\chi \frac{\epsilon}{k} \widetilde{Z''^2} \frac{f(Z)}{\int_0^1 f(Z) P(Z) dZ}, \quad f(Z) = e^{-2[erfc^{-1}(2Z)]^2} \,. \tag{11}$$

Here, $C_\chi \sim 2$ is a model constant and $erfc^{-1}$ represents the inverse of the complementary error function.

5 Computational Methodology

The structured-grid multi-block parallel solvers [15] using finite-volume methods are based on general non-orthogonal coordinates and employ a fully co-located storage arrangement for all transport properties. The calculation procedures incorporate various difference schemes. Different upwind-based convective schemes are implemented explicitly in the sense of deferred correction. High-order bounded and unbounded schemes have been embedded to approximate the advective volume-face fluxes, among which are Van Leer's TVD/MUSCL scheme and the QUICK scheme of Leonard. The second-order central difference approximation scheme is generally retained for the diffusion terms.

As LES requires precise numerical methods, improved numerical schemes have been incorporated in NATURLES [7] to reduce the inherent numerical diffusion and dispersion of the original Roe scheme [14] i.e. a $\beta - \gamma$ modified Roe solver with a Turkel low-Mach number preconditioning and a 3rd order Runge-Kutta time integration. The code NATURLES is also parallelized using a domain decomposition technique and a message passing library.

6   Discussion of Results

6.1   Flows without Combustion

The ORACLES configuration [3] can be considered as a challenging flow for LES as it involves various difficult flow features such as fully turbulent channel flows, spatial mixing layers and backward-facing steps that have never been studied by LES in a single simulation at such a high Reynolds number (≈ 25000), which is based on the height of the channel inlet (H) and the mean streamwise component U_0 of the velocity measured at the channel inlet's centerline. According to Abbot and Kline [1], the geometry of the double backward-facing step corresponds to an area expansion ratio of 1.84, which leads to an asymmetry of the inert mean velocity field. The experimental analysis of inert and combustion flows within this configuration can be found in [3], where the experimental conditions are well defined and a variety of results are available for comparison. The inert case has been simulated by LES and RANS using various turbulence models.

Computation with LES

The ORACLES configuration and its inflow boundary conditions are shown in Fig. 3. A particular effort was necessary in order to get the inflow conditions for the combustion chamber as close as possible to fully developed turbulent channel flow. The experimental conditions were accurately reproduced in terms of pressure gradient/skin friction velocity in both channels. For LES, the mean velocity profiles cannot be imposed at the entrance. Thus two different channel flows are computed in order to impose independent instantaneous flows in each channel. Consequently, the flow through a reference plane of each channel is used as inflow conditions for the combustion chamber. The 3D finite-element mesh used consists of one million cells, one simulation requires several days on a Linux cluster of 8 DEC Alpha 500. Comparison of the phase-averaged quantities is limited here to mean velocity $\langle U \rangle$ and turbulent kinetic energy $\langle k \rangle$ profiles. Results are

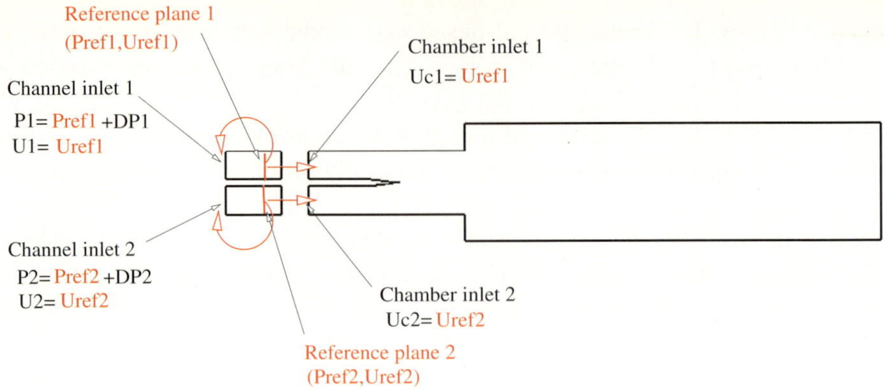

Figure 3 ORACLES configuration and inflow boundary conditions.

shown in Fig. 4 concerning the averaged longitudinal velocity profile in the upper channel expressed in terms of wall coordinates. As the first grid point is placed at $y^+ \approx 0.5$,
the viscous sub-layer ($y^+ \leq 10$) is well resolved by the calculation, which proves that the

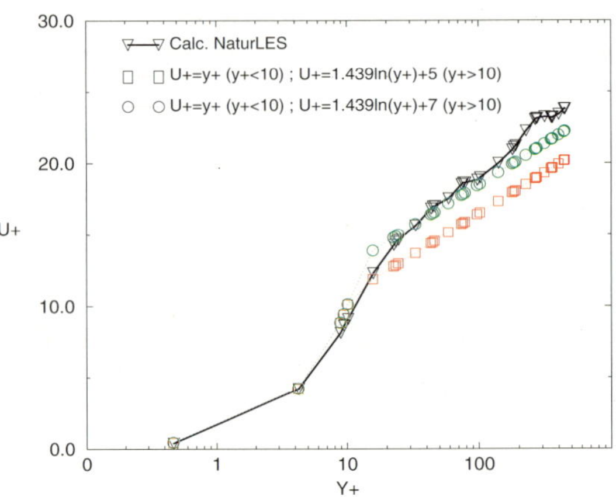

Figure 4   Upper channel's mean longitudinal velocity profile in wall coordinates. Comparison
with logarithmic laws.

grid is fine enough in this region. In the log-layer, the velocity is slightly overestimated
($U^+ = 2.439ln(y^+) + 7$ as compared to the theoretical law $U^+ = 2.439ln(y^+) + 5$),
which can be explained by a small overestimation of the mean velocity near the centerline of each channel. Fig. 5 depicts the profiles of the mean velocity at the stations:
x = -170 mm, x = 0 mm (sudden expansion region), x = 50, 110 mm (behind the sudden expansion). These results show that LES predicts the asymmetry of the mean flow
observed experimentally.

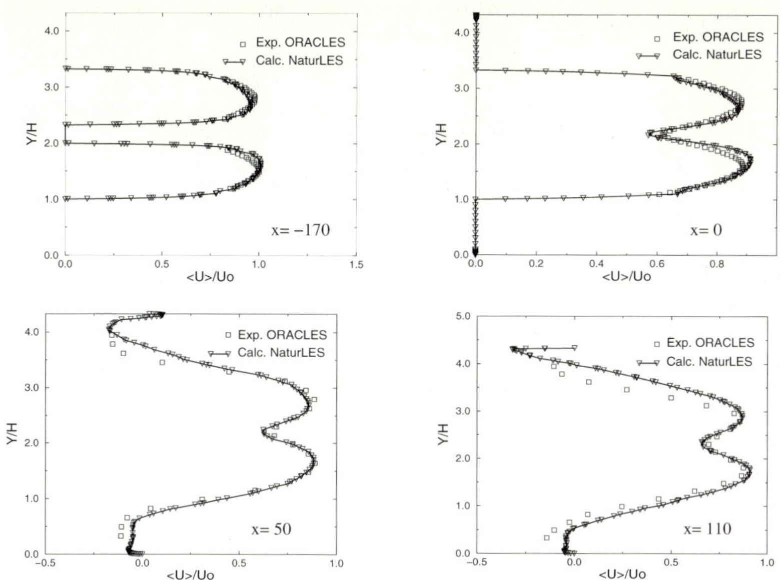

Figure 5 LES of the inert flow in the ORACLES configuration: profiles of mean longitudinal velocity $\langle U \rangle$ at various stations.

Computation using RANS

The same configuration was computed with various turbulence models, namely, the standard $k - \epsilon$ model, the Wilcox $k - \omega$ model, EASM and RSM [10], respectively. The dimensions of the computational domain were set as follows: the inflow and outflow planes are 10 H upstream and 30 H downstream of the step, respectively. A non-uniform grid distribution was used. The velocity inflow condition in the combustion chamber is set as close as possible to fully developed turbulent channel flow. According to the experiment, the velocity distributions of both inlets are slightly different, which is reproduced in the computation. k and ϵ profiles were specified using uniform distributions corresponding to a free-stream turbulence intensity of $T_u = 3\%$ and $\nu_t = 10\nu$.

Fig. 6 shows the velocity and turbulent kinetic energy fields computed with EASM. The asymmetry of the inert mean velocity and the turbulent kinetic energy is clearly predicted. The use of other turbulence models displays a similar behaviour.

Comparisons of the velocity and turbulent kinetic energy profiles are shown in Fig. 7. At the station x = -170 mm near the inlet, where the inlet conditions have a strong influence on the $\langle U \rangle$ distribution, all computations produce almost the same profiles. In the mid-plane of the channel, the turbulence energy is underestimated.

At the station x = 0 mm, the velocity and turbulent kinetic energy profiles are well captured and all models produce similar results. The maximum velocity is slightly overpredicted, corresponding to an underestimated turbulent kinetic energy value. At the mid-plane, the two inflows are not yet fully mixed, causing the velocity profiles being underestimated.

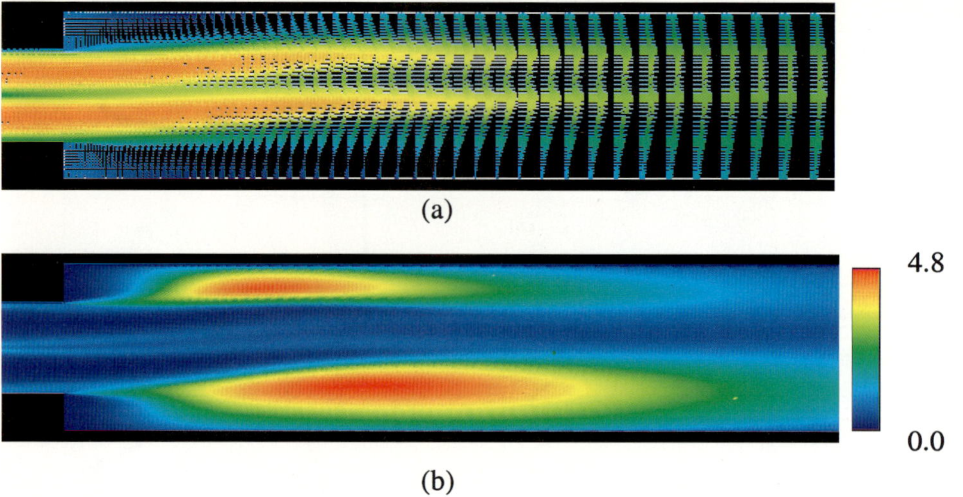

Figure 6 EASM results of the inert flow in the ORACLES configuration: the velocity (a) and the dimensionless turbulent kinetic energy (b) fields.

At the station x = 50 mm, the maximum turbulent kinetic energy value is underestimated and the velocity at the mid-plane is smaller than in the experimental data. Also shown in these figures is that none of these models can accurately reproduce the velocity profiles in the near wall region. Compared to the other models, the RSM shows a poor performance. The reason may be that no wall effect is taken into account in the high-Re RSM application. However, the RSM produces the best prediction of the turbulent kinetic energy. Furthermore, all computational profiles in the outer region are in close agreement with the measurement.

As seen from the results at the stations x = 190 and 250 mm, the flow is clearly asymmetric. The RSM predicts the best results of the turbulent kinetic energy here too, while the $k - \omega$ model gives the lowest turbulent kinetic energy value. The Wilcox k-ω model reproduces the best velocity profiles in the lower region, but in the upper region the profiles display a larger difference to the experimental data than the other models. In general, EASM has a better behaviour.

6.2   Reactive Flows

In order to assess the aforementioned turbulence models concerning the prediction of diffusion flames, a 2D axisymmetric flow in a combustion chamber has been selected. The flame is stabilized by a bluff body separating the fuel jet (methane) and the oxidant stream (air). The experimental conditions are well defined and a variety of results are available for comparison [8]. Because of density variations in the reacting flow, the Navier-Stokes equations are considered in their averaged, mass-weighted conservative form. The effects of velocity divergence ($\partial u_m/\partial x_m \neq 0$) are absorbed in the redefinition of the strain–rate tensor, which results in a trace–free strain-rate tensor for both

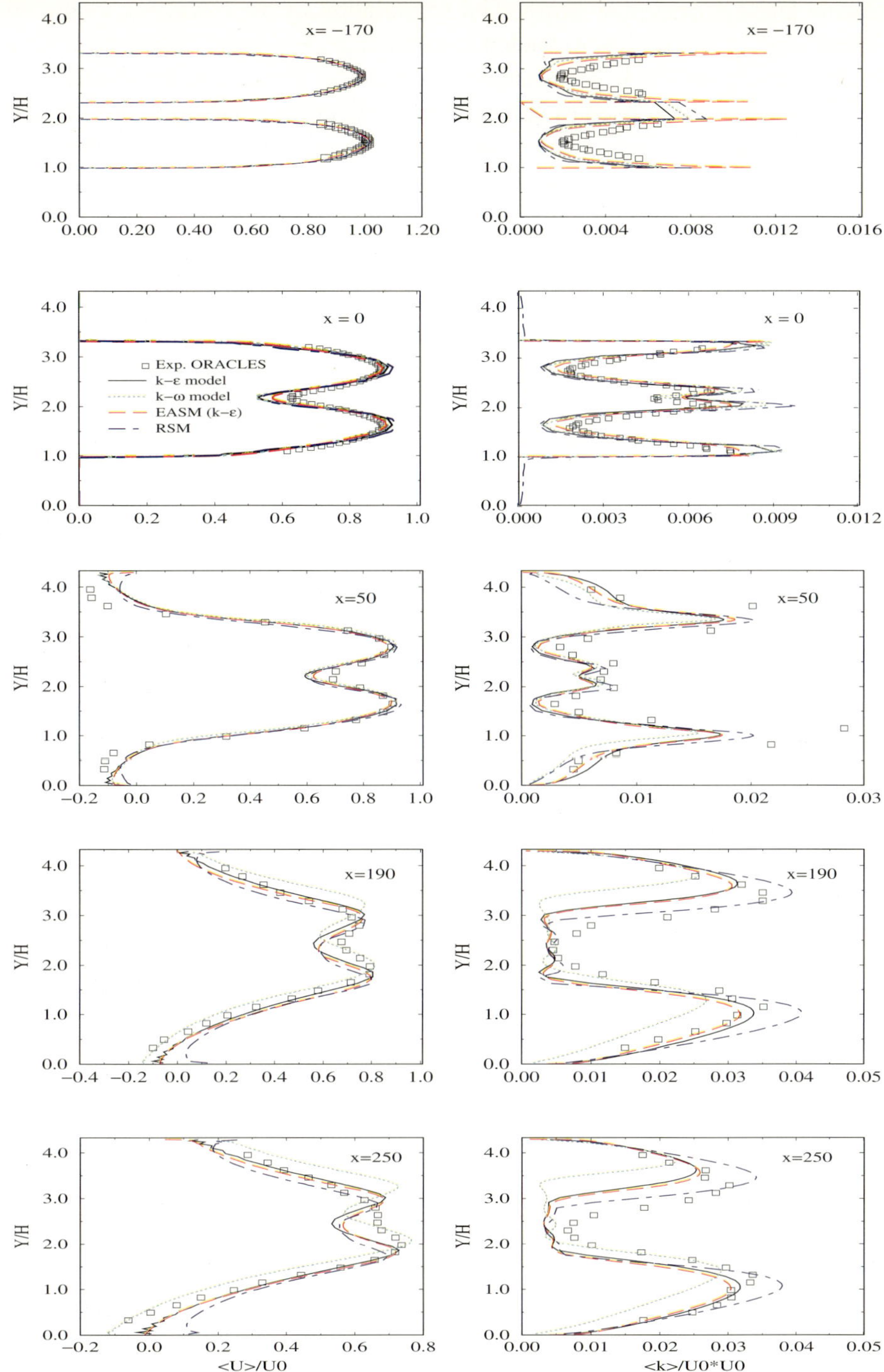

Figure 7 RANS results of the inert flow in the ORACLES configuration: profiles of mean longitudinal velocity (left) and turbulent kinetic energy (right) at various stations.

incompressible and compressible flows. The formulation of the EASM is unaltered. A detailed description of the EASM extended to compressible flows can be found in [16].

The burner consists of a 5.4 mm diameter methane jet located in the center of a 50 mm diameter cylinder which itself is surrounded by an outer tube supplying coaxial air (Fig. 8). The axial bulk velocity in the central methane jet is 21 m/s corresponding to a Reynolds number of Re=7000. The axial bulk velocity for the coaxial air is set to 25 m/s. All calculations are performed on the same Cartesian non-equidistant grid. Three turbulence models, namely, the standard $k - \epsilon$ model, EASM and RSM are investigated. In

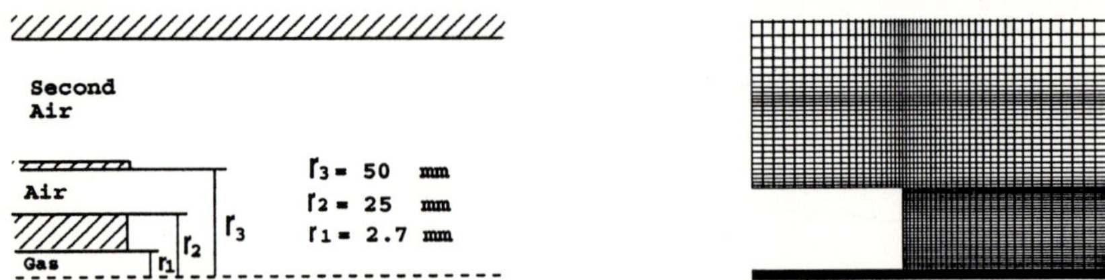

Figure 8   Inlet region and burner dimension (left) and grid (right) for a combustion chamber [8].

order to evaluate the potential and limits of these turbulence models, we first consider an inert configuration. The axial centerline profiles of the streamwise velocity, the turbulent kinetic energy and the mixture fraction are displayed in Fig. 9. In general, the distribution of the turbulent kinetic energy is predicted. Only using EASM, the position of the two peaks in the profiles for the turbulent kinetic energy can be approximately detected from the computation. The centerline profile of the mean mixture fraction is found to be

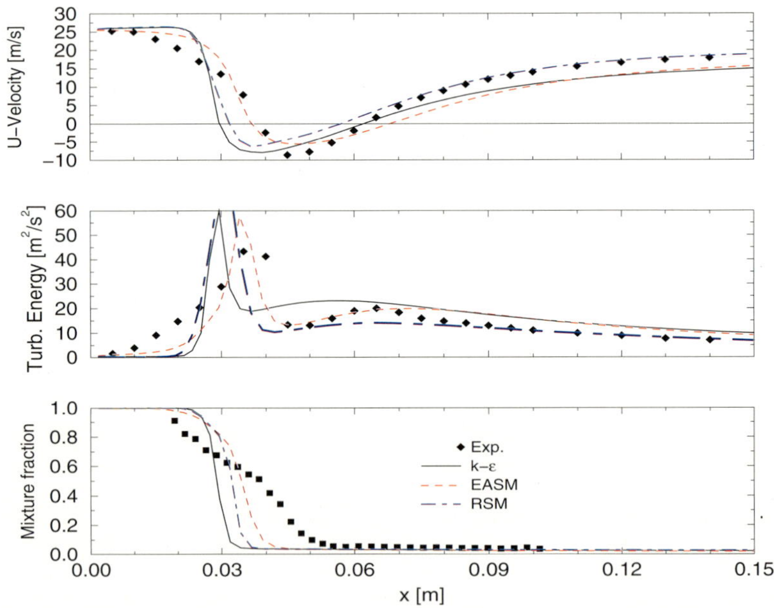

Figure 9   Axial centerline profiles of mean axial velocity, turbulent kinetic energy and mean mixture fraction for the inert case.

156

very sensitive to the exact location of the flow inversion, but it is not accurately predicted by any model. The radial profiles of the mean velocity and the turbulent kinetic energy are shown in Fig. 10. EASM has a good behaviour compared to the other models.

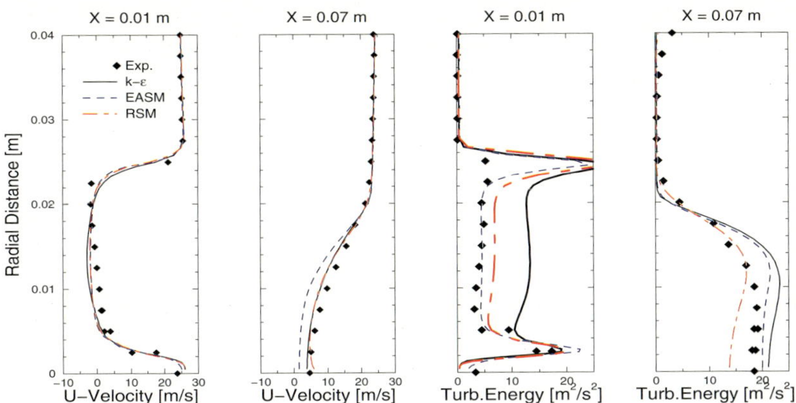

Figure 10 Radial profiles of mean axial velocity and turbulent kinetic energy for the inert case at the stations x = 0.01, 0.07 m.

For reacting flow, the results using the fast combustion model with presumed PDF are shown in Figs. 11-12. Looking at the centerline profiles of the mean axial velocity in the flame, the flow inversion in the inert case is replaced by a mere flow stagnation caused by the drop of density due to the temperature increase. All models are able to

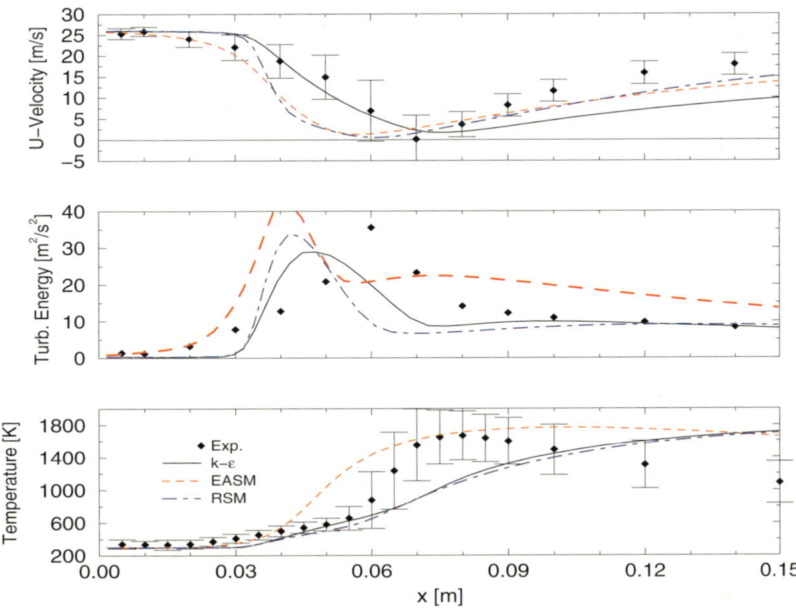

Figure 11 Axial centerline profiles of mean axial velocity, turbulent kinetic energy and mean temperature for the reacting case using fast combustion model.

157

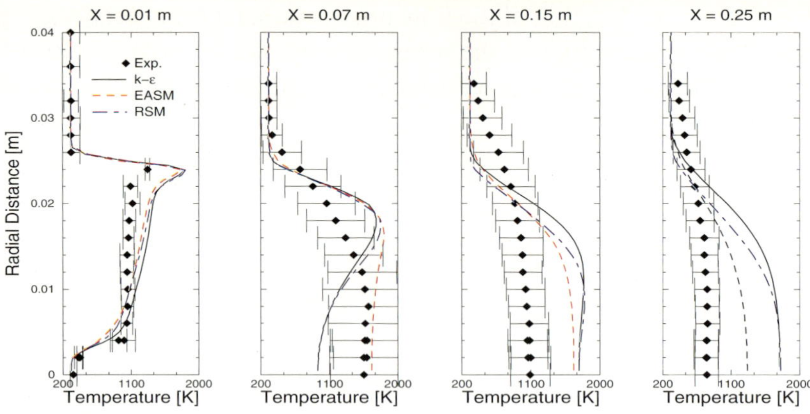

Figure 12 Radial profiles of mean temperature for the reacting case using fast combustion model at stations x = 0.01, 0.07, 0.15, 0.25 m.

predict this trend at least qualitatively. Detailed analysis indicates that the standard $k - \epsilon$ model reproduces better the experimentally observed velocity profiles before the stagnation point, but downstream from this point the velocity profile is clearly underestimated. Concerning the axial temperature behaviour and the radial temperature distributions at various stations, standard $k - \epsilon$ model and RSM fail to predict the temperature behaviour.

Fig. 13 depicts the centerline profiles of the mean velocity, the turbulent kinetic energy and the temperature using the flamelet model. Concerning the velocity profiles, all models display a similar behaviour. Downstream of the stagnation point, the RSM

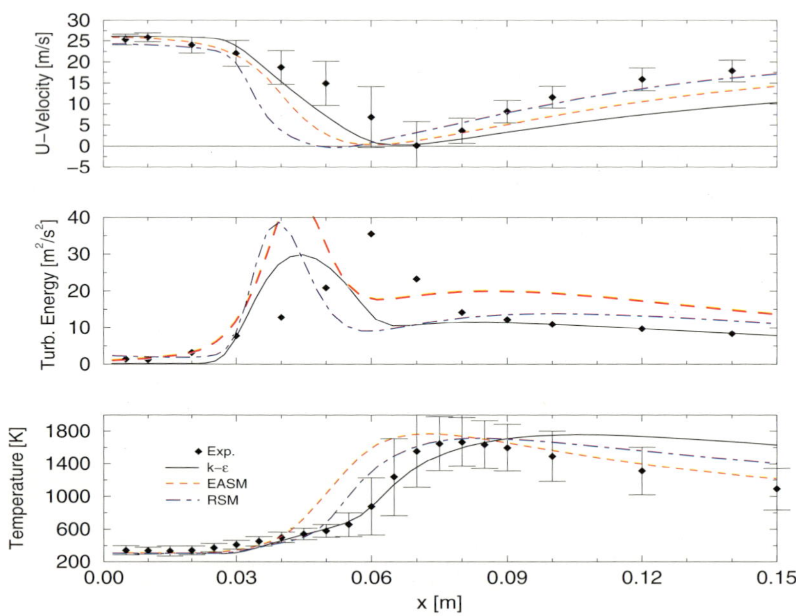

Figure 13 Axial centerline profiles of mean axial velocity, turbulent kinetic energy and mean temperature for the reacting case using the flamelet combustion model.

158

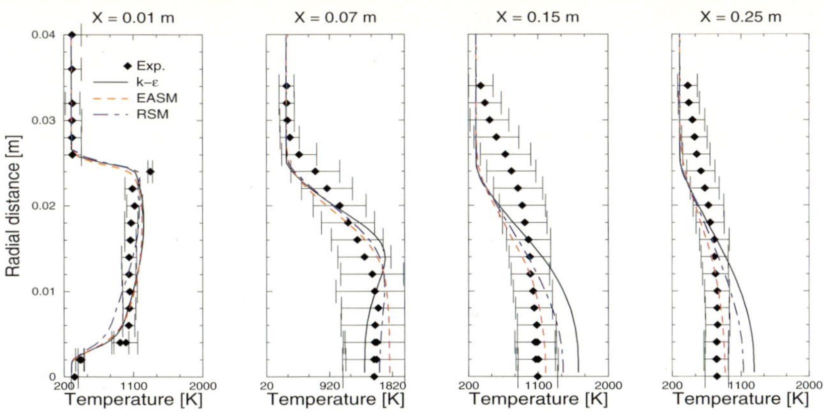

Figure 14 Radial profiles of mean temperature for the reacting case using flamelet combustion model at stations x = 0.01, 0.07, 0.15, 0.25 m.

reproduces better the experimentally observed velocity profiles. EASM produces more turbulent kinetic energy than the standard $k - \epsilon$ model and RSM, however, none of models can predict the peak position of turbulent kinetic energy. Fig. 14 shows the radial distribution of the temperature at various stations. The flamelet model predicts the temperature profiles better than the fast chemistry model, in conjunction with both RSM and standard $k - \epsilon$ model. However, both models fail to capture the trailing temperature decrease, too. EASM, in contrast, captures the trailing temperature decrease.

7   Conclusion

The inert flow inside the ORACLES configuration has been extensively studied by LES and RANS using different turbulence models. All computations predict the asymmetry. Compared to the other turbulence models, RSM produces the best results concerning the turbulent kinetic energy, but shows poor behaviour in the wall region. The investigations of the methane combustion indicate the sensitivity of turbulence models applied. It has been found that EASM has a better performance in the prediction of diffusion flames using both fast chemistry and flamelet models. The standard $k - \epsilon$ model and RSM show better behaviour in connection with flamelet model than with the fast chemistry model. In the future, the improved low-Reynolds-number high-order turbulence models will be adopted to study these configurations.

Acknowledgment

The authors would like to thank the CNRS and the DFG for grants under which the project "Simulation of the Stabilization Process in Flames" is supported, embedded in the French-German project "Numerische Strömungssimulation – Simulation Numérique d'Ecoulements".

References

[1] D. Abbot and S. Kline. "Experimental investigation of a subsonic turbulent flow over single and double backward-facing steps". In J. Basic Eng., ASME, series D, Volume 84, 1962, pp. 317-325.

[2] R. Abid, C. Rumsey and T. Gatski. "Prediction of nonequlibrium turbulent flows with explicit algebraic stress models". In AIAA Journal, Volume 33, 1995, pp.2026-2031.

[3] M. Besson, P. Bruel J. Champion and B. Deshaies. "Inert and combustion flows developing over a plane symmetric expansion: experimental analysis of the main flow characteristics". In 37 AIAA Aerospace Science meeting and Exhibit, 99-0412, January, 1999.

[4] H. C. Chen and C. Poatel. "Near-wall turbulence models for complex flows including separation". In AIAA Journal, Volume 26, 1999, pp. 641-649.

[5] T. Craft, B. E. Launder and K. A. Suga. "A non-linear eddy-viscosity model including sensitivity to stress anisotropy". In Proc. 10th Symp. on Turbulent Shear Flows, Pennsylvania State University, 1995, 23-19.

[6] A. Dervieux. "Steady Euler simulations using unstructured meshes". VKI, Lectures series, 1985, 1884-04.

[7] L. Duchamp de Lageneste. "Simulation des grandes échelles d'écoulements compressibles et reactifs sur maillages non-structurés". PhD thesis, Ecole Centrale de Lyon, 1999.

[8] D. Garreton and O. Simonin Eds. First Workshop on Aerodynamics of Steady State Combustion Chambers and Furnace. Chatou, France, 1994.

[9] Y. Guo, S. Fu, T. Rung and F. Thiele. "Improvement and application of low-Reynolds-number high-order turbulence models". In Proc. 8th Int. Symposium on Computational Fluid Dynamics, Bremen, Germany, Sept. 5-10, 1999.

[10] B. E. Launder, G. Reece and W. Rodi. "Progress in the development of a Reynolds stress turbulence closure". In J. Fluid Mech., Volume 68, 1975, pp. 537-566.

[11] F. S. Lien, W. L. Chen and M. A. Leschziner. "Low-Reynolds number eddy-viscosity modelling based on non-linear stress-strain/vorticity relations". In Engineering Turbulence Modelling and Eperiments 3, Flsevier, 1996, pp.91-100.

[12] A. Page and M. Buffat. A parallel unstructured solver for 3d turbulent reacting flows. In Proc. Int. Conf. on Numerical Methods for Fluids. Swansee, England, 1997.

[13] N. Peters. Four Lectures on Turbulent Combustion, ERCOFTAC Summer School, Sept. 15-19, 1997, Aachen, Germany.

[14] P. Roe. "A numerical method for solving the equations of compressible viscous flows". In AIAA Journal, Volume 20, 1982, pp. 1275-1281.

[15] T. Rung, H. Lübcke, M. Franke, L. Xue, F. Thiele and S. Fu. "Assessment of Explicit Algebraic Stress Models in Transonic Flows", In Proc. 4th Int. Symposium on engineering Turbulence Modelling and Measurements, Corsica, France, 1999, pp. 659-668

[16] T. Rung, F. Thiele and S. Fu. "On the realisability of non-linear stress-strain relationships for Reynolds-stress closures" In J. Flow Turbulence and Combustion, Volume 60, 1999, pp. 333-359.

[17] C. G. Speziale and S. Sarkar and T. B. Gatski. "Modelling the pressure–strain correlation of turbulence: an invariant dynamical systems approach". In J. Fluid Mech., Volume 227, 1991, pp. 245-272.

[18] J. Warnatz and U. Mass. Technische Verbrennung. Springer-Verlag, 1993.

Modeling partially premixed turbulent combustion

M. Herrmann[+], M. Chen[+], B. Binninger[+], N. Peters[+],
V. Favier*, J. Réveillon* and L. Vervisch*

[+]Institut für Technische Mechanik, RWTH-Aachen
Templergraben 64, 52056 Aachen, Germany.

*INSA and Université de Rouen, UMR-6614-CNRS-CORIA
Avenue de l'Université, B.P. 8
76801 Saint Etienne du Rouvray Cdx, France.

Summary

This paper investigates combustion in the partially premixed regime that may be observed in non-premixed systems. Direct numerical simulation (DNS) is first utilized to further investigate the properties of diffusion flames at quenching along with the development of edge flames. DNS of spray flames are also performed and reveal the crucial role played by partially premixed combustion when liquid fuel injection is used. The ultimate product of this work is a flamelet model for partially premixed turbulent combustion that combines the flamelet models for non-premixed and premixed combustion. In addition, a new model for the turbulent burning velocity in partially premixed flows is proposed. The model is used to simulate the stabilization process of turbulent methane/air and propane/air jet diffusion flames.

1   Introduction

Many combustion systems operate with fuel and oxidizer not perfectly premixed before entering the combustion chamber. This is particularly the case in devices where liquid injection of the fuel is retained leading to nonpremixed or to partially premixed flames (Diesel and aircraft engines, furnaces, gas turbines...). New technologies for these nonpremixed systems imply the precise determination of the position in the turbulent flow where combustion starts, incorporating liquid fuel injection and the accurate control of pollutants emission. The development of numerical models for nonpremixed turbulent flames is therefore an important issue of combustion.

Partial premixing is observed in these nonpremixed flames when fuel and oxidizer have mixed without burning. This mixing can occur before ignition or after local quenching, sometimes the reactants also mix with products to insure stabilization, as in bluff-body type burners. In most of the devices where a spray of liquid fuel is injected, partially premixed flame propagation will be observed, as in a gasoline direct injection engine or in aircraft engines. Even in premixed systems, the premixing of the reactants is not always perfect and some partial premixing may be observed. Our objective is to investigate and model partially premixed combustion in nonpremixed systems.

First, direct numerical simulation (DNS) is used to understand the properties of diffusion flames submitted to partial premixing. The cases of diffusion flame quenching and edge flames are discussed along with DNS of spray flames, where partial premixing is found to play a crucial role. Then, a new combustion model combining a level set method ($G-$equation) with flamelet modeling is proposed and applied to the calculations of partially premixed combustion observed in turbulent lifted flames.

2 Edge flames in quenching of trailing diffusion flames

Finite rate chemistry and quenching effects in diffusion flames have been the subject of multiple studies motivated by the need to understand intermediate combustion regimes observed during transition from fast chemistry to fast micromixing and quenching of the reaction zones [1]. Local flow conditions promoting quenching of diffusive-reactive layers prohibit stabilization or development of combustion. Partial extinction also modifies pollutants formation. The development of new strategies for Reynolds Averaged Simulation (RANS) or Large Eddy Simulation (LES) of turbulent combustion implies a precise knowledge of both quenching mechanisms and flame structure bordering quenched locations [2, 3].

¿From numerical simulations of diffusion flame quenching, it is shown that using a scalar dissipation rate one may distinguish between conditions leading to transition from burning to quenching, and, conditions observed at a quenched location featuring different flame properties, related to partially premixed combustion.

Diffusion flames are usually parameterized using conserved scalars as the mixture fraction Z ($Z = 0$ in pure oxidizer and $Z = 1$ in pure fuel). Fundamentals of diffusion flame reveals that regions where the chemical reaction is frozen (flame quenching) appear when the local Damköhler number (Da) falls below a critical value Da_q. The number Da is defined as the ratio of a diffusive time τ_χ to a chemical time [4]. τ_χ may be estimated from the scalar dissipation rate $\chi = D|\nabla Z|^2$ as $\tau_\chi \approx \chi^{-1}$, where D is a diffusion coefficient. Solutions of planar flame problems for a variety of χ, called flamelet library, provide a quenching scalar dissipation rate χ_q^*, a critical value widely used in turbulent combustion modeling [1].

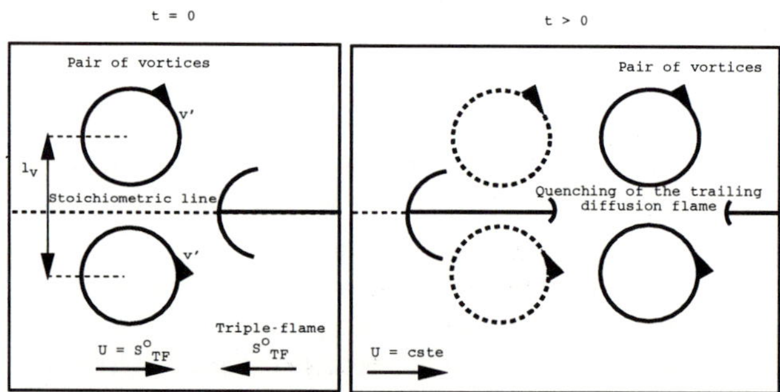

Figure 1 Schematic of the simulations of two-dimensional diffusion flame quenching.

To study flamelet quenching without curvature of the stoichiometric line, we have further analyzed a Direct Numerical Simulation (DNS) database initially developed to focus on triple flame / vortices interaction [5]. For strong vortices, the trailing diffusion flame is submitted high

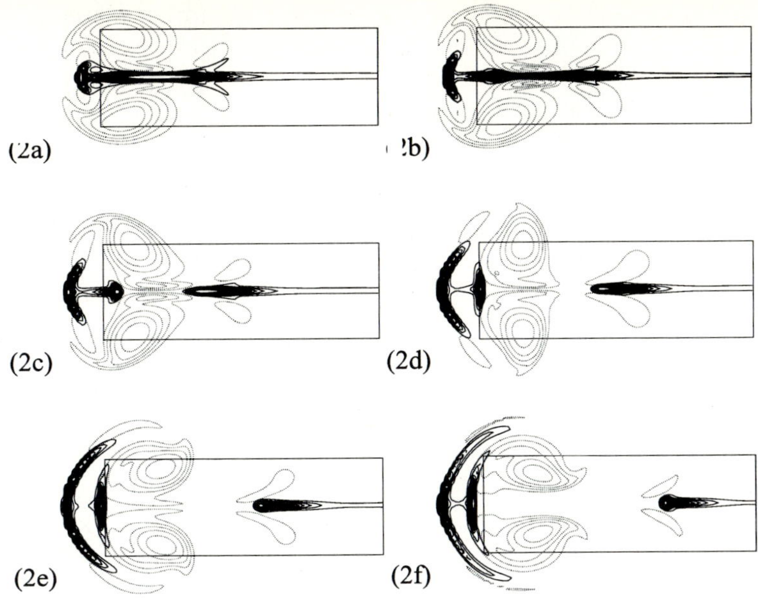

Figure 2 Isolines of burning rate (line) and vorticity (dotted line) for six successive times: squeezed diffusion flame (cases 2a,b) and edge flame (cases 2c,d,e,f). The frame denotes the domain $\mathcal{D}$.

levels of strain and scalar dissipation rates leading to quenching of the two-dimensional planar diffusion flamelet (Fig. 1). We focus on this part of the simulation where quenching appears. The diffusion flame is initially stabilized within a two-dimensional domain following a procedure proposed in [6].

A laminar flamelet library has been constructed using the same chemistry and transport than in the full Navier Stokes simulation. The reference quenching scalar dissipation rate χ_q^*, its corresponding quenching length $\delta_q^* = \sqrt{D/\chi_q^*}$ and the scalar dissipation rate in the unperturbed trailing diffusion flame χ_{Diff} may be used as control parameters. The vortices are characterized by their radius R, velocity v' and the length l_v (Fig. 1). The reported simulations have been performed using a representative condition where diffusion flame quenching is found:

$$\frac{\chi_q^*}{\chi_{Diff}} \approx 100 \quad , \quad \frac{l_v}{\delta_q^*} \approx 12 \quad , \quad \frac{R}{\delta_q^*} \approx 5 \quad , \quad \frac{v'}{\chi_q^* \delta_q^*} \approx 40 \tag{2.1}$$

Varying these numbers changes the streamwise position of quenching, but does not fundamentally modify the presented results.

The studied sequence may be decomposed in three stages. The planar diffusion flame is first squeezed by the vortices (Fig. 2a-b), then quenching occurs and edge flames are formed (Fig. 2c-f). Latter, premixed kernels develop at the reaction zone extremities (Fig. 2f).

After quenching, the overall flame is organized in three main parts (Fig. 3a). The partially premixed front of the initial triple flame has been transported upstream, its response in mixture

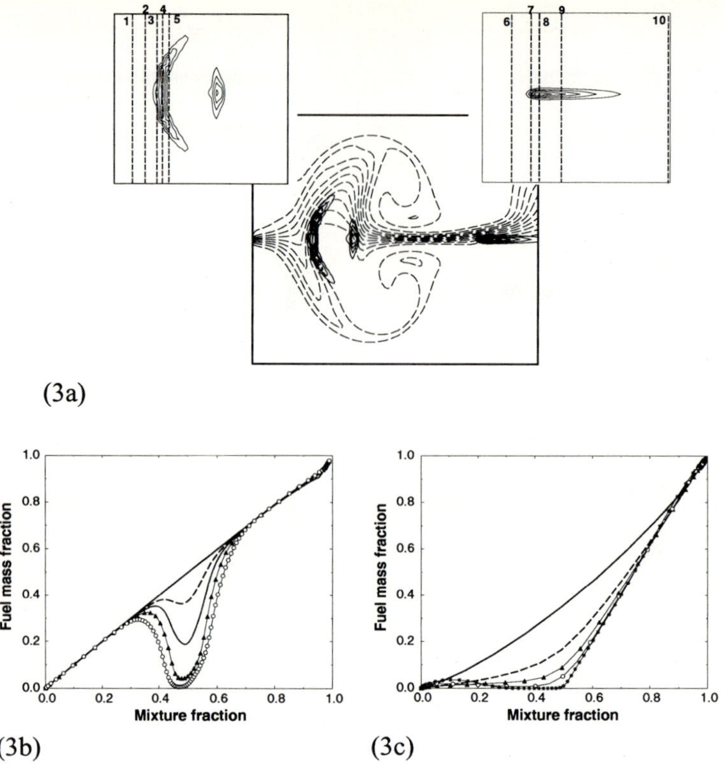

(3a)

(3b)   (3c)

Figure 3 (3a): Isolines of burning rate (lines) and fuel mass fraction (dashed lines) corresponding to the case (2d). (3b): Profiles of fuel mass fraction versus mixture fraction for the triple flame. At position 1 in Fig. 3a: line , 2: dashed line, 3: dotted line, 4: triangles, 5: circles. (3c): Profiles of fuel mass fraction versus mixture fraction for the edge-flame. At position 6 in Fig. 3a: line, 7: dashed line, 8: triangles, 9: circles, 10: stars

fraction space partly evolves on mixing lines ($Y_F(Z) = Y_{F,o}Z$ and $Y_O(Z) = Y_{O,o}(1 - Z)$) before combustion in a partially premixed regime (Fig. 3b). $Y_{F,o}$ and $Y_{O,o}$ are free stream mass fractions of fuel and oxidizer respectively. Two reaction zone extremities result from the quenching of the trailing diffusion flame, one attached to the main body of the initial triple flame (on the left), the other (on the right) features an edge flame [7] followed by the diffusion flame. For these edge flames the flame structure in mixture fraction space reveals combustion in a diffusive regime (Fig. 3c). In what follows, the scope of the study is restricted to a zone $\mathcal{D}$ of the computational domain where only edge flame combustion is concerned. Thus, the partially premixed front of the initial triple flame is not considered. For the simulated case, the vortices do not carry any hot gases and products remaining from their interaction with the partially premixed front (see Fig. 3c). The triple flame is only used to stabilize a two-dimensional planar steady diffusion flame, subsequently submitted to an unsteady increase of χ, free stream conditions being fuel and air.

The time evolution of averaged quantities, $\overline{Q}$, obtained from integration over the domain $\mathcal{D}$ are first discussed. $\overline{Q}$ is computed as:

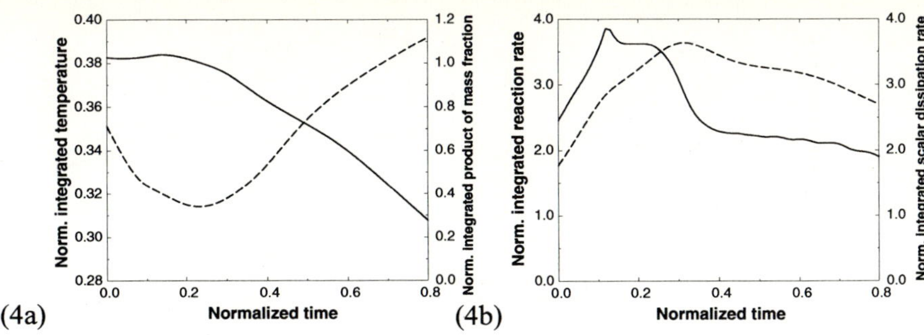

(4a)   (4b)

Figure 4 Time evolution of averaged temperature $\overline{T}$ (4a, line) and product of mass fractions $\overline{Y_F Y_O}$ (4a, dashed line), burning rate $\overline{\dot{\omega}_T}$ (4b, line) and scalar dissipation rate $\overline{\chi}$ (4b, dashed line). Averaging is defined by Eq. (2.2).

$$\overline{Q}(t) = \frac{\int_{\mathcal{D}} Q(x,y,t)\, dx\, dy}{L \int_{\partial \mathcal{D}} Q(\infty,y,t = t_o)\, dy} \tag{2.2}$$

Properties in the trailing diffusion flame were chosen to normalize $\overline{Q}(t)$ over the domain $\mathcal{D}$ of streamwise length L.

When the flame starts to be pinched, the vorticity field modifies the mixture and brings up the amount of heat released. The averaged temperature $\overline{T}$ increases to reach a plateau (Fig. 4a). The increase of the averaged scalar dissipation rate $\overline{\chi}$ is accompanied by an increase of the mean burning rate $\overline{\dot{\omega}_T}$ (Fig. 4b). However, when $\overline{\chi}$ reaches its maximum, $\overline{\dot{\omega}_T}$ suddenly drops, indicating local quenching of the diffusion flame. ¿From this instant in time a hole exists in the flame. $\overline{\chi}$ decays slowly and $\overline{\dot{\omega}_T}$ is almost constant, but combustion is not organized in a pure diffusive regime. This constant burning rate goes with a growth of the product of the fuel and oxidizer mass fractions $\overline{Y_F Y_O}$, indicating partial premixing of the reactants at the quenched point (Fig. 4a). One therefore observes the development of combustion in a partially premixed regime, at the zones bordering the quenched locations, with a burning rate larger than the one of pure diffusion combustion.

To further analyze flame quenching, the responses of the burning rate and of the temperature are collected along the stoichiometric line and plotted versus the inverse of the scalar dissipation rate. Four successive times are considered on Fig. (5), the results are also compared with the reference one-dimensional flamelet library. Despite the existence of effects of unsteadiness, before quenching and during the first phase of the creation of the hole in the flame, an interesting agreement is observed between the flamelet and the real flame response (Fig. 5a-c). The maximum burning rate increases with χ till the quenching point is reached, i.e. for $\chi \approx \chi_q^*$ the flamelet quenching scalar dissipation rate. Once a reaction zone extremity exists, the edge flame cannot sustain values of χ greater than $\chi_{qEd} = \chi_q^* - \Delta\chi$ leading to a scalar dissipation rate at the reaction zone extremity χ_{qEd} smaller than χ_q^* (Fig. 5b-d). In the trailing burning zone, however, $\dot{\omega}_T(\chi)$ perfectly follows the flamelet behavior. The overall flamelet response being shifted toward lower values of χ.

These results suggest that one should distinguish between the amount of scalar dissipation rate necessary to quench a diffusion flame χ_q, and, χ_{qEd}, the value measured at the extremity of a reaction zone that is geometrically defined as a quenching location. In our simulations, the level

165

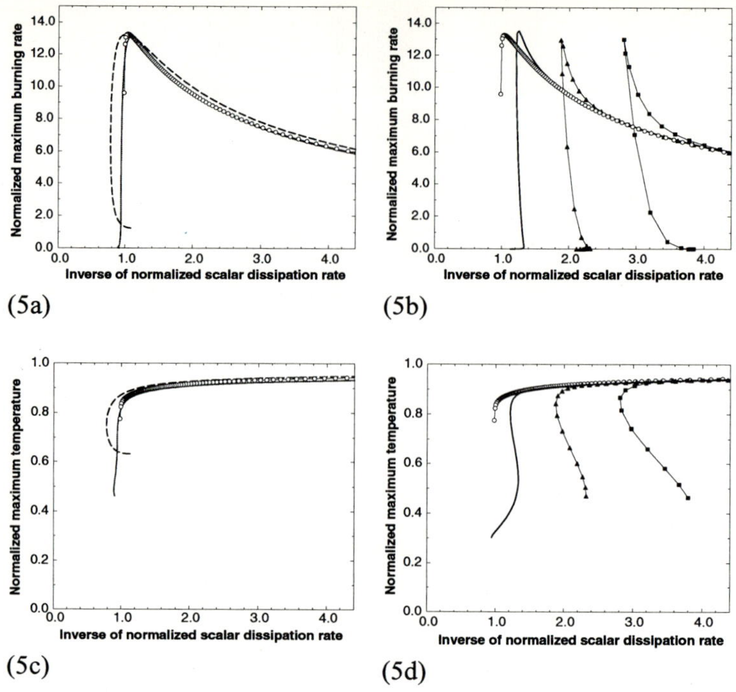

Figure 5 DNS and one-dimensional flamelet theory (circles). For succesive times, comparison of the normalized maximum burning rate (5a,b) and temperature (5c,d) responses versus the inverse of normalized scalar dissipation rate. The one-dimensional strained diffusion flame (cases 5a,c) and two-dimensional edge-flame (premixing, cases 5b,d) behaviors are related to the corresponding DNS times : (2b) dashed line, (2c) dotted line, (2d) line, (2e) triangles and (2f) squares. The burning rate and the temperature are made nondimensional by their values in the initial trailing diffusion flame and the scalar dissipation rate by the reference quenching scalar dissipation rate χ_q^*.

of χ that should be applied to transition from burning to quenching is found to be of the order of χ_q^* predicted by flamelet theory. In particular when $\chi > \chi_q^*$, quenching is always observed. However, once a hole exists in the reaction zone, due to multi-dimensional fluxes of heat and species, χ_{qEd}, the level of scalar dissipation rate necessary to maintain quenching, is always lower than χ_q^*.

Except when unsteadiness was explicitly quantified using oscillating one-di-mensional flamelets, in DNS or in experiments of nonpremixed turbulent flames, the exact distinction between χ_q and χ_{qEd}, the two scalar dissipation rate "at quenching", is non-trivial, but is needed to carefully described strong finite rate chemistry effects.

3   Partially premixed combustion in spray flames

Many industrial devices dedicated to propulsion systems or energy transformation involve the injection of liquid fuel with a gaseous oxidizer to feed the combustion chamber. To develop accurate numerical models of turbulent combustion in such devices, it is important to carefully understand the physical phenomena controlling the interactions between spray turbulent mixing and combustion. To this end, direct numerical simulations (DNS) of a turbulent spray flame are performed. The gaseous phase is captured in an Eulerian context in association with a Lagrangian solver for dispersed vaporizing droplets.

Previous studies have described vaporizing turbulent spray in decaying [8, 9] or forced [10, 11] homogeneous turbulence. We report DNS results of a burning jet surrounded by a forced preheated coflow of air. For this configuration, it is shown that turbulent combustion mainly occurs in a partially premixed regime.

3.1   Numerical procedure

The continuous phase is a fully compressible Newtonian fluid following the equation of state for perfect gases. DNS allows us to solve exactly all the scales of the flow from the Kolmogorov up to the integral scale.

The modeling of the liquid phase [12] includes several simplifications. The droplets are assumed to be spherical without any motion in the liquid core and droplet-droplet interaction are neglected. Time and space dependent vaporization rate of droplets is computed following local properties of the gas phase (temperature, pressure, gaseous fuel mass fraction). Local saturation properties are determined through the Clausius-Clapeyron relation. A modified convective Sherwood number accounting for convection of the gas around droplets is retained.

Spatial derivatives of Eulerian equations are determined by a 6th order Pade scheme [13]. Time advancement is computed with a third order Runge Kutta method used with a minimal data storage algorithm [14]. Figure 6 shows the computational configuration and the injection profile, Fig. 7 is a snapshot of the mixture fraction levels at a given time.

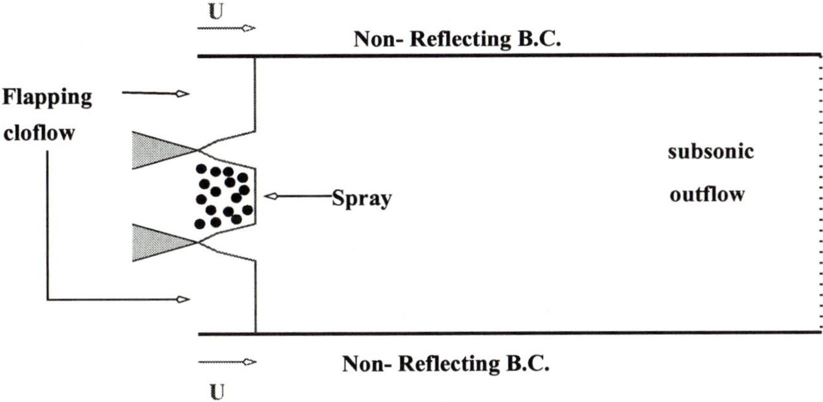

Figure 6   Computational domain.

3.2   Flame structure at stabilization

Figure 8 is a snapshot of the flame at a representative time. The stoichiometric ratio between oxidizer and fuel corresponds to n-heptane.

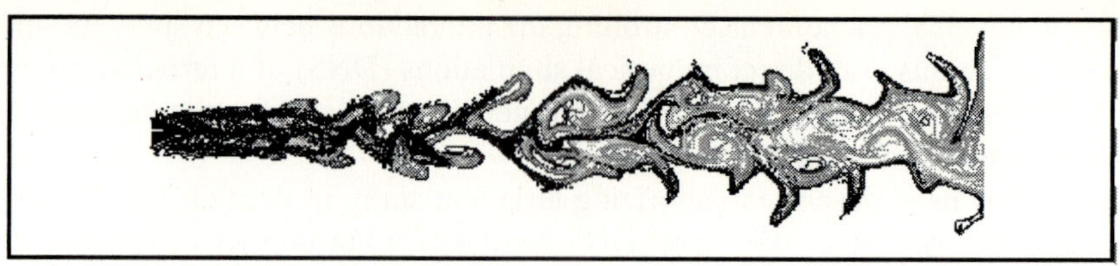

Figure 7 Snapshot of the fuel mass fraction (isolines) and the vaporizing particles (dots), pure mixing case.

Combustion starts with triple points surrounding the spray. Three flames emerge from each triple point: A rapidly vanishing lean premixed flame on the oxidizer side, a trailing diffusion flame and a rich premixed flame, both in the continuity of the triple point. Combustion is therefore organized in an hybrid regime. The large mean vaporization time of the droplets implies the presence of the partially premixed (rich) flame along the whole droplets trajectories (Fig. 8).

Figure 9 presents the flame at two successive times. The droplets are also shown, some are crossing the computational domain without being totally vaporized. A variety of flame topologies is observed, pockets of burning droplets behind the diffusion flame (A') or multiple interaction between premixed and diffusion flames (B'). Local extinctions may be detected as well. These DNS illustrate the need for numerical models capturing both diffusion and premixed combustion. Such a turbulent combustion model is now discussed.

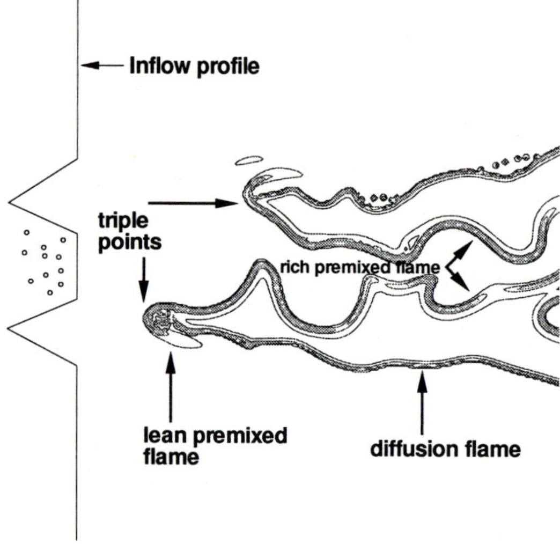

Figure 8 Structure of the turbulent flame attached to the spray.

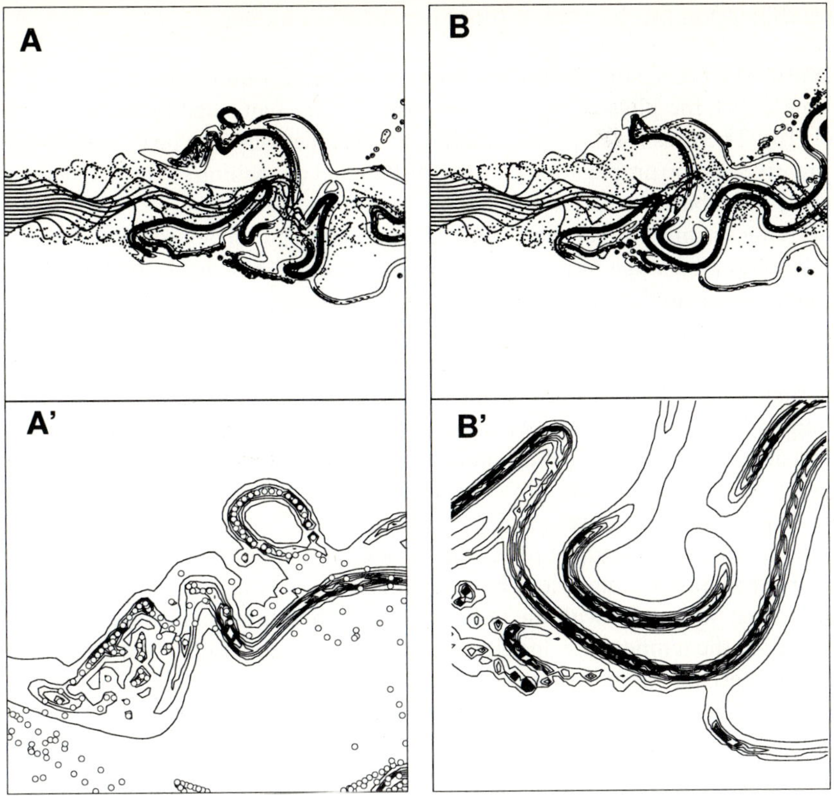

Figure 9 Two snapshots (A and B) of the energy heat release (isolines) and the vaporizing droplets (dots). Local zooms show a burning ring of droplets (A') and flame complex interactions and extinctions (B'). (Droplets size is not representative of the computation.)

4 Modeling of partially premixed turbulent combustion

We will present a flamelet model for partially premixed turbulent combustion that combines the flamelet models for non-premixed and premixed combustion. In addition, a new model for the turbulent burning velocity in partially premixed flows is proposed. It is based on a formulation for a conditional turbulent burning velocity, which depends on mixture fraction. The effect of partially premixing is taken into account by using the presumed pdf approach in terms of the mixture fraction. Mean scalar quantities on both sides of the premixed flame front are calculated using presumed pdfs on flamelet libraries. From a computational point of view, the model has the advantage that the calculation of the chemical processes can be decoupled from the flow calculation, allowing for simulations of realistic configurations, yet retaining detailed chemistry. The model is used to simulate the stabilization process of turbulent methane/air and propane/air jet diffusion flames. The calculated lift-off heights compare favorably with experimental data from various authors.

4.1 The stabilization mechanism of lifted jet diffusion flames

Research on lifted jet diffusion flames has been conducted for more than 50 years [15]. Despite this long time effort, the physical mechanisms of turbulent flame stabilization are still not well understood [16]. Theories for the flame stabilization mechanism may be divided into three categories: a) premixed flame propagation [17, 18], b) flamelet quenching [19] and c) flame extinction due to large-scale turbulent structures [20].

The underlying assumption for the premixed flame propagation approach is that fuel and oxidizer are fully premixed at the base of a lifted diffusion flame and that stabilization occurs at the position, where the mean flow velocity at the contour of mean stoichiometric mixture is equal to the burning velocity of a stoichiometric premixed turbulent flame [17, 18]. In contrast, Peters and Williams [19] proposed that diffusion flamelet extinction is responsible for flame stabilization. They argue that there is insufficient residence time below the flame base to achieve spatial and temporal uniformity of the mixture. Although there is little doubt that diffusion flame quenching is responsible for the lift-off of an initially attached flame, detailed experimental analysis conducted over the last fifteen years do not confirm the flamelet quenching hypothesis for flame stabilization [1]. Finally, Broadwell et al. [20] propose that hot combustion products are carried by large-scale turbulent structures to the edge of the jet, where they re-enter the jet and ignite the combustible mixture. In their view, both lift-off and blow-out occurs, when the re-entrained products are mixed so rapidly with the unburnt jet fluid that there is insufficient time to initiate the reaction before the temperature and the radical concentration drop below some critical value. In his review of these different approaches, Pitts [16] comes to the conclusion that none of these theories can satisfactorily predict lift-off and blow-out behaviour.

In recent years, triple flames have attracted much interest, because it is believed that they may play a crucial role in many partially premixed combustion situations including the stabilization mechanisms of turbulent jet flames. Liñán [21] and Kioni et al. [22] have shown theoretically that in laminar flows lifted flames are stabilized by a triple flame configuration. Veynante et al. [23] and Favier and Vervisch [5] have demonstrated that triple flames are able to survive strong interactions with vortices by adjusting their structure to a new transient environment thus being more robust than pure diffusion flames.

Here, we propose a flamelet model for partially premixed turbulent combustion, that is based on the premixed flame propagation mechanism, but which will take the triple flame structure as a key element of the partially premixed situation into account. Flamelet models [24, 25, 26, 27] have been very useful in combining turbulence and non-equilibrium chemistry. The advantage of the flamelet concept is that the calculation of the chemistry can be separated from the calculation of the turbulent flow field.

4.2 The flamelet model for partially premixed turbulent combustion

At the base of the lifted turbulent diffusion flame fuel and oxidizer are partially premixed. The instantaneous surface of stoichiometric mixture separates lean and rich regions. When a flame propagates through the inhomogeneous fluctuating mixture of fuel and oxidizer an instantaneous flame front separates burnt and unburnt gases. Thus a formulation for both premixed and non-premixed combustion has to be used. For this purpose, the flamelet model of non-premixed combustion [28] will be combined with the flamelet model for premixed combustion [29].

The mixing of fuel and oxidizer in the turbulent flow field is described by the transport equations of the mean mixture fraction $\tilde{Z}$ and the variance $\widetilde{Z''^2}$

$$\frac{\partial \left(\bar{\rho}\tilde{Z}\right)}{\partial t} + \nabla \cdot \left(\bar{\rho}\tilde{\mathbf{v}}\tilde{Z}\right) = \nabla \cdot \left[\frac{\mu_t}{\mathrm{Sc}_{\tilde{Z}}}\nabla\tilde{Z}\right] \tag{4.3}$$

$$\frac{\partial \left(\bar{\rho}\widetilde{Z''^2}\right)}{\partial t} + \nabla \cdot \left(\bar{\rho}\tilde{\mathbf{v}}\widetilde{Z''^2}\right) = \nabla \cdot \left[\frac{\mu_t}{\mathrm{Sc}_{\widetilde{Z''^2}}}\nabla\widetilde{Z''^2}\right] + \frac{2\mu_t}{\mathrm{Sc}_{\widetilde{Z''^2}}}\left(\nabla\tilde{Z}\right)^2 - \bar{\rho}\tilde{\chi}\,, \tag{4.4}$$

where the Schmidt numbers $\mathrm{Sc}_{\tilde{Z}}$ and $\mathrm{Sc}_{\widetilde{Z''^2}}$ are chosen as 0.7 and the scalar dissipation rate $\tilde{\chi}$ is modeled as

$$\tilde{\chi} = c_\chi \frac{\tilde{\varepsilon}}{\tilde{k}}\widetilde{Z''^2}\,, \qquad c_\chi = 2.0\,. \tag{4.5}$$

Here $\tilde{k}$ is the turbulent kinetic energy and $\tilde{\varepsilon}$ its dissipation rate.

In order to describe premixed combustion, the level set approach [30] based on the G-equation is introduced. The scalar G is equal to the constant G_0 at the location of the instantaneous premixed flame front. Thus the surface $G(\mathbf{x},t) = G_0$ divides the flow field into the regions of burnt gas, where $G(\mathbf{x},t) > G_0$, and unburnt gas, where $G(\mathbf{x},t) < G_0$. The equation for the mean location of the turbulent flame front then reads [29]

$$\frac{\partial \left(\bar{\rho}\widetilde{G}\right)}{\partial t} + \nabla \cdot \left(\bar{\rho}\tilde{\mathbf{v}}\widetilde{G}\right) = \bar{\rho}s_{T,p}\left|\nabla\widetilde{G}\right| - \bar{\rho}D_t\tilde{\kappa}\left|\nabla\widetilde{G}\right|\,, \tag{4.6}$$

where $\tilde{\kappa}$ is the curvature of the mean flame front and D_t the turbulent diffusivity, which can be determined from the integral length scale ℓ and the fluctuation velocity v'

$$D_t = a_4\ell v'\,, \qquad a_4 = 0.78\,. \tag{4.7}$$

In addition, the turbulent flame brush thickness $l_{F,t}$ can be determined from the variance of G by the simple relation

$$l_{F,t} = \left.\frac{(\widetilde{G''^2})^{1/2}}{\left|\nabla\widetilde{G}\right|}\right|_{\tilde{G}=G_0}\,, \tag{4.8}$$

evaluated at the location of the mean premixed flame front $\tilde{G} = G_0$. The equation for the variance of G is [29]

$$\frac{\partial \left(\bar{\rho}\widetilde{G''^2}\right)}{\partial t} + \nabla \cdot \left(\bar{\rho}\tilde{\mathbf{v}}\widetilde{G''^2}\right) = \nabla_\parallel \cdot \left(\bar{\rho}D_t\nabla_\parallel\widetilde{G''^2}\right) + 2\bar{\rho}D_t\left(\nabla\tilde{G}\right)^2 - c_s\bar{\rho}\frac{\tilde{\varepsilon}}{\tilde{k}}\widetilde{G''^2}\,, \tag{4.9}$$

where $\nabla_\parallel$ denotes differentiation only tangential to the mean flame front. Using eq. (4.8) and (4.9) it is shown in [1] that for large times the turbulent flame brush thickness $l_{F,t}$ of a one-dimensional unsteady flame is proportional to the integral length scale

$$l_{F,t} = b_2\ell\,, \qquad b_2 = 1.78\,. \tag{4.10}$$

What remains is the determination of the turbulent partially premixed burning velocity $s_{T,p}$ in eq. (4.6). In order to model this quantity, we follow in essence the assumption that fuel and oxidizer are locally premixed, such that the partially premixed flame propagates through a stratified, though locally premixed environment. For premixed turbulent combustion the turbulent burning velocity s_T can be determined from [29]

$$\frac{s_T - s_L}{v'} = -\frac{a_4 b_3^2}{2b_1}\text{Da} + \left[\left(\frac{a_4 b_3^2}{2b_1}\text{Da}\right)^2 + a_4 b_3^2 \text{Da}\right]^{1/2} , \qquad (4.11)$$

where s_L is the laminar burning velocity of a plane flame, $\text{Da} = s_L \ell/(v' l_F)$ is the Damköhler number and ℓ and l_F are the integral length scale and the laminar flame thickness, v' is the turbulence intensity and $a_4 = 0.78$, $b_1 = 2.0$ and $b_3 = 1.0$ are constants derived from turbulence modeling.

Let us, for illustration purposes, consider a stationary laminar triple flame in a constant velocity field. The leading edge of such a flame, called the triple point, propagates along a surface that is in the vicinity of the stoichiometric mixture. On the lean side of that surface there is a lean premixed branch and on the rich side there is a rich premixed branch, both propagating with a lower burning velocity. Behind the triple point a diffusion flame develops into which unburnt intermediates like H_2 and CO diffuse from the rich premixed flame branch and the left-over oxygen diffuses from the lean premixed flame branch. The premixed branches are inclined in such a way that, while the normal burning velocity decreases as one moves downstream on the lean and on the rich branch, its projection onto the oncoming flow direction has to be equal to the oncoming flow velocity. This indicates, that each part of a triple flame, parameterized in terms of the mixture fraction, contributes to the propagation velocity of the whole structure in a similar way. Therefore they can be considered separately. A conditional turbulent Damköhler number $\text{Da}(Z)$ can then be introduced into eq. (4.11) to determine the conditional burning velocity $s_T(Z)$ as

$$s_T(Z) = s_L(Z) + v' f\left\{\text{Da}(Z)\right\} , \qquad (4.12)$$

where $f\{\ \}$ represents the right hand side of eq. (4.11) and $\text{Da}(Z)$ is defined as

$$\text{Da}(Z) = \frac{s_L(Z)\ell}{v' l_F(Z)} = \frac{s_L^2(Z)\ell}{v' D} . \qquad (4.13)$$

In the second part of eq. (4.13) the laminar flame thickness $l_F(Z)$ has been replaced by $l_F(Z) = D/s_L(Z)$, where the laminar diffusivity D has been assumed mixture fraction independent. Using a presumed pdf approach, the mean turbulent burning velocity of a partially premixed flame $s_{T,p}$ can then be determined from

$$(\bar{\rho} s_{T,p}) = \int_0^1 \rho(Z) s_T(Z) P(Z) dZ , \qquad (4.14)$$

where $P(Z)$ is chosen to be a beta-pdf. If $s_T(Z)$ is defined with respect to the unburnt mixture, $\rho(Z)$ is to be evaluated there. If it was assumed that turbulent partially premixed flame propagation proceeds by an ensemble of laminar triple flamelets, the laminar burning velocity $s_L(Z)$ should be the velocity normal to the premixed flame surface of a laminar triple flame. In the present paper, however, the laminar burning velocity $s_L(Z)$ is taken as that of a premixed flame in a homogeneous mixture with mixture fraction Z. In Plessing et.al. [31] these two burning velocities were found to be in qualitative good agreement justifying the chosen approach as a first approximation.

The set of equations (4.3), (4.4), (4.6) and (4.10) represents the flamelet model for partially premixed turbulent combustion used in this paper, while eqs. (4.12) and (4.14) model the turbulent partially premixed burning velocity needed in eq. (4.6).

4.3   The Numerical Method

In order to simulate turbulent partially premixed combustion, the flamelet model described above
has been implemented into the FLUENT code [32]. In addition to the conservation equations of
mass and momentum, an equation for the mean total enthalpy $\tilde{h}$ is solved

$$\frac{\partial\left(\bar{\rho}\tilde{h}\right)}{\partial t} + \nabla\cdot\left(\bar{\rho}\tilde{\mathbf{v}}\tilde{h}\right) = \frac{d\bar{p}}{dt} + \nabla\cdot\left(\frac{\bar{\rho}\nu_t}{\mathrm{Pr}_t}\nabla\tilde{h}\right) \;, \tag{4.15}$$

replacing the original energy equation of the FLUENT code. Here Pr_t is the turbulent Prandtl
number, which is chosen as 0.7.

To avoid numerical difficulties, the scalar function $\tilde{G}$ is calculated as a distance function,
meaning that away from the mean flame front, a re-initialization procedure of the $\tilde{G}$ field using
$\left|\nabla\tilde{G}\right| = 1$ has to be performed. This is achieved using an algorithm proposed by Sussman et al.
[33]. The turbulence is described by a standard $\tilde{k}$-$\tilde{\varepsilon}$ model, which includes buoyancy effects and
a round jet correction.

In order to describe the scalar fields, a flame sheet model is adopted as a first approximation.
This model does not resolve the laminar premixed flame structure, but rather replaces it by a
jump. The dependence of the scalar field on the mixture fraction, however, is taken into account
by calculating the diffusion flamelet structure. Thus, there are two possible states for the diffusion
flamelet, either burning for $G > G_0$ or non-burning for $G < G_0$. For the burning flamelets the
mass fractions of the chemical species are determined by using a steady state flamelet library
with the conditional scalar dissipation rate χ_{st} as a parameter. In the burnt gas the mean mass
fractions are calculated using a presumed pdf approach

$$\tilde{Y}_{i,b}\left(\tilde{Z},\widetilde{Z''^2},\tilde{\chi}_{st}\right) = \int_0^1 Y_i(Z,\chi_{st})P(Z)dZ \;. \tag{4.16}$$

Here $Y_i(Z,\chi_{st})$ is determined from a library of burning diffusion flamelets, setting the conditional scalar dissipation rate χ_{st} of the flamelets equal to the conditional mean scalar dissipation
rate $\tilde{\chi}_{st}$, both defined at stoichiometric mixture. The latter can be calculated from

$$\tilde{\chi}_{st} = \frac{\tilde{\chi}f(Z_{st})}{\int_0^1 f(Z)P(Z)dZ} \quad , \quad f(Z) = \exp\left(-2\left[\mathrm{erfc}^{-1}(2Z)\right]^2\right) , \tag{4.17}$$

where the mean value $\tilde{\chi}$ is determined by eq. (4.5). A beta function pdf is used in eq. (4.16) and
(4.17). The boundary conditions for the diffusion flamelets are pure air ($Z = 0$) and pure fuel
($Z = 1$). In the unburnt gas all mass fractions are zero except those of fuel and oxidizer. These
mass fractions, being linear in mixture fraction, are evaluated from

$$\tilde{Y}_{F,u} = Y_{F,1}\tilde{Z} \quad , \quad \tilde{Y}_{Ox,u} = Y_{Ox,2}(1 - \tilde{Z}) . \tag{4.18}$$

Within the turbulent flame brush the average mass fractions are determined from the weighted
sum

$$\tilde{Y}_i = p_b\tilde{Y}_{i,b} + (1 - p_b)\tilde{Y}_{i,u} . \tag{4.19}$$

Here p_b denotes the probability of finding burnt gas

$$p_b = p_b(G > G_0) = \int_{G=G_0}^{\infty} \frac{1}{\sqrt{2\pi\widetilde{G''^2}}} \exp\left(-\frac{(G - \tilde{G})^2}{2\widetilde{G''^2}}\right) dG , \tag{4.20}$$

where a Gaussian distribution is assumed for the pdf of G and $\widetilde{G''^2}$ is determined from eq.(4.8) and (4.10). This approach differs from the one proposed in [27] in that here the turbulent flame brush is numerically resolved whereas in [27] it was treated as a numerical discontinuity requiring an in cell reconstruction scheme using jump conditions over the turbulent flame front. The preferred method depends on the number of grid points that resolve the turbulent flame brush. If it is reasonably well resolved, which should be the case since the integral length scale ℓ (4.10) typically has to be resolved by the turbulence model, than the method described here should be used, otherwise the method presented in [27] should be employed.

The mean temperature $\tilde{T}$ can then be calculated from eq. (4.15) and (4.19) using

$$\sum_{i=1}^{n} \tilde{Y}_i h_i(\tilde{T}) = \tilde{h}, \tag{4.21}$$

where the specific enthalpies h_i are taken from NASA polynomials. Figure 10 summarizes the computational steps in the simulation of partially premixed turbulent combustion using the proposed flamelet approach.

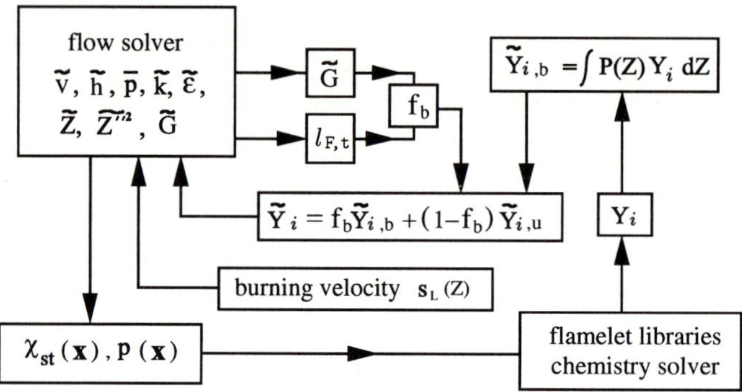

Figure 10 The code structure for the flamelet model of partially premixed turbulent combustion

4.4 Numerical Results

The model presented above will be used to calculate methane/air and propane/air turbulent jet flames for a wide variety of fuel nozzle exit velocities and different fuel nozzle diameters. The results will be compared to experimental data of Kalghatgi [34], Miake-Lye and Hammer [35], Donnerhack and Peters [36] and Røkke [37].

Turbulent methane/air jet flames

In the experiments to be considered, a fuel stream of pure methane was injected into ambient air through a nozzle with a diameter $D = 4$ mm or $D = 8$ mm. In the calculated cases with a nozzle diameter of 8 mm, the mean fuel exit velocity $\tilde{u}_0$ is varied from 40 m/s to 100 m/s, whereas it is varied from 20 m/s to 50 m/s for the cases with a nozzle diameter of 4 mm. The fuel exit velocity profile is assumed to follow the 1/7 power law, the turbulent intensity is set equal to 10 percent of the inlet flow velocity and the integral length scale of the turbulent inflow is set equal to the nozzle diameter. Fuel and air temperatures are both 298 K and the ambient pressure is 1 bar.

For the cases with a nozzle diameter of 8 mm, the simulations have been performed for a domain of 1000 mm × 400 mm axial × radial length with 191 × 77 non-equidistant computational grid cells. For the case with the nozzle diameter of 4 mm, the domain size is 440 mm × 190 mm axial × radial length with the same number of grid cells.

The mass fractions Y_i of the laminar diffusion flamelets are determined by using a steady flamelet library with the scalar dissipation rate χ_{st} as the parameter. The flamelet library is produced by the RIF code [38], in which the chemistry of methane/air diffusion flames is described by a detailed chemical mechanism involving 354 chemical reactions among 30 chemical species. For the methane/air flame, the stoichiometric mixture fraction is $Z_{st} = 0.055$.

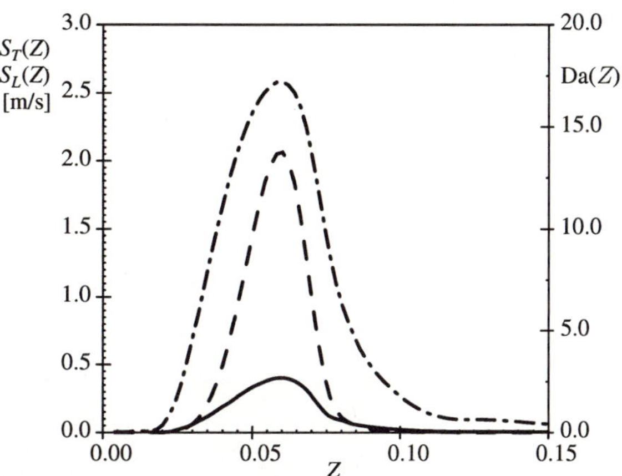

Figure 11 Quantities at the lift-off height. The laminar burning velocity (——), the conditional Damköhler number (- - - -), and the conditional turbulent burning velocity (— · — · — · —), evaluated for $v' = 1.41$ m/s and $\ell = 8.48$ mm as a function of the mixture fraction.

In order to initialize the simulation, the cold flow is calculated at first for the different fuel exit velocities, using eq. (4.18) to determine the mean chemical mass fractions. Then the mixture is ignited at a downstream location by initialization of the G-field in such a way that $\tilde{G} = G_0 \pm |\mathbf{x} - \mathbf{x}_0|$. After ignition, the flame front propagates until it finally reaches a steady state, stabilizing at the lift-off height H. Since the mean curvature term in eq. (4.6) was found to be small it was neglected in the following. In Fig. 11 the laminar burning velocity $s_L(Z)$ for eq. (4.12) taken from [39, 31], the conditional turbulent burning velocity $s_T(Z)$ from eq. (4.12), and the Damköhler number Da(Z) are shown as a function of the mixture fraction Z, evaluated for $\ell = 8.48 \cdot 10^{-3}$ m and $v' = 1.41$ m/s. These quantities are calculated at the lift-off height of the jet flame with a fuel exit velocity of $\tilde{u}_0 = 80$ m/s and a fuel nozzle diameter of $D = 8$ mm.

Figure 12 shows the mean flame fronts $\tilde{G} = G_0$ for different fuel exit velocities with a fuel nozzle diameter of 4 mm after stabilization has been reached. It can be seen that the mean shape of the lifted diffusion flame is similar to that of a laminar triple flame. The stabilization points are found to be located on the lean side and, in the case of low fuel exit velocities, near the iso-line of stoichiometric mixture.

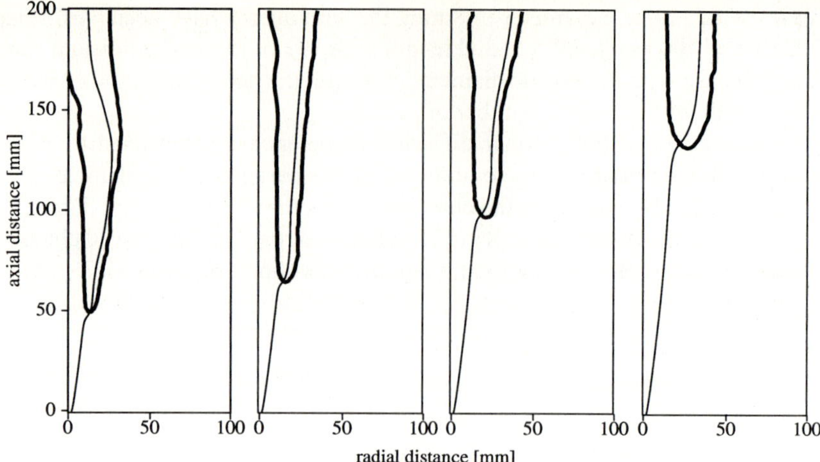

Figure 12 The mean shape of the turbulent flame front (thick line) for methane/air jet flames at fuel nozzle exit velocities of u_0 =20, 30, 40, 50 m/s (from left to right) for a fuel nozzle diameter of $D = 4$ mm. Thin lines are iso-lines of the mean mixture fraction at stoichiometric mixture.

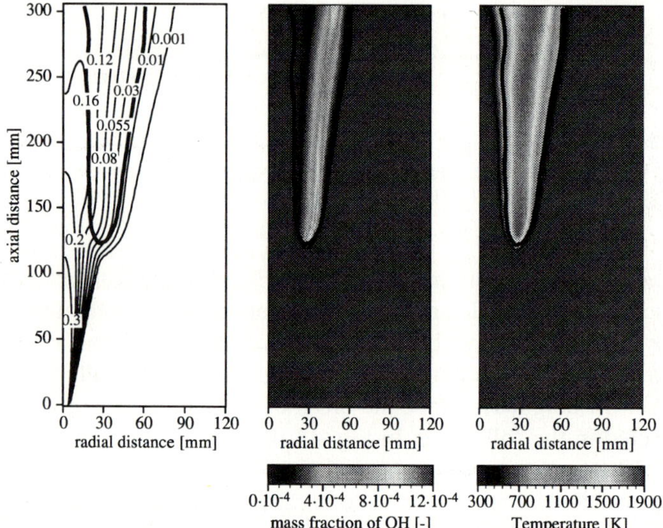

Figure 13 Results for the methane/air jet flame with a fuel nozzle exit velocity of $u_0 = 40$ m/s and $D = 8$ mm. The left picture shows iso-lines of the mixture fraction (thin lines) and the shape of the mean turbulent flame front (thick line), the middle picture the mean OH mass fraction and the right picture the mean temperature distribution.

Figure 13 shows blow-ups of the stabilization region for an exit velocity of $u_0 = 40$ m/s and an exit diameter of $D = 8$ mm. In the left picture, thin lines denote iso-contours of mixture fraction and the thick line represents the mean flame front contour. The expansion at the flame front deflects the stream lines and thereby the mixture fraction iso-lines at the flame base. Stabilization occurs in this case slightly on the lean side at $\tilde{Z} = 0.05$. The middle picture shows the mean OH distribution while the mean temperature distribution is shown on the right side. The location of the maximum OH concentration marks the location of the trailing mean diffusion flame.

The calculated non-dimensional lift-off heights H/D are shown as a function of the fuel exit velocity $\tilde{u}_0$ in Fig. 14. It can be seen that the predicted lift-off heights are in good agreement with the experimental data of Kalghatgi [34], Miake-Lye and Hammer [35] and Donnerhack and Peters [36].

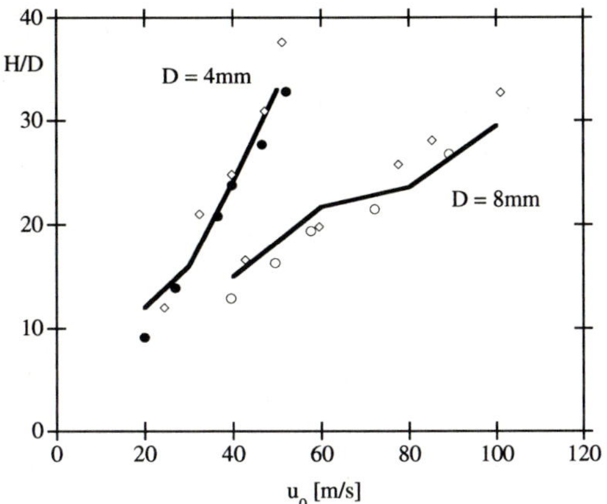

Figure 14 Normalized lift-off heights H/D of methane/air jet diffusion flames for $D = 4$ mm and $D = 8$ mm. Comparison of flamelet model (———) with experimental data by Kalghatgi (o) [34], Miake-Lye and Hammer (●) [35] and Donnerhack and Peters (◇) [36].

Turbulent propane/air jet flames

The simulations are carried out according to the configuration and experimental conditions given by Røkke [37]. Pure propane is injected into the ambient air through a nozzle with a diameter of $D = 6$ mm. In the calculated cases, the mean fuel exit velocity $\tilde{u}_0$ is varied from 20 m/s to 120 m/s. The turbulent intensity is assumed to be 10 percent of the inlet flow velocity and the integral length scale of the turbulent inflow is assumed to be equal to the nozzle diameter. Fuel and air temperatures are both 293 K and ambient pressure is 1 bar. The simulations have been performed for a domain of 440 mm × 190 mm axial × radial length with 191 × 77 non-equidistant computational grid cells. The laminar diffusion flamelets are calculated by the RIF code, in which the chemistry of propane/air diffusion flames is described by a detailed chemical mechanism, involving 36 chemical species. The mixture fraction at stoichiometric mixture is $Z_{st} = 0.0601$. The laminar burning velocity $s_L(Z)$ of the unstretched premixed propane/air flame is obtained from [40].

Figure 15 shows the calculated values of H/D compared to the measured data given by Røkke [37] and Kalghatgi [34]. It can be seen that the calculated lift-off heights are in good agreement with the experimental data of Røkke [37], whereas there is a slight discrepancy with the data of Kalghatgi [34].

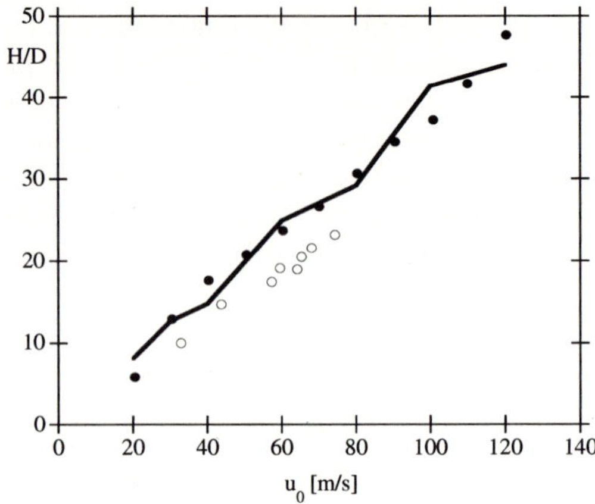

Figure 15 Normalized lift-off heights H/D of propane/air jet diffusion flames for $D = 6$ mm. Comparison of flamelet model (——) with experimental data by Røkke ($\bullet$) [37] and Kalghatgi ($\circ$) [34].

References

[1] N. Peters. Turbulent combustion. Cambridge University Press, 2000.

[2] V. Nayagam, R. Balasubramaniam, and P. D. Ronney. Diffusion flame holes. *Combustion theory and modelling*, 3(4):727–742, 1999.

[3] L. Vervisch and T. Poinsot. Direct numerical simulation of non-premixed turbulent flame. *Annu. Rev. Fluid Mech.*, 30:655–692, 1998.

[4] A. Liñán. The asymptotic structure of counterflow diffusion flames for large activation energies. *Acta Astronautica*, 1007(1), 1974.

[5] V. Favier and L. Vervisch. Effects of unsteadiness in edge-flames in liftoff in non-premixed turbulent combustion. In *Twenty-Seventh Symposium (International) on Combustion*, pages 1239–1245, Pittsburgh, PA, 1998. The Combustion Institute.

[6] G. Ruetsch, L. Vervisch, and A. Linan. Effects of heat release on triple flames. *Physics of Fluids*, 7, 1995.

[7] J. Buckmaster and R. Weber. Edge-flame holding. In *Proceedings of the 26th Symposium (International) on Combustion*, Pittsburgh, 1996. The Combustion Institute.

[8] J. Réveillon and L. Vervisch. Accounting for spray vaporization in non-premixed turbulent combustion modeling: A single droplet model (SDM). *Combustion and Flame*, 121(1):75–90, 2000.

[9] J. Réveillon, K.N.C. Bray, and L. Vervisch. Dns study of spray vaporization and turbulent micromixing. In *36st Aerospace Sciences Meeting and Exhibit AIAA Paper 98-1028*, Reno, NV, January 1998.

[10] F. Mashayek. Direct numerical simulation of evaporating droplet dispersion in forced low mach number turbulence. *Int. J. Heat Mass Transfer*, 41(17):2601–2617, 1998.

[11] F. Mashayek. Simulations of reacting droplets dispersed in isotropic turbulence. *AIAA Journal*, 37(11):1420–1425, 1999.

[12] G. M. Faeth. Evaporation and combustion of sprays. *Prog. Energy Combust. Sci.*, 9:1–76, 1983.

[13] S. K. Lele. Compact finite difference schemes with spectral like resolution. *J. Comput. Phys.*, 103:16–42, 1992.

[14] A. A. Wray. Minimal storage time-advancement schemes for spectral methods. Technical report, Center for Turbulence Research, Stanford University, 1990.

[15] K. Wohl, N. M. Kapp, and C. Gazley. The stability of open flames. In *Third Symposium on Combustion, Flame and Explosion Phenomena*, pages 3–21, 1949.

[16] W. M. Pitts. Assessment of theories for the behaviour and blowout of lifted turbulent jet diffusion flames. In *Twenty-Second Symposium (International) on Combustion*, pages 809–816, Pittsburgh, PA, 1988. The Combustion Institute.

[17] L. Vanquickenborne and A. Van Tiggelen. The stabilization mechanism of lifted diffusion flames. *Combust. Flame*, 10:59–69, 1966.

[18] H. Eikhoff, B. Lenze, and W. Leukel. Experimental investigation on the stabilization mechanism of jet diffusion flames. In *Twentieth Symposium (International) on Combustion*, pages 311–318, Pittsburgh, PA, 1985. The Combustion Institute.

[19] N. Peters and F. A. Williams. Lift-off characteristics of turbulent jet diffusion flames. *AIAA Journal*, 21:423–429, 1983.

[20] J. E. Broadwell, W. J. A. Dahm, and M.G. Mungal. Blowout of turbulent diffusion flames. In *Twentieth Symposium (International) on Combustion*, page 303, Pittsburgh, PA, 1985. The Combustion Institute.

[21] A. Liñán. Ignition and flame spread in laminar mixing layers. In J. Buckmaster, T. L. Jackson, and A. Kumar, editors, *Combustion in High-Speed Flows*, pages 461–176. Kluwer Academic, Dordrecht, 1994.

[22] P. N. Kioni, B. Rogg, K. N. C. Bray, and A. Linan. Flame spread in laminar mixing layers: The triple flame. *Combust. Flame*, 95:276–290, 1993.

[23] D. Veynante, L. Vervisch, T. Poinsot, A. Linan, and G. Ruetsch. Triple flame structure and diffusion flame stabilization. In *Proceedings of the Summer Program 1994*, pages 55–73. Center for Turbulence Research, 1994.

[24] N. Peters. Laminar flamelet concepts in turbulent combustion. In *Twenty-First Symposium (International) on Combustion*, pages 1231–1250, Pittsburgh, PA, 1986. The Combustion Institute.

[25] M. Wirth and N. Peters. Turbulent premixed combustion: A flamelet formulation and spectral analysis in theory an IC–engine experiments. In *Twenty-Forth Symposium (International) on Combustion*, pages 493–501, Pittsburgh, PA, 1992. The Combustion Institute.

[26] C. M. Müller, H. Breitbach, and N. Peters. Partially premixed turbulent flame propagation in jet flames. In *Twenty-Fifth Symposium (International) on Combustion*, pages 1099–1106, Pittsburgh, PA, 1994. The Combustion Institute.

[27] V. Favier, L. Vervisch, M. Herrmann, P. Terhoeven, B. Binninger, and N. Peters. Numerical simulation of combustion in partially premixed turbulent flows. In Ernst Heinrich Hirschel, editor, *Numerical Flow Simulation*, Notes on Numerical Fluid Mechanics, pages 203–221. Vieweg, 1998.

[28] N. Peters. Laminar diffusion flamelet models in non-premixed turbulent combustion. *Prog. Energy Combust. Sci.*, 10:319–339, 1984.

[29] N. Peters. The turbulent burning velocity for large-scale and small-scale turbulence. *J. Fluid Mech.*, 384:107–132, 1999.

[30] J. A. Sethian. *Level Set Methods*. Cambridge University Press, Cambridge, 1996.

[31] T. Plessing, P. Terhoeven, and N. Peters. An experimental and numerical study on a laminar triple flame. *Combust. Flame*, 115:335, 1998.

[32] Fluent Europe. *FLUENT User's Guide, Version 4.4*, Jan. 1996.

[33] M. Sussman, P. Smereka, and S. Osher. A level set approach for computing solutions to incompressible two-phase flow. *J. Comp. Phys.*, 114:146–159, 1994.

[34] G. T. Kalghatgi. Lift-off height and visible lenghts of vertical turbulent jet diffusion flames in still air. *Combust. Sci. and Tech.*, 41:17, 1984.

[35] R. C. Miake-Lye and J. A. Hammer. Lifted turbulent jet flames: A stability criterion based on the jet large-scale structure. In *Twenty-Second Symposium (International) on Combustion*, pages 817–824, Pittsburgh, PA, 1992. The Combustion Institute.

[36] S. Donnerhack and N. Peters. Stabilization heights in lifted methane-air jet diffusion flames diluted with nitrogen. *Combust. Sci. and Tech.*, 41:101–108, 1984.

[37] N. A. Røkke. A study of partially premixed unconfined propane flames. *Combust. Flame*, 97:88–106, 1994.

[38] H. Barths, H. Pitsch, G. Paczko, and N. Peters. *RIF User Guide*. ITM, RWTH-Aachen, url: http://www.flamelets.com/RifUG.pdf, 1998.

[39] F. Mauss and N. Peters. Reduced kinetic mechanisms for premixed methane-air flames. In N. Peters and B. Rogg, editors, *Reduced Kinetic Mechanisms for Applications in Combustion Systems, Lecture Notes in Physics*, volume m 4, pages 58–75, Berlin, 1993. Springer Verlag.

[40] H. Pitsch. *FlameMaster, A C++ Program for 0D and 1D Flame Calculation*. ITM, RWTH-Aachen, 1993.

Development of a Parallel Unstructured Multigrid Solver for Laminar Flame Simulations with Detailed Chemistry and Transport

S. Paxion[1], R. Baron[2], A. Gordner[1], N. Neuss[1], P. Bastian[1],
D. Thévenin[2] and G. Wittum[1]

[1] Universität Heidelberg, I.W.R., im Neuenheimer Feld 368, D-69120 Heidelberg, Germany
E-mails : sebastien.paxion@iwr.uni-heidelberg.de, peter.bastian@iwr.uni-heidelberg.de

[2]École Centrale Paris, E.M2.C, Grande Voie des Vignes, F-92295 Châtenay-Malabry, France
E-mails : baron@em2c.ecp.fr, thevenin@em2c.ecp.fr

Summary

We develop a computer code for steady laminar flame simulations at low Mach numbers, with detailed models for chemical and molecular transport properties. The so called *UG-C* code is based on the *UG* library developed at IWR, Heidelberg. With a view to reducing computational time as much as possible, we combine an appropriate low Mach number modelisation with implicit time integration, Krylov-Newton and multigrid preconditioning, on unstructured, dynamically refined grids, with storage optimization of sparse matrices. The code runs on distributed memory parallel machines, with several load balancing algorithms. We show applications to diffusion hydrogen/air and premixed methane/air flames. Multigrid acceleration has been obtained for sufficiently fine grids.

1 Motivations and Model Problem

Accurate simulations of complex reactive flows remain difficult and costly, even today. These simulations are, however, becoming so important from the scientific, economical and ecological point of view, that there is a real need for prediction tools, complementary to experimental investigations. Applications range from fundamental understanding of flame structures on simple and fully controllable laboratory burners, to accurate performance prediction of industrial combustion systems, with respect to pollutant emissions, heat release, safety issues, etc. It is still today impossible to meet all these requirements for all kind of combustion problems at all flow conditions. Depending on the problem addressed, emphasis is either laid on hydrodynamics (high Reynolds numbers, turbulent models or large eddy combustion models [1]), or acoustics (acoustic-chemical coupling, flame instabilities), or chemistry (detailed reaction mechanisms). The combination of all detailed aspects still remains restricted to DNS computations [2][3] and is at present limited to simple geometries and fairly low Reynolds numbers.

This paper relates to an ongoing DFG-CNRS project where it has been decided to focus mainly on the chemical aspect, putting aside for the time being any turbulent and acoustic issues. This work consisted in developing a computer code for steady laminar flame simulations in gaseous media, at low Mach numbers [4][5][6][7][8]. Detailed models are taken into account to describe reactive mechanisms, thermodynamic properties of the mixture and molecular diffusion of chemical species. A detailed thorough description of the model problem and the numerical methods, as well as a detailed presentation of the results and validation test cases can be found in [9]. We restrict in this paper to the major features of our computer code and solving strategy (see Section 2) and to the latest results (see Section 3). This section is concerned with derivation and approximations leading to equations of our model problem.

1.1 Chemical modelisation and thermodynamics

The reactive mixture consists of a multicomponent gas of K different chemical species, interacting and possibly reacting with each other. Identical molecules at different quantum degeneracy level may be regarded as different species (to account for effects like spontaneous emission). The overall chemical mechanism is modeled by I elementary reactions involving no more than 3 reactants and 3 products. If each elementary reaction (index i) is formally written

$$\sum_{k=1}^{K} \nu'_{ki} \mathcal{X}_k \rightleftharpoons \sum_{k=1}^{K} \nu''_{ki} \mathcal{X}_k \ ,$$

with ν'_{ki} and ν''_{ki} as stœchiometric coefficients (0 for most species or 1 or 2), then the molar production rate of one species k per unit volume is

$$\dot{\omega}_k = \sum_{i=1}^{I} (\nu''_{ki} - \nu'_{ki}) \, q_i \ \text{with} \ q_i = k_{f_i} \prod_{k=1}^{K} [\mathcal{X}_k]^{\nu'_{ki}} - k_{r_i} \prod_{k=1}^{K} [\mathcal{X}_k]^{\nu''_{ki}} \ ,$$

where $[\mathcal{X}_k]$ is the concentration of species k and the forward and backward rates of progress of each reaction are related to each other via the reaction equilibrium constant, and are modeled by an Arrhenius law:

$$k_{f_i} = A_i T^{\beta_i} \exp \left(\frac{-E_i}{RT} \right) \ \text{and} \ k_{r_i} = \frac{k_{f_i}}{\text{Ke}_i} \ . \tag{1}$$

The equilibrium constant Ke_i is theoretically expressed from the standard enthalpy and entropy, and is mainly a function of temperature. On table 1 are listed all reactions considered and their related Arrhenius coefficients A_i, β_i and E_i, taken from [10] for all methane/air flame computations presented in this paper. The chemical scheme used for the hydrogen/air computations is taken from [11].

Each species k is given by its mass fraction Y_k, which represents the mass of species k per unit mass of the mixture. It is more sensible and usual in gazeous combustion to consider the specific enthalpy than the internal energy as the former is conserved through chemical processes at constant pressure and with adiabatic boundaries. The enthalpy h of the mixture is expressed from the species enthalpies, themselves being functions of temperature: temperature only, and the standard enthalpy at standard temperature (T_0):

$$h = \sum_{k=1}^{K} Y_k h_k \ \text{and for each species,} \ h_k(T) = h_k^0(T_0) + \int_{T_0}^{T} c_{pk} \, (T') \, \mathrm{d}T' \ ,$$

where $h_k^0(T_0)$ and C_{pk} are the standard enthalpy (at standard temperature T_0) and the specific heat at constant pressure of species k.

Table 1: Chemical scheme for methane/air flames, from [10]. Some reaction require the interaction with any third body, M, which remains unchanged. An efficiency factor (1 by default) may be given for some species to either enhance their efficiency as a third body, or decrease it, or even forbid it (e.g. line 5).

species : CH_4, CH_3, CH_2O, HCO, CO_2, CO, H_2, H, O_2, O, OH, HO_2, H_2O and N_2 (dilutant)

Elementary reactions	A_i (cgs)	β_i	E_i (cgs)	Elementary reactions	A_i (cgs)	β_i	E_i (cgs)
				$2OH \rightleftharpoons O+H_2O$	5.750×10^{12}	0.0	775
$OH+H_2 \rightleftharpoons H_2O+H$	1.170×10^9	1.3	3626.3	$CO+OH \rightleftharpoons CO_2+H$	1.500×10^7	1.3	-765
$H+O_2 \rightleftharpoons OH+O$	1.420×10^{14}	0.0	16393	$CO+O+M \rightleftharpoons CO_2+M$	5.400×10^{15}	0.0	4570
$O+H_2 \rightleftharpoons OH+H$	1.800×10^{10}	1.0	8902	$N_2/0.44/$ $O_2/0.35/$ $CO/0.74/$ $CO_2/1.47/$ $H_2O/6.5/$			
$H+O_2+M \rightleftharpoons HO_2+M$	3.610×10^{17}	-0.72	0	$CO+H+M \rightleftharpoons HCO+M$	5.000×10^{14}	0.0	1500
$N_2/0.44/$ $O_2/0.35/$ $CO/0.74/$ $CO_2/1.47/$ $H_2O/6.5/$				$N_2/0.44/$ $O_2/0.35/$ $CO/0.74/$ $CO_2/1.47/$ $H_2O/6.5/$			
$H+HO_2 \rightleftharpoons 2OH$	1.400×10^{14}	0.0	1073	$CH_4+O \rightleftharpoons OH+CH_3$	4.070×10^{14}	0.0	13988
$H+HO_2 \rightleftharpoons O+H_2O$	1.000×10^{13}	0.0	1073	$CH_4+H \rightleftharpoons CH_3+H_2$	7.240×10^{14}	0.0	15081
$H+HO_2 \rightleftharpoons H_2+O_2$	1.250×10^{13}	0.0	0	$CH_4+OH \rightleftharpoons H_2O+CH_3$	1.550×10^6	2.13	2444
$OH+HO_2 \rightleftharpoons H_2O+O_2$	7.500×10^{12}	0.0	0	$CH_4+M \rightleftharpoons CH_3+H+M$	4.680×10^{17}	0.0	93210
$O+HO_2 \rightleftharpoons O_2+OH$	9.1054×10^{12}	0.061	765.82	$CH_3+O \rightleftharpoons CH_2O+H$	6.020×10^{13}	0.0	0
$H+H+H_2 \rightleftharpoons H_2+H_2$	9.200×10^{16}	-0.6	0	$CH_2O+O \rightleftharpoons HCO+OH$	1.820×10^{13}	0.0	3080
$H+H+N_2 \rightleftharpoons H_2+N_2$	1.000×10^{18}	-1.0	0	$CH_2O+H \rightleftharpoons HCO+H_2$	3.310×10^{14}	0.0	10511
$H+H+O_2 \rightleftharpoons H_2+O_2$	1.000×10^{18}	-1.0	0	$CH_2O+OH \rightleftharpoons HCO+H_2O$	7.580×10^{12}	0.0	143
$H+H+H_2O \rightleftharpoons H_2+H_2O$	6.000×10^{19}	-1.25	0	$HCO+O_2 \rightleftharpoons CO+HO_2$	3.000×10^{12}	0.0	0
$H+H+CO \rightleftharpoons H_2+CO$	1.000×10^{18}	-1.0	0	$HCO+H \rightleftharpoons CO+H_2$	4.000×10^{13}	0.0	0
$H+H+CO_2 \rightleftharpoons H_2+CO_2$	5.490×10^{20}	-2.0	0	$HCO+OH \rightleftharpoons CO+H_2O$	5.000×10^{12}	0.0	0
$CH_4+H+H \rightleftharpoons CH_4+H_2$	5.490×10^{20}	-2.0	0	$HCO+O \rightleftharpoons CO+OH$	1.000×10^{13}	0.0	0
$H+OH+M \rightleftharpoons H_2O+M$	1.600×10^{22}	-2.0	0	$CH_2O+CH_3 \rightleftharpoons HCO+CH_4$	2.230×10^{13}	0.0	5146
$H_2O/5/$				$CH_3+OH \rightleftharpoons CH_2O+H_2$	3.980×10^{12}	0.0	0
$H+O+M \rightleftharpoons OH+M$	6.200×10^{16}	-0.6	0	$CH_3+HO_2 \rightleftharpoons CH_4+O_2$	1.020×10^{12}	0.0	397
$H_2O/5/$				$HO_2+CO \rightleftharpoons CO_2+OH$	1.500×10^{14}	0.0	23645

The thermodynamical and chemical description of the reactive mixture is complete, being given the chemical scheme (such as table 1), and, for each species, the standard enthalpy and entropy, and the specific heats at constant pressure as a function of temperature. In practice, the latter is a fourth order polynomial fit of experimental measurements (given with its 5 coefficients), then leading to a fifth order polynomial for the specific enthalpy. Our computations are based on the JANAF thermodynamical database and use the CHEMKIN [12] processing library.

1.2   Detailed models for molecular diffusion

Although the macroscopic balance equations for mass, momentum and enthalpy for the gaseous mixture may be derived from the Onsager's theory for multiple coexistent continua, this approach does not give any information regarding molecular transport properties and how the diffusion coefficients should be related to the macroscopic variables and the mixture composition. The only rigorous analysis is via the kinetic theory of multicomponent gases [13]. The non-dimensionalized multi-component Boltzmann equation is perturbed for small values of the Knudsen number K_n (ratio of the mean free path of particles to the characteristic hydrodynamic length) around the Maxwellian velocity distribution. The rigorous asymptotic analysis of Chapman and Enskog [14] shows that velocity distributions can be expanded to integer series of K_n, the zeroth (Maxwellian distribution) and first orders of which leading respectively to the Eulerian and Navier-Stokes regimes. All macroscopic variables and diffusion coefficients are fully determined by integrals of microscopic properties weighted by the velocity distribution function, over the whole velocity range. Expanding the integrand in series of Sonine polynomials and integrating leads to diffusion coefficients being related to temperature and the collision integrals of particles, which in turn can be evaluated assuming potential interaction laws. No explicit formulations can be obtained for multicomponent gases. Linear systems of equations, with size depending on the number of terms which are kept in the truncated Sonine expansion, must be solved instead. This is performed with the **EGLIB** library developed by Ern and Giovangigli, for evaluating the transport coefficients [15][16][17]. Optimized iterative methods are used to solve to some accuracy these linear kinetic systems, based on a thorough mathematical analysis of their properties [18].

Molecular transport then takes the form of diffusion velocities V_k each being defined as the difference between the mean velocity of species k and the mean velocity of the mixture. They are related to the species molar fractions X_k (ratio of the number of moles of species k to number of moles of the mixture, related to mass fractions via molar weights) by:

$$Y_k \boldsymbol{V}_k = -\sum_{\ell=1}^{K} \tilde{D}_{k\ell} \boldsymbol{\nabla} X_\ell \approx -\tilde{\Upsilon}_k \boldsymbol{\nabla} X_k + Y_k \boldsymbol{V}_{\mathrm{corr}} \; ,$$

where the second formulation is a cheaper approximation neglecting interspecies coupling, and requires the addition of a correction diffusion velocity $\boldsymbol{V}_{\mathrm{corr}}$ to enforce the $\sum_{k=1}^{K} Y_k \boldsymbol{V}_k = \boldsymbol{0}$ condition. Both formulations are implemented but all computations presented use the second diffusion vector flux formulation. Coupling between species diffusion and thermal diffusion, known as the Soret effect, has been neglected in this formulation. Such effect can however become noticeable at some combustion regimes, as shown by Ern *et. al.* [19] and will be implemented in the future. Pressure gradients have also been neglected for the computation of the diffusion velocities, which is justified within the low-Mach number approach presented in the next section.

1.3   Low-Mach number approximation of the governing equations

In flows at low Mach numbers M, acoustic waves travel much faster than entropy waves, with a velocity ratio of about M^{-1}. Due to numerical stability criteria, the time step is dictated by acoustic speed and is therefore unnecessary small when a stationary solution is sought. Furthermore, detailed mathematical analyses reveal that the compressible Euler equations exhibit a singular limit as M tends to zero, as the coupling between pressure and density vanishes. All this results in very ill-conditioned systems of equations when low speed flows are solved with compressible equations. To circumvent this problem, the incompressible fluid assumption is commonly used in non reactive fluid dynamics (possibly with the Boussinesq approximation for buoyancy driven flows), which is obviously not of practical interest for exothermal reactive flows whereby density may vary up to an order of magnitude. Nevertheless, it is observed that pressure variations through laminar flames at low Mach numbers are always of the order of magnitude of a few Pascals and stem for the very largest part from hydrodynamical and not from compressibility effects. Stated differently, density increase only result from heat release due to chemical reactions, and from changes in the mixture composition, but not from local fluid compression. Temperature and density vary in opposite directions, such that their effects on the ideal gas law compensate. These physical observations motivates the decomposition of pressure into a bulk background uniform thermodynamic pressure p_u and a hydrodynamic $\tilde{p}$ term:

$$p\left(\boldsymbol{x}, t\right) = p_u\left(t\right) + \tilde{p}\left(\boldsymbol{x}, t\right) .$$

If acoustic waves may propagate in the gas mixture, then an additional acoustic pressure term has to be considered. Since we do not address cases of acoustic/flame interactions, we assume acoustic waves are either inexistent or of negligible effect on the flame structure and the flow.

Giovangigli [20][21] shows from an asymptotic analysis of aerothermochemical structures of laminar flames [22], that hydrodynamic pressure fluctuations lie within the Mach number squared of the overall thermodynamic pressure. The so-called isobaric flame equations, as recalled hereafter, can then be derived from the fully compressible ones after a simple scale analysis of different terms, whereby pressure has previously been replaced by its decomposition. This yields the following set of balance equations, written in conservative form, for specific mixture and species mass, momentum and enthalpy:

$$\frac{\partial \rho}{\partial t} + \boldsymbol{\nabla} \cdot (\rho \boldsymbol{v}) = 0 \ , \tag{2}$$

$$\frac{\partial (\rho Y_k)}{\partial t} + \boldsymbol{\nabla} \cdot (\rho Y_k \boldsymbol{v}) = -\boldsymbol{\nabla} \cdot (\rho Y_k \boldsymbol{V}_k) + W_k \dot{\omega}_k \ , \tag{3}$$

$$\frac{\partial (\rho \boldsymbol{v})}{\partial t} + \boldsymbol{\nabla} \cdot (\rho \boldsymbol{v} \boldsymbol{v}) = -\boldsymbol{\nabla} \tilde{p} + \boldsymbol{\nabla} \cdot \left\{ \mu \left(\boldsymbol{\nabla} \boldsymbol{v} + (\boldsymbol{\nabla} \boldsymbol{v})^T \right) \right\} \ , \tag{4}$$

$$\frac{\partial (\rho h)}{\partial t} + \boldsymbol{\nabla} \cdot (\rho h \boldsymbol{v}) = -\boldsymbol{\nabla} \cdot \left[-\lambda \boldsymbol{\nabla} T + \sum_{k=1}^{K} (\rho h_k Y_k \boldsymbol{V}_k) \right] + \frac{\mathrm{d} p_u}{\mathrm{d} t} \ , \tag{5}$$

with the addition of the ideal gas law,

$$\rho = \frac{p_u W (Y_1, Y_2, \cdots, Y_K)}{RT} \ . \tag{6}$$

All chemical and diffusional properties appear decoupled from pressure fluctuations and are computed with p_u. Acoustic wave are canceled via the decoupling between density ρ and $\tilde{p}$. Moreover, energy viscous dissipation is of lower order of magnitude, as well as compressibility effects, which both disappear in the enthalpy balance, except from a possible bulk compression (if combustion is to be simulated in a confined domain). In our simulations, the domain is at constant thermodynamic equilibrium with the ambient atmospheric pressure, and hence $p_u (t) = p_{\text{atmospheric}}$. The mathematical properties of equations (2) to (5) are similar to those of purely incompressible uniform density flows, namely parabolic-elliptic. The same stabilization methods and solving algorithms behave similarly. Boundary conditions are much simpler (of Dirichlet or Neumann type, for each primitive variable), in comparison with the usual NSCBS non-reflecting conditions required by full compressible combustion simulations [23].

This approach is now widely used in the low Mach combustion community [24] [25] [26], though it has to be pointed out that, to our knowledge, rigorous justifications of the isobaric flame equations with detailed chemical and diffusional models are still lacking. Majda and Sethian [27] proved the asymptotic consistency of the pressure decomposition and derived the equations from a rigorous asymptotic expansion but their analysis is restricted to premixed, simple diffusion and chemistry flames with adiabatic slip wall conditions. Karlin *et al.* [28] suggest that a validity criterion for applicability of isobaric flame equations to laminar flame computations be that $M \ll Da^{-1}$, where the Damköhler number Da is the ratio of chemical to hydrodynamical characteristic time scales. This criterion results from numerical computations of laminar flames in confined pipes, but with a simple Arrhenius low to describe chemistry.

It is however reasonable to assume that this low-Mach approach is valid to some accuracy for the very low-Mach number flame computations presented in this paper. Comparison with experimental results have eventually given some confidence in this respect. A *pseudo-conservative* equation for temperature can easily be derived from (3) and (5),

$$\frac{\partial(\rho T)}{\partial t} + \boldsymbol{\nabla} \cdot (\rho T \boldsymbol{v}) = \frac{1}{c_p}\frac{dp_u}{dt} - \frac{1}{c_p}\boldsymbol{\nabla} \cdot (-\lambda \boldsymbol{\nabla} T)$$

$$\underbrace{- \frac{1}{c_p}\left(\sum_{k=1}^{K} \rho Y_k \boldsymbol{V}_k c_{p_k}\right) \cdot \boldsymbol{\nabla} T}_{(*)} - \frac{1}{c_p}\sum_{k=1}^{K} h_k W_k \dot{\omega}_k \ . \tag{7}$$

We have seen from 1D flame computations that term $(*)$ has a negligible effect. We have omitted it, as previously done by other authors [29][24]. Our model problem to be solved finally consists of equations (2) (4) (7) and $K-1$ equations of type (3), bearing in mind that the last species (nitrogen) is a dilutant, such that $Y_{N_2} = 1-\sum_{k=1}^{K-1} Y_k$. They are solved for the primitive variables $\tilde{p}$, $\boldsymbol{v}$, T and $Y_{k\in\{1,..,K-1\}}$.

2   The *UG-C* Code

The equations are discretised with a cell vertex finite volume method on conforming unstructured grids. All primitive variables are stored at the grid nodes. Such a collocated discretization is not stable with respect to velocity-pressure coupling. Stabilization is performed by the FIELDS method proposed by Schneider *et. al.*[30][31] initially for fully incompressible flows, which we have extended to the case of our incompressible dilatable flows: it consists in a physically motivated interpolation of the velocity at the integration points where convective and diffusive fluxes are constructed; the integration point velocities are regarded as new degrees of freedom, and are related to nodal values of pressure and velocity by discretising, on the current grid cell, the momentum equation, with an approximate finite difference scheme. Analysis for the Stokes equation shows some similarities with stabilization methods based on artificial compressibility. This discretization is second order in space.

Time discretization is of first order implicit type. Its particularity resides in the explicit treatment of density and viscosity fields: time marching from time index n to $n+1$ is performed with viscosity at time n and density ρ at times $n-1$ and n, which, for instance for the transient term of the temperature equation, writes:

$$\frac{\left(\rho^n T^{n+1} - \rho^{n-1}T^n\right)}{\text{timestep}} = \text{convection/diffusion/source terms at time } n+1, \text{ with } \rho^n \ .$$

The Navier-Stokes equations (2) and (4) then appear fully decoupled at time $n+1$ from the thermo-reactive ones (3) and (7), and are marched in time separately but each sub-system with full implicit integration. Linearisation is achieved by fixed point iterations for the momentum convection term, and with an approximate-Newton method for non-linearities in the chemical and diffusional terms of the thermo-reactive sub-system. The Jacobian matrix is computed by first order numerical differentiation for chemical sources and analytically for other convection/diffusion contributions, after neglecting variations of diffusion coefficients. The sparsity pattern of the Jacobian consists of full diagonal blocks, whereby all species and temperature are coupled via chemical sources, and a single diagonal for off-diagonal blocks. Its storage in the memory is optimized for this structure [32], which, in comparison to a full Jacobian, results in a size reduction

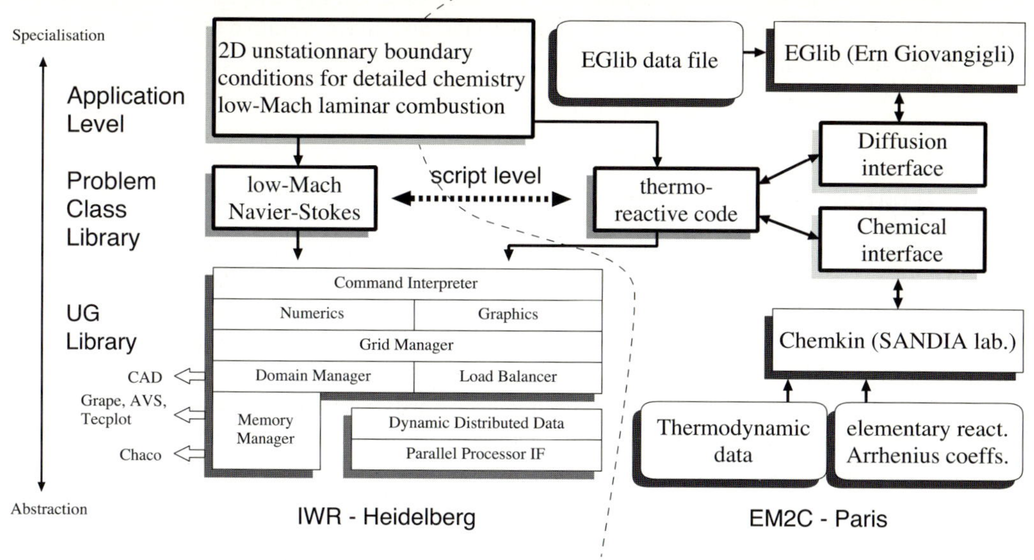

Figure 1: Overall structure of the *UG-C* code and the *UG* library. Development distribution between both German and French institutes is indicated by the dashed line.

factor that tends to 7 in 2D (15 in 3D) for quadrilaterals as the number of species increases. All matrix/vector operations are equally optimized, giving an overall speed-up of more than 4 for one iteration of the linear solver for our practical cases.

2.1 Implementation in the parallel *UG* numerical toolbox

The *UG-C* code is designed as an application of the *UG* library, developed by the German side of the project at IWR, Heidelberg. The *UG* library is a modular, C-coded, numerical toolbox [33][34] aimed at both the development of simulation codes for a wide range of PDE problems (compressible and incompressible flows, structural mechanics, porous media and density driven flows,etc.) and at fundamental investigations of multigrid methods on simpler model problems. The Navier-Stokes module was derived in Germany from an already existing incompressible Navier-Stokes application [35], whereas the thermo-reactive module with interfaces to CHEMKIN and EGLIB has been developed in France. The overall *UG-C* code structure along with the *UG* subsystems is shown in Figure 1: it is based on different levels of abstraction: the *UG library* itself, independent of the PDE problem to be solved; the *problem-class library*, concerned with discretization, using abstract grid elements description, independent of space dimensions (2D or 3D) and element types (triangles, quadrangles, tetrahedra ...); finally the *application level* implements the boundary value problem (description of geometries and boundary conditions), the grid and any data files, and the definition of the solving strategy. The *UG-C* code automatically inherits all features of the *UG* library, among which the most important for our applications are: dynamic local refinement and coarsening of the multi-level grid; portability on distributed memory parallel machines with various spectral and geometrical load-balancing algorithms, specifically designed for multilevel-grid partitioning; a command interpreter with an object oriented environment for combining iterative methods together with discretizations for building and interactively controlling the whole solving strategy during run time; graphical tools; grid generators; and a large variety of linear and non-linear iterative methods coded in an application independent fashion.

2.2   Dynamic local multigrid

The linearized Navier-Stokes and thermo-reactive problems are solved by the Bi-CGSTAB method [36], preconditioned by multigrid V-cycles with a Gauss-Seidel smoother. Special multigrid methods have been devised in *UG* for problems which require strong local mesh refinements. A good compromise between algorithm complexity and robustness in this context is achieved by adding new refined grid levels, overlapping each other, which may only be local (in regions where a high spatial resolution is needed), hence not necessarily covering the whole computational domain. They are kept conform and slightly extended, as explained in Figure 2. It is shown in [37] that smoothing must be performed on all 'red', 'green' and 'yellow' elements, the defect being previously set to zero for nodes lying on the local grid boundary. In the right grid of Figure 2, white elements (thin lines) are not used and do not exist. We call *surface grid* the set of all elements that are not refined, from all levels.

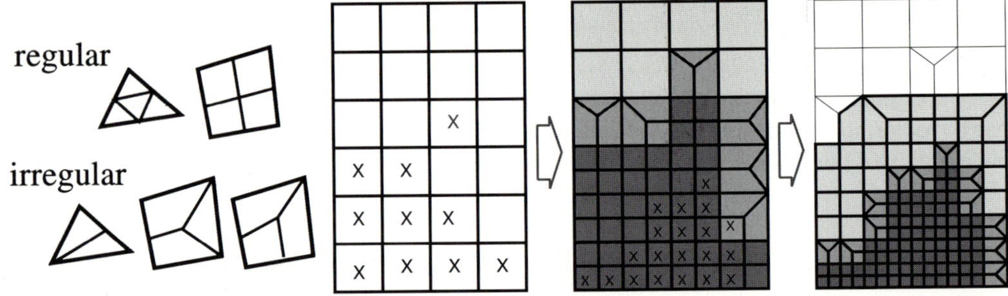

Figure 2: The grid is built from the level 0 grid (left) by successive local refinements: first, marked elements (crosses) are refined according to regular rules (red or dark grey); a new grid level (middle) is made on top of the base grid; it is made conforming by adding irregular elements (green or middle grey) and is furthermore surrounded by two rows of copied elements (yellow or light grey) as required by the local multigrid method. Further refinement of irregular elements is forbidden to avoid strongly skewed elements. In case one such green element is flagged for refinement (e.g. on middle grid), the father element (left grid) is re-refined according to a regular rule before further refinement may proceed (see resulting right grid).

Laminar flames feature sharp fronts in a very narrow region in space where most of the chemical activity takes place. The flame properties, such as its propagation speed (which determines its location) and its heat release, are strongly dependent on some very unstable intermediate species, which have to be solved to a high resolution. This leads to strong requirement on the size of the grid elements, which, inside the flame, is often more than 50 times smaller than in the far field. Our general solving procedure consists in a rough flame initialization (from 1D flame database) and makes use of dynamic grid refinement and coarsening to track the flame front. Finally, the grid is further refined to achieve high spatial resolution at stationary convergence, as illustrated in Figure 3.

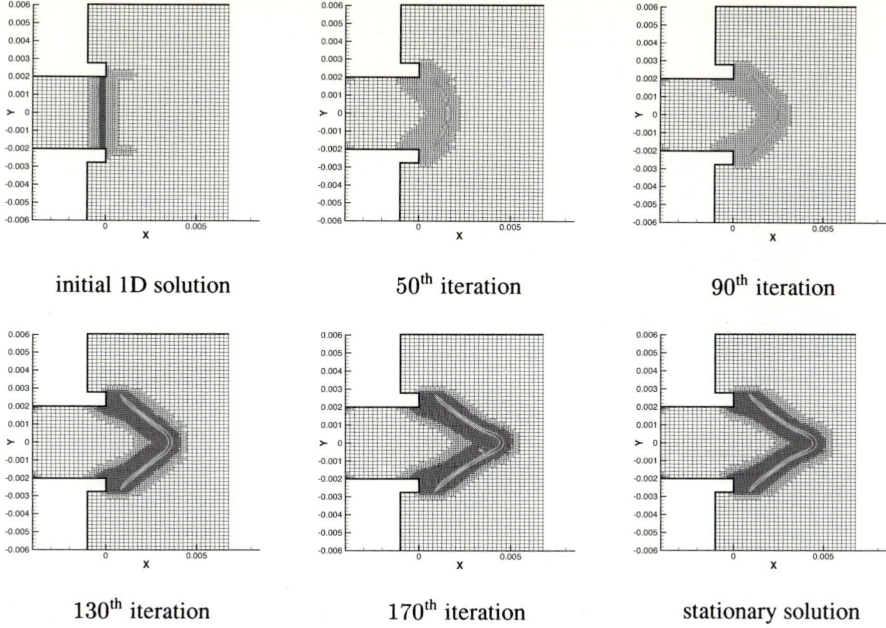

initial 1D solution   50th iteration   90th iteration

130th iteration   170th iteration   stationary solution

Figure 3: Solution history for a methane/air premixed flame computation on the Bunsen configuration of Somers and de Goey [38]. The stationary solution is reached after 190 time iterations and 10 CPU hours on a 128Mb Sun Ultra-1 workstation. The final grid has about 10000 surface elements.

3 Validations and Applications

The *UG-C* code has already been used to simulate premixed and non-premixed laminar flames, on various configurations ranging from fundamental laboratory burners to industrial-like burners. In all cases, the numerical procedure is similar and proved its efficiency in terms of robustness, computing time and required resources. We present below the two major validation simulations, corresponding to two extreme configurations: a hydrogen/air diffusion flame and a methane/air premixed flame.

3.1 A counterflow diffusion flame

This is a fundamental configuration that has been intensively studied in the French institute, experimentally, theoretically and numerically [39][40][41]. The experimental set-up developed for many years by C. Rolon is represented schematically in its diffusion hydrogen/air configuration in Figure 4: the fuel (hydrogen diluted with nitrogen) and oxidizer (oxygen present in the air) are injected against each other in a counterflow. Such flows are well understood and theoretical asymptotic analysis can describe their structures (assuming infinite injector diameter) and relate some properties, like the strain rate at the stagnation point, to the injector mass flow rates. If the strain rate lies below the extinction limit, which depends on the composition of the fuel and oxidizer mixtures, then a diffusion flame may exist in the vicinity of the stagnation line, close to the line where fuel and oxidizer are at stoechiometric concentrations. The burned gases are convect-

189

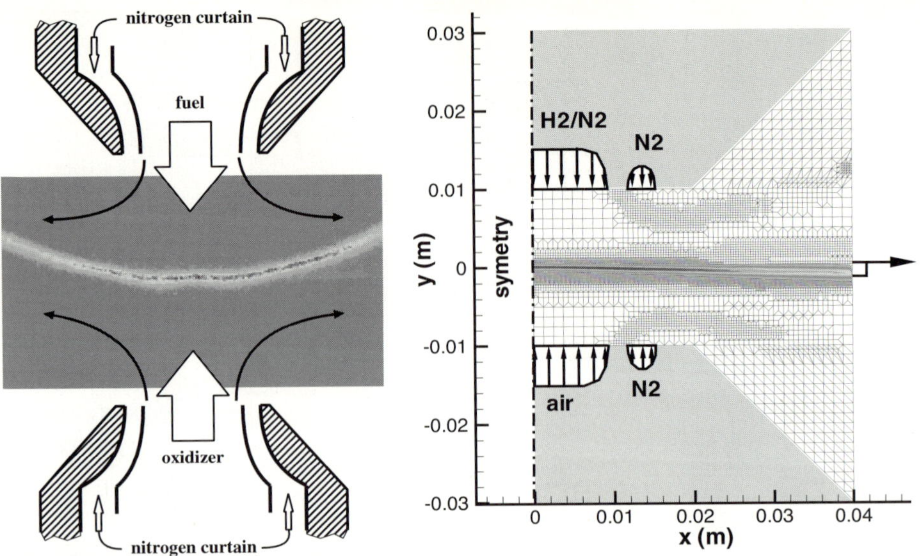

Figure 4: Diffusion hydrogen/air flames in a counterflow: schematic view of the experimental axisymmetrical burner (not to scales) with induced fluorescence laser image of the OH radical (left), and 2D computational configuration and grid with OH mass fraction (right). Buoyancy is responsible for the flame curvature (left) and is not taken into account in our computations.

ed away along the stagnation line. Both injector outlets are moreover surrounded by a nitrogen curtain to isolate the flame from external hydrodynamical disturbances and to avoid a secondary diffusion flame between the fuel jet and the ambient air. Such counterflow flames are stable and easy to control and to characterize, which makes valuable test cases for code validations.

We have computed a similar flame but in a two-dimensional planar counterflow (see results in Figure 5). Intensive use of local grid refinement and coarsening is made to capture the mixing layers and all intermediate species in the flame front. The computation has been conducted in 2 steps: first, a solution has been obtained on the final 8-level, 15 000 node grid represented in Figure 4 after 10 hours on a single SUN Ultra-10 workstation; second, the solution has been carried further on a much finer grid of about 120000 nodes on a Cray T3E parallel computer with 8 processors. It is believed that grid convergence has been reached: the 1D plots of Figure 6 show solution profiles along the symmetry line, with symbols corresponding to grid nodes.

One-dimensional equations can be derived to describe the flow and flame structure along the symmetry axis, assuming that the injector diameter is of infinite length and that the solution is self-similar. A 1D code has been developed at the French institute [42] and validated for many years by comparison with experimental results [43]. It includes two different configurations: either axisymmetrical (like the experimental set-up) or planar two-dimensional counterflow. We have used the latter configuration to validate our results as the *UG-C* code does not yet comprise an axisymmetrical formulation of the equation. The agreement is satisfactory (see Figure 6), being aware of the approximations inherent to the 1D formulation. We are now working on an axisymmetrical version of *UG-C* which will soon allow direct validations on experiments. The gravity effects will also be included with a view to reproduce the observed flame curvature.

The effect of the multigrid method used as a preconditioner of BiCGSTAB has been investigated for this configuration. It is compared to a simple Gauss-Seidel preconditioner for the

190

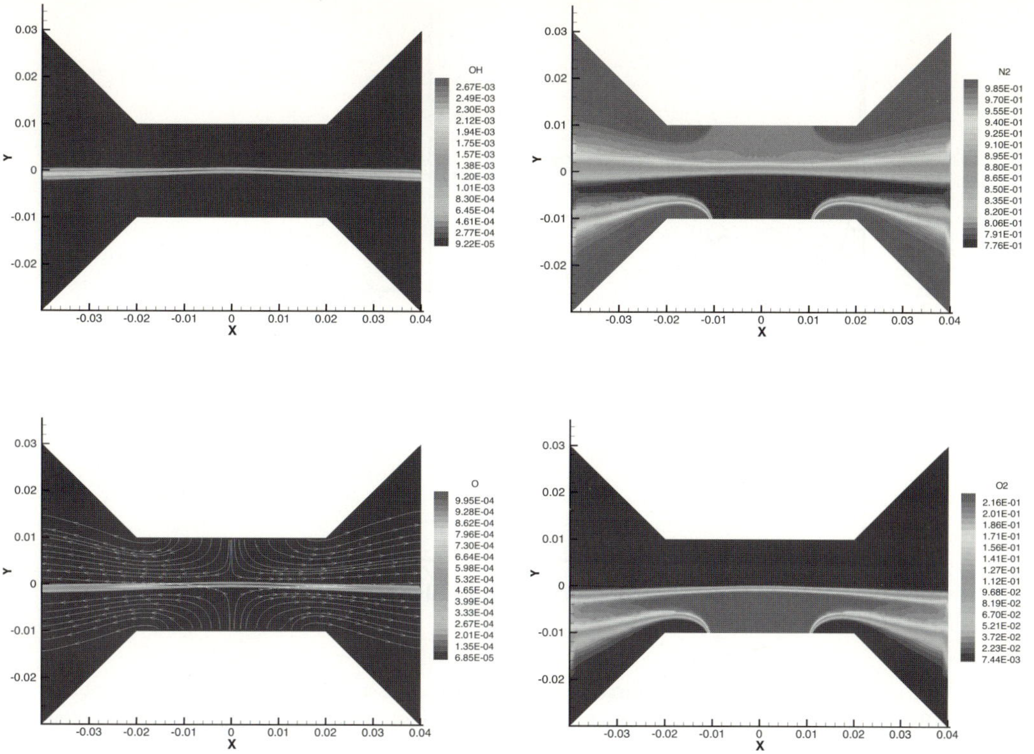

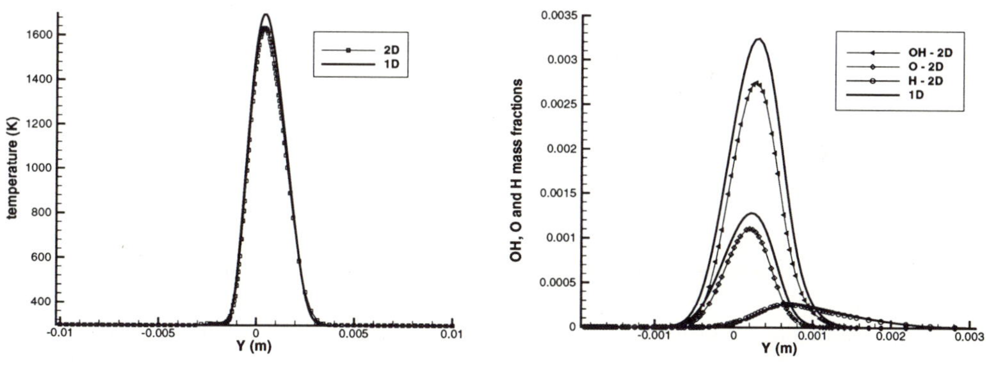

streamlines and O mass fraction O_2 mass fraction

Figure 5: 2D hydrogen/air diffusion flame in a counterflow

Figure 6: Comparison between 2D and 1D approximation computations of diffusion hydrogen/air flames along the center symmetry line, for temperature and mass fractions of some intermediate species.

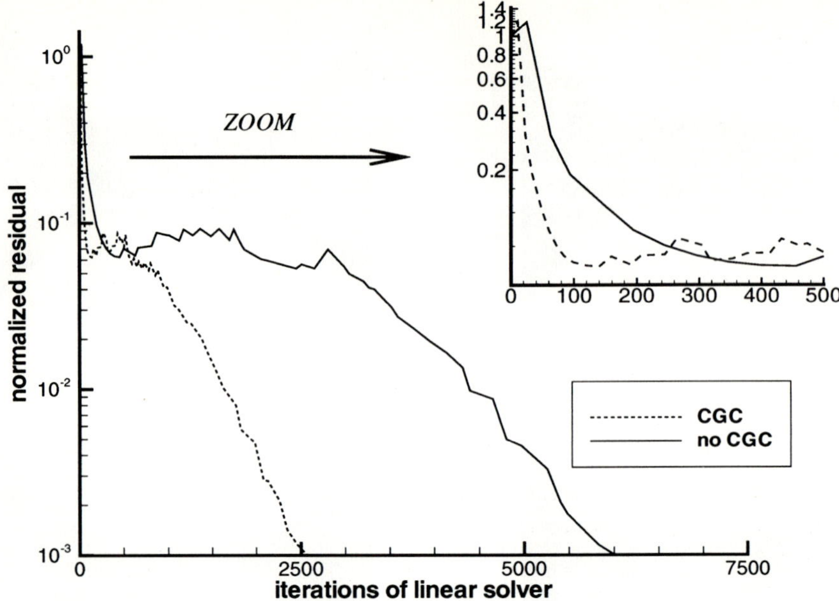

Figure 7: Convergence history of the normalized global residual as a function of the cumulated number of BiCGSTAB iterations. Comparison between multigrid with Gauss Seidel and single grid Gauss-Seidel (no Coarse Grid Correction) preconditioners.

second computation step described previously, see Figure 7. To our knowledge, effectiveness of multigrid methods had not previously been shown for laminar flames with finite rate detailed chemical models. Our literature review on this topic (refer to [9] for a detailed discussion) found only applications of multigrid methods to either hypersonic detailed chemistry reactive flows, or simple chemistry turbulent flame, or flame-sheet modelisations of diffusion laminar flames. But none of these problems feature stiff intermediate species as in low-Mach number laminar flames. From our numerical tests, it is observed that preconditioning with coarse grid correction reduces the total computation time by a factor of about 2, provided that the grid be fine enough for the stiffest species to be represented on at least 2 grid levels.

3.2   A premixed methane flame

The performances of the *UG-C* code are also demonstrated for more realistic configurations close to industrial combustion systems: we simulated two flame configurations of the TOPDEC burner: it is an idealized two-dimensional industrial burner, being currently experimentally and numerically studied as a European BRITE-EURAM project [44], in collaboration with Gaz de France, British Gas, Gasunie and several European universities. The computational domains for both cases are shown in Figure 8: the experimental burner consists of 7 rows of 2D injectors, which, for the middle injector, can be approximated by an infinite number of injectors hence reducing the computational domain to one half of the injector and symmetry boundary conditions. A slender body with a triangular section is injected in the injector for test case # 5, but both configurations are operated at the same conditions. A wealth of experimental data are available at the

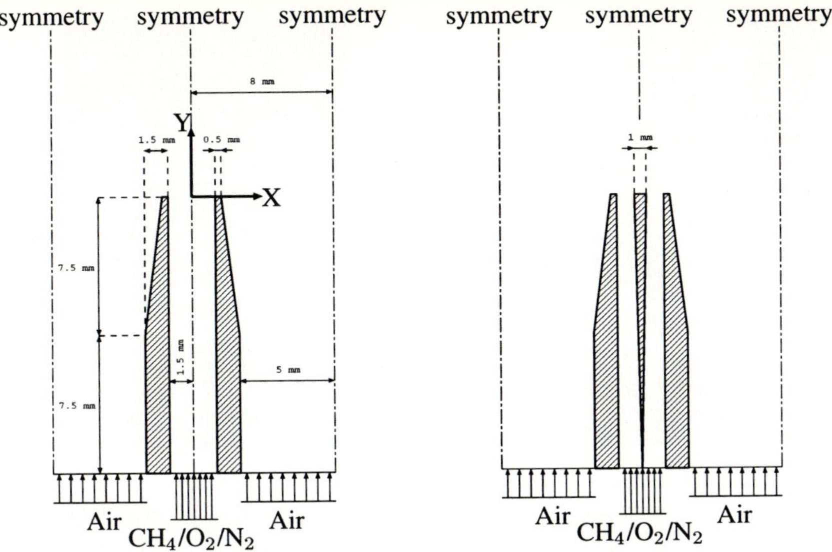

Figure 8: Computational configuration for the TOPDEC burner, configurations # 4 (left) and # 5 (right). The total computational domain (truncated here) extends 6 cm above the injector tip. Computations are performed on half of the domain (all results are shown with symmetrical reconstruction).

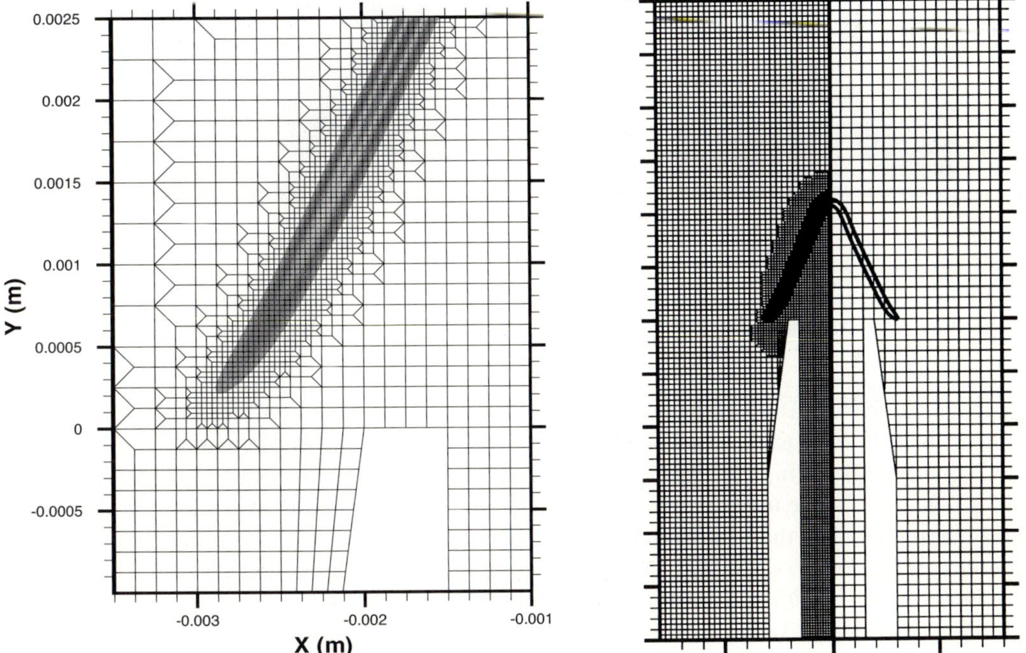

Figure 9: Grid at convergence for the TOPDEC #4 configuration: finest surface grid (15200 nodes) and base level grid (2500 nodes) with flame position given by the isoline $Y_{HCO} = 5.10^{-6}$ (right). Close-up on the injector tip showing the 3 local grid levels, with HCO mass fraction (left)

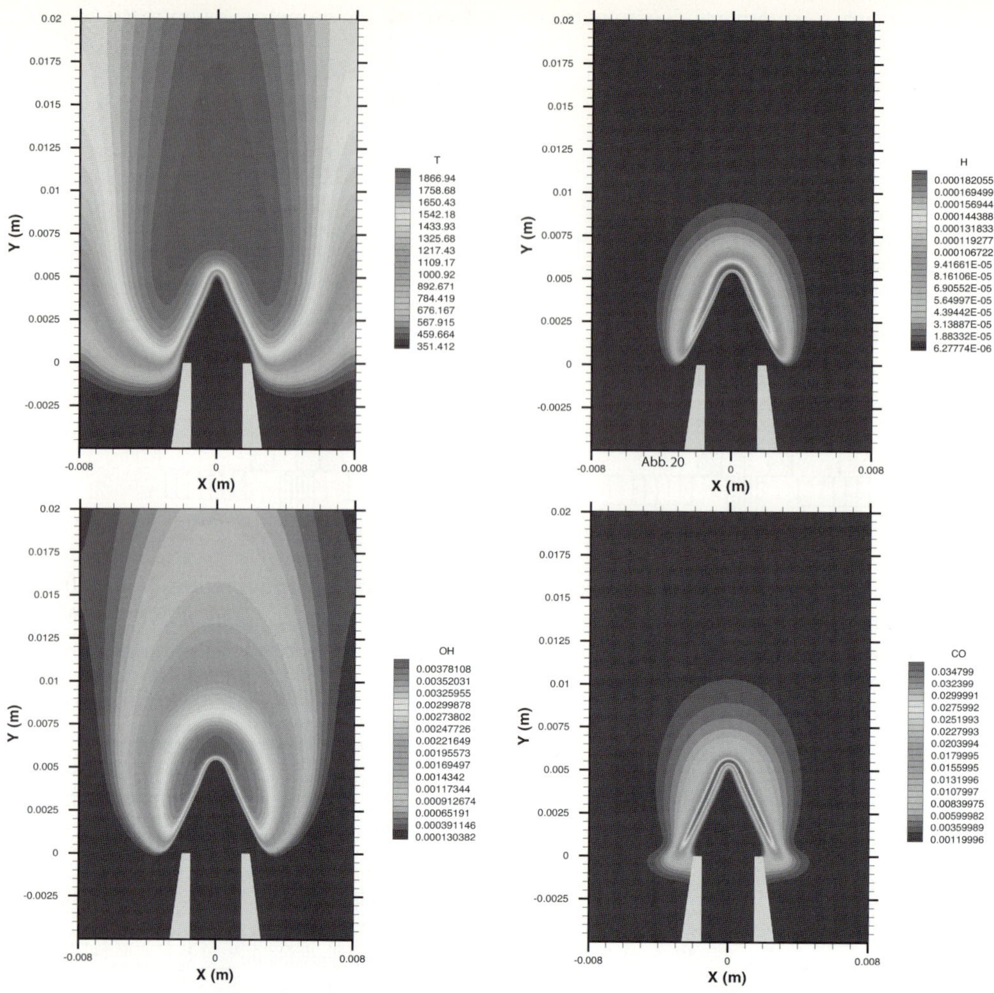

Figure 10: Methane/air premixed flame on the TOPDEC burner: from left to bottom and left to right, temperature, mass fractions of H, OH and CO.

French institute [45]. Both cases that we present here use methane as a fuel which is premixed with air in lean proportions (equivalence ratio of 0.83, with, at the injector inlet, $Y_{CH_4} = 0.046$ and $Y_{O_2} = 0.222$) and injected at 0.98 m.s^{-1}. A secondary air flow is injected between the injectors at 0.05 m.s^{-1}. The total computation from a rough initial solution takes about 20 hours on a SUN Ultra-5 with less than 180 Mb and a final surface grid of about 15200 nodes (see Figure 9). Results (see Figure 10) show reasonable agreement with experimental images of the flame front (see Figure 11). The *UG-C* code also proved its ability to predict subtle influences of the thermal boundary conditions, like those observed on the second test case, see Figure 12: two completely different flame configurations are then possible, depending on the temperature of the burner. The second V-shaped flame is the only one that is stable experimentally, but the first flame has already been observed before the burner warms up.

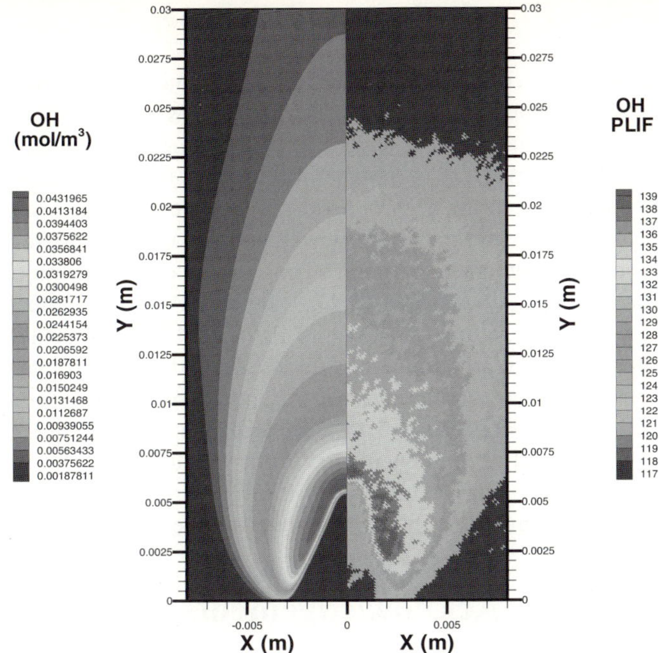

Figure 11: Comparison between computed OH mole fraction (left) and experimental Planar Laser Induced Fluorescence image (right)

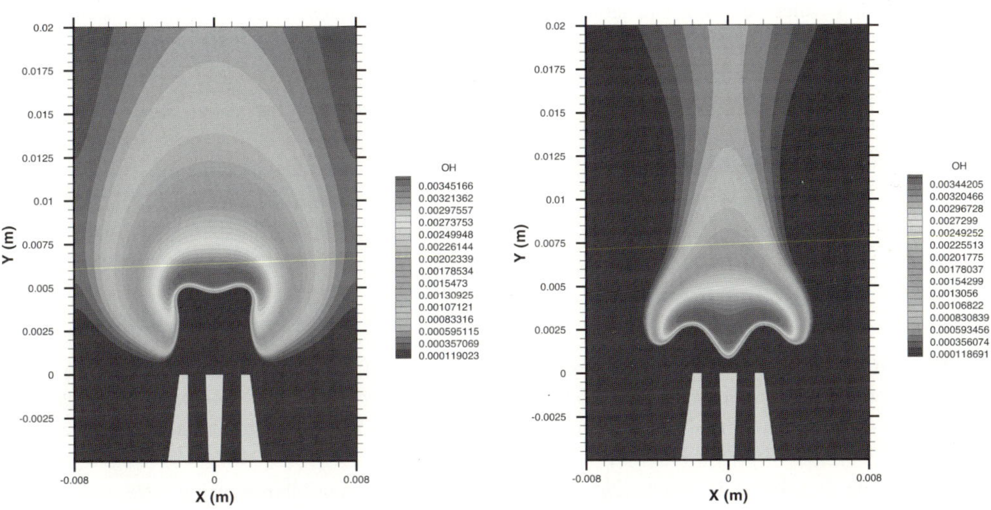

Figure 12: Influence of thermal boundary conditions for the TOPDEC #5 configuration: for cold walls (298 K) the flame is detached from the inserted body (left) whereas it takes a V-shape if temperature is set to 500 K on the upper face of the inserted body (right) which is closer to experimental observations.

Conclusions and Prospects

We have developed the *UG-C* code aimed at simulations of laminar flames with detailed models for molecular diffusion and chemical reactions. Most features of the *UG* library have proved efficient for this physical problem among which: local dynamic grid refinements, robust multigrid/Krylov solvers and the portability on parallel machines. Multigrid acceleration is observed for diffusion flames provided that the highest-level grid be fine enough with respect to the stiffest species present in the flame front. The first validations have been performed for diffusion hydrogen/air and premixed methane/air flames. The interest of such simulations as a prediction tool is demonstrated on the industrial TOPDEC burner. A further reduction in computing time is expected from the future implementation of the ILDM/FPI chemical kinetics reduction method. This will then enable larger simulations (more detailed chemical kinetic schemes, three dimensions) and other applications (chemical vapor deposition reactors, burner optimizations). Current developments are also concerned with the realization of the first 3D computations on a 80-PC-cluster that has recently been constructed at the German institute. An axisymmetrical discretization is also being implemented, which will allow direct validations of the code with experimental results for the counterflow flame that are available at the French institute.

Acknowledgements

This work has been mostly financed by a Ph.D. grant from DGA, France. A prerequisite for its success was that French and German participants could meet on a regular basis. This would not have been possible without support of the French-German project "Numerische Strömungssimulation - Simulation numérique d'écoulements" between the DFG and the CNRS. The authors also wish to thank V. Reichenberger, S. Lang, from the German institute, and S. Belhalfaoui, L. Brenez from the French institute for their help and fruitfull discussions on either software and parallelisation issues, or physical and experimental aspects. We finally thank the I.D.R.I.S., computer center of CNRS, for access to the Cray T3E on which all parallel computations presented in this paper has been performed.

References

[1] Thévenin, D., Thibaut, D., Piana, J., Veynante, D. and Candel, S., Progress in direct and large-eddy simulations of turbulent combustion, *Computation and Visualization of Three-Dimensional Vortical and Turbulent Flows*, (Friedrich, R., and Bontoux, P., Eds.), Notes in Numerical Fluid Mechanics, Vieweg Verlag, 64:263-275, 1998.

[2] Thévenin, D. and Baron, R., Investigation of turbulent non-premixed flames using DNS with detailed chemistry, *Direct and Large Eddy Simulation III*, (Voke, P.R., Sandham, N.D. and Kleiser, L., Eds.), Kluwer Academic Publishers, pp.323-334, 1999.

[3] Candel, S., Thévenin, D., Darabiha, N. and Veynante, D., Progress in numerical combustion, *Combustion Science and Technology*, 149:297-337, 1999.

[4] S. Paxion, A. Gordner, D. Thévenin, P. Bastian, G. Wittum, and S. Candel. Development of a Parallel Multigrid Solver to Investigate Low Mach Number Reactive Flows Using Detailed Chemistry.

In *7th Colloquium of the French-German Research Program on Numerical Flow Simulation (Turbulence/Chemistry group)*, Karlsruhe, 1998.

[5] S.C. Paxion, D. Thévenin, P. Bastian, and S. Candel. Using Multigrid Methods for the Computation of Reactive Flows at Low Mach Numbers. In C.-H. Bruneau, editor, *Sixteenth International Conference on Numerical Methods in Fluid Dynamics*, pages 536–541, Arcachon, France, 1998. Springer-Verlag.

[6] S.C. Paxion, D. Thévenin, P. Bastian, and S. Candel. Towards Low-Mach Number Multigrid Simulations on Distributed-Memory Parallel Computers Using Detailed Chemistry. In *7th International Conference on Numerical Combustion*, page 57, York, 1998.

[7] S.C. Paxion, A. Gordner, N. Neuss, P. Bastian, and D. Thévenin. Parallel Multigrid Computations of Steady Laminar Flames at Low Mach Numbers with Detailed Chemistry. In *8th International Symposium on Computational Fluid Dynamics*, Bremen, 1999.

[8] S. Paxion, A. Gordner, R. Baron, D. Thévenin, and P. Bastian. Development of a Parallel Multigrid Solver to Investigate Low Mach Number Reactive Flows Using Detailed Chemistry. In *8th Colloquium of the French-German Research Program on Numerical Flow Simulation*, Berlin, 1999.

[9] S. Paxion. *Développement d'un solveur multigrille non-structuré parallèle pour la simulation de flammes laminaires en chimie et transport complexes.* PhD thesis, École Centrale Paris, 1999.

[10] T.P. Coffee. Kinetic Mechanisms for Premixed, Laminar, Steady State Methane/Air Flames. *Combust. Flame*, 55:161–170, 1984.

[11] U. Maas and J. Warnatz. Ignition Processes in Carbon-Monoxide-Hydrogen-Oxygen Mixtures. In *22nd Symposium (International) on Combustion*, pages 1695–1704. The Combustion Institute, 1988.

[12] R.J. Kee, F.M. Rupley, and J.A. Miller. CHEMKIN-II: a Fortran Chemical Kinetics Package for the Analysis of Gas Phase Chemical Kinetics. Technical Report SAND89-8009B, SANDIA, 1991.

[13] J.O. Hirschfelder, C.F. Curtiss, and R.B. Bird. *Molecular Theory of Gases and Liquids.* John Wiley & Sons, New York, 1954.

[14] S. Chapman and T.G. Cowling. *The Mathematical Theory of Non-Uniform Gases.* Cambridge University Press, 1970.

[15] A. Ern and V. Giovangigli. Fast and Accurate Multicomponent Property Evaluation. *Journal of Computational Physics*, 120:105–116, 1995.

[16] A. Ern and V. Giovangigli. Optimized Transport Algorithms for Flame Codes. *Combust. Sci. Tech.*, 118:387–395, 1996.

[17] A. Ern and V. Giovangigli. EGLIB: A General-Purpose Fortran Library for Multicomponent Transport Property Evaluation. Technical report, CERMICS, 1997.

[18] A. Ern and V. Giovangigli. *Multicomponent Transport Algorithms*, volume m24 of *Lecture Notes in Physics*. Springer-Verlag, Heidelberg, 1994.

[19] A. Ern and V. Giovangigli. Thermal Diffusion Effects in Hydrogen-air and Methane-air Flames. *Combust. Theory Modelling*, 2:349–372, 1998.

[20] V. Giovangigli. *Structure et extinction de flammes laminaires prémélangées.* PhD thesis, École Centrale Paris, 1988.

[21] V. Giovangigli. Un exemple d'écoulement réactif à cinétique chimique complexe : les flammes laminaires. In *Ecole de Printemps de Mécanique des Fluides Numérique*, Aussois, 1989.

[22] F.A. Williams. *Combustion Theory.* Addison-Wesley, 2nd edition, 1985.

[23] T. Poinsot and S.K. Lele. Boundary Conditions for Direct Simulations of Compressible Viscous Reacting Flows. *Journal of Computational Physics*, 101(1):104–129, 1992.

[24] M. Braack. *An Adaptive Finite Element Method for Reactive-Flow Problems.* PhD thesis, Universität Heidelberg, 1998.

[25] M. Lai, J.B. Bell, and P. Colella. A Projection Method for Combustion in the Zero Mach Number Limit. *AIAA Paper*, (93-3369), 1993.

[26] R.B. Pember, L.H. Howell, J.B. Bell, P. Colella, W.Y. Crutchfield, W.A. Fiveland, and J.P. Jessee. An Adaptative Projection Method for Unsteady, Low-Mach Number Combustion. *Combust. Sci. Tech.*, 140:123–168, 1998.

[27] A. Majda and J. Sethian. The Derivation and Numerical Solution of the Equations for Zero Mach Number Combustion. *Combust. Sci. Tech.*, 42:185–205, 1985.

[28] V. Karlin, G. Makhviladze, and J. Roberts. Numerical Algorithms for Premixed Flames in Closed Channels. In C.-H. Bruneau, editor, *Sixteenth International Conference on Numerical Methods in Fluid Dynamics*, pages 500–505, Arcachon, France, 1998. Springer-Verlag.

[29] L.P.H. de Goey, L.M.T. Somers, W.M.M.L. Bosch, and R.M.M. Mallens. Modeling of the Small Scale Structure of Flat Burner-Stabilized Flames. *Combust. Sci. Tech.*, 104:387–400, 1995.

[30] G.E. Schneider and M.J. Raw. Control Volume Finite-Element Method for Heat Transfer and Fluid Flow Using Colocated Variables. *Num. Heat Transfer*, 11:363–390, 1987.

[31] G.E. Schneider and S.M.H. Karimian. Advances in Control-Volume-Based Finite-Element Methods for Compressible Flows. *Computational Mechanics*, (14):431–446, 1994.

[32] N. Neuss. A New Sparse Matrix Storage Method for Adaptative Solving of Large Systems of Reaction-Diffusion-Transport Equations. Technical Report 99-04, IWR, Heidelberg University, 1999.

[33] P. Bastian, K. Birken, K. Johannsen, N. Neuss, H. Rentz-Reichert, and C. Wieners. UG - a Flexible Software Toolbox for Solving Partial Differential Equations. *Comp. Vis. Sc.*, 1:27–40, 1997.

[34] P. Bastian, K. Birken, K. Johannsen, S. Lang, V. Reichenberger, C. Wieners, G. Wittum and C. Wrobel: Parallel Solutions of Partial Differential Equations with Adaptive Multigrid Methods on Unstructured Grids. High Performance Computing in Science and Engineering, to appear in 2000.

[35] H. Rentz-Reichert. *Robuste Mehrgitterverfahren zur Lösung der inkompressiblen Navier-Stokes Gleichung: Ein Vergleich.* PhD thesis, Institut für Computeranwendungen, Universität Stuttgart, 1996.

[36] H.A. Van der Vorst. Bi-CGSTAB : a Fast and Smoothly Converging Variant of Bi-CG for the Solution of Nonsymmetric Linear Systems. *SIAM Journal of Sci. Statist. Comput.*, 13(n1):631–644, 1992.

[37] P. Bastian. *Parallele adaptive Mehrgitterverfahren.* PhD thesis, Ruprecht-Karls-Universität Heidelberg, 1994.

[38] L.M.T. Somers and L.P.H. de Goey. A Numerical Study of a Premixed Flame on a Slit Burner. *Combust. Sci. Tech.*, 108:121–132, 1995.

[39] T. Dagusé, A. Soufiani, N. Darabiha and J.C. Rolon: Structure of diffusion and premixed laminar counterflow diffusion flames including molecular radiative transfer, *Combust. Explos. Shock Waves*, 29:311-315, 1993.

[40] F. Aguerre, N. Darabiha, J.C. Rolon and S. Candel: Experimental and numerical study of transient laminar counterflow flames, *Combust. Explos. Shock Waves*, 29:306-310, 1993.

[41] F. Lacas, N. Darabiha, J.C. Rolon and S. Candel: Laminar counterflow spray diffusion flames - a comparison between experimental results and complex chemistry computations, *Combust. Sci. Tech.*, 90:35-60, 1993.

[42] F. Aguerre. *Étude expérimentale et numérique des flammes laminaires étirées stationnaires et instationnaires.* PhD thesis, École Centrale Paris, 1994.

[43] T. Croonenbroek. *Diagnostics optiques appliqués aux milieux réactifs . Application aux flammes laminaires étirées à contre-courant.* PhD thesis, Paris VI, 1996.

[44] M. Perrin, A. Garnaud, F. Lasagni, S. Hasko, M. Fairweather, H. Levinsky, J.-C. Rolon, J.-P. Martin, J. Rolon, A. Soufiani, H. Volpp, T. Dreier, and R. Linstedt. TOPDEC Project: New Tools and Methodology for the Design of Natural Gas Domestic Burners and Boilers. In *International Gas Research Conference*, San Diego, 1998.

[45] P. Miquel, N. Larass, M. Perrin, F. Lasagni, M. Beghi, S. Hasko, M. Fairweather, G. Hargrave, G. Sherwood, H. Levinsky, A. Mokhov, H. de Vries, J.-P. Martin, J.-C. Rolon, L. Brenez, P. Scouflaire, D. Shin, G. Peiter, T. Dreier, and H. Volpp. Detailed Measurements in an Idealized and a Practical Natural Gas Household Boiler. In *International Gas Research Conference*, San Diego, 1998.

Numerical Methods for weakly compressible reactive Flows

Th. Schneider[1,2], R. Klein[1,2,3], R. Fortenbach[4] and C.-D. Munz[4]

[1]FB Mathematik und Informatik, Freie Universität Berlin, Arnimallee 2-6,
14195 Berlin, Germany
[2]Konrad-Zuse-Zentrum für Informationstechnik, Takustraße 7,
14195 Berlin, Germany
[3]Potsdamer Institut für Klimafolgeforschung, Telegrafenberg C4,
14412 Potsdam, Germany
[4]Institut für Aerodynamik und Gasdynamik, Universität Stuttgart, Pfaffenwaldring 21,
70550 Stuttgart, Germany

Summary

A numerical method is outlined for reactive and non reactive flows in the regime of a vanishing Mach number. Laminar flames are treated as reactive discountinuities by which burnt and unburnt gas is separated. This approach requires to explicitly prescribe the laminar burning velocity s_l which is the speed of the flame relative to the unburnt gas. A level set representation of the flame front is used to describe the dynamical behaviour of the front and to obtain geometric information of the front such as normal vectors and local curvature etc. in an easy way. As the laminar burning velocity vanishes ($s_l \mapsto 0$) the front becomes a material interface which can be treated by the proposed method as well. The finite volume method is fully conservative and is based upon a projection method. Special attention was paid to design a very simple but efficient numerical scheme.

1   Governing equations

We consider a set of governing equations for the zero Mach number limit. Details can be found in *Matalon et al.* [18], *Schochet* [20], *Klainerman et al.* [13], *Majda et al.* [16] and *Klein* [14]. In case of a reactive flow only two species are considered, namely the burnt and unburnt one. Laminar flames are treated as reactive discontinuities. Two pressure variables are introduced : the thermodynamic pressure P_0 and the hydrodynamic pressure $p^{(2)}$. An asymptotic analysis yields spatial homogenity of P_0 ($\nabla P_0 = 0$). $p^{(2)}$ acts as a balance of force agent that forces the velocity field to obey a divergence constraint which results from the energy equation, as it will be shown later in this section.

Since a finte volume method is used to solve the governing equations it is natural to consider the integral formulation of the governing equations for an arbitrary control volume V which are :

$$\nabla P_0 = 0,$$

$$\frac{d}{dt} \int_V \rho \, dV + \oint_{\partial V} \rho \vec{v} \cdot \vec{n} dA = 0,$$

$$\frac{d}{dt} \int_V \rho \vec{v} \, dV + \oint_{\partial V} (\rho \vec{v} \circ \vec{v}) \cdot \vec{n} dA = -\oint_{\partial V} p^{(2)} \, \vec{n} dA + \frac{1}{Fr^2} \int_V \rho \vec{g} dV$$

$$+ \frac{1}{Re} \oint_{\partial V} \left(S - \frac{2}{3}(\nabla \cdot \vec{v}) \, I \right) \cdot \vec{n} dA \,,$$

"

$$\frac{d}{dt} \int_V \rho E \, dV + \oint_{\partial V} (\rho E + P_0)\, \vec{v} \cdot \vec{n} dA \quad = \quad \frac{\gamma}{\gamma - 1} \frac{1}{RePr} \oint_{\partial V} \nabla \left(\frac{P_0}{\rho} \right) \cdot \vec{n} dA \, ,$$

$$\frac{d}{dt} \int_V \rho Y \, dV + \oint_{\partial V} \rho Y \vec{v} \cdot \vec{n} dA \quad = \quad \int_V \rho \omega_Y \, dV \, ,$$

$$P_0 \quad = \quad \rho T \, . \tag{1.1}$$

Where ω_Y represents chemical production, Y denotes mass fraction of the unburnt species and S is the shear tensor.

1.1 Divergence constraint

In case of a vanishing Mach number, two species (burnt and unburnt) and a constant heat release Q the total specific inner energy is given by:

$$\rho E = \frac{P_0}{\gamma - 1} + \rho Y Q \, . \tag{1.2}$$

Therefore, the energy equation can be written as follows:

$$\frac{|V|}{\gamma - 1} \frac{dP_0}{dt} + \frac{\gamma P_0}{\gamma - 1} \oint_{\partial V} \vec{v} \cdot \vec{n} dA = \quad - \int_V \rho \omega_Y Q dV$$

$$+ \quad \frac{\gamma P_0}{\gamma - 1} \frac{1}{RePr} \oint_{\partial V} \nabla \left(\frac{1}{\rho} \right) \cdot \vec{n} dA. \tag{1.3}$$

Straighforward algebraic manipulation of the above equation yields a divergence constraint :

$$\oint_{\partial V} \vec{v} \cdot \vec{n} dA \quad = \quad -\frac{1}{\gamma P_0} \frac{dP_0}{dt} |V| - \frac{\gamma - 1}{\gamma P_0} \int_V \rho \omega_Y Q dV + \frac{1}{RePr} \oint_{\partial V} \nabla \left(\frac{1}{\rho} \right) \cdot \vec{n} dA. \tag{1.4}$$

The temporal change of the background pressure is determined by integration the latter equation over the complete domain Ω:

$$\frac{1}{\gamma P_0} \frac{dP_0}{dt} = \quad - \quad \frac{1}{|\Omega|} \oint_{\partial \Omega} \vec{v} \cdot \vec{n} dA - \frac{1}{|\Omega|} \frac{\gamma - 1}{\gamma P_0} \int_\Omega \rho \omega_Y Q dV$$

$$+ \quad \frac{1}{|\Omega| P_0} \frac{1}{RePr} \oint_{\partial \Omega} \nabla T \cdot \vec{n} dA. \tag{1.5}$$

The first and last term on the right hand side of equation 1.5 are given by boundary condtions. Therefore, the second term remains to be determined which is achieved by using a level set representation of the flame front as it will be explained in the next section.

1.2   Level set method

In a level set method an additional scalar field G is introduced; its contour line or surface where $G = 0$ represents the flame front. The evolution in time is determined by a transport equation that was first given by *Markstein* [17]:

$$\frac{\partial}{\partial t} G + (\vec{v} + s_l \vec{n}) \cdot \nabla G = 0, \ \vec{n} = \frac{\nabla G}{|\nabla G|} \ . \tag{1.6}$$

s_l is the laminar burning velocity which is assumed to be a function of the unburnt state and the local front curvature κ.

Strictly speaking only the iso line or iso surface where $G = 0$ is physically meaningful, which can be taken into account when designing a numerical method to solve the G transport equation to reduce the numerical effort.

The second term on the right hand side of equation 1.4 can be expressed in terms of G:

$$\int_V \rho \omega_Y Q dV = - \int_V \rho_u s_l \delta(G) |\nabla G| Q dV. \tag{1.7}$$

ρ_u represents the local unburnt density next to the front.

1.3   Jump conditions

Since laminar flames are considered as reactive discontinuities the *Rankine-Hugoniot* jump conditions hold across the discontinuity. Under the assumption of a constant heat release they are given by:

$$\begin{aligned}
[\![\rho]\!] &= -\frac{P_0 [\![T]\!]}{T_u (T_u + [\![T]\!])} \ , \\
[\![v_n]\!] &= \frac{s_l}{\rho_b} [\![\rho]\!] \ , \\
[\![v_t]\!] &= 0 \ , \\
[\![p^{(2)}]\!] &= \frac{s_l^2 \rho_u}{\rho_b} [\![\rho]\!] \ , \\
[\![P_0]\!] &= 0 \ .
\end{aligned} \tag{1.8}$$

Where $[\![\phi]\!]$ denotes a difference $[\![\phi]\!] = \phi_b - \phi_u$.

2   Fully conservative Finite Volume Method

The underlying method to solve the governing equations is a projection method by *Chorin* [9] and *Teman* [27]. Major modifications were introduced by *Bell et al.* [4]. In a conservative finite volume method that is based upon a collocated discretisation two kinds of velocity fields exist. The first one is the field of the velocities normal to the cell interface that are associated with the convective fluxes; these velocities will be referred to as interface velocities in the following. The second one is corresponding to the volume averaged momentum within the grid cell itself; from now on, the term cell-centered velocities will be used instead.

Therefore, two different control volumes are considered to formulate the divergence in each case. For the interface velocities the control volume V_h is the cell itself. A staggered control volume $\overline{V}_h$ is used for approximating the divergence of the cell-centered velocities. For the two dimensional case the conrol volumes are illustrated below.

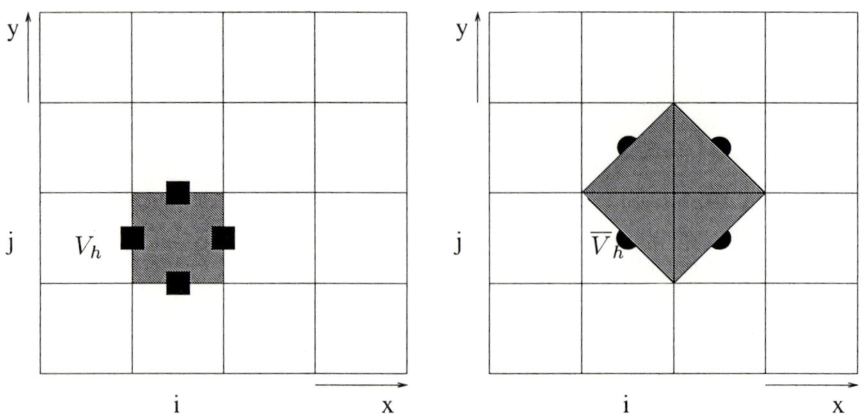

Figure 1 Sketch of control volumes $\overline{V}_h$ and V_h

The corresponding discrete second order approximations are:

$$\frac{1}{|\overline{V}_h|} \oint_{\partial \overline{V}_h} \vec{v} \cdot \vec{n} dA \approx \frac{1}{2\Delta x \Delta y} \left[\; (u_{i+1,j+1} + u_{i+1,j} - u_{i,j+1} - u_{i,j}) \Delta y \right.$$

$$\left. + \; (v_{i+1,j+1} + v_{i,j+1} - v_{i+1,j} - v_{i,j}) \Delta x \; \right],$$

$$\frac{1}{|V_h|} \oint_{\partial V_h} \vec{v} \cdot \vec{n} dA \approx \frac{1}{\Delta x \Delta y} \left[\; (u_{i+1/2,j} - u_{i-1/2,j}) \Delta y \right.$$

$$\left. + \; (v_{i,j+1/2} - v_{i,j-1/2}) \Delta x \; \right]. \tag{2.9}$$

In the present method projection operators are chosen such that both velocity fields comply with their corresponding divergence constraint in an exact way. In this respect the method is different from *approxmaitve* projection methods by *Almgren et al.* [1]. They sacrifice exact compliance of the velocity field with its divergence constraint in favor of a convenient pressure discretisation that avoids local grid decoupling and thus suppresses spurious pressure oscillations. A detailed description of the numerical algorithm that performes four projections by requiring only two solutions of elliptic poisson equations per time step will be given by *Schneider et al.* in [24].

2.1   Efficient numerical solution of poisson problems

As explained in the previous subsection, two projection steps are performed during a time step. Each of them requires a solution of a poisson equation. Therefore, the overall efficiency of the method mainly depends on how efficient these poisson problems can be solved numerically. As in [21] a preconditioned conjugate gradient (PCG) method is used. A multigrid f-cycle serves as a preconditioner. Grid tranfer operators are constructed in an algebraic multigrid fashion that was proposed by *Kickinger* in [15]. Linear coarse grid operator are obatined by the Galerkin product and a Gauß-Seidel algorithm is used as a smoother involving two pre- and post smoothing steps. Numerical tests indicate that this solver requires a numerical effort which is proportional to the number of unknowns of the linear system.

2.2 Level Set based Front Tracking

Sussman et al. used a level set approach to represent material interfaces in incompressible two phase flows. One major problem they reported is conservation of mass. Several strategies were developed to overcome this drawback. *Schneider et al.* presented at [22] a fully conservative level set based tracking method.

Smilianoski, *Moser* and *Klein* introduced in [25] a capturing tracking hybrid scheme algorithm to compute flows in a compressible regime to simulate deflagration to detonation transitions. Their key idea is to divide all grid cells into mixed and non mixed cells each time step. Mixed are intersected by the front during a time step. The level set approach is used to obtain information about the front topology. *Terhoeven* extended this is idea to the zero Mach number limit in [28].

3   Numerical Flow Simulations

In this section we dicuss three simulations that represent flow problems which can be treated with the numerical method described above. The first example is a non-reactive flow which is used to demonstrate that effects of global compression and expansion are treated properly by the method such that thermoacoustic effects can be resolved. The second example shows the capability of the method to *survive* the limit of a vanishing laminar burning velocity s_l where the flame front becomes a material interface as $s_l \mapsto 0$. Thus, the flow is non reactive in this case as well. The last example is a bunch of reactive flow computations to illustrate the problems that remain to be solved.

3.1 Thermoacoustic Refrigerator

Previous numerical investigations of a thermoacoustic refrigerator were performed by *Worlikar et al.* in [29] and [30]. As sketched in fig. 2 a thermoacoustic refrigerator basically consists of a resonance tube filled with gas, an acoustic driver (e.g. a loudspeaker), two heat exchangers and a stack of plates. The acoustic driver generates a standing wave within the tube; due to continous compression and expansion of the gas temperature gradients in the gas as well as in the plates are established which are tangential to the plate surface. A fluid particle receives heat during expansion from plate right from its mean position. When the gas compressed the temperature of the particle rises and it moves to left. At a certain position it releases heat to the plate. As a consequence of this the right half of the plate is cooled and the left half is heated by the gas. Provided a proper choice of the design parameters heat can be transported from the cold heat exchanger to the hot heat exchanger. The impact of numerical flow simulations on this problem is to assist determining the design parameters of the device such as plate length, blockage ratio etc.

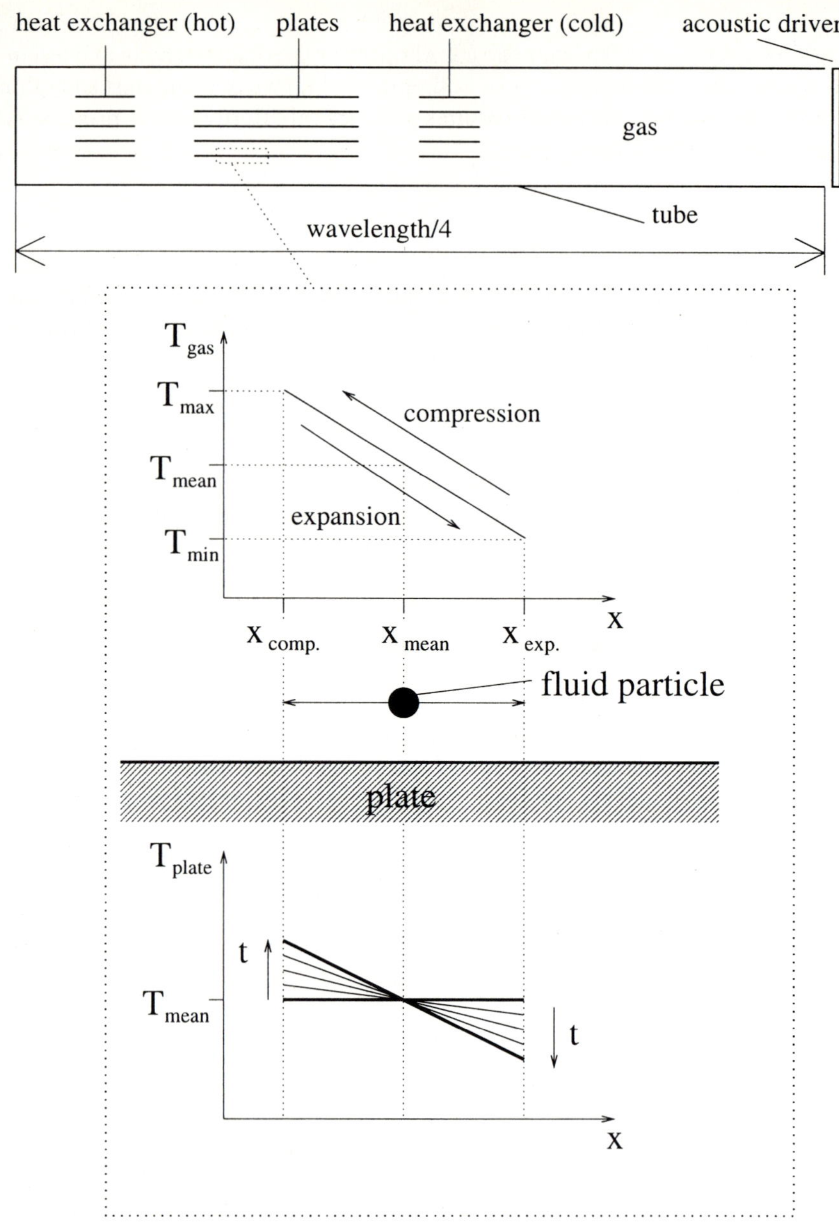

Figure 2 Basic principle of a thermoacoustic refrigerator

The flow is non-reactive, vicous effects and heat conduction are accounted for. The flow problem is characterized by two length scales which differ by orders of magnitudes. The hydrodynamic scale is given by the particle displacement; whereas the acoustic scale is given by the acoustic wavelength itself. The reference length l_{ref} in the problem is given by the distance between the plates of the stack; the reference time is $t_{ref} = \omega^{-1}$. The Reynolds number of the problem is

200, the Prandtl number 0.67 (Properties of Helium were considered). Since the Mach number is small $M \approx O(10^{-3})$ the zero Mach number regime discribes the problem in a proper way. The modelling approach is to resolve the two-dimensional zero Mach number flow in the neighbourhood of the stack and accounting for acoustic effects by imposing suitable boundary conditions at the inlet and outlet at the same time. Making extensive use of geometric symmetries the integrational domain can be reduced as sketched in fig. 3.

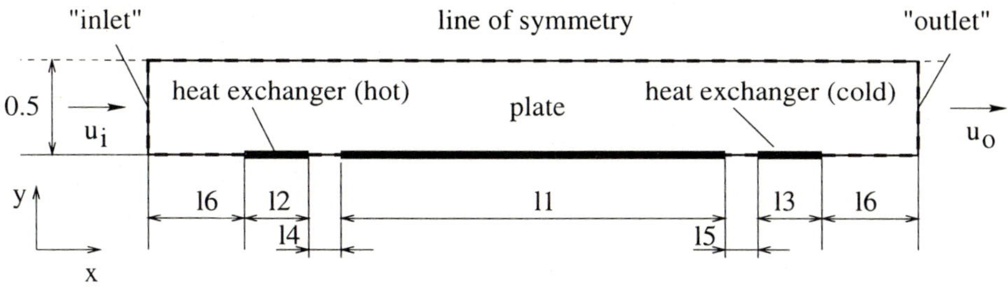

Figure 3   Integration domain

Both heat exchangers are modelled as bodies of a constant temperatures ($T_h = 1.05, T_c = 0.95$). The boundary conditions at the in- (suffix $_i$) and outlet (suffix $_o$) are :

$$u_i(t) = \hat{u}_i \cos(t), \quad v_i = 0, \quad u_o(t) = \hat{u}_o \cos(t), \quad v_o = 0. \tag{3.10}$$

To determine the normal velocity amplitudes at the in- and outlet $\hat{u}_i, \hat{u}_o$ the energy equation is considered :

$$\frac{1}{\gamma P_0} \frac{dP_0}{dt} = -\frac{1}{|\Omega|} \oint_{\partial\Omega} \vec{v} \cdot \vec{n} dA + \frac{1}{|\Omega| P_0} \frac{1}{RePr} \oint_{\partial\Omega} \nabla T \cdot \vec{n} dA . \tag{3.11}$$

When neglecting the second term on the right hand side which is reasonable due the magnitude of the Reynolds number the energy eqution reduces to an ordinary differential equation if the boundary conditions eq. 3.10 are taken into account. This ordinary differential equation be integrated with relative ease :

$$\rightsquigarrow P_0(t) = P_{0,0} \exp\left(-\gamma \frac{l_{\mathrm{ref}}}{|\Omega|} (\hat{u}_o - \hat{u}_i) \sin(t)\right) . \tag{3.12}$$

$\exp()$ denotes the exponential function. The thermodynamic pressure P_0 obviously only depends on the difference of the amplitudes of the normal velocities. In addition to the governing equations a heat conduction equation for the plates has to be solved :

$$\frac{\partial}{\partial t} T = \frac{1}{Fo} \left(\frac{\partial^2}{\partial x^2} T + \frac{2\kappa}{\delta} \frac{\partial T}{\partial y}\bigg|_{\mathrm{Gas}}\right), \delta = 0.1 * l_{\mathrm{ref}} . \tag{3.13}$$

In the plates temperature variations in the y-direction are small and therefore neglected. δ is a nominal plate thickness. A *Crank-Nicholson* scheme is used to solve the heat conduction equation.

Five different cases were chosen to investigate the influence of length of the heat exchangers on the heat flux that out of the cold heat exchanger. Complete geometric information of all cases is listed in the table below.

Table 1 Geometric data of the cases investigated, $Dr = \Delta P_0/P_0 = 1.4\%$, $Re = 200$ and $Pr = 0.67$

case	l_1	l_2 bzw. l_3	l_4 bzw. l_5	$\hat{u}_o$	$\hat{u}_i$	total length	n_x	n_y
1	15	0.5	0.25	1.03519	0.79297	28.5	684	12
2	15	1.0	0.25	1.03884	0.78818	29.5	708	12
3	15	1.5	0.25	1.04247	0.78337	30.5	732	12
4	15	2.0	0.25	1.04608	0.77854	31.5	756	12
5	15	3.0	0.25	1.05324	0.76884	33.5	804	12

In order to reduce the time to reach a kind periodical steady state a non constant density distribution is set as initial data. This density field is given by

$$
\rho(x,y,t=0) = \begin{cases} \rho_h & x \leq l_h, \\ \rho_h + \frac{\rho_c - \rho_h}{l_c - l_h}(x - l_h) & l_h < x < l_c \\ \rho_c & x \geq l_h. \end{cases} \tag{3.14}
$$

were $\rho_h = \dfrac{P_0}{T_h}$, $\rho_c = \dfrac{P_0}{T_c}$, $l_h = l_6 + l_2$, $l_c = l_6 + l_2 + l_4 + l_1 + l_5$.

As it can be seen from fig. 4 after 15 cycles the heat flux into the cold heat exchanger indicates to be close to a steady state. A comparison of the integrated heat fluxes that are obtained with the present method and the method of *Worlikar et al.* (see [30]) however show a difference that increases as the length of the heat exchanger increases.

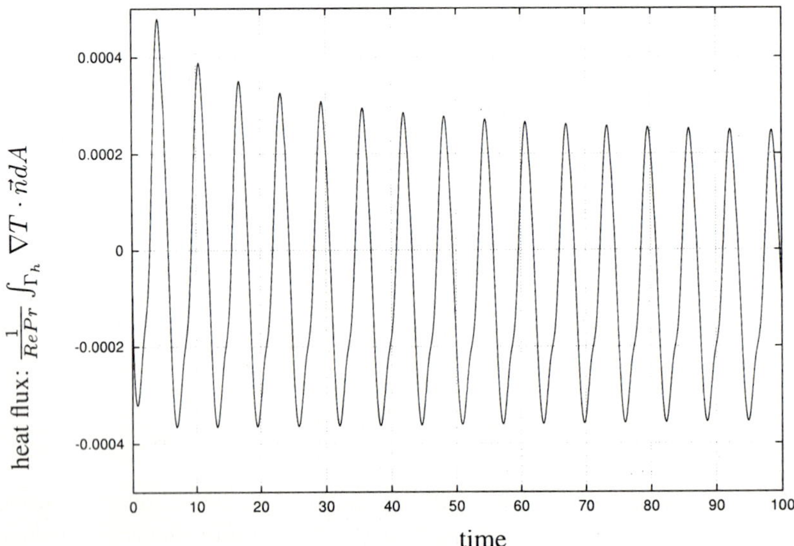

Figure 4 Temporal evolution of heat **into** cold heat exchanger, case 1

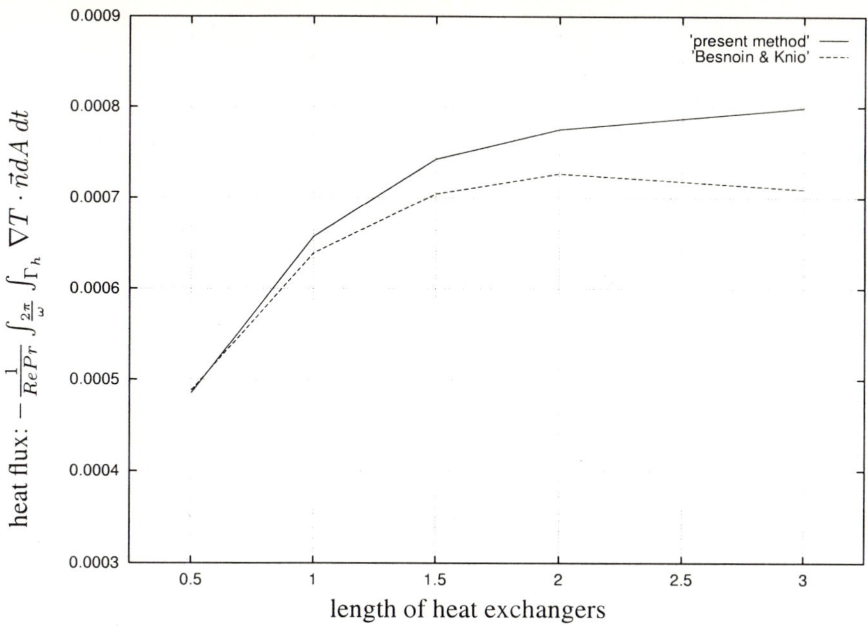

Figure 5 Integrated heat flux **out of** the cold heat exchanger during 60th cycle, cases 1-5; present method (solid line), data of *Besnoin* und *Knio* (dashed line)

3.2 Interface Tracking in Two-Phase Flow

If a level representation of an interface is used in combination with a finite volume method to compute incompressible two-phase flows one difficulty arises due to the redundancy of the information the level set provides and the continuity equation. In the discrete case there is a difference in the information that is obtained from the level set and the density distribution which is in the order of the truncation errors of the scheme used. Therefore, *Schneider et al.* proposed [23] a way to synchronize the level set and the continuity equation by applying a normal correction velocity to the level set in order to control the relative error. A similiar idea was presented by *Bourlioux* [7]. When solving the continuity equation an elliptic flux correction is applied to ensure that the volume fraction f in a cell remains bounded : $0 < f < 1$. This is a correction which is of the magnitude of the truncation errors of the scheme. Therefore, the overall order of the method is neither effected by the elliptic correction nor by the slight normal movement of the level set, that introduced to aviod an error accumulation. A complete description of this synchronisation can be found in [23].

To illustrate the potential of this idea an incompressible rising bubble flow is considered. Rising bubble flows have been experimentally investigated by *Bhaga et al.*[6] and *Hnat et al.* [12].

The reference length scale is the initial diameter of the bubble. The density ratio is 714:1, the viscosity ratio 6667:1 the charactistic numbers are : Reynolds number $Re = 9.7$, Froude number $Fr = \sqrt{0.78}$, Weber number $W = 7.6$. The integrational domain consists of 100×200 grid cells. The initial shape of the front is a circle located at (0.5,0.5) with a radius of 0.5. The integration starts at t=0 and stopped at t=4.0.

In fig. 6 the relative error of the level set representation of the front is shown. It indicates that an

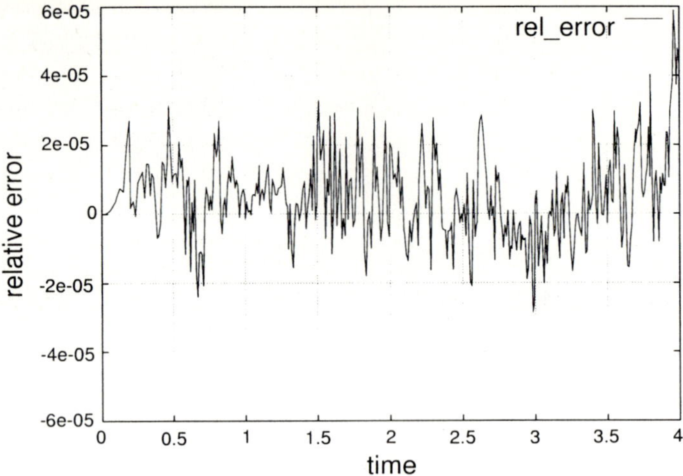

Figure 6 Temporal evolution of relative heavy mass error $(m_s(G) - m_s)/m_s$

accumulation of the relative error is avoided by the synchronisation. In fig. 7,8 the interface, the velocity field and instantaneous streamlines are shown at different times.

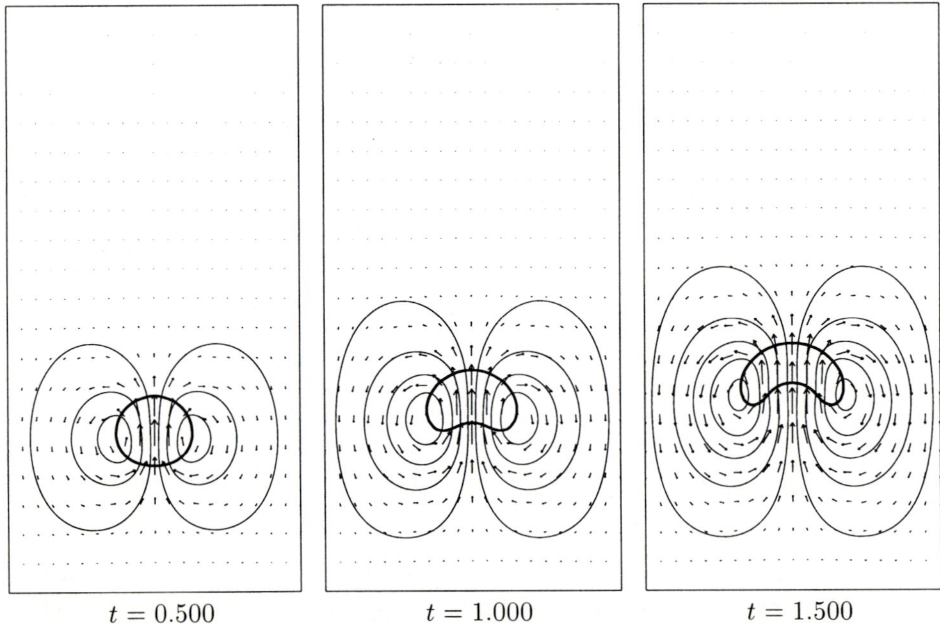

Figure 7 Rising bubble at different times: material interface (thick line), 10 instantaneous streamlines [-0.1667,0.1667] (thin lines) and velocity field (arrows), density ratio $\rho_h/\rho_l = 714$, viscosity ratio $\mu_h/\mu_l = 6667$, Reynolds number $Re = 9.7$, Froude number $Fr = \sqrt{0.78}$, Weber number $W = 7.6$

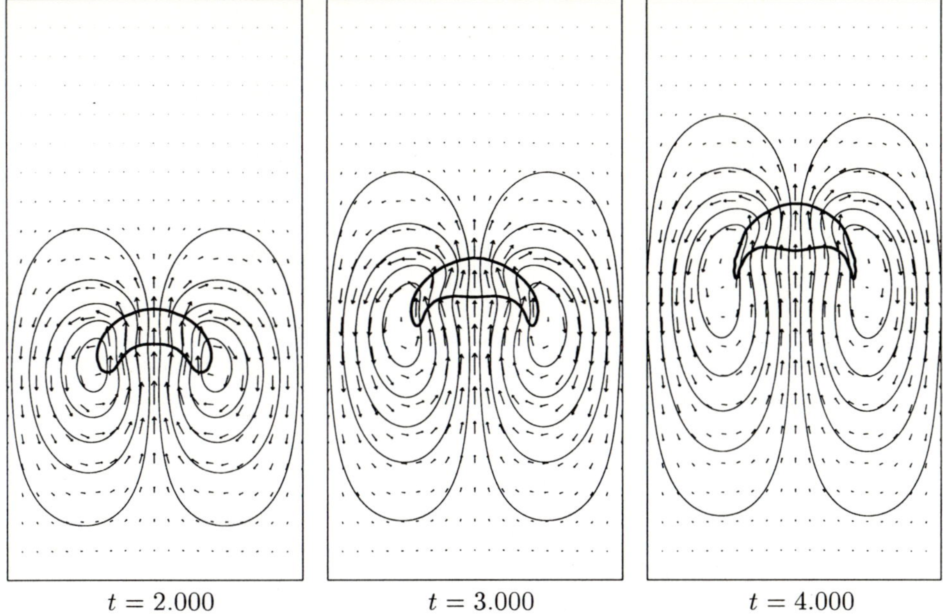

$t = 2.000$ $t = 3.000$ $t = 4.000$

Figure 8 Rising bubble at different times: material interface (thick line), 10 instantaneous stream-lines [-0.1667,0.1667] (thin lines) and velocity field (arrows), density ratio $\rho_h/\rho_l = 714$, viscosity ratio $\mu_h/\mu_l = 6667$, Reynolds number $Re = 9.7$, Froude number $Fr = \sqrt{0.78}$, Weber number $W = 7.6$

3.3 Flame Tracking

Three different cases are presented to demonstrate the capabilities of the method and to illustrate tasks that remain to be solved. All cases are initialized with a density ratio of unburnt to burnt density of 5. In all cases a simple burning law was used $s_l = 0.1/\rho_u$ as well as a constant heat release $Q = 4\gamma/(\gamma - 1)$. For each case to different figures are shown : the velocity field in combination with the front and a pseudocolor plot showing the density field.

Circular Flame

The initial front is a circle located at $(0.5,0.5)$ with radius of 0.2. The domain is a unit square consisting of 65×65 grid cells. Both figures (9, 10) show reasonable results. The symmetry of the problem is maintained. In the next case a collision of two initially circular flames was tested.

Flame Collision

In this case two circular flames are initialized. Each flame has an initial radius of 0.1. The centers are located at $(0.35,0.5)$ and $(0.65,0.5)$. As it can be seen from the figure below at $t = 0.4$ the solution is not symmetric any longer . The integrational domain is the same as in the previous case. The topology change of the front is resolved by the level set in correct way.

One major problem can be seen from fig. 12. At locations where the front developed a sharp

209

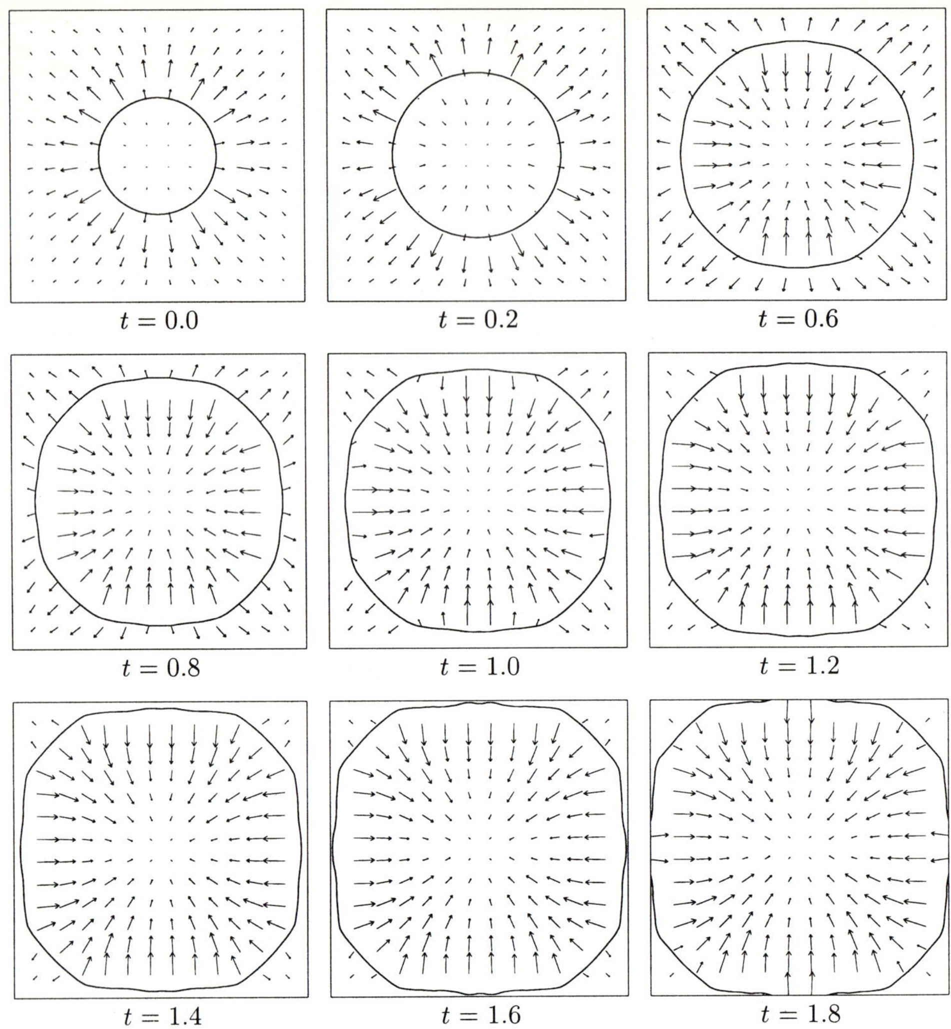

Figure 9 Two-dimensional circular flame : flame front (line) and velocity field (arrows); burning law $s_l(\mathbf{u}_u) = 0.1/\rho_u$, heat release $Q = 4\gamma/(\gamma - 1)$, 65×65 grid cells

cusp a trail of high densities remains within the burnt gas which is unphysical. The reason for this behaviour is relateted to the level set representation of the front. Under certain circumstances a front may not be detected in a cell since piecewise linear functions are used to represent the level set function. Therefore, the local divergence of the velocities at the cell surface of this particular cell is to small which implies that not sufficient mass is transported away from this particular cell.

Tulip Flame

In the last case a rectangular domain was chosen with a length to width ratio of 4:1 consisting of

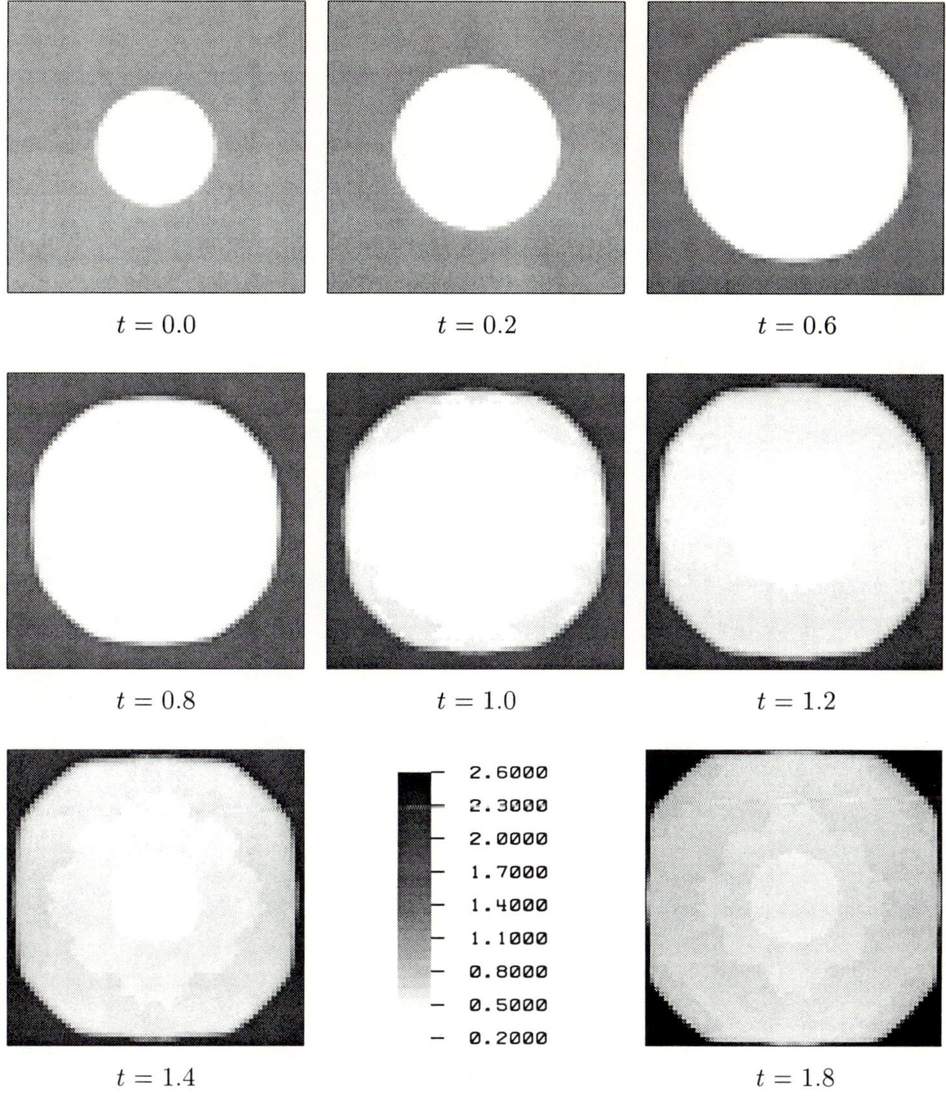

Figure 10 Two-dimensional circular flame : density field; burning law $s_l(\mathbf{u}_u) = 0.1/\rho_u$, heat release $Q = 4\gamma/(\gamma - 1)$, 65×65 grid cells

128×32 grid cells. Again the shape of the intial front is circular, the center is located at (0.5,0.5) the radius is 0.25. The results of this case show the evolution of the wellknown tulip flame. At time $t = 4.0$ two small cusps appear in the center that have merged at $t = 8.0$. The case shows the same large density trails in the neighbourhood of cusps. The reason for this are the same as in the previous case.

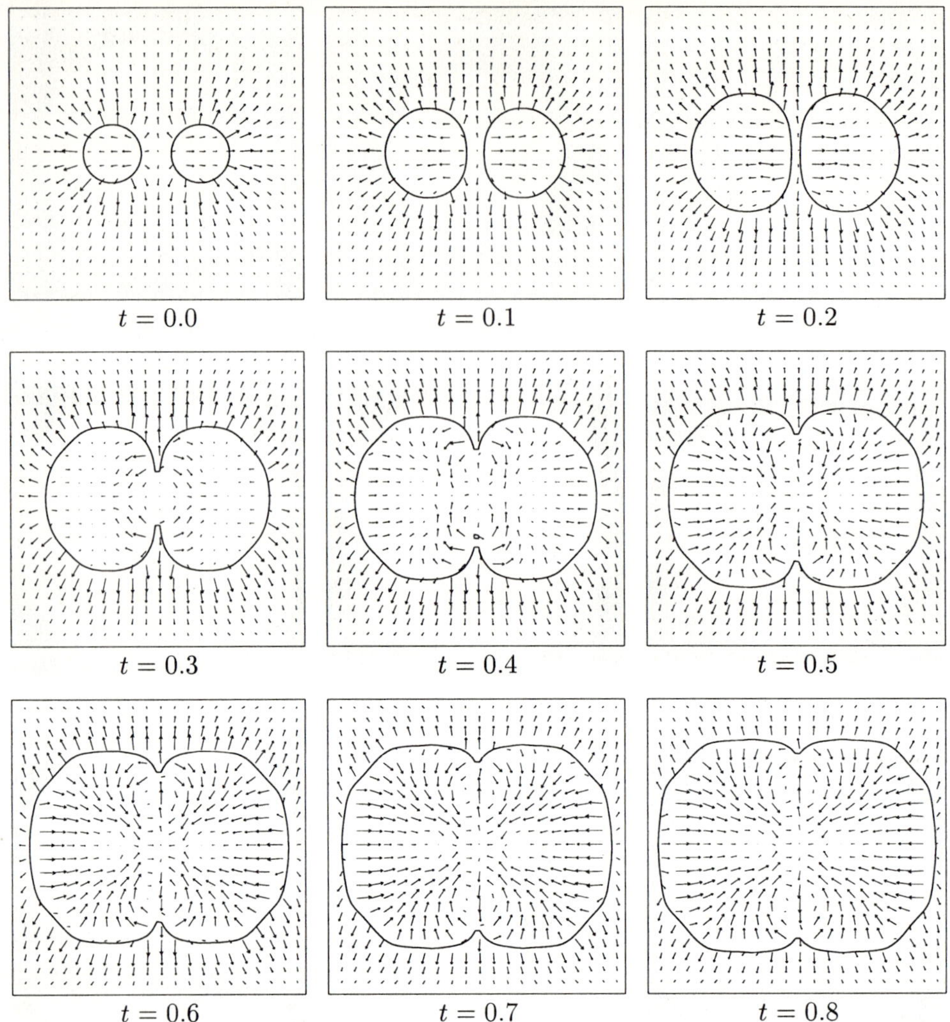

Figure 11 Two-dimensional flame collision : flame front (line) and velocity field(arrows); burning law $s_l(\mathbf{u}_u) = 0.1/\rho_u$, heat release $Q = 4\gamma/(\gamma - 1)$, 65×65 grid cells

4   Conclusions and future work

The examples presented in the last section show the applicability of the numerical method to a wide range of problems. Nevertheless, the treatment of mixed cells which are cells that are intersected by the front must be improved in order to avoid such detection problems that lead to unreasonable effects in the density field as discussed in the last two cases.

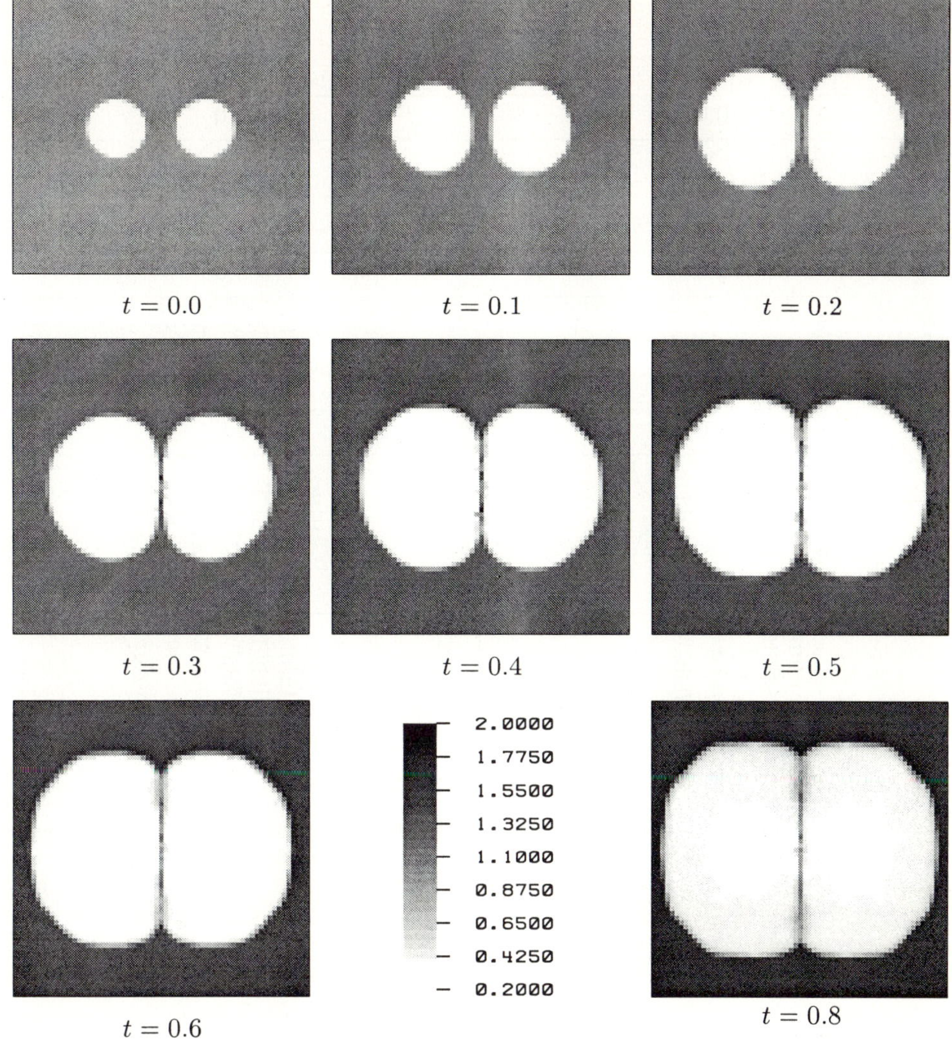

Figure 12 Two-dimensional flame collision : density field; burning law $s_l(\mathbf{u}_u) = 0.1/\rho_u$, heat release $Q = 4\gamma/(\gamma - 1)$, 65×65 grid cells

Acknowledgements

The authors wish to thank the DFG (Deutsche Forschungsgemeinschaft) for a grant under which the project "Numerical Methods for weakly compressible reactive flows" (KL 611-5) is supported, embedded in the French-German project "Numerische Strömungssimulation - Simulation Numerique d'Ecoulements". Furthermore, the authors are indebted to Dr. O. M. Knio from Johns Hopkins University Baltimore, Maryland for advice and helpful discussions concerning the thermoacoustic part of the work presented. Finally, TS and RK would like to thank the DAAD (Deutscher Akademischer Austauschdienst) for generous financial support that enabled cooperation with Dr. Knio.

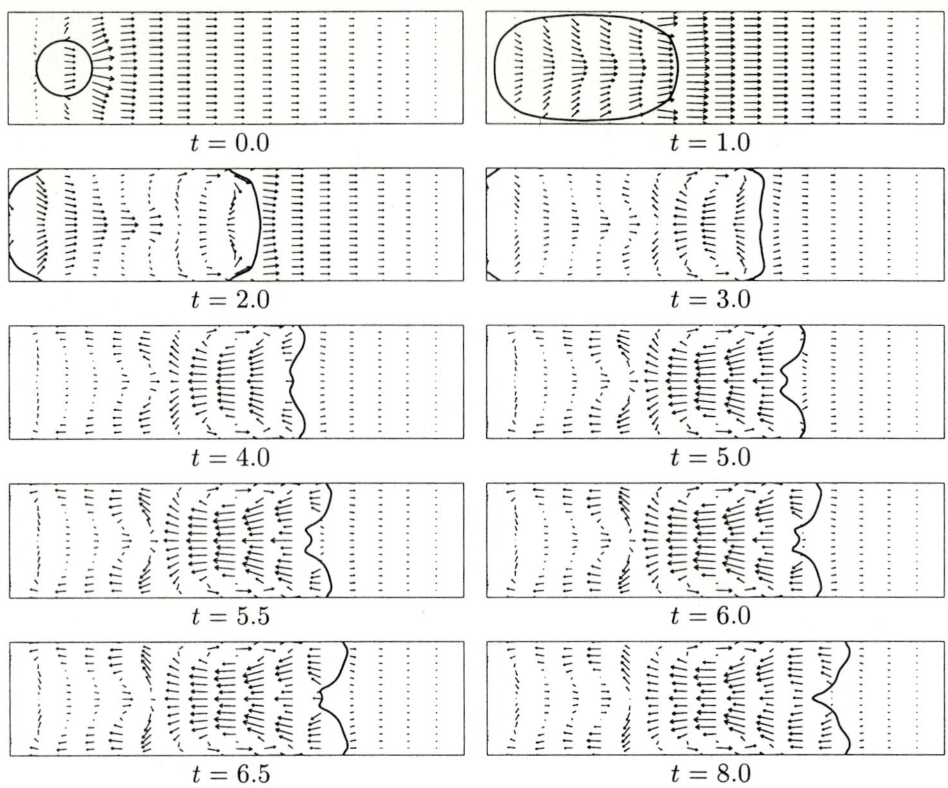

Figure 13 Two-dimensional tulip flame: flame front (line) and velocity field (arrows); burning law: $s_l(\mathbf{u}_u) = 0.1/\rho_u$, heat release $Q = 4\gamma/(\gamma - 1)$, 128×32 grid cells

References

[1] ALMGREN, A. S., BELL, J. B. AND SZYMCZAK, W. G., *A Numerical Method for the Incompressible Navier-Stokes Equation Based on an Approximate Projection*, SIAM Journal of Scientific Computing, **17(2)**, 358-369, 1996

[2] ALMGREN, A. S., BELL, J. B., COLELLA, P., HOWELL, L. H. AND WELCOME, M. L., *A Conservative Adaptive Projection Method for the Variable Density Incompressible Navier-Stokes Equations*, Journal of Computational Physics, **142**, 1-46, 1998

[3] ALMGREN, A. S., BELL, J. B. AND CRUTCHFIELD, W. Y., *Approximate Projection Methods: Part I. Inviscid Analysis*, eingereicht zur Veröffentlichung in SIAM Journal on Scientific Computing, erhältlich als Vorabdruck: Ernest Orlando Lawrence Berkeley National Laboratory LBNL-43374, Mai 1999

[4] BELL, J. B., COLELLA, P. AND GLAZ, H. M., *A second order projection method for the incompressible Navier-Stokes equations*, Journal of Computational Physics, **85**, 257-283, 1989

214

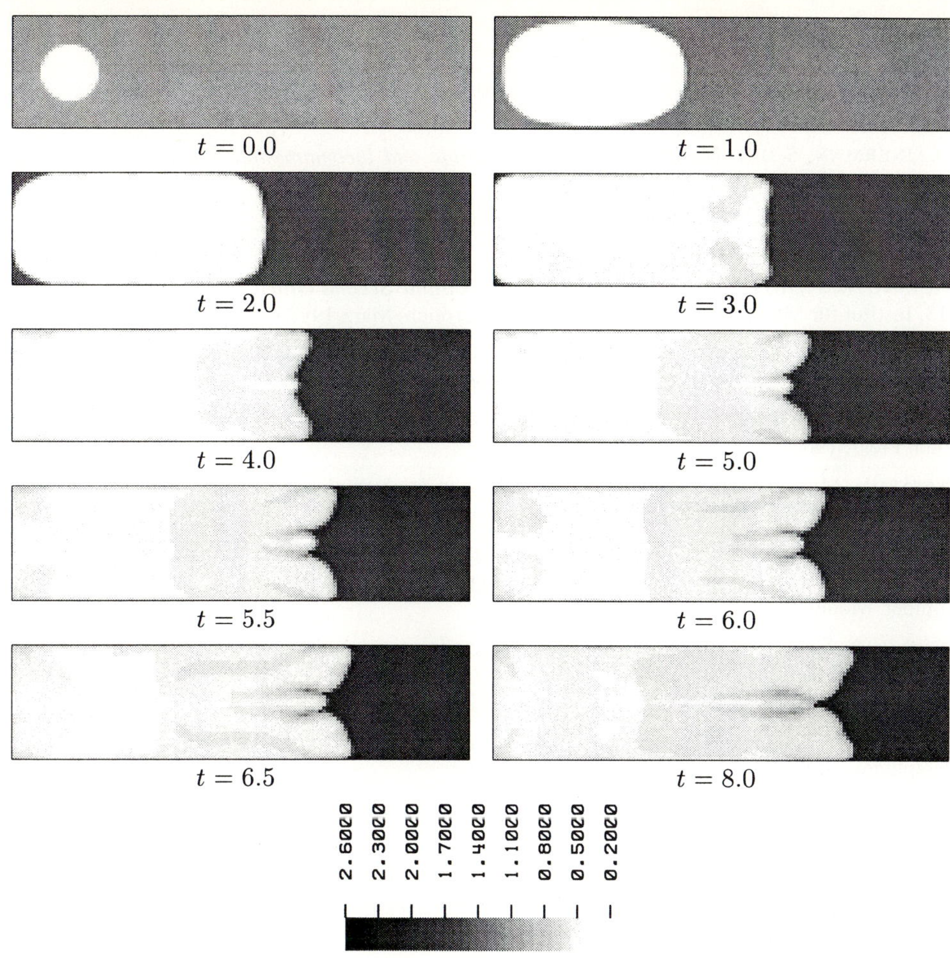

Figure 14 Two-dimensional tulip flame: density field; burning law : $s_l(\mathbf{u}_u) = 0.1/\rho_u$, heat release $Q = 4\gamma/(\gamma - 1)$, 128×32 grid cells

[5] BELL, J. B. AND MARCUS, D., *A second order projection method for variable density flows*, Journal of Computational Physics, **101**, 334-348, 1992

[6] BHAGA, D. AND WEBER, M. E., *Bubbles in viscous liquids: shapes, wakes and velocities*, Journal of Fluid Mechanics, **105**, 61-85, 1981

[7] BOURLIOUX, A., *A Coupled Level-Set Volume-Of-Fluid Algorithm for Tracking Material Interfaces*, 3rd Annual Conference of the CFD Society of Canada, June 25-27, 1995, Banff, Alberta, Canada

[8] CHORIN, A. J., *A Numerical Method for Solving Incompressible Viscous Flow Problems*, Journal of Computational Physics, **2**, 12-26, 1967

[9] CHORIN, A. J., *Numerical solution of the Navier-Stokes equations*, Math. Comput., **22**, 745-762, October 1968

[10] CHOPP, D. L., *Computing Minimal Surfaces via Level Set Curvature Flow*, Journal of Computational Physics, **106**, 77-91, 1993

[11] CLANET, C. AND SEARBY, G., *On the tulip flame phenomenon*, Combustion Science and Technology, **105**, 225-238, 1996

[12] HNAT, J. G. AND BUCKMASTER, J. D., *Spherical cap bubbles and skirt formation*, The Physics of Fluids, **19**, 2:182, 1976

[13] KLAINERMAN, S UND MAJDA, A. J., *Compressible and Incompressible Fluids*, Commun Pure Appl. Math, **35**, 629, 1982

[14] KLEIN, R., *Semi-Implicit Extension of a Godunov-Type Scheme Based on Low Mach Number Asymptotics I: One-Dimensional Flow*, Journal of Computational Physics, **121**, 213-237, 1995

[15] KICKINGER, F., *Algebraic Multigrid for Discrete Elliptic Second Order Problems*, Institutsbericht 513, Institut für Mathematik, Universität Linz, Österreich, März 1997

[16] MAJDA, A. UND SETHIAN, J., *The Derivation and Numerical Solution of the Equations for Zero Mach number Combustion*, Combustion Science and Technology, **42**, 185-205, 1985

[17] MARKSTEIN, G. H., *Nonsteady Flame Propagation*, Kapitel B, *Theory of Flame Propagation*, Pergamon Press, 1964

[18] MATALON, M. UND MATKOWSKY, B. J., *Flames as gasdynamic discontinuities*, Journal of Fluid Mechanics, **124**, 239-259, 1982

[19] MUNZ, C.-D., ROLLER, S., KLEIN, R. UND GERATZ, K.J. *Extension of Incompressible Flow Solvers to the Weakly Compressible Regime*, eingereicht zur Veröffentlichung bei Theoretic. and Comput. Modelling, 1997

[20] SCHOCHET, S., *Asymptotics for symmetric hyperbolic systems with a large parameter*, Journal of Differential Equations, **75**, 1-27, 1988

[21] SCHNEIDER, T, BOTTA, N, GERATZ, K. J. UND KLEIN, R., *Extension of Finite Volume Compressible Flow Solvers to Multi-dimensional, Variable Density zero Mach Number Flow*, Journal of Computational Physics, **155**, 248-286, 1999

[22] SCHNEIDER, TH., KLEIN, R., BESNOIN, E. UND KNIO, O. M., *Computational analysis of a thermoacoustic refrigerator*, Joint meeting: ASA/EAA/DEGA, 1999, Berlin, Deutschland

[23] SCHNEIDER, TH. UND KLEIN, R., *Overcoming mass losses in Level Set-based interface tracking schemes*, Second International Symposium on Finite Volumes for Complex Applications -Problems and Perspectives-, 19.-22. Juli, 1999, Duisburg, Deutschland

[24] SCHNEIDER, TH., TERHOEVEN, P AND KLEIN R., *A cell centered fully conservative projection method for zero Mach number variable density flows*, in preparation for submission to the Journal of Computational Physics

[25] SMILJANOVSKI, V., MOSER, V. UND KLEIN, R., A Capturing-Tracking Hybrid Scheme for Deflagration Discontinuities, Journal of Combustion Theory and Modelling, 2(1):183-215, 1997

[26] SUSSMAN, M., SMEREKA, P. UND OSHER, S., *A Level Set Approach for Computing Solutions to Incompressible Two-Phase Flow*, Journal of Computational Physics, **114**, 146-159, 1994

[27] TEMAM, R., *Sur l'approximation de la solution des equations de Navier-Stokes par le methode des fractionaire ii.*, Arch. Rational Mech. Anal.,**33**, 377-385, 1969

[28] TERHOEVEN, P., *Ein numerisches Verfahren zur Berechnung von Flammenfronten bei kleiner Mach-Zahl*, Dissertation, RWTH-Aachen, Deutschland, 1998

[29] WORLIKAR, S. A. UND KNIO O. M., *Numerical Simulation of a thermoacoustic refrigerator - I. Unsteady Adiabatic Flow around the Stack*, Journal of Computational Physics, **127**, 424-451, 1996

[30] WORLIKAR, S. A., KNIO O. M. UND KLEIN, R., *Numerical Simulation of a thermoacoustic refrigerator - II. Stratified Flow around the Stack*, Journal of Computational Physics, **144**, 299-324, 1998

Numerical Simulation of Shock Wave Interaction Effects on Supersonic Mixing Layer Growth

U. Wepler, Ch. Huhn, W. Koschel
Jet Propulsion Laboratory, Aachen University of Technology
D-52062 Aachen, Germany

Summary

This work is motivated by the need of the development of improved numerical tools to investigate the flow field in a supersonic combustion ramjet engine. For the application of supersonic combustion in air breathing propulsion systems the mixing process is a major problem. The injected fuel has to be mixed with air before reaction takes place. Several experiments show a decrease of the spreading rate of a supersonic mixing layer with increasing convective Mach number. Aim of this work is to investigate the influence of shock / mixing layer interactions on the mixing process by means of computational fluid dynamics. Experiments carried out at the DLR Space Propulsion Institute predict an enhancement of the mixing layer growth due to shock waves. The mixing process is mainly determined by turbulent structures in the flow field. For the turbulent mixing a modified $k - \varepsilon$ turbulence model is applied. It includes modifications introduced by Sarkar to take compressibility effects into account. For the computation of the flow field a Finite Element code on unstructured triangular grids is applied. A brief overview of the numerical method used for solving the Navier-Stokes equations will be presented. The code is applied to a supersonic combustion chamber test case of the above mentioned DLR experiment. In this test case a hydrogen - air mixing layer with and without shock wave interaction is studied. A comparison between numerical and experimental data will be made and discussed in detail.

1   Introduction

In recent years the development of an air breathing hypersonic vehicle has received considerable attention. This is a complex task and requires innovative research in many technical areas such as aerodynamics, propulsion, structures and materials. The development of such a vehicle is strongly dependent on the development of efficient propulsion systems, capable of producing sufficient thrust for acceleration. For hypersonic flight within the atmosphere a supersonic combustion ramjet (Scramjet) is envisioned to be a viable propulsion system [1]. The work reported here is concerned with the computational analysis of the scramjet flow field.

The flow field in the scramjet combustor is highly complex. The premise for a efficient combustion is an optimal mixing of fuel and oxydizer. Therefore the behaviour of shear and mixing layers has been investigated extensively [2] [3] [4], mainly as a result of the observed reduced growth rate of compressible mixing layers. The combustion efficiency is reduced due to the limited growth rate in turbulent mixing. To obtain higher efficiencies a fast and effective mixing is necessary. One possibility to enhance the turbulent mixing in a mixing layer is to apply a shock interaction. This has been investigated by experiments at the DLR Space Propulsion Institute [5] [6] and is now subject of numerical computations at the Aachen University of Technology.

For solving the Navier-Stokes equations a Finite-Element solver is used, which has been developed at the Jet Propulsion Laboratory in Aachen in the last decade [7] [8]. For the computation of the flowfield unstructured grids are used. A solution dependent adaptive remeshing technique is applied to improve accuracy and to study local flow phenomena.

2   Governing Equations

For a compressible flow the Navier-Stokes equations along with equations for energy and species continuity governing multi species flow with chemical reactions are considered in their Favre-averaged conservative form

$$\frac{\partial U}{\partial t} + \frac{\partial}{\partial x_i} F_i^k = \frac{\partial}{\partial x_i} F_i^d + Q \,, \tag{2.1}$$

where U denotes the averaged conservative variables

$$U = \left[\bar{\rho} \,, \overline{\rho} \widetilde{Y_n} \,, \overline{\rho} \widetilde{u_i} \,, \overline{\rho} \widetilde{E}\right]^T \,. \tag{2.2}$$

The convective and diffusive fluxes are

$$F^k = \begin{pmatrix} \overline{\rho} \widetilde{u_i} \\ \overline{\rho} \widetilde{Y_n} \widetilde{u_i} \\ \overline{\rho} \widetilde{u_i} \widetilde{u_j} + \overline{p} \delta_{ij} \\ \left(\overline{\rho} \widetilde{E} + \overline{p}\right) \widetilde{u_i} \end{pmatrix} \tag{2.3}$$

and

$$F_i^d = \begin{pmatrix} 0 \\ \overline{J_{i_n}} - \overline{\rho u_i'' Y_n''} \\ \widetilde{\sigma_{ij}} - \overline{\rho u_i'' u_j''} \\ \left(\widetilde{\sigma_{ij}} - \overline{\rho u_i'' u_j''}\right) \widetilde{u_i} + \widetilde{q_i} + \overline{\rho u_i'' h''} \end{pmatrix} \,. \tag{2.4}$$

Finally Q represents a source term

$$Q = \left[0 \,, \overline{\rho} \widetilde{\dot{\omega}} \,, 0 \,, 0\right]^T \,. \tag{2.5}$$

The total specific energy $\widetilde{E}$ is given by

$$\widetilde{E} = \widetilde{e} + \frac{1}{2} \widetilde{u_i} \widetilde{u_i} + k \,, \tag{2.6}$$

introducing the mass averaged turbulent kinetic energy $k := \frac{1}{2} \frac{\overline{\rho u_i'' u_i''}}{\overline{\rho}}$. The internal energy $\widetilde{e}$ calculates to

218

$$\widetilde{e} = \sum_{n=1}^{N_{sp}} \left(\widetilde{Y}_n \int_0^T c_{vn}\, dT + Y_n h_n^0 \right) . \tag{2.7}$$

The mass flux $\overline{J_{i_n}}$ is given by

$$\overline{J_{i_n}} = \overline{\rho} D_n \frac{\partial \widetilde{Y}_n}{\partial x_i} = \frac{\mu}{Sc_n} \frac{\partial \widetilde{Y}_n}{\partial x_i} \tag{2.8}$$

and the turbulent diffusion flux $\overline{\rho u_i'' Y_n''}$ is calculated from

$$\overline{\rho u_i'' Y_n''} = -\frac{\mu_t}{Sc_{t_n}} \frac{\partial \widetilde{Y}_n}{\partial x_j} . \tag{2.9}$$

The Schmidt numbers are assumed to be constant for each species ($Sc = 0.9$ and $Sc_t = 0.7$). The Newton stress tensor $\widetilde{\sigma_{ij}}$ is a linear function of the velocity gradients tensor with the molecular viscosity μ being the proportionality constant

$$\widetilde{\sigma_{ij}} = \mu \cdot \left(\left(\frac{\partial \widetilde{u}_i}{\partial x_j} + \frac{\partial \widetilde{u}_j}{\partial x_i} \right) - \frac{2}{3} \frac{\partial \widetilde{u}_k}{\partial x_k} \delta_{ij} \right) \tag{2.10}$$

and the turbulent stress tensor $\overline{\rho u_i'' u_j''}$ is obtained by

$$\overline{\rho u_i'' u_j''} = \mu_t \left(\frac{\partial \widetilde{u}_i}{\partial x_j} + \frac{\partial \widetilde{u}_j}{\partial x_i} - \frac{2}{3} \frac{\partial \widetilde{u}_k}{\partial x_k} \delta_{ij} \right) - \frac{2}{3} \overline{\rho} k\, \delta_{ij} . \tag{2.11}$$

The turbulent viscosity μ_t is determined as

$$\mu_t = C_\mu \frac{\overline{\rho} k^2}{\varepsilon} . \tag{2.12}$$

The temperature dependency of the molecular viscosity μ is considered by Sutherland's empirical equation

$$\mu = \mu_{ref} \cdot \left(\frac{T}{T_{ref}} \right)^{1.5} \cdot \left(\frac{T_{ref} + S}{T + S} \right) \tag{2.13}$$

with

$$\begin{aligned}
\mu_{ref} &= 1.711 \cdot 10^{-5} \tfrac{Ns}{m^2} \\
T_{ref} &= 273.15\,K \\
S &= 110.56\,K \quad .
\end{aligned}$$

The heat flux $\widetilde{q}_i$ and its turbulent transport $\overline{\rho u_i'' h''}$ is denoted as

$$\widetilde{q}_i = -c_p \frac{\mu}{Pr} \frac{\partial \overline{T}}{\partial x_i} - \sum_{n=1}^{N_{sp}} \frac{\mu}{Sc_n} h_n(\overline{T}) \frac{\partial \widetilde{Y}_n}{\partial x_i} \tag{2.14}$$

$$\overline{\rho u_i'' h''} = \frac{\mu_t}{Pr_t} \frac{\partial H_T}{\partial x_j} \tag{2.15}$$

assuming constant Prandtl numbers ($Pr = 0.72$ and $Pr_t = 0.9$). The equation set described above is completed by addition of the thermal equation of state

$$\overline{p} = \rho R \overline{T} = \rho R \sum_{n=1}^{N_{sp}} \frac{\widetilde{Y}_n}{\mathcal{M}_n} \overline{T} \,. \tag{2.16}$$

There are the turbulent energy k and the dissipation rate ε, representing the influence of turbulence, and the chemical source term $\widetilde{\omega}$, representing the influence of combustion, left to close the system. Since in this presentation only non-reacting flows are regarded, the chemical source term $\widetilde{\omega}$ is neglected. A detailed description of the combustion modelling can be found in reference [8]. Turbulence modelling will be considered in the following section.

3 Turbulence Modelling

The turbulence is modelled by the standard $k - \varepsilon$ model for high Reynolds number suggested by Launder and Spalding [9] with the addition of the Sarkar model [10] to take compressibility effects into account. The transport equations are

$$\frac{\partial U_{k\varepsilon}}{\partial t} + \frac{\partial}{\partial x_i} F^k_{i_{k\varepsilon}} = \frac{\partial}{\partial x_i} F^d_{i_{k\varepsilon}} + Q_{k\varepsilon} \,, \tag{3.17}$$

with the conservative variables

$$U_{k\varepsilon} = \begin{pmatrix} \overline{\rho} k \\ \overline{\rho} \varepsilon_s \end{pmatrix} \,. \tag{3.18}$$

where ε_s is the solenoidal dissipation rate, which is unaffected by compressibility. The turbulent fluxes

$$F^k_{i_{k\varepsilon}} = \begin{pmatrix} \overline{\rho \widetilde{u}_i k} \\ \overline{\rho \widetilde{u}_i \varepsilon_s} \end{pmatrix} \tag{3.19}$$

and

$$F^d_{i_{k\varepsilon}} = \begin{pmatrix} \left(\mu + \frac{\mu_t}{\sigma_k} \right) \frac{\partial k}{\partial x_i} \\ \left(\mu + \frac{\mu_t}{\sigma_\varepsilon} \right) \frac{\partial \varepsilon}{\partial x_i} \end{pmatrix} \tag{3.20}$$

contain the usual convective and diffusive terms. The turbulent source term is determined as

$$Q_{k\varepsilon} = \begin{pmatrix} P_t - \overline{p} \left(\varepsilon_s + \varepsilon_d \right) \\ C_{\varepsilon 1} P_t \frac{\varepsilon_s}{k} - C_{\varepsilon 2} \overline{\rho} \frac{\varepsilon_s^2}{k} \end{pmatrix} \,, \tag{3.21}$$

where P_t is the turbulent production term

$$P_t = \mu_t \left(\frac{\partial \widetilde{u}_i}{\partial x_j} + \frac{\partial \widetilde{u}_j}{\partial x_i} - \frac{2}{3} \frac{\partial \widetilde{u}_k}{\partial x_k} \delta_{ij} \right) - \frac{2}{3} \overline{\rho} k \frac{\partial \widetilde{u}_k}{\partial x_k} \delta_{ij} \,. \tag{3.22}$$

According to Sarkar's model the dilatation dissipation ε_d is assumed to be proportional to ε_s

$$\varepsilon_d = \alpha_1 M_t^2 \varepsilon_s \quad \text{with} \quad \alpha_1 = 1 \,. \tag{3.23}$$

Here M_t is the the local turbulent Mach number defined as

$$M_t = \sqrt{\frac{2k}{a^2}} \tag{3.24}$$

with a equal to the speed of sound.

220

4 Grid Generation

The triangulation of arbitrary shaped computational domains with embedded interior boundaries is accomplished by an automatic mesh generation scheme following the generalised advancing front method [11]. Points and elements are introduced simultaneously and therefore significant changes in the local mesh structure are possible. The shape of the triangles is controlled by the parameters: element size, element stretching and stretching direction. A spatial distribution of these mesh parameters is provided by elements when generating an initial mesh.

The automatic mesh generation scheme described above offers the ability of applying some method of adaptive remeshing to improve the solution quality in a computationally efficient manner. During the analysis of a certain flow problem the computational grid is repeatedly adapted by completely regenerating the mesh based upon information provided by the computed solution on the present grid. The new mesh is constructed using the automatic mesh generation scheme allowing a significant variation in element size and stretching of the elements in the vicinity of one–dimensional flow features. The initial computational mesh is now acting as a background grid providing the spatial distribution of the mesh parameters. In order to determine the nodal values of these mesh parameters in an optimal manner it is necessary to apply some method of error estimation. The second derivatives of a certain scalar key variable, e.g. density, mixture fraction or Mach number, are used to give some indication of error magnitude and direction. The condition of a uniformly distributed error indicator within the domain leads to the optimal nodal values of the mesh parameters.

After the mesh generation a two–stage adaptive smoothing process is applied. In the first stage the element connectivity is optimised by swapping the element edges. The mesh is relaxed to the optimum case with each node being surrounded by six elements. In the subsequent stage the nodes are moved with regard to the prescribed distribution of the mesh parameters using a spring system analogy.

5 Numerical Method

The transport equations are integrated by the application of the Galerkin weighted residual method in combination with a Runge-Kutta type time marching scheme [12][7]. The spatial discretisation of the eq.(2.1) and eq.(3.17) is denoted as

$$\int_\Omega \frac{\Delta \hat{U}}{\Delta t} N d\Omega - \int_\Omega \frac{\partial N_J}{\partial x_k} \hat{F}_k d\Omega \;\; + \int_\Gamma N_J \hat{F}_k n_k d\Gamma$$

$$= \int_\Omega \hat{H} N d\Omega \tag{5.25}$$

where Ω represents the computational domain and Γ it's boundary. Applying linear approximation functions N

$$\hat{U} = \sum_{J=1}^{3} N_J \cdot U_J \,, \quad \hat{F}_J = \sum_{J=1}^{3} N_J \cdot F_J \tag{5.26}$$

for the triangular elements the integral equation can be discretized as follows

$$\sum_e \int_{\Omega_e} N_j N_i d\Omega_e \cdot \frac{\Delta U_i}{\Delta t} =$$

$$\sum_e \left[\int_{\Omega_e} \frac{\partial N_j}{\partial x_k} N_i d\Omega_e \cdot F_{k_i} \quad - \int_{\Gamma_e} N_j N_i d\Gamma_e \cdot F_{k_i} n_k \right.$$

$$\left. + \int_{\Omega_e} N_j N_i d\Omega_e \cdot H_j \right] . \tag{5.27}$$

The discretisation above can be regarded as a node-centred scheme with an arithmetic averaging of the fluxes normal to the edges of each element.

Time integration is achieved by a 5-step Runge-Kutta scheme

$$U_J^{(\nu+1)} = U_j^n \quad + \quad \alpha_\nu \Delta t_J \left[M_{LJ}^{-1} R_J \left(U^\nu \right) \right.$$

$$+ \quad \left. \left(D_J^{(2)} - D_J^{(4)} \right)^\gamma \right] \tag{5.28}$$

$$U_J^{(n+1)} = U_j^{(5)} \qquad ,$$

with $\nu = 0, \cdots, 4$ and $\gamma = \min(\nu, 1)$ and $R_J \left(U^n \right)$ being the right hand side of eq.(5.27). The Runge-Kutta coefficients α_ν correspond to maximum stability reducing the time accuracy to second order. In order to avoid spurious oscillation and to achieve a stationary solution a second-order and a fourth-order damping term are required for the flow equations. $D^{(2)}$ represents a nonlinear shock-capturing term [13]

$$D_J^{(2)} = c_d^{(2)} M_{LJ}^{-1} \sum_e \frac{S_e}{\Delta t_e} \left(M_e - M_{L_e} \right) U^n , \tag{5.29}$$

where $c_d^{(2)}$ is a constant diffusion coefficient and S_e is a dimensionless pressure switch that varies between 0 and 1 from smooth flow to near discontinuous flow

$$S_e = \left| \sum_e \frac{\left(M_e - M_{L_e} \right) p}{\left| \left(M_e - M_{L_e} \right) p \right|} \right| . \tag{5.30}$$

$D_J^{(4)}$ denotes a linear fourth-order high frequency damping term also published in [13]

$$D_J^{(4)} = M_{LJ}^{-1} \sum_e \frac{\chi^{(4)}}{\Delta t_e} \left(M_e - M_{L_e} \right)$$

$$\left[M_L^{-1} \sum_e \left(M_e - M_{L_e} \right) \cdot U^n \right] . \tag{5.31}$$

In order to reduce overshoots near shock waves strong contributions of $D_J^{(4)}$ are suppressed by defining

$$\chi^{(4)} = \max \left(0, c_d^{(4)} - c_d^{(2)} S_e \right) \tag{5.32}$$

with $c_d^{(4)}$ as a user-specified constant.

6   Results and Comparison

First computations were performed for plane compressible shear/mixing layers at different supersonic flow conditions [14]. For this case, a lot of different experimental results[3][2] are available showing the dependence of the relative shear layer growth $\frac{C_{\delta'}}{C_{\delta'_0}}$ on the convective Mach number

$$Ma_c = \frac{|U_1 - U_2|}{a_1 + a_2} . \tag{6.33}$$

For comparison with data from literature the compressible growth rate $C_{\delta'}$ is normalized with the incompressible growth rate $C_{\delta'_0}$. Figure 1 shows the well-known behaviour of reduced shear layer growth with increasing convective Mach number at supersonic conditions. The calculations have been done without and with the presented modification of the turbulence model. The results of the numerical simulation fit only if the compressibility correction for the $k - \varepsilon$ model is applied. This example shows the applicability of the modified turbulence model to compressible shear layers.

In the following the code was applied to a supersonic combustion chamber flow test case. This test case was investigated experimentally at the DLR Space Propulsion Institute [6] [5]. Herein the injection of hydrogen in air at supersonic conditions is considered without combustion. Computations were performed with and without shock interaction. The geometry of the computational domain is illustrated in fig.2. The test section is $50mm$ in height at the inlet and $150mm$ in width. The upper and lower wall have a divergence with an angle of 3 degrees for the reference case without shock interaction. The shock wave itself was generated by a wedge in the upper wall of the test section. For the test case presented here a inclination angle of 7 degrees was chosen. The shock/mixing layer interaction was positioned at $x0 = 60mm$ behind the inlet. The hydrogen injector ($4mm$ in height) divides the inlet in two symmetrical parts.

Computations have been made for the different convective Mach numbers, which are achieved by varying the temperature for the H_2 and the air.

First simulations have been been made for the mixing layer without shock interaction in order to get an insight view of the mixing process. Figure 3 presents the solution adapted grid for the convective Mach number $Ma_c = 0.9$. It already reveals a first impression of the complex flow pattern.

The contour plot of the density in fig.4 gives a more detailed view of the flow field. Due to to the divergence of the channel the flow is accelerating. Expansion fans generated at the upper and lower edge of the inlet can be detected. These expansion waves show little influence on the mixing layer. The mixing layer thickness is increasing, where they hit the hydrogen jet.

Figure 5 presents the density distribution of the calculation with the same inflow conditions, but with shock interaction. The flow field shows similar phenomena as for the before mentioned reference test case. The difference can be observed where the oblique shock hits the hydrogen jet. In the vicinity of the shock the mixing layer is obviously getting thicker. Also the main flow direction downstream the shock is slightly turned downward.

The development of the mixing layer in axial direction can be seen from the profiles of mean velocity and mass fraction hydrogen in fig.6 and fig.7 respectively. In the reference case the profiles are nearly symmetric, whereas in the case of the shock interaction the profiles are slightly shifted to the lower side. Again the increase of the mixing layer due to the oblique shock interaction is obvious. Refering to these profiles the mixing layer growth rate is determined using the following definition

$$C_{\delta,} = \frac{\partial \delta_Y}{\partial x} \cdot \frac{U_1 + U_2}{|U_1 - U_2|} \cdot \tag{6.34}$$

The mixing layer width δ_Y is taken where the mass fraction of hydrogen is down at 1% of the initial value.

For comparison with experimental data the growth rate is normalized with the incompressible growth rate

$$C_{\delta_0'} = C_{\delta,} \, (Ma_c = 0) \, . \tag{6.35}$$

The results are presented in fig.8. The normalized mixing layer growth rate is plotted against the convective Mach number for the reference case and a shock interaction test case. This figure reveals the general behaviour of a reduction of the growth rate with increasing convective Mach number. But it shows also that the growth rate for the shock interaction test case tends to higher values.

In fig.9a) the comparison of the results of the mixing layer growth rate without shock interaction to the experimental data taken from Nuding [5] is made. A good fit of both the computational and experimental data is given by the function

$$\frac{dc}{di} = 0.7 \cdot e^{-2.5 \cdot Ma_c^2} + 0.3 \, , \tag{6.36}$$

which was published in [5]. Equation (6.36) represents the compressibilty effect on the mixing layer growth rate and is also used as a reference for the mixing layer studies with shock interaction.

In fig.9b) finally the results of the mixing layer growth rate with shock interaction are compared. Here the data are plotted against the fit function (eq. (6.36)). In both cases (reference case and shock interaction test case) the calculated values for the mixing layer growth agree very well with the experimentally obtained data.

7   Conclusion

A finite-element solver on unstructured grids for compressible flow has been presented. The code has been successfully applied to calculate non-reacting scramjet combustion chamber flows. The obtained results for the mixing layer growth depending on the convective Mach number are in good agreement with experimental data. Furthermore the code has been used to investigate the influence of shock wave interaction on the mixing layers. The results clearly show an enhancement of the mixing layer growth due to the shock interaction.

Further computations will be carried out in order to get a more detailled insight view of the mixing layer behaviour. The dependence on different parameters like convective Mach number, shock strength or upstream interaction position will be evaluated. In addition the computational work will be extended to reacting mixing layers.

Acknowlodgement

This research project is funded by the Deutsche Forschungsgemeinschaft (DFG) within the frame of the French-German priority programme on "Numerical Flow Simulation". The support of the DFG is acknowledged.

References

[1] Heiser, W.H. and Pratt, D.T. *Hypersonic Airbreathing Propulsion*. AIAA Education Series, 1994.

[2] Elliott, G.S., Samimiy, M. and Arnette, S.A. Study of compressible mixing layers using filtered rayleigh scattering based visualizations. *AIAA-Journal*, 30:2567–2569, 1992.

[3] Papamoschou, D. and Roschko A. The compressible turbulent shear layer: an experimental study. *Journal of Fluid Mechanics*, 197:453–477, 1988.

[4] Kline, S.J., Cantwell, B.J. and Lilly, G.M. AFSOR-HTTM Stanford conference on complex turbulent flows1981. Technical report, Stanford University, CA, 1981.

[5] J.-R. Nuding. Interaction of compressible shear layers with shock waves: an experimental study, part i. *AIAA-96-4515*, 1996.

[6] Brummund, U. and Mesnier, B. Experimental study of compressible mixing layers. In *14th International Symposium on Airbreathing Engines, Florenz, Italy*. ISABE, Sept. 5-10. 1999.

[7] W. Rick. *Adaptive Galerkin Finite Elemente Verfahren zur numerischen Strömungssimlation auf unstrukturierten Netzen*. PhD thesis, Institut für Strahlantriebe, RWTH Aachen, 1994.

[8] S. Sasse. *Numerischen Simulation von nichtreaktiven und reaktiven turbulenten Überschallströmungen in Staustrahltriebwerken*. PhD thesis, Institut für Strahlantriebe, RWTH Aachen, 1998.

[9] B.E. Launder. *Mathematical Models of Turbulence*. Academic Press, 1972.

[10] Sarkar, S., Erlebacher, G., Hussaini, M. Y., Kreiss, H. O. The analysis and modeling of dilatation terms in compressible turbulence. *ICASE Report No. 89-79*, December 1989.

[11] Bikker, S., Greza, H., Koschel, W. Parallel computing and multigrid solution on adaptive unstructured meshes. *Notes on Numerical Fluid Mechanics*, 47:9–16, 1994.

[12] D.J. Mavriples. Accurate multigrid solution of the euler equations on unstructured and adaptive meshes. *AIAA-Journal*, 28, 1990.

[13] Morgan, K., Peraire, J. Finite element methods for compressible flows. In *Lecture Series for Fluid Dynamics*, 1987.

[14] Wepler, U., Koschel, W., Stoukov, A., Vandromme, D., et al. Numerical simulation of turbulent high speed flows. *Notes on Numerical Fluid Mechanics*, 66:278–297, 1998.

Figures

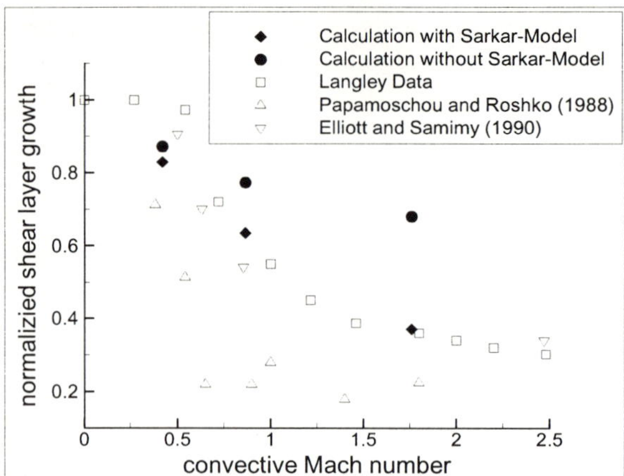

Figure 1: Normalized growth rate of compressible shear layers; comparison of experimental and numerical results

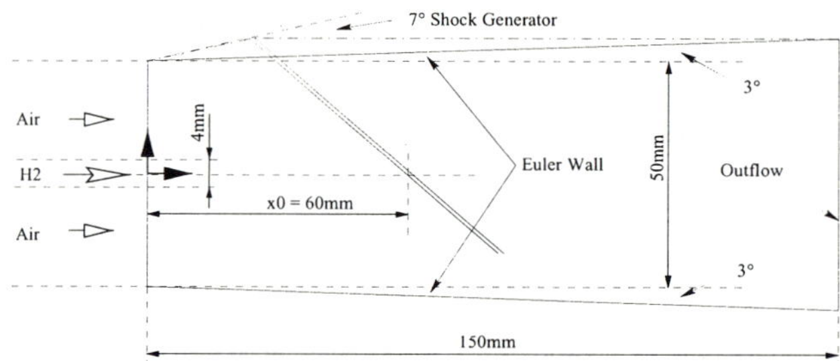

Figure 2: Boundary of computational domain

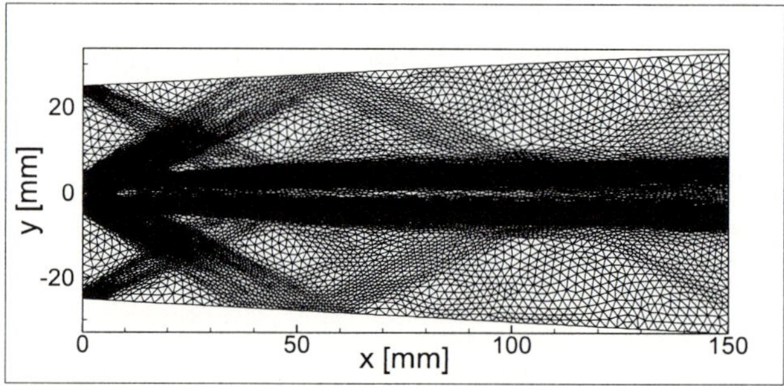

Figure 3: Solution adapted grid for the computation of the reference test case $Ma_c = 0.9$

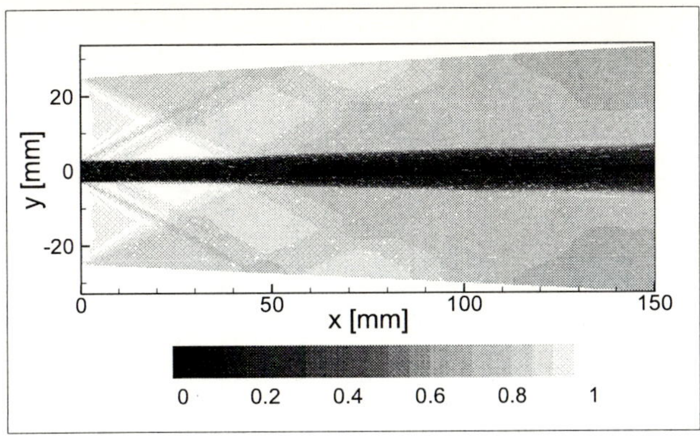

Figure 4: Computed density $[kg/m^3]$ distribution at a convective Mach number $Ma_c = 0.9$

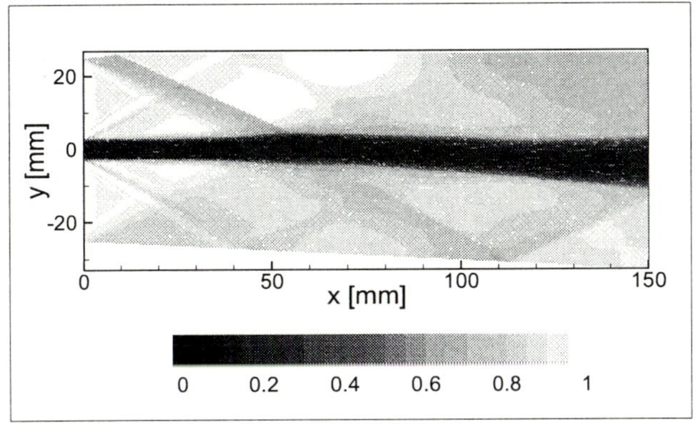

Figure 5: Computed density $[kg/m^3]$ contour plot for shock / shear layer interaction $[Ma_c = 0.9$, wedge 7°, $x_0 = 60mm]$

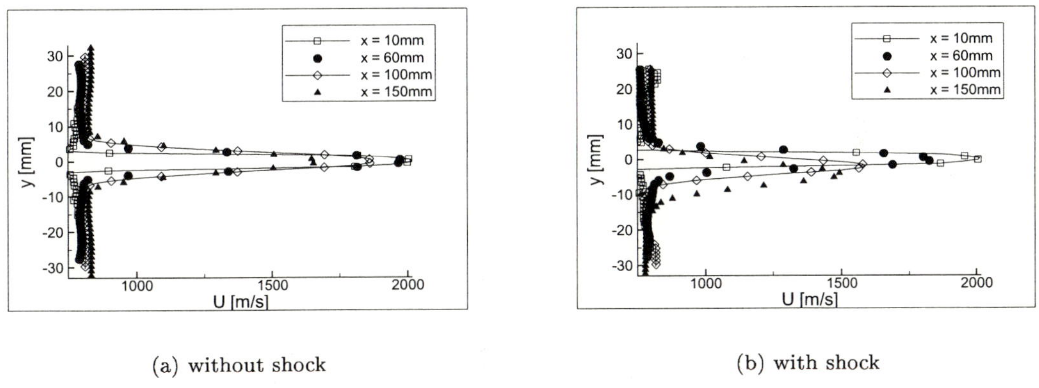

(a) without shock   (b) with shock

Figure 6: Comparison of computed velocity profiles for the test case $Ma_c = 0.9$ (a) without shock interaction and (b) with shock interaction [wedge 7°, $x_0 = 60mm$]

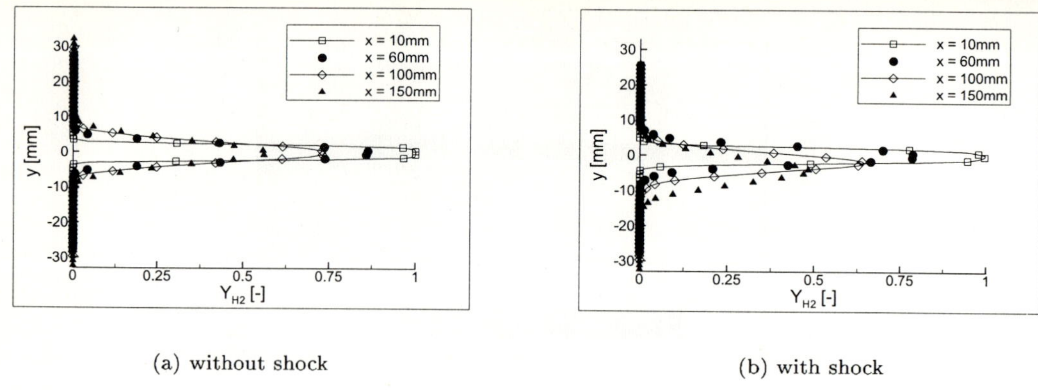

(a) without shock   (b) with shock

Figure 7: Comparison of computed profiles of the hydrogen mass fraction for test case $Ma_c = 0.9$ (a) without shock interaction and (b) with shock interaction [wedge 7°, $x_0 = 60mm$]

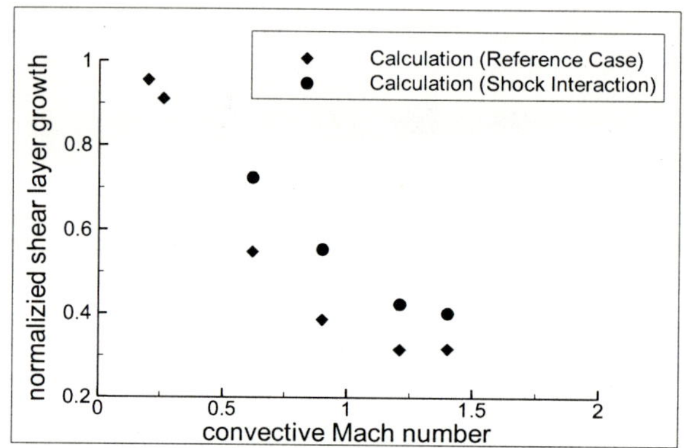

Figure 8: Calculated shear layer growth (with and without shock interaction)

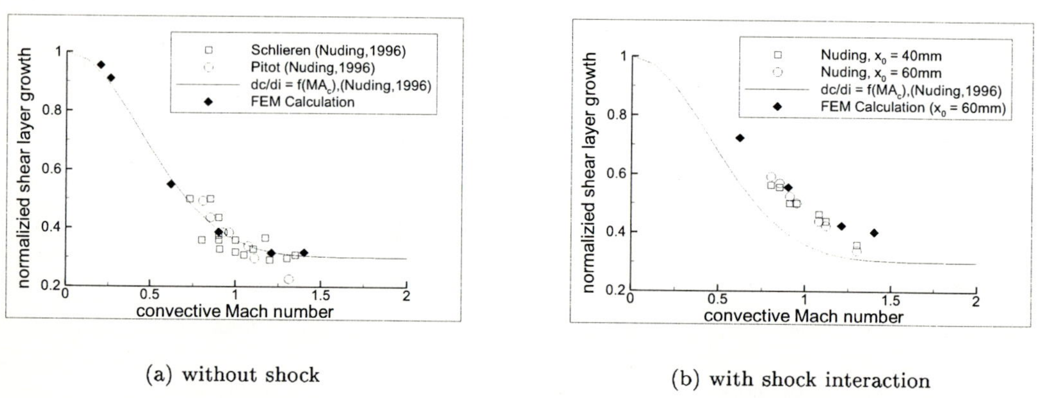

(a) without shock   (b) with shock interaction

Figure 9: Comparison of calculated and measured growth rate ((a)without and (b) with shock interaction)

IV.   TURBULENT FLOWS

Large Eddy Simulation of Flow around Circular Cylinders on Structured and Unstructured Grids, II

J. Fröhlich[1], W. Rodi[1], J.P. Bertoglio[2], U. Bieder[3], H. Touil[2]

[1] Inst. für Hydromechanik, Universität Karlsruhe, 76128 Karlsruhe, Germany

[2] LMFA, URA CNRS 263, Ecole Centrale de Lyon, 69131 Ecully, France

[3] CEA–Grenoble, 17 rue des Martyrs, 38054 Grenoble, France

Dedicated to R. Peyret

Summary

The paper presents LES computations of subcritical flow around circular cylinders. Uniform upstream conditions have been investigated with $Re = 3900$ and $Re = 140000$. The first case is considered for comparison between results obtained by the authors from earlier structured and unstructured computations and new results from a recently developed unstructured French code. Spanwise shear modifies the flow in substantial points and has been considered at $Re = 6250$. Time signals have been recorded and analyzed by means of Fourier and wavelet techniques. These reveal important differences between the flow with and without shear. Furthermore, work on the LES methodology is reported. We present a model for the laminar boundary layer employing an integral method and discuss the impact of streamwise variations of the grid on the computed resolved flow.

1 Introduction

The present paper is a follow–up of [17] concerned with the second phase of a common research project between IFH Karlsruhe, EDF Chatou, and LMFA Lyon on LES for the flow around cylinders.

Several studies have shown that Large Eddy Simulation (LES) is an appropriate simulation technique for bluff body flows. It generally yields better results than Reynolds–averaged models and on the other hand is affordable also at high Reynolds numbers when a Direct Numerical Simulation is too costly. Reviews on the subject can be found in [33], [34]. The flow around a subcritical circular cylinder has been considered in several papers on LES starting with the seminal work in [3]. This type of flow, due to its transitional character, is very sensitive to even small disturbancies. This is well–known from experiments [45] and has also been highlighted from a numerical point of view by the recent grid–refinement studies in [22]. Hence, the configuration is a severe test case. It is presumably more difficult than most industrial applications in similar geometry where the upstream turbulence level is generally higher such as for the tube bundle flow investigated by means of an LES on unstructured grids in [36].

The present paper is outlined as follows: We first report on recent computations for the circular cylinder at $Re = 3900$ focusing on the application of the new unstructured code developed by the French partners. Subsequently, computations at higher Reynolds numbers are reported. We then outline new model boundary conditions particularly suited for bluff–body flows. Finally, the flow around a cylinder in uniform shear flow is considered focusing on the analysis of the complicated wake structure by means of wavelet techniques.

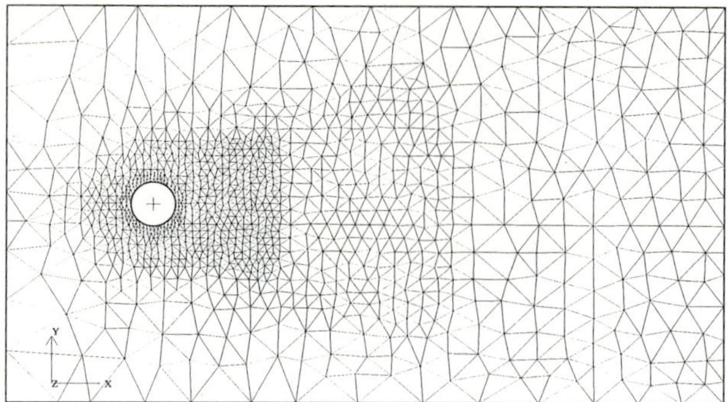

Figure 1 Cut in the $x-y$-plane through the unstructured grid with 300000 tetrahedra used for the computation with PRICELES.

2 Circular cylinder at $Re = 3900$

As an industrial partner, EDF experiences more than do academics the economical pressure on its research and development activities. This different attitude, with all its pros and cons, is one of the reasons for carrying out common research projects with partners from both areas to a mutual benefit. Recently, management as well as technical issues led EDF to abandon their code N3S for research on LES in favour of a new object–oriented Finite Element code, PRICELES (Platform Rapide Industrielle CEA EDF de LES) being developed together with CEA. At the same time the discretization scheme was revisited. This issue is more involved and less standardized with unstructured discretizations compared to structured or quasi–structured ones classically used for LES. Rollet–Miet et al. [36] proved that the P1/P1 element with suitable damping of oscillations is better suited for LES than classical Finite Elements. Further tests [5] however showed that in some cases a relatively fine grid is required to obtain satisfactory results. A new scheme was therefore developed, a P1–non-conforming/P1–bubble element, which mimics the staggered grid arrangement in structured Finite Volume codes [6].

The flow around a circular cylinder at $Re = 3900$ which had been computed with the structured Finite Volume code LESOCC [8] of IFH and N3S of EDF as reported in [17] was therefore revisited as one of the test cases for PRICELES [4]. Preliminary studies have shown that with PRICELES, more so than with N3S, regularity of the grid is an important issue. Hence, the previous Finite Element mesh was replaced by a new one with more degrees of freedom (600000 velocity nodes), lower aspect ratios, and smoother variation of the step size. A cross section is depicted in Fig. 1. Spanwise periodicity was replaced by a free–slip condition while increasing the respective extent of the calculation domain from $L_z = \pi$ to $L_z = 5$. (Note that everywhere in the present text distances are normalized with the cylinder diameter and velocities with the mean free–stream velocity.) All further characteristics were taken over from [17].

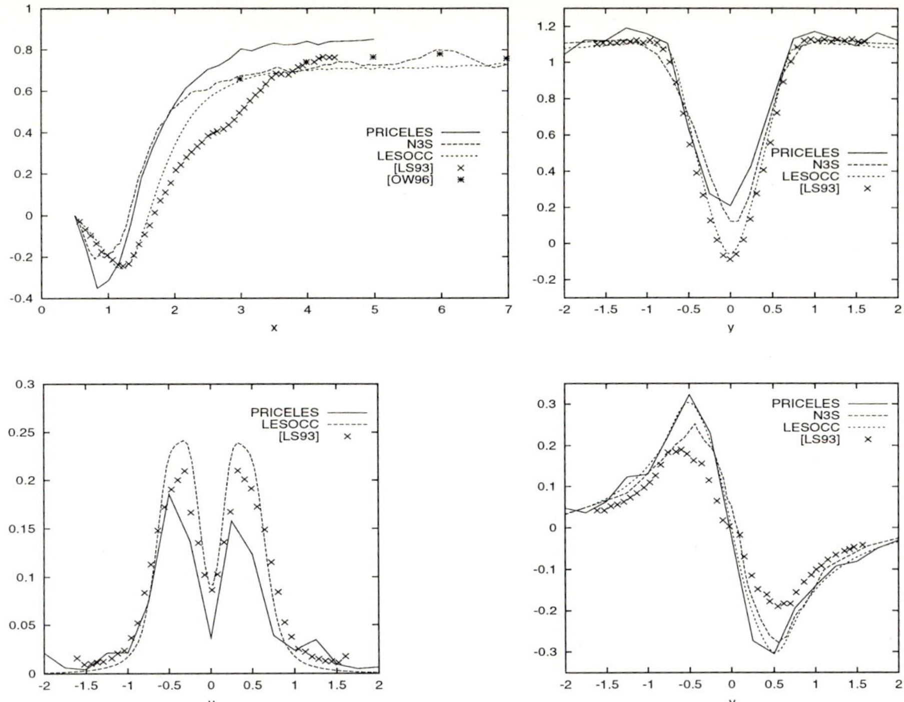

Figure 2 Cylinder flow at Re=3900: The new results obtained with PRICELES compared to the results of LESOCC and N3S (LRUN3 and NRUN1 in [17], respectively) as well as to the experimental data of [23] as reported in [3] and [27]. Top: Profiles of the mean streamwise velocity field at $y = 0$ (left) and $x = 1.54$ (right) Bottom: Profiles of the resolved fluctuations $\langle u'u' \rangle$ (left) and the mean normal velocity at $x = 1.54$ (right). The points where values from PRICELES were stored have no relation to the grid employed.

The computed instantaneous flow exhibits the features known from earlier computations and experiments such as [43]. Flow visualizations [4] show the typical vortex roll–up and braid–structures resulting from the formation of secondary spanwise vortices. The computation was performed for 7 shedding cycles, 3 of which were used for averaging. This amount is definitely too small for the statistics in the wake to converge with only some saturation close to the cylinder being observed. Nevertheless, important information on the behaviour of the code could be gained. Fig. 2 displays profiles of the average velocity field in the wake and for $\langle u'u' \rangle$ in comparison to the results in [17]. The recirculation length is obtained too short compared to the experiment [23]. In light of the discussion in [22] this is likely to result from insufficient resolution of the separating laminar shear layers and would certainly be improved by refining the grid in this region. In [22] it is even conjectured that the experiment of [23] suffered from disturbances and that ideally the recirculation region is larger reflecting a later transition of the shear layers.

Fig. 3 shows spectra from different LES of the considered flow as well as from the experiment of [27]. Fig. 3a taken from [24] illustrates the better representation of small scales when replacing an upwind scheme with a central scheme. Fig. 3b displays spectra from the computation with the structured Finite Volume code LESOCC using central differences, LRUN3 in [17], which have not been reported previously. It shows that with

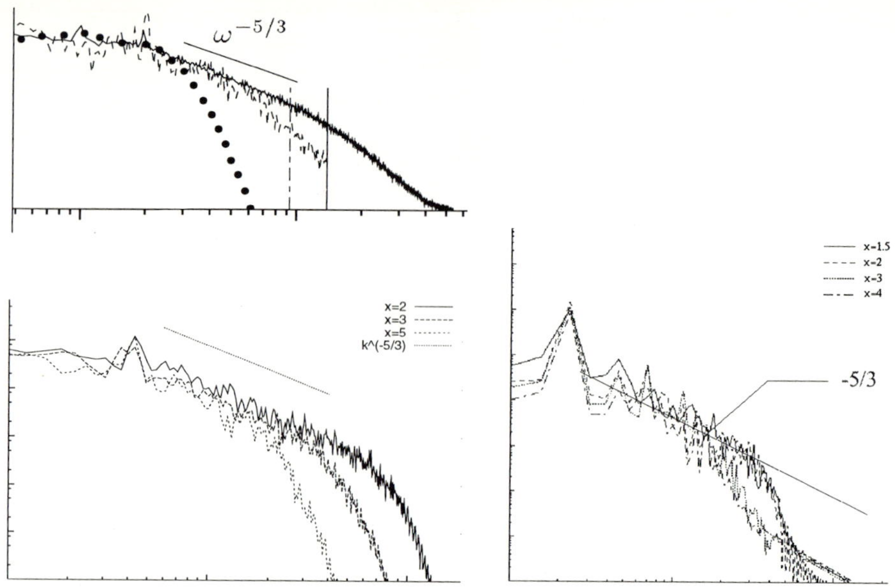

Figure 3 Spectra in the wake of the cylinder at $Re = 3900$ on the centerline. a) One-dimensional energy spectra E_{11} at $x = 5$ as compiled in [24]: – Experiment of [27], • LES of [3] with upwind-biased scheme, - - - LES with central/Fourier scheme of [24]. b) E_{11} in the computation with LESOCC, LRUN3 of [17], at $x = 2,3,5$. c) Spectra of turbulent kinetic energy obtained with PRICELESS at $x = 1.5,2,3,4$ [4]. Note that for reasons of flow topology E_{11} exhibits a maximum at $2St$ while the kinetic energy has a maximum at St. Tic marks have been droped because of different normalization of the axes.

increasing distance from the cylinder in the wake the spectra decay faster. This is partly due to the physical decay of the turbulence as reported in [27] but mainly due to the coarsening of the grid in streamwise direction. The extent of the inertial range at $x = 5$ in LRUN3 is comparable to the one in [24]. Fig. 3c displays energy spectra from the computation with PRICELES discussed above. Since the grid is coarser, in particular beyond $x = 3$ (cf. Fig.1), the inertial range extends less far.

Table 1 Computations for the flow around a circular cylinder at $Re = 140000$.

	RUN1	RUN2	RUN3	Experiments
$N_r \times N_\theta \times N_z$ L_z/D	$166 \times 206 \times 64$ 1	$166 \times 206 \times 64$ 4	$182 \times 306 \times 96$ 4	
St	0.217	0.208	0.207	0.179 [9], 0.2 [37]
C_D	1.157	1.246	0.91	1.237 [9], 1.2 [37]
C_{bp}	−1.33	−1.40	−0.93	−1.21[9], −1.34[40]
θ_{sep}	$93.5°$	$89.7°$	$88.6°$	$79° \pm 1°$ [39]
L_r/D	0.42	0.59	0.70	0.5 ± 0.05 [9]

3   Circular cylinder at $Re = 140000$

The computation of the cylinder flow at $Re = 140000$ is substantially more demanding than at the lower Reynolds number considered before. The flow is subcritical, just below the drag crisis, so that the thin laminar boundary layer has to be resolved if a no–slip condition is applied. Since the boundary layer thickness is proportional to $\sqrt{Re}$ this results in a factor of 6 in the resolution requirements compared to $Re = 3900$ [17]. Also, in azimuthal direction a much finer grid has to be employed. Due to the refined grid the critical time step decreases which further increases the cost. Consequently, we have to

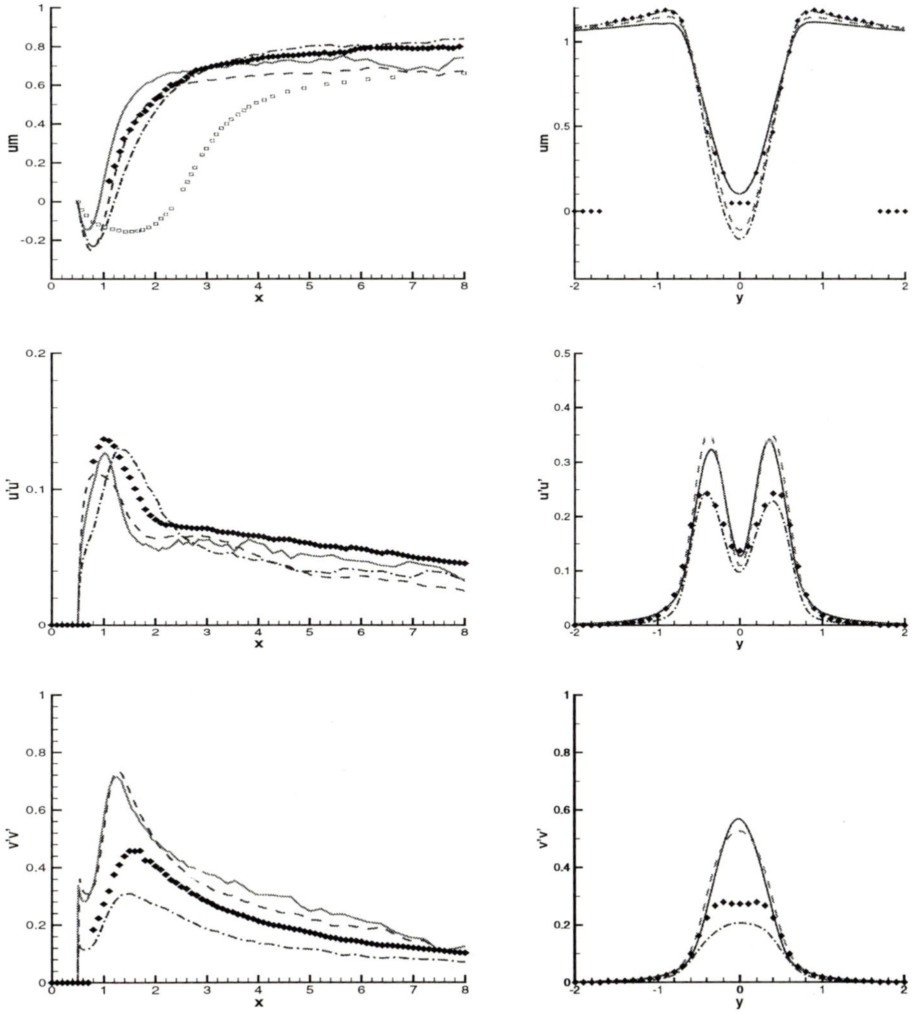

Figure 4   LES of flow around cylinder at $Re = 140000$. Mean streamwise velocity u and Reynolds stresses $\langle u'u'\rangle$, $\langle v'v'\rangle$ along the centerline $y = 0$ (left) and on a line $x = 1$ (right). In all plots ——— RUN 1, $---$ RUN 2, $-\cdot-\cdot-$ Run 3. Comparison with experimental results of [9] (full diamond) and an unsteady $k - \varepsilon$ computation of [15] (light square).

be content with fewer computations and less averaging in time. Two–dimensional studies were performed in order to construct grids that meet best the requirements with the available resources. To enhance the resolution of the boundary layer any stretching in azimuthal direction was dropped and the number of points in this direction increased. In [17] we have presented a first LES for this case with $L_z/D = 1$. Two further computations have now been performed with $L_z/D = 4$, motivated by the observed correlation in z and studies of the square cylinder [35]. They are summarized in Table 1. Other computations are available in [7] and [42]. In [42] the authors observe that very long averages of the order of 40 shedding cycles might be required for fully converged statistics. Due to limited resources we could only average over 10–14 cycles. Nevertheless, the performed computations give valuable information on trends and sensitivities. In [7] no superiority of the dynamic model with respect to the Smagorinsky model used in the present LES has been observed.

Figure 4 reports profiles along the centerline $y = 0$ and $x = 1$ for various quantities. It is apparent that in RUN1 the vortex shedding is too strong. As a consequence the maxima of the mean streamwise velocity are less pronounced and the v–fluctuations are over–predicted. Increasing the domain size to $L_z = 4$ with the same number of grid points, i.e. a coarser grid in z, yields only minor changes. An exception is the mean centerline velocity which levels off at a substantially lower value. The third run has been made with an increased resolution in all directions. The resolution of the shear layer is now substantially better which is reflected by the improved result for $\langle \bar{u} \rangle (x = 1)$ outside the recirculation. The Reynolds stresses are substantially lower which is the right tendency (the reported stresses are the resolved ones to which a subgrid scale contribution has to be added). On the other hand the values for $C_D, - C_{bp}$ are too small. We conjecture that the discrepancy results from insufficient resolution of the separating shear layer. The importance of this issue has recently been highlighted in [22] for $Re = 3900$. As in [7], we observe too large separation angles compared to the experiments which is also of relevance for the downstream shear layer. This occurs although the radial resolution in our LES is $\Delta r_1 = 0.00017$ compared to 0.0004 in [7]. The cited computations in the literature have mostly been made with a smaller spanwise domain size. Our results show that increasing L_z does not modify the result substantially. The resolution in radial and azimuthal direction rather appears to be the limiting factor in the present cases. On the other hand, substantial refinement in these directions in [7] did not yield a clear uniform improvement of the result. Interestingly, the best agreement with experiments was obtained with $L_z/D = 1$. We shall continue to work on modelling issues as discussed below in order to possibly obtain an improved and more cost–effective LES of this case.

4   Circular cylinder in presence of shear

4.1   Physics of the flow

Uniform upstream conditions constitute the classical setting for the study of the flow around cylinders. In practice, however, the upstream flow is often non–uniform. Of particular interest is a gradient in spanwise direction as it disrupts the spanwise symmetry of the configuration and generates qualitatively new features [45]. Applications are in the area of buildings placed in the earth's boundary layer, vehicle aero- and hydrodynamics and industrial flows.

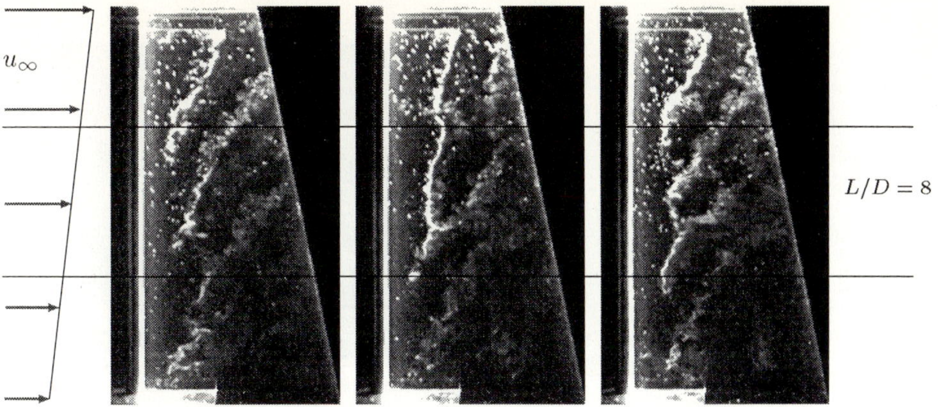

Figure 5 Flow visualization of the vortex shedding behind a circular cylinder in uniform shear flow at $Re_m = 6250$, pictures from [20]. Light regions show hydrogen bubbles trapped in like–sign vortices crossing a laser sheet in the centerplane. There is no temporal correlation between the snapshots. The aspect ratio is $L/D = 20.1$ in the experiment. The aspect ratio $L/D = 8$ of the LES is indicated by two lines.

Reviews on a circular cylinder in uniform spanwise shear are given in [18], [45]. The characteristic parameter is the dimensionless gradient $\beta = du_\infty/dz$, where, as above, cylinder diameter D and average free–stream velocity U_m are used as reference quantities. The gradient causes a variation in the stagnation pressure along the span which in turn generates a secondary flow in front of the body and even more so in the rear (cf. Fig. 7 below). This yields a complicated three–dimensional flow structure interacting with the vortex shedding. An important finding in the experiments with $\beta \neq 0$ such as [32] is that the Strouhal number $St = fD/U_\infty$ does not adjust continuously to the free stream velocity. Rather, the dominant frequency exhibits cells of uniformity. The origin of this behaviour is not entirely clear up to now. Woo et al. [44] have conducted experiments with slotted end plates which inhibit the secondary flow to a smaller extent than the solid end plates in other experiments. They found decreased cell structure. Balasubramanian et al. [2] analyzed these results and conjectured that the strength of frequency cells is closely related to end conditions. The interaction between the three–dimensional average flow and the instantaneous shedding process is not yet understood [45].

4.2 Computational setup

Until now, cylinders in shear flow have mainly been investigated experimentally. RANS models are expected to have similar problems as for uniform upstream flow. Here, LES has a high potential. Recent investigations by means of DNS have been conducted for Reynolds numbers $Re_m = DU_m/\nu$ up to 200 [26],[25].

The present computations deal with the flow at a substantially higher Reynolds number in a configuration which parallels the experiments by Kappler and Rodi [20], [21]. Fig. 5 displays the oblique vortex shedding generated by the shear flow. Characteristic is the occurence of Y–shaped vortex dislocations with the points of junction propagating from the high–speed end to the low–speed end. The experiment in Fig. 5 has been conducted at $Re_m = 6250$, $\beta = 0.04$ and an aspect ratio of $L/D = 20.1$. Since an LES for this geometry

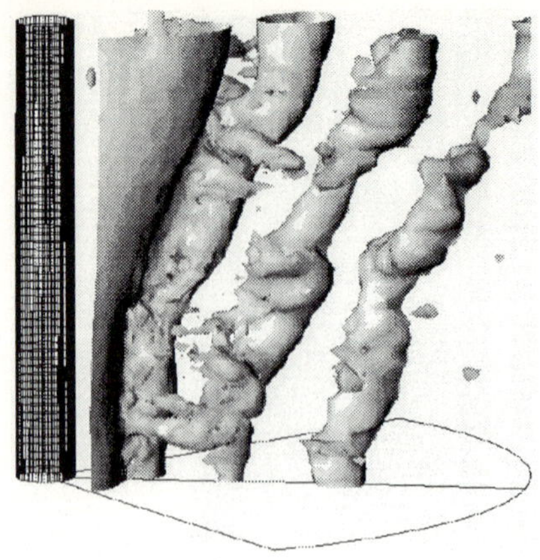

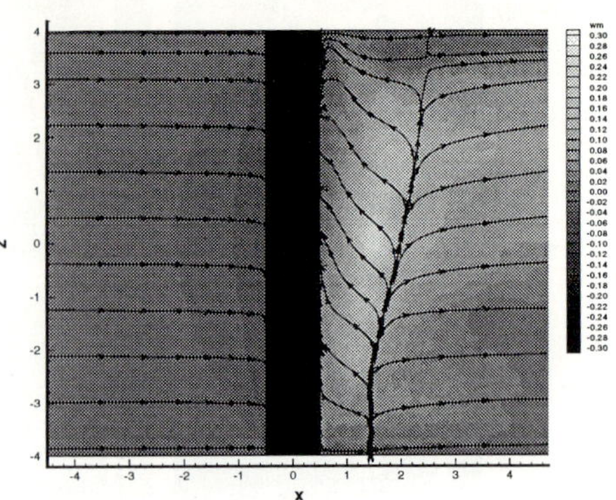

Figure 6 Instantaneous iso–pressure surface $p = -0.6$ in the rear of the cylinder. In angular and radial direction only part of the domain is represented.

Figure 7 Average flow field in the center plane for cylinder in shear flow: streamlines in the $x - z$ plane and average spanwise velocity depicted by grey scale.

was unaffordable with the available resources the aspect ratio was reduced to $L/D = 8$. The code LESOCC was used with the same numerical scheme as above. This is motivated by the typical extent of the frequency cells mentioned above of around $4D$, so that ideally 2 cells would be generated in this domain. This was confirmed in a later experiment [21] under the same conditions but with $L/D = 8$. Guided by the experience from the earlier LES at $Re = 3900$, an O-grid of diameter $30D$ was employed with $178 \times 176 \times 174$ internal cells in radial, azimuthal and spanwise direction, respectively. Points were clustered in the wake and at the cylinder surface. Laminar inflow was prescribed with $u_\infty = 1 + \beta z$ and a convective outflow condition was used. At $z = \pm 4$ a free slip condition was imposed. It represents at best the effects of the end plates with sidelength $8D$ in the experiment. The effect of the employed Smagorinsky subgrid–scale model is only small since $\nu_t/\nu \leq 1.5$. After a transient phase, the computation has been conducted for 220 time units, i.e. 46 shedding cycles. Time signals were collected at $x = 2.5, y = 0.5, z = 0.4n$, where $n = -10 \ldots 10$ (the outermost at $z = \pm 3.9$), similar to the locations in the experiment. Another computation with $\beta = 0$ has been performed under exactly the same conditions for comparison. Discussion and results in addition to the ones presented below can be found in [16].

4.3   Computed flow field

Figure 6 shows a calculated pressure surface of the instantaneous flow. Closely behind the cylinder the vortex core is parallel to the cylinder axis. Further downstream the von Karman vortices become oblique due to the average shear. The angle which is formed is clearly visible. Braid–like structures connecting the von Karman vortices can be discerned as well.

Figure 7 depicts the computed secondary mean flow discussed above. The spanwise component $\langle \overline{w} \rangle$ increases up to 0.26. The streamlines of the average flow in the center-

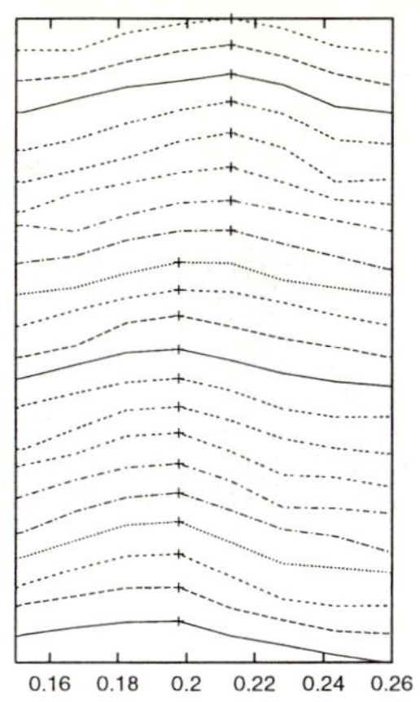
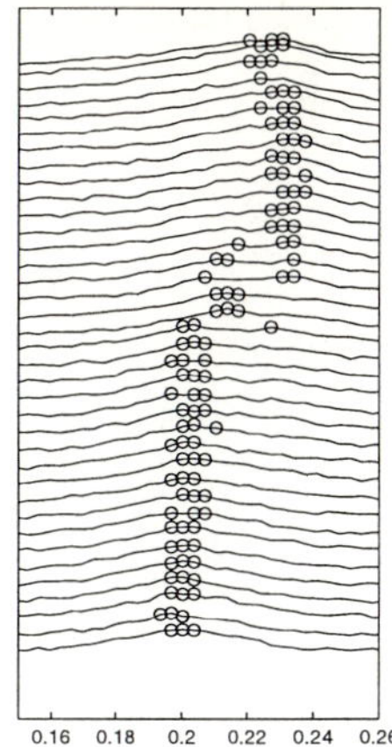

Figure 8 Spectra of the v–velocity signal. Left: results from the present LES with maxima marked by crosses, right: experimental data from [21]. The frequency axis is normalized to represent the Strouhal number. The spectra at different locations have been shifted upwards according to their z–coordinate.

plane highlight the complex three–dimensional character of the flow. They barely display uniformity in z around mid–span and exhibit substantial variations in z at the upper and lower end. Also note the secondary recirculation zone formed at the upper end behind the cylinder.

4.4 Fourier analysis of time signals

The spectra of the velocity signals recorded in the wake have been computed. They exhibit a peak at the dominant shedding frequency and an inertial range over about one decade [16]. Figure 8 shows a zoom around the dominant frequency and compares the LES result to the experimental data for the same configuration [21]. The same cells as in the experiment are observed.

Since, unlike in the experiment, in the LES the time signals have been stored simultaneously at the different locations the correlation coefficient ρ_{ij} between two signals at z_i and z_j can be determined. We have defined a correlation length $l(z_i) = \Delta z \sum_j \rho_{ij}$, $\Delta z = 0.4$, to display this information in Figure 9. In the case without shear l is larger than half the domain size around mid–span and decreases away from it due to the presence of the boundaries. With $\beta = 0.04$ three distinct regions are observed with almost constant value. In particular, the correlation drops towards mid–span where the jump in the shedding frequency occurs in Figure 8. The lower spanwise correlation observed here results in a lower amplitude of the fluctuations of the lift coefficient. Indeed, these are only half as large for $\beta = 0.04$ compared to $\beta = 0$ [16].

4.5 Wavelet analysis of time signals

The above spectra and correlations constitute valuable statistical information but give only an indirect picture of instantaneous mechanisms. This requires time–local averaging as performed with the continuous wavelet transform (CWT) [19], [14]. Further quantities can be defined based on this approach to address particular issues as detailed below.

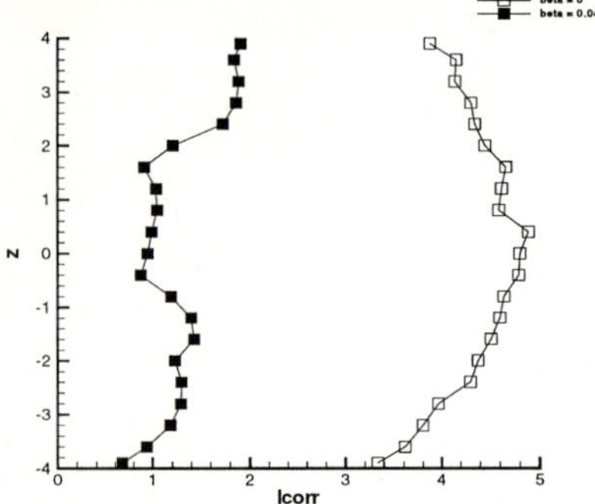
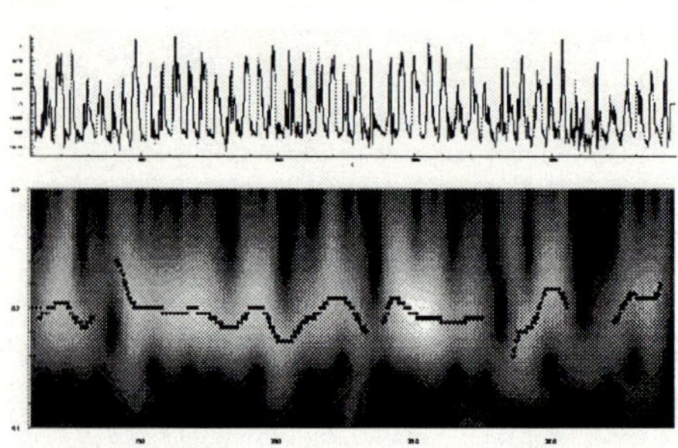

Figure 9 Correlation length l as defined in the text at different spanwise positions. Open symbols: $\beta = 0$, full symbols: $\beta = 0.04$.

Figure 10 Time signal of v at midspan with $\beta = 0.04$ and its wavelet analysis. The ordinate s is scaled to correspond to the Strouhal number and covers the range $[0.1, 0.3]$. The ridge is only retained if $|W(s_R,t)|$ is larger than 80% of the average of $|W|$ along the ridge.

Any function $\psi(t)$ with Fourier transform $\widehat{\psi}$ and $c_\psi = \int_\infty^\infty |\widehat{\psi}(\omega)|^2/|\omega| d\omega < \infty$ can be used to define a wavelet transform

$$W_f(s,t) = \int_{-\infty}^{\infty} f(t')\, \psi_{s,t}^*(t')dt'. \tag{4.1}$$

Here, dilated and translated versions of ψ are denoted

$$\psi_{s,\tau}(t) = \sqrt{s}\, \psi\left(s(t-\tau)\right) \tag{4.2}$$

while the asterisk stands for the complex conjugate. Oscillating signals are conveniently analyzed with complex wavelet such as the one of Morlet with

$$\psi(t) = \pi^{-1/4}(e^{i\omega_0 t} - e^{-\omega_0^2/2})e^{-t^2/2} \tag{4.3}$$

and $\omega_0 = 5$. The transform W then also is a complex value. The scale parameter s plays the same role as the frequency in Fourier analysis and is normalized here so as to correspond to the Strouhal number at which $|\widehat{\psi_{s,\tau}}(\omega)|$ attains its maximum. The value $|W(s,t)|^2$ hence represents the energy of the signal f at time t around the frequency s. If s_R represents the scale number at which $|W(s,t)|^2$ has a maximum for a given time t, the curve $s_R(t)$, the so–called ridge, is an instantaneous analogon of the time–averaged dominant frequency above.

In the following we concentrate on the normal velocity component as it is most directly related to the shedding process. Figure 10 displays a v–signal and its CWT together with the ridge. The frequency range has been chosen in the vicinity of the Strouhal number. It is obvious, that at irregular instances an interrupt or defect in the vortex shedding causes the energy on scales $St \approx 0.2$ to collapse. We therefore retain the ridge only if $|W|$ exceeds a threshold value. A result of this investigation is that the instantaneous shedding frequency hardly ever remains constant for more than 3–4 cycles but rather fluctuates, as reflected by the undulations of the ridge, with a typical period of 5–6 cycles. Without shear ($\beta = 0$) these undulations are much less pronounced in amplitude and have a slightly shorter period, cf. Figure 12.

240

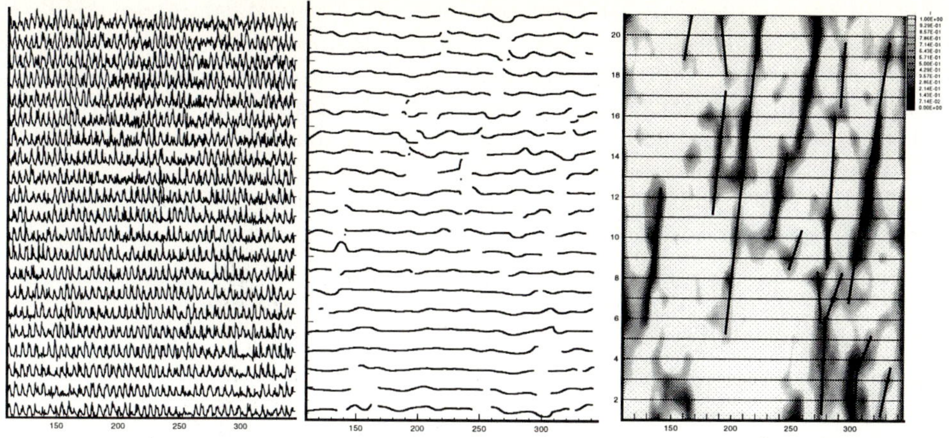

Figure 11 Signals, ridges and intermittency coefficient for $\beta = 0.04$. The horizontal axis is time, the vertical axis is the spanwise position of the signal. Left: v–signals. Middle: ridges as discussed in the text. The shift in the vertical coordinate s between the ridges is 0.15. Right: Intermittency factor I for $St = 0.2$. The continuous distribution between the signals results from the interpolation of the graphics tool. To bring out more clearly the minima of I the gray scale has been squeezed to the lower third of the range in each case. The straight lines have been inserted by hand.

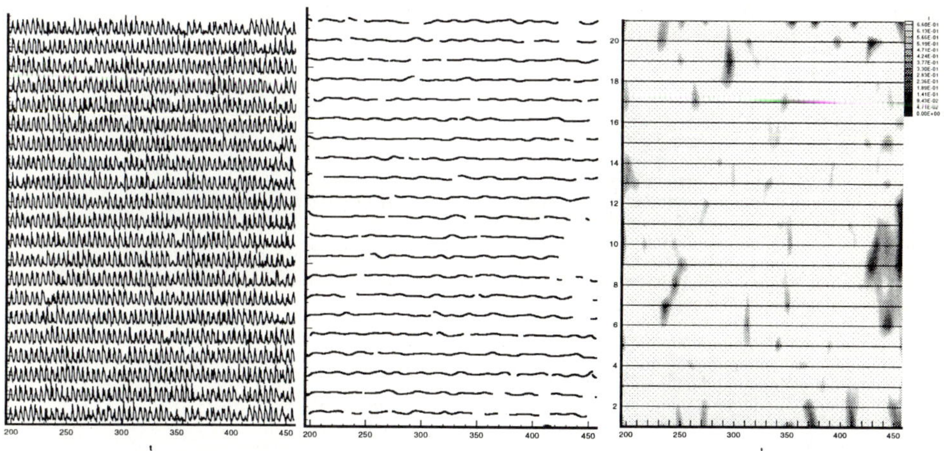

Figure 12 Signals, ridges and intermittency coefficient for $\beta = 0$. For caption see Figure 11.

The defects mentioned above are related to an instantaneous decrease in $|W|$. Since W is a smooth function (cf. Figure 10) s can be kept constant when addressing this issue so that the intermittency coefficient [14]

$$I(s,t) = \frac{|W_f(s,t)|^2}{\langle |W_f|^2 \rangle} \tag{4.4}$$

with $\langle . \rangle$ denoting time averaging is an ideal quantity for their detection. Figure 11 and 12 display $I(0.2,t)$ for all signals. It turns out that with $\beta = 0.04$ defects are stronger than without shear. Furthermore, they propagate from the low–speed end to the high–speed end. Lines have been inserted in the figure by hand in order to estimate their speed yielding a non–dimensional average of about 0.16. The original signals are displayed also. Although here and there defects can be detected by hand, the intermittency coefficient gives a simple and robust means for their determination.

Finally, the CWT can be employed to investigate which events contribute to the Reynolds stresses. The idea is based on the observation that for a complex wavelet [28]

$$\int_{-\infty}^{\infty} f(t)\, g(t)\, dt = \frac{1}{c_\psi} \int_{0}^{\infty} \int_{-\infty}^{\infty} C_{fg}(s,t)\, dt\, ds \tag{4.5}$$

where $C_{fg}(s,t) = R\{W_f(s,t)\, W_g^*(s,t)\}$ is the so–called wavelet co-scalogram and R designates the real part. Now, if f and g are different components of the velocity vector the total Reynolds stress can be obtained by integrating over C_{fg}. Hence, $C_{fg}(s,t)$ indicates which scales and times contribute to the Reynolds stress. Figure 13 shows as an example the vw–correlation at midspan. The intermittent generation of the Reynolds stress is visible. Furthermore, there appears to be a drastic change in the phase around $s = 0.2$ in the case with shear compared to the one without shear.

In [16] a similar method is proposed to investigate two–point correlations and phase shifts taking signals at different location for f and g in (4.5). It turns out that the observed defects are accompanied by large phase differences between adjacent signals. Further investigations along the above lines should be carried out to fully exploit the potential of the approach.

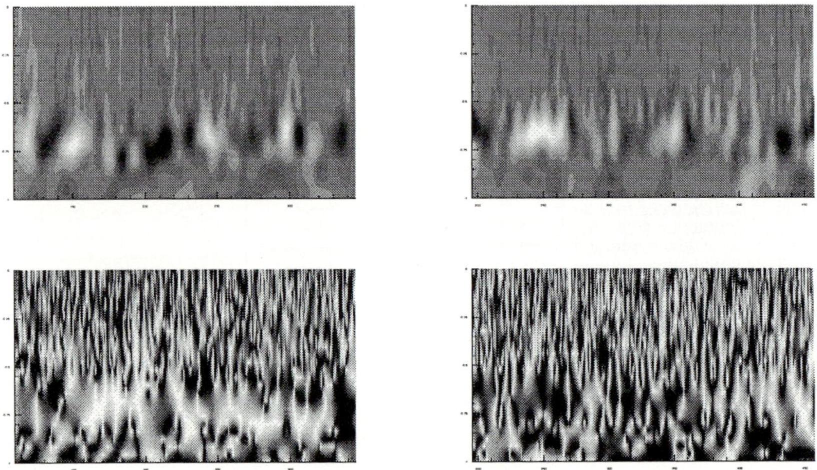

Figure 13 Instantaneous and scalewise correlation between the $v-$ and the $w-$signals at midspan. Horizontal axis is time, vertical axis is $\log(s)$, $s = 0.1 \ldots 1$. Left: $\beta = 0.04$, right: $\beta = 0$. Top: C_{vw} as defined in the text, scale from -0.0009 to 0.0008. bottom: angle between W_v and W_w in absolute value, light is $90°$, dark is $0°$.

5 Boundary condition by means of an integral method

The computations of flow around circular cylinders described above were made for Reynolds numbers in the sub–critical regime where the boundary layer along the cylinder wall remains laminar until separation. Transition to turbulence occurs after separation in the shear layers. Resolving the boundary layer becomes increasingly difficult as the Reynolds number increases and requires a finer and finer mesh. We have therefore developed a model for the windward boundary layer and implemented it in the LESOCC code. It is based on an integral method and allows a coarser discretization normal to the wall yielding savings in CPU time and memory.

The laminar boundary layer is unsteady as it is shifted around the cylinder by the vortex shedding. This motion, however, is small and slow compared to the internal time scale of the boundary layer: Dwyer, McCrosky [13] measured an amplitude of 3.7° for the oscillation of the stagnation point for $Re = 1.06 \cdot 10^5$ and in [31] the criterion $St < 14$ is derived for the second property to hold (here, $St \approx 0.2$). The instantaneous velocity profiles in the boundary layer at a certain distance from the stagnation point hence correspond to the steady state profiles at this distance from the average stagnation point and the laminar boundary layer is amenable to a quasi–steady description.

Integral methods represent a boundary layer in terms of integrals over its thickness such as the displacement thickness δ^*, or the momentum thickness θ. The method of Thwaites [10] is one of them based on the empirical relation

$$\frac{u_e}{\nu} \frac{d\theta^2}{ds} = F(\lambda) = A + B\lambda \quad ; \quad A = 0.45 \quad ; \quad B = 6 \quad , \tag{5.6}$$

where $\lambda = \theta^2 \frac{du_e}{ds}/\nu$, s the tangential coordinate, and u_e the external velocity in the direction of s. Integration in closed form is possible yielding

$$\theta^2(s) = \frac{1}{u_e^B} \int_{s_0}^{s} A \nu u_e^{B-1} ds + u_e^B(s_0) \theta^2(s_0) \quad . \tag{5.7}$$

The last term vanishes if s_0 is a stagnation point. Eq. (5.6) can be used for $\lambda \in [-0.1; 0.1]$ with $\lambda = -0.09$ corresponding to separation. The shear stress is subsequently obtained by an empirical correlation for $S(\lambda) = c_f u_e \theta/(2\nu)$, and δ^* is obtained from a similar correlation for the shape factor $H(\lambda)$ [10].

The implemented algorithm proceeds as follows: a one–dimensional problem along a line of constant z is generated. This is straightforward due to the structured O–grid. Actually, the method is applied to the mean flow so that only one such problem has to be solved. Next, the stagnation point s_0 is determined as a zero crossing of the tangential velocity. Then u_e in (5.7) is replaced by $\sqrt{2(p_w - p_0)/\rho}$. Using the wall pressure is advantageous as it avoids to determine the location of u_e, particularly difficult for medium Reynolds numbers. Eq. (5.7) is then integrated from s_0 until $\lambda < 0.095$, i.e. shortly beyond separation where θ and δ^* drastically increase.

The basis for using the above method in an LES is Schumann's relation [38]

$$\tau_w = \frac{\langle \tau_w \rangle}{\langle \overline{u}_1 \rangle} \overline{u}_1 \quad , \tag{5.8}$$

where $\overline{u}_1$ is the tangential velocity at the wall–adjacent grid point. The integral method applied to the mean flow yields $\langle \tau_w \rangle$. With (5.8), local variations are accounted for by the relation to $\overline{u}_1$. Eq. (5.8) furthermore allows to blend this method with a no–slip condition which can be put in the same form. This is performed depending on the resolution of the boundary layer, measured by the coefficient $\Delta r_1/\delta^*$. If the latter is small, the radial resolution is good and a no–slip condition is adequate (Fig. 14).

Tests have been performed in the two–dimensional setting for $Re = 140000$. Recall that the physics of the wake are substantially different in this case compared to 3d, but the windward laminar boundary layer remains almost unaffected. Figure 14 shows the computed boundary layer thickness until separation and the resulting wall–shear stress along the circumference of the cylinder. Three–dimensional LES with this model will be performed in the near future. Note that with (5.8) blending to other wall models can easily be performed by choosing a different way to determine $\langle \tau_w \rangle$ [16]. The method can also be extended to turbulent boundary layers when replacing the method of Thwaites by another, appropriate integral method.

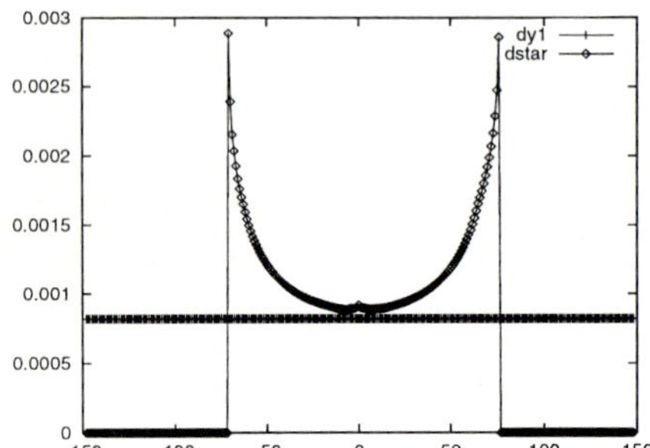 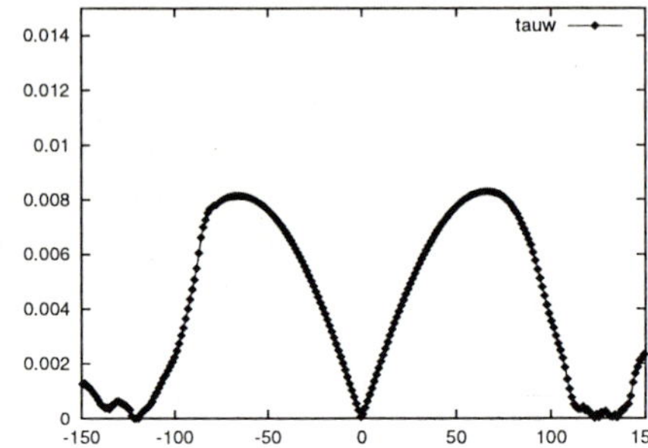

Figure 14   Determination of the wall–shear stress on the windward face with Thwaites' method as described in the text. Left: δ^* and Δr_1, the size of the wall adjacent cell in radial direction, versus the azimuthal angle. Right: computed wall–shear stress with blending to a no–slip condition used.

6   Investigation of a variable filter size

In complex flows the mesh size of the employed grid cannot be kept constant as envisaged in the classical concept of LES. This also applies to the above computations of the flow around cylinders requiring substantial refinement near the surface of the body. The cutoff scale then varies in space and interacts with the turbulence modelling. This can be appreciated best when considering a flow with a statistically homogeneous direction. Varying the mesh size and hence the filter width in this direction introduces a statistical inhomogeneity of the resolved flow. To isolate the effect we consider developed channel flow and apply a streamwise non–uniform grid. In [17] LMFA contributed a computation in which an abrupt change of the mesh size was modelled. We have now performed a closer comparative study of this phenonenom by means of plane channel flow at $Re_b = 10935$ for which a DNS is reported in [1]. An LES using the LESOCC code was perfomed on a grid with a gradually varying streamwise step size and streamwise periodicity (Figure 15). It contains two regions of constant step size, $\Delta x^+ = 220$ and $\Delta x^+ = 30$, respectively, limited by two intermediate zones with 5% geometric stretching. The dynamic model is used with averaging in spanwise direction and temporal relaxation [8]. The Werner-Wengle wall model is used to bridge the distance between the first point at $y_1^+ = 30$ and the wall, furthermore $\Delta z^+ = 30$. For the same grid LMFA performed a computation with the statistical two–point closure S.C.I.T. [30] which has, compared to the study reported in [17], now been generalized for anisotropic turbulence [41] and is briefly described in Appendix A. The S.C.I.T. method is usually applied to model the entire spectrum, but for the application to the problem of subgrid–scale modelling a spectral cut off $K < K_c$ is introduced. The contribution $K > K_c$ is then modelled by a spectral eddy viscosity

$$\nu_{t_c} = \left(C_1 + C_2 \left(\frac{K}{K_c} \right)^{C_3} \right) \sqrt{\frac{E(K_c)}{K_c}} \tag{6.9}$$

with $C_1 = 0.267$, $C_2 = 0.4724$, $C_3 = 3.742$ as proposed in [11]. To represent the size of the computational grid the cutoff wavenumber K_c is set proportional to $1/\Delta_x$.

Figure 15 shows $< u'/u_\tau >^2$ on a cut along the centerline of the channel. The stream-wise mesh size of the grid is marked by crosses. It is obvious that both methods, although entirely different in their approach, give a very similar result. In accordance with the fil-tering underlying an LES the resolved stress is smaller in regions of the coarse grid and higher in regions of the fine grid. Instructive, however is the observed relaxation in zones of constant mesh size after the stretching or shrinking. When the grid is coarsened, the effect of filter becomes more and more pronounced. There, the energy loss appears to be roughly proportional to the mesh size. The settling in the subsequent area of constant mesh size is a little less pronounced with the LES method than with S.C.I.T. When the grid is refined energy has to be supplied to the newly added resolved fine scales by the turbulence cascade. This process is slow compared to the time it takes a particle to advance into a region of finer grid, although the stretching rate is only 5%. The resolved stress increases slower than it decreased before and a substantial relaxation is observed. In fact, the zone of constant fine grid has to be larger to achieve complete relaxation, but this would have increased the cost of the LES substantially. It should be recalled that in an a priori test when filtering a fully resolved flow field with a variable mesh size the stresses would follow exactly the step size of the grid. This is not the case here due to the dynamic process described above. It is illustrative to mirror the distribution which corresponds to a reversed flow direction (Figure 15 right) to support the above remarks.

The performed computations demonstrate that coarsening the grid in streamwise di-rection has much less effect than refining the grid. Indeed, the latter appears to be detrimental to current subgrid-scale modelling. In the above bluff–body flows we are only faced with the former, since the refinement of the grid takes place in the laminar part of the flow. Research will be pursued in order to possibly develop a correction to standard subgrid-scale models which compensates the observed phenomenon.

7   Concluding remarks

The present paper assembles various activities of the cooperation between the German and the French partners. They concern the numerical aspect in that a structured and two unstructured codes have been employed for the same flow. In particular, the reso-lution requirement which is extremly demanding for the higher Reynolds number case is discussed. A proposal for a boundary layer model to be integrated in such an LES

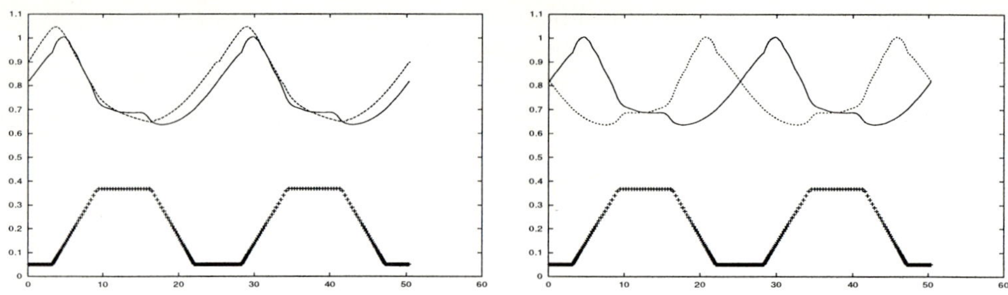

Figure 15 Reynolds stress $< u'/u_\tau >^2$ on the centerline of a channel with variable streamwise grid spacing as described in the text (+). The computation was performed for one period $[0, 25.1]$, which is repeated in the figure to enhance clarity. Right: comparison between results of S.C.I.T. ($---$) and the LES performed with LESOCC (——). Left: plot of LES result (——) together with the same data for a reversed flow direction ($---$).

of a subcritical cylinder has been made to alleviate this problem. The approach can be extended for turbulent boundary layers as well. A more physical study has been undertaken with the computation of the flow around a cylinder in uniform shear flow at a much higher Reynolds number than considered in the current literature. For this purpose, analyzing techniques have been developed which take into account variations in the instantanous character of the signal. They reveal defects in the regular vortex shedding which propagate in spanwise directions and appear more frequently in the presence of spanwise shear. The implemented algorithms can partly also be applied to the available experimental time signals, and this will be done in future work. Finally, the model study of streamwise grid refinement highlights the methodological problems this may create. Although this issue is pertinent to many cases where LES is applied to complex flows it has to our knowledge not been investigated in detail up to now. We will continue work on this issue as it seems to be desirable that an SGS model should account for such effects.

Acknowledgments

The authors thank the Deutsche Forschungsgemeinschaft (DFG) and the Centre National de la Recherche Scientifique (CNRS) for the support of this work within the French–German Research Programme 'Numerical Flow Simulation'. The first author is particularly gratefull to R. Peyret for the continuous interest, support and the gratifying collaboration during and after the time he worked in his group in Nice.

Appendix A: Turbulence model employed by LMFA

The S.C.I.T. model (for Simplified Closure for Inhomogeneous Turbulence) [30] is based on a statistical spectral approach, that is to say on a description of turbulence by correlations at two points. It is known that two–point models directly take into account information on different length scales, up to the Kolmogorov scale. Consequently no ϵ equation is required. Before being applied to real flows, the complex formulations of two–point closures for inhomogeneous turbulence must however be simplified. This is the approach followed when developing the S.C.I.T. model.

A first version of the S.C.I.T. model (S.C.I.T.0) was proposed in [30]. The basic quantity in S.C.I.T.0 is the turbulent kinetic energy spectrum $E(K,\vec{X},t)$ and consequently, although a detailed spectral information was retained, the anisotropic properties of turbulence are only grossly accounted for. The next stage was to extend the approach to a tensorial description of the turbulent spectra

$$\varphi_{ij}(K,\vec{X},t) = \int_{|\vec{K}| \leq K} \Phi_{ij}(\vec{K},\vec{X},t)\,d\sigma(\vec{K}) \quad , \tag{8.10}$$

with $\Phi_{ij}(\vec{K},\vec{X},t)$ being the Fourier transform on the correlation $\langle u_i(\vec{x}-\vec{r}/2)\,u_j(\vec{x}+\vec{r}/2)\rangle$ and K the modulus of $\vec{K}$. It results in a new version of the model (S.C.I.T.1) [41] which is applied here. The equation for φ_{ij} reads:

$$\left(\frac{\partial}{\partial t} + \langle U_e \rangle \frac{\partial}{\partial X_l}\right)\varphi_{ij}(\vec{X},K,t) = \left(-2\nu K^2 + \frac{\nu}{2}\nabla^2\right)\varphi_{ij}(\vec{X},K,t) \tag{8.11}$$
$$+P_{ij}(\vec{X},K,t) + p_{ij}^L(\vec{X},K,t)$$
$$+p_{ij}^{LW}(\vec{X},K,t) + t_{ij}^L(\vec{X},K,t)$$
$$+t_{ij}(\vec{X},K,t) + D_{ij}(\vec{X},K,t) \quad ,$$

in which the first term of the right hand side is a viscous contribution which can be expressed exactly. The second term P_{ij} is also a term which is not requiring a closure (production term). The other terms $p_{ij}^L, p_{ij}^{LW}, t_{ij}^L, t_{ij}$ and D_{ij}, respectively, stand for the rapid part of the pressure–strain spectrum, the echo term associated with wall effects, the linear transfer, the non linear transfer and the inhomogeneous transport term. These terms were closed introducing various assumptions described in [41]. One of the basic ingredients in the model is the use of the EDQNM theory to express the transfer term [29]. The relation between S.C.I.T.0 and S.C.I.T.1 is illustrated by the relation $E(K,\vec{X},t) = 1/2\varphi_{ii}(K,\vec{X},t)$. Correspondingly, in S.C.I.T.0 only a transport equation for E is solved.

The computation procedure consists in discretizing the K-space, and in solving at each wave length K the six equations for φ_{ij} on the physical computational domain. Then, one gets the information required to close the averaged Navier-Stokes equations, which are solved together with (8.11):

$$\langle u_i u_j \rangle = \int_0^\infty \varphi_{ij}(K,\vec{X})\,dK \tag{8.12}$$

Concerning the near wall treatment, a spectral "infrared cut-off", depending on the distance to the wall is imposed as in [30]. For the pressure strain correlation, an echo term is introduced. The boundary conditions for the different turbulent quantities are deduced from wall functions. A log law is assumed and a hyperbolic tangent profile is used to match with the viscous sublayer [12].

References

[1] AGARD, Neuilly–sur–Seine, France. *A selection of test cases for the validation of Large Eddy Simulations of turbulent flows*, 1998.

[2] S. Balusubramanian, F.L. Haan Jr., A.A. Szewczyk, and R.A. Skop. On the existence of a critical shear parameter for cellular vortex shedding from cylinders in nonuniform flow. *J. Fluids & Struct.*, 12:3–16, 1998.

[3] P. Beaudan and P. Moin. Numerical experiments on the flow past a circular cylinder at sub–critical Reynolds number. Technical Report TF-62, Stanford University, 1994.

[4] U. Bieder. PRICELES: Large Eddy Simulation of the very near wake of a circular cylinder. Technical Report DRN/DTP/SMTH/LATA/99-73, CEA Grenoble, 1999.

[5] U. Bieder. PRICELES: Tests of the numerical scheme. Technical Report DRN/DTP/SMTH/LATA/99-52, CEA Grenoble, 1999.

[6] U. Bieder, Ph. Emonot, and D. Laurence. PRICELES: Summary of the numerical scheme. Technical Report DRN/DTP/SMTH/LATA/98-50, CEA Grenoble, 1998.

[7] M. Breuer. A challenging test case for Large Eddy Simulation: High Reynolds number circular cylinder flow. In S. Banerjee and J.K. Eaton, editors, *Turbulence and Shear Flow Phenomena – 1*, pages 735–740, New York, 1999. Begell house inc.

[8] M. Breuer and W. Rodi. Large eddy simulation of complex turbulent flows of practical interest. In E.H. Hirschel, editor, *Flow simulation with high performance computers II*, volume 52 of *Notes on Numerical Fluid Mechanics*, pages 258–274. Vieweg, Braunschweig, 1996.

[9] B. Cantwell and D. Coles. An experimental study of entrainment and transport in the turbulent near wake of a circular cylinder. *J. Fluid Mech.*, 136:321–374, 1983.

[10] T. Cebeci and P. Bradshaw. *Physical and Computational Aspects of Convective Heat Transfer*. Springer Verlag, New York, 1984.

[11] J.P. Chollet and M. Lesieur. Parameterization of small scales of three dimensional isotropic turbulence. *J. Atmos. Sci.*, 38:2747–2757, 1981.

[12] P. Debaty. *Performances des modèles de turbulence au second ordre appliqués à des configurations axisymétriques simulées par éléments finis*. PhD thesis, École Centrale de Lyon, 1994.

[13] H. Dwyer and W.J. McCroskey. Oscillating flow over a cylinder at large Reynolds number. *J. Fluid Mech.*, 61:753–767, 1973.

[14] M. Farge, N. Kevlahan, V. Perrier, and E. Goirand. Wavelets and turbulence. *Proceedings of the IEEE, Special Issue on Wavelets*, 84:639–669, 1996.

[15] R. Franke. *Numerische Berechnung der instationären Wirbelablösung hinter zylindrischen Körpern*. PhD thesis, Universität Karlsruhe, 1991.

[16] J. Fröhlich. LES of vortex shedding past circular cylinders. to appear in Proceedings of ECCOMAS 2000, Barcelona, 11–14 September, 2000.

[17] J. Fröhlich, W. Rodi, Ph. Kessler, S. Parpais, J.P. Bertoglio, and D. Laurence. Large eddy simulation of flow around circular cylinders on structured and unstructured grids. In E.H. Hirschel, editor, *Notes on Numerical Fluid Mechanics*, volume 66, pages 319–338. Vieweg, 1998.

[18] O.M. Griffin. Vortex shedding from bluff bodies in a shear flow: a review. *Trans. ASME: J. Fluids Eng.*, 107:298–306, 1985.

[19] A. Grossman, R. Kronland-Marinet, and J. Morlet. Reading and understanding continuous wavelet transforms. In J.M. Combes, A. Grosman, and P. Tchamitchian, editors, *Wavelets, Time–Frequency Methods and Phase Space*, pages 2–20. Springer, 1989.

[20] M. Kappler and W. Rodi. Experimentelle Untersuchung der Strömung um Kreiszylinder mit ausgeprägten dreidimensionalen Effekten. Technical Report Ro-558/15-1, Institut für Hydromechanik, Universität Karlsruhe, 1998.

[21] M. Kappler, W. Rodi, and O. Badran. Experiments on the flow past a long circular cylinder in a shear flow. In *accepted for presentation at 4th International Colloquium on Bluff Body Aerodynamics & Applications, September 2000, Bochum*, 2000.

[22] A. G. Kravchenko and P. Moin. Numerical studies of flow over a circular cylinder at $Re_D = 3900$. *Phys. Fluids*, 12:403–417, 2000.

[23] L.M. Lourenco and C. Shih. Characteristics of the plane turbulent near wake of a circular cylinder. A particle image velocimetry study. (data taken from Beaudan,Moin(1994)), 1993.

[24] R. Mittal. Progress of LES of flow past a circular cylinder. In *Annual Research Briefs 1996*, pages 233–241. Center for Turbulence Research, 1996.

[25] A. Mukhopadhyay, P. Venugopal, and S. P. Vanka. Numerical study of vortex shedding from a circular cylinder in linear shear flow. *J. Fluids Eng.*, 121:460–468, 1999.

[26] D.J. Newmann and G.E. Karniadakis. A direct numerical simulation of flow past a freely vibrating cable. *J. Fluid Mech.*, 344:95–136, 1997.

[27] L. Ong and J. Wallace. The velocity field of the turbulent very near wake of a circular cylinder. *Experiments in Fluids*, 20:441–453, 1996.

[28] M Onorato, M V Salvetti, G Buresti, and P Petagna. Application of a wavelet cross-correlation analysis to DNS velocity signals. *Eur. J. Mech. B/Fluids*, 16(4):575–598, 1997.

[29] S.A. Orszag. Analytical theories of turbulence. *J. Fluid Mech.*, 41:363–386, 1970.

[30] S. Parpais and J. P. Bertoglio. A spectral closure for inhomogeneous turbulence applied to turbulent confined flow. Kluwer Academic editor, ETC VI, July 1996.

[31] D.E. Paxson and R.E. Mayle. Velocity measurements on the forward portion of a cylinder. *J. Fluids Eng.*, 112:243–245, 1990.

[32] R.D. Peltzer and D.M. Rooney. The effects of roughness and shear on vortex shedding cell lengths behind a circular cylinder. *Trans. ASME: J. Fluids Eng.*, 107:61–66, 1985.

[33] W. Rodi. Comparison of LES and RANS calculations of the flow around bluff bodies. *J. Wind Ind. Aerodyn.*, 69-71:55–75, 1997.

[34] W. Rodi. Large–Eddy Simulation of the flow around bluff bodies. In B. Launder and N. Sandham, editors, *Closure Strategies for Turbulent and Transitional Flows*, volume to appear. Cambridge University Press, 2000.

[35] W. Rodi, J.H. Ferziger, M. Breuer, and M. Pourquié. Status of large eddy simulation: Results of a workshop. *J. Fluid Eng.*, 119:248–262, 1997.

[36] P. Rollet-Miet, D. Laurence, and J.H. Ferziger. LES and RANS of turbulent flow in tube bundles. *Int. J. Heat Fluid Flow*, 20:241–254, 1999.

[37] G. Schewe. On the force fluctuations acting on a circular cylinder in crossflow from sub-critical up to transcritical Reynolds numbers. *J. Fluid Mech.*, pages 265–285, 1983.

[38] U. Schumann. Subgrid scale model for finite difference simulations of turbulent flows in plane channels and annuli. *J. Comput. Phys.*, 18:376–404, 1975.

[39] J. Son and T.J. Hanratty. Velocity gradients at the wall for flow around a cylinder at Reynolds numbers from 5×10^3 to 10^5. *J. Fluid Mech.*, 35:353–368, 1969.

[40] S. Szepessy and P.W. Bearman. Aspect ratio and end plate effects on vortex shedding from a circular cylinder. *J. Fluid Mech.*, 234:191–217, 1992.

[41] H. Touil, S. Parpais, and J. P. Bertoglio. A spectral model for anisotropic inhomogeneous turbulence. 13th Australasian Fluid Mechanics Conference, Monash University, Melbourne, Australia, December 1998.

[42] A. Travin, M. Shur, M. Strelets, and Ph. Spallart. Detached–eddy simulations past a circular cylinder. submitted, 2000.

[43] T. Wei and C.R. Smith. Secondary vortices in the wake of circular cylinders. *J. Fluid Mech.*, 169:513–533, 1986.

[44] H.G.C. Woo, J.E. Cermak, and J.A. Peterka. Secondary flows and vortex formation around a circular cylinder in constant-shear flow. *J. Fluid Mech.*, 204:523–542, 1989.

[45] M.M. Zdravkovich. *Flow Around Circular Cylinders*. Oxford University Press, 1997.

DIRECT NUMERICAL AND STATISTICAL SIMULATION OF TURBULENT BOUNDARY LAYER FLOWS WITH PRESSURE GRADIENT

T.J. Hüttl[1], G.B. Deng[2], M. Manhart[1], J. Piquet[2], R. Friedrich[1]
[1] Lehrstuhl für Fluidmechanik, TU München
Boltzmannstr. 15, D-85748 Garching, Germany
[2] Laboratoire de Mécanique des Fluides, Ecole Centrale de Nantes
1, Rue de la Noë, F-44072 Nantes Cedex, France

Summary

A zonal grid approach has been developed as a means to reduce the large number of grid points required for direct numerical simulation (DNS) of boundary layers. The case of a boundary layer flow with zero pressure gradient shows that computational resources can be saved by zonal DNS compared to a full fine grid DNS and that the accuracy of the results is increased as compared to a completely coarse grid. The zonal grid approach is also used for DNS of boundary layer flow with adverse pressure gradient. DNS results are compared with experimental data. In order to evaluate the performance of existing turbulence models, RANS simulations have been performed for the same boundary layer under adverse pressure gradient. Several models, ranging from two-equation models to Reynolds stress transport models have been tested and the results are compared with DNS and experimental data.

1   Introduction

The identification of weaknesses of turbulence models for the Reynolds Averaged Navier-Stokes equations (RANS) and for large eddy simulations (LES) is one focus of the current turbulence research. A DNS can provide highly reliable data bases for this goal, because all relevant turbulent length and time scales are resolved. Due to the limitations of computational power, only low or moderate Reynolds numbers can be investigated by DNS up to now. Although the computational power has been continuously increasing during the last years, a DNS of a complex flow, like a separated turbulent boundary layer, at a moderate Reynolds number requires the use of the fastest available computers and the most efficient algorithms currently available. A promising way to save computer resources in a DNS is to use a locally refined computational grid in certain regions of the flow and a coarse grid elsewhere.

The use of locally refined (zonal) grids in DNS is a relatively new approach. Kravchenko *et al.* [6] performed a zonal grid DNS of turbulent plane channel flow using a combined B-spline/spectral method. Later Kravchenko and Moin [7] extended the method to the flow around a cylinder. Sullivan *et al.* [24] used zonal grids to perform an LES of a planetary boundary layer. Like Kravchenko *et al.* [6] they used spectral interpolation at the boundary between coarse and fine grid. Spectral interpolation delivers highly accurate results but has the disadvantage of being restricted to relatively simple geometries with periodic boundaries. A first step to overcome this drawback has been made by Manhart [11, 12, 13, 14] in zonal DNS of turbulent plane channel flow and turbulent boundary layers, using the finite volume code MGLET, a well-tested code for LES and DNS of turbulent flows in complex geometries (Werner and Wengle [26] and Manhart and Wengle [16]).

The purpose of the current research is to extend the zonal grid approach to the DNS of turbulent boundary layers with pressure gradients. Then, the DNS data is used to test and improve turbulence models. In the present paper the application of zonal DNS to an adverse pressure gradient boundary layer is evaluated and compared with a zero pressure gradient boundary layer. The adverse pressure gradient boundary layer is used for comparison with RANS models and with experimental data of Watmuff [25].

2 Numerical method

2.1 The basic scheme of the DNS Code

The code MGLET is based on a finite volume formulation of the Navier-Stokes equations on a staggered Cartesian non-equidistant grid. The spatial discretization is of second order (central) for the convective and diffusive terms. For the time advancement of the momentum equations an explicit second-order time step (leapfrog with time-lagged diffusion term) is used.

The Poisson equation for the pressure is solved by an iterative point-wise velocity-pressure iteration like that described in Hirt et al. [5]. It can be used as a single-grid iteration or as a smoother in a multigrid cycle. The maximum divergence is chosen in order to keep the maximal velocity error below $\Delta u_{max} \leq 10^{-5} u_\infty$ (according to the relation $div_{max} - \Delta u_{max}/\Delta x_{min} \cdot l/u_\infty$). Here, u_∞ and l are the characteristic velocity and length scales, respectively.

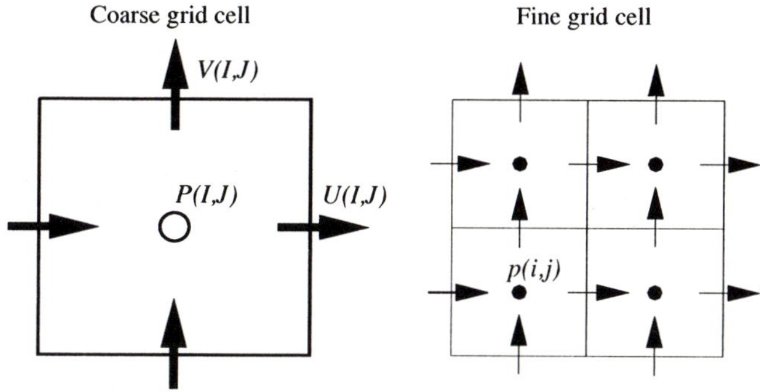

Figure 1: Refinement strategy of the grid with staggered variable arrangement.

2.2 Zonal grid algorithm

The refinement for the local grids is done by dividing one coarse grid cell into 8 fine grid cells (Figure 1). The coarse and the fine grid are arranged in an overlapping way, so that the coarse grid is defined globally (global grid) and the fine grid is defined only locally (zonal grid). Each second cell-face of the local fine grid lies exactly on a coarse grid cell-face. The overlapping of the grids allows for flexible handling of grid refinement. One can start the simulation on a coarse grid and later, if in some region of the computation the resolution is not sufficient, a locally refined grid can easily be inserted.

In our approach, the coarse-grid and the fine-grid solutions are fully coupled. The coupling is achieved by transfering the fine-grid solution in the overlap region to the coarse grid. This so-called restriction is done at certain steps within the solution algorithm. We use averaging over four cell faces for the velocities and averaging over 8 grid cells for the pressure restriction. The solution on the coarse-grid level in the non-overlapping region serves as a boundary condition for the fine grid. The interpolation of the coarse-grid variables to the fine-grid boundary points is done by a first-order interpolation in order to provide conservation of mass and momentum fluxes. If the interpolation were not conservative, the turbulent fluctuations would be strongly damped near the grid interface [11]. For solving the Poisson equation on both levels, we use the pressure correction on the coarse grid as a new pressure estimate for the fine grid in a multigrid cycle.

The advancement from time-level n to time-level $n + 1$ is done by the following steps:

1. explicit time step on both grid-levels independently
2. restriction of the fine-grid solution to the coarse grid
3. solving the Poisson equation for the pressure on the coarse grid
4. prolongation of the pressure correction from coarse grid to fine grid
5. setting velocity boundary conditions for the fine grid on the intergrid boundary
6. solving the Poisson equation for the pressure on the fine grid
7. restriction of the fine-grid solution to the coarse grid

The steps 2 to 6 in the algorithm are part of the multigrid cycle for solving the Poisson equation on both grid levels. Depending on the pressure boundary condition for the zonal grid and on the smoother for the Poisson equation, the multigrid cycle has to be performed iteratively.

2.3   Configuration and boundary conditions

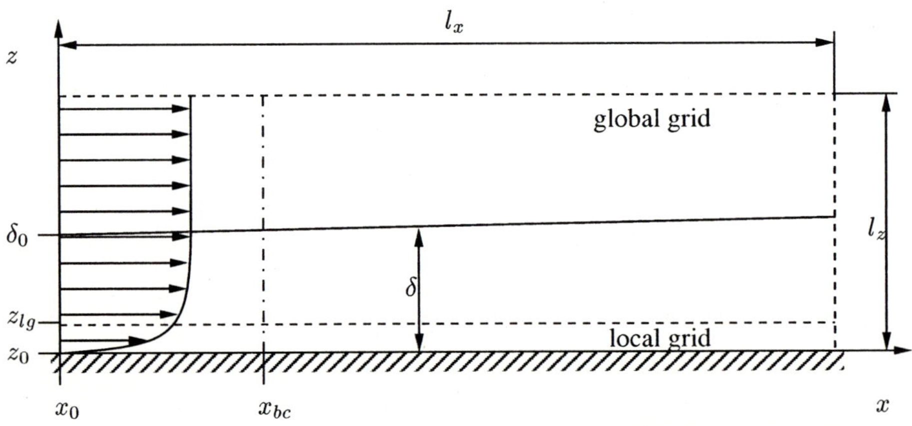

Figure 2: Geometry of the boundary layer simulations (not to scale).

The geometry, used for the simulations of turbulent boundary layers, is sketched in Figure 2. The streamwise, spanwise and wall-normal directions are denoted by x, y and z, respectively.

The global grid defines the whole computational box with dimensions l_x, l_y and l_z. In addition to that, a locally refined grid is used between the wall and the wall-normal position z_{lg}. Manhart [11] showed for zonal DNS of channel flow, that z_{lg} should preferably lie in the logarithmic region of the mean streamwise velocity profile.

At the inlet plane, time dependent boundary conditions are needed to prescribe turbulent velocity fluctuations. They are generated by taking fluctuations $\vec{u}''$ from a position x_{bc} downstream and adding them to a mean velocity profile:

$$\vec{u}_0\left(x=0,y,z,t\right) = \langle\vec{u}_0\rangle\left(z\right) + \vec{u}''\left(x=x_{bc},y,z,t\right). \tag{2.1}$$

For $\langle\vec{u}_0\rangle\left(z\right)$, a time-averaged velocity profile of a zero-pressure gradient boundary layer simulation of Spalart [22] is taken that matches the Reynolds number desired. The fluctuations are computed by using space-averaging in the homogeneous spanwise direction. An exponential damping function f_d at a wall distance higher than the inlet boundary layer thickness δ_0 prevents the boundary layer from growing in time:

$$f_d = \left(e^{\left(\frac{\delta_0-z}{\delta_0}\right)}\right)^4, \quad \text{for } z > \delta_0. \tag{2.2}$$

This procedure gives similar results than the method proposed by Richter et al. [20] for LES of boundary layers and the method of Lund et al. [9, 10] applied to a zero-pressure gradient boundary layer (see section 3). In the exit plane and at the top surface, a Neumann condition for the velocities and a Dirichlet condition for the pressure is used. The velocity derivatives normal to the boundary are set to zero by the formulations $u(N+1) = u(N)$, $v(N+1) = v(N)$ and $w(N+1) = w(N)$. Here N is the index of the last cell of the computational domain and $N+1$ is the index of an auxiliary cell outside. The pressure at the outflow boundary is set to zero: $p(N+1) = 0$. Near the exit plane a negative effect of this simple formulation on the flow quantities is visible up to about two boundary layer thicknesses upstream. Therefore, a sponge layer will be introduced in the future, in which the solution is forced to attain a desired state. At the upper boundary, $p(N+1) = 0$ is imposed, if a streamwise zero pressure gradient boundary layer is computed. In the case of a pressure gradient in streamwise direction, $p(N+1)$ is a function of the streamwise position thus introducing the pressure gradient. At the wall, impermeability and no-slip conditions are realized and in spanwise direction periodic boundary conditions are used.

2.4 Computational effort

The computational requirements of direct numerical simulations are extremely high. Therefore, the simulations have been performed on two high performance computers. The one, a Cray T3E-900, is a massively parallel computer (HLRS computing center, Stuttgart/Germany). The other, a Fujitsu VPP700, is a vector-parallel computer (LRZ, Leibniz Computing Center, Munich/Germany). Both computers are well suited for large-scale flow computations. For single processor runs of the code MGLET, the minimum requirement of core memory is 12 words per grid point. Parallel computations have additional requirements in order to keep neighbouring information [13]. The necessary CPU-time depends on the architecture of the computer. On the VPP700 vector machine $1.8 \cdot 10^{-6}$ seconds per time step and grid point are needed and a maximum MFLOP-rate of 1000 per processor can be achieved. On the Cray T3E-900 the required CPU-time per time step and grid point is about $2.5 \cdot 10^{-5}$ at 70 MFLOPS per processor. Therefore, about 10 to 15 times more processors are needed on the Cray T3E to get the performance of the VPP700. If the problem size is large enough, parallelisation gives a linear speed-up to more than 200 processors on the Cray T3E-900, [12].

2.5 Numerical method for RANS simulations

In order to evaluate the performance of existing turbulence models in the prediction of boundary layer flow under adverse pressure gradient, several RANS simulations have been performed. For these simulations, the HORUS code is used, which solves the Reynolds averaged Navier-Stokes equations by using a finite difference method with a body-fitted structured grid. A cell-centered layout is used in which the pressure, turbulence and velocity unknowns share the same location. The momentum and continuity equations are coupled through the PISO procedure and several implicit second order accurate schemes are implemented for the space and time discretizations [2]. Preconditioned conjugate gradient solvers (CGS, CGSTAB) are used to solve the linear systems. A performance comparison between the HORUS code and MGLET is presented in [15] for DNS of the minimal flow unit.

Many turbulence models are implemented in the HORUS code, ranging from linear and non-linear two-equation models to models requiring the solution of Reynolds stress transport equations. To avoid any wall-function boundary conditions which turn out to be unacceptable when three-dimensional flows are considered, near-wall low-Reynolds number treatments are systematically implemented in the turbulence models.

3 Zero pressure gradient boundary layer flow at low Reynolds number

3.1 Testing inflow conditions

The spatially developing boundary layer with zero pressure gradient in streamwise direction has been computed to find an optimal way of specifying turbulent inflow conditions. Several test cases have been performed at low Reynolds number in order to keep the number of grid points and therefore the computational costs small. The Reynolds number was $Re_\theta = 300$, based on the momentum thickness at the inlet. A computational domain with the dimensions $l_x/\delta_0 = 15.0$, $l_y/\delta_0 = 3.5$, $l_z/\delta_0 = 4.5$, resolved with $n_x = 256$, $n_y = 144$, $n_z = 96$ grid points, is used. The mesh sizes in wall units are $\Delta x^+ = 9$, $\Delta y^+ = 4$, $\Delta z^+_{min} = 0.5$ The number of time steps required for the statistics is in the range of $n_t = 200\,000$.

For the two test cases ZPGL1 and ZPGL2, the treatment of the boundary condition at the inlet plane, described in section 2.3 was used and the streamwise position x_{bc}, where the fluctuations are taken from, was varied. In the case of ZPGL1 the position $x_{bc} = 5\delta_0$ was used; in ZPGL2 it was $x_{bc} = 10\delta_0$. The results are compared with the approach proposed by Lund *et al.* [9, 10] (case ZPGL3).

3.2 Numerical results

The skin friction coefficients $c_f = \tau_w/(\frac{\rho}{2}u_\infty^2)$ of all the three runs show a smooth streamwise development after the inlet with the exception that the run ZPGL2 immediately after the inlet exhibits jumps to a somewhat higher value than that prescribed at the inlet plane. In Figure 3, c_f is plotted versus the momentum thickness Reynolds number Re_θ and compared with some results from the literature [1, 18, 19, 22]. One can see that the run ZPG2 shows the most realistic

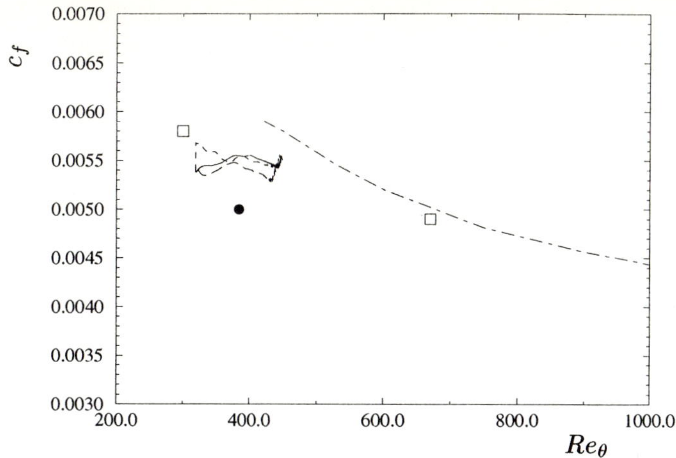

Figure 3: Skin-friction coefficient c_f versus Reynolds number based on momentum thickness Re_θ: ——— ZPGL1, - - - - - ZPGL2, – – – – ZPGL3, —·—· Coles [1], ● Na and Moin [18, 19], □ Spalart [22].

behaviour over the complete streamwise development that is directly linked to the development of Re_θ through the streamwise growth of the boundary layer.

The test of the inflow boundary condition in cases ZPGL1 and ZPGL2 shows that it is possible to prescribe turbulent fluctuations at the inflow plane with a simple algorithm and to yield results which are in fair agreement with those of Lund *et al.* [9, 10], if the position x_{bc} is chosen far enough downstream ($x_{bc} = 10\delta_0$). From a theoretical point of view, the treatment of Lund *et al.* is more sophisticated and perhaps more promising. On the other hand, it is much more difficult to apply to boundary layer flow with pressure gradient.

4 Zero pressure-gradient boundary-layer flow at moderate Reynolds number

4.1 Validation of the zonal grid approach

In order to validate the zonal grid approach for the case of a developing turbulent boundary layer, three direct numerical simulations of a zero pressure gradient boundary layer have been performed at different grid resolutions. Case ZPGM1 only uses a coarse grid. Case ZPGM2 uses a coarse grid as the global grid and a locally refined grid between the wall and $z_{lg}^+ = 60$. In the third case ZPGM3, a fully refined global grid is used. The number of grid points in each direction n_x, n_y, n_z, the total number of grid points $n_{tot} = n_x \cdot n_y \cdot n_z$ and the mesh sizes in wall units Δx^+, Δy^+, Δz^+ are shown in Table 1. For all of these simulations, the same computational box with dimensions $l_x/\delta_0 = 40.96$, $l_y/\delta_0 = 3.2$ and $l_z/\delta_0 = 3.8$ is used and the Reynolds number based on the inlet momentum thickness is $Re_\theta = 670$. Over $n_t = 200\,000$ time steps have been done per case in order to achieve correct second-order statistics. For the boundary condition at the inlet plane, the fluctuations are taken from a position $x_{bc} = 10\delta_0$ downstream.

Table 1: Parameters of the grids used for zero pressure gradient boundary layer flow simulations at a moderate Reynolds number $Re_\theta = 670$ and computational requirements of the code on 8 processors of a Fujitsu VPP700.

Case	ZPGM1	ZPGM2		ZPGM3
		local	global	
n_x	512	1024	512	1024
n_y	80	160	80	160
n_z	96	32	96	192
n_{tot}	$3.9 \cdot 10^6$	$3.9 \cdot 10^6$	$5.2 \cdot 10^6$	$31.4 \cdot 10^6$
Δx^+	26	13	26	13
Δy^+	13	6.5	13	6.5
Δz^+_{min}	3.2	1.6	3.2	1.6
memory [words]	$52 \cdot 10^6$	$125 \cdot 10^6$		$396 \cdot 10^6$
CPU-time/Δt	$1.5sec$	$2.5sec$		$8.0sec$

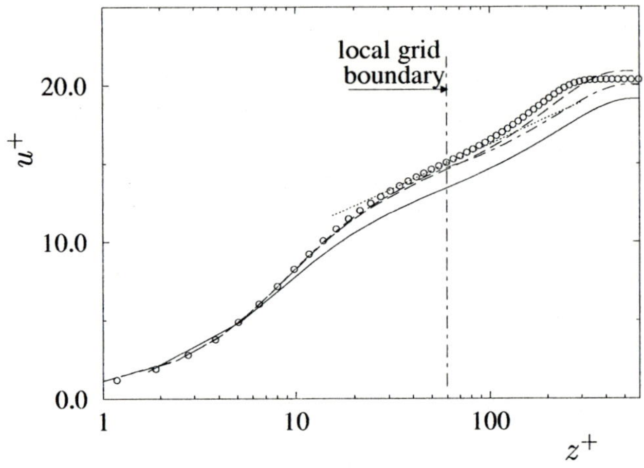

Figure 4: Time-averaged velocity profiles non-dimensionalized with inner coordinates: ——— ZPGM1, —·—· ZPGM2, – – – – ZPGM3, · · · · · · $u^+ = ln(z^+)/0.41 + 5.0$, ○ Spalart ($Re_\theta = 670$) [22].

4.2   Numerical results

The computational requirements, also listed in Table 1, demonstrate how memory and CPU-time can be saved by a zonal grid approach compared to a full grid computation. In the case ZPGM2, only 32% of the memory of a full fine grid is required and the zonal DNS runs 3.2 times faster than case ZPGM3. The achieved accuracy improvement of zonal DNS compared to the coarse grid (ZPG1) can be seen in Figure 4. Here, the mean streamwise velocity profiles, averaged in time and spanwise direction and taken at the position $x = 10\delta_0$ downstream of the inlet, are compared with the result of Spalart's DNS, performed with a spectral method. Up to a wall distance of about $z^+ = 100$ a remarkable agreement between the locally refined and the globally fine grid runs can be observed, while case ZPGM1 differs strongly from the other cases. A high

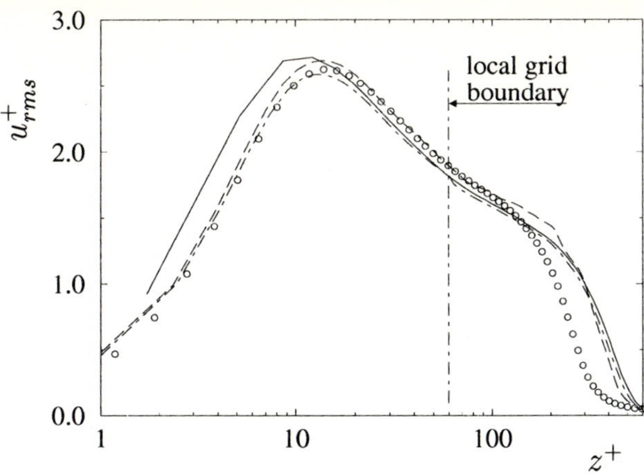

Figure 5: RMS values of streamwise velocity fluctuations non-dimensionalized with inner co-ordinates: ——— ZPGM1, —·—· ZPGM2, – – – – ZPGM3, ○ Spalart ($Re_\theta = 670$) [22].

near-wall resolution is required to reproduce the logarithmic law $u^+ = ln(z^+)/0.41 + 5.0$. Only in the far field, where the grid of case ZPGM2 is too coarse, its curve differs from that one of case ZPGM3. In Figure 5 the RMS values of the streamwise velocity fluctuations are compared with DNS data of Spalart [22]. Near the wall ($z^+ \leq 20$) the case ZPGM1 yields values of u_{rms}, that are clearly too high. On the other hand, the values are slightly too low in the logarithmic layer. The locally refined grid improves the situation in the near-wall region, but not in the logarithmic region where the grid interface is located. An effect of the interface on the curve is hardly visible. The best results in the log-region are obtained with the globally fine grid. In the outer region all computations provide higher turbulence levels than Spalart's results [22]. This is a consequence of the higher Reynolds number at the position $x = 10\delta_0$. It can be stated that by using a locally refined grid in the near-wall region, a significant improvement of the solution can be achieved. The computational cost is much lower than for a fully refined grid.

5   DNS of adverse pressure gradient boundary layer flow

5.1   Configuration

For the simulation of the adverse pressure gradient boundary layer (case APG1), we used the same grid as for case ZPGM2 with local refinement near the wall. The new case APG1 only differs from ZPGM2 in the description of a variable pressure at the upper boundary of the computational domain. The pressure distribution in streamwise direction has been derived from Bernoulli's equation using a $u_\infty(x)$ which meets the experiment of Watmuff [25]. Na and Moin [18] also computed this well documented experiment by DNS, but chose a different way to realize the pressure gradient at the upper boundary.

In the experiment of Watmuff [25], the upper wall of the wind tunnel has been adjusted to accelerate the flow before the adverse pressure gradient (deceleration) has been applied. In our simulation, we let a zero pressure gradient boundary layer develop for 5 inlet boundary layer

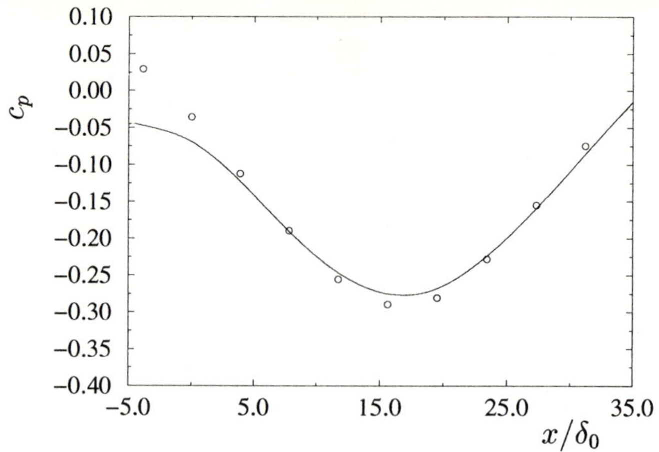

Figure 6: Streamwise development of the wall pressure coefficient c_p: ——— APG1, ○ Watmuff [25].

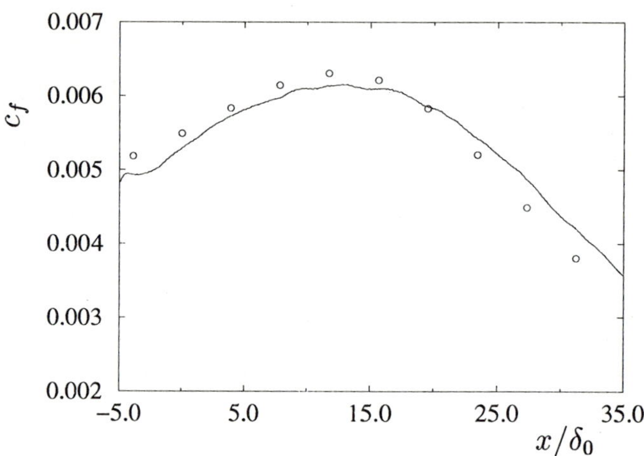

Figure 7: Streamwise development of the friction coefficient c_f: ——— APG1, ○ Watmuff [25].

thicknesses δ_0 before we apply the favourable pressure gradient of the experimental position $x = 0.4m$ at the simulations' top surface at $x = 0$.

5.2 Numerical results

The streamwise development of the wall pressure coefficient $c_p = (p - p_0)/(\frac{\varrho}{2}u_0^2)$ is drawn in Figure 6. u_0 denotes the freestream velocity $u_0 = u_\infty(x = 0)$. The influence of the developing zone $(-5 \le x/\delta_0 < 0)$ leads to a smooth onset of the pressure gradient at the wall.

The decreasing pressure is connected with a rising skin friction coefficient and the rising pressure leads to a decreasing in c_f (Figure 7). Compared to the experimental results, this effect seems to be delayed in the simulation. For the favourable pressure gradient region $(x/\delta_0 \le 15.0)$

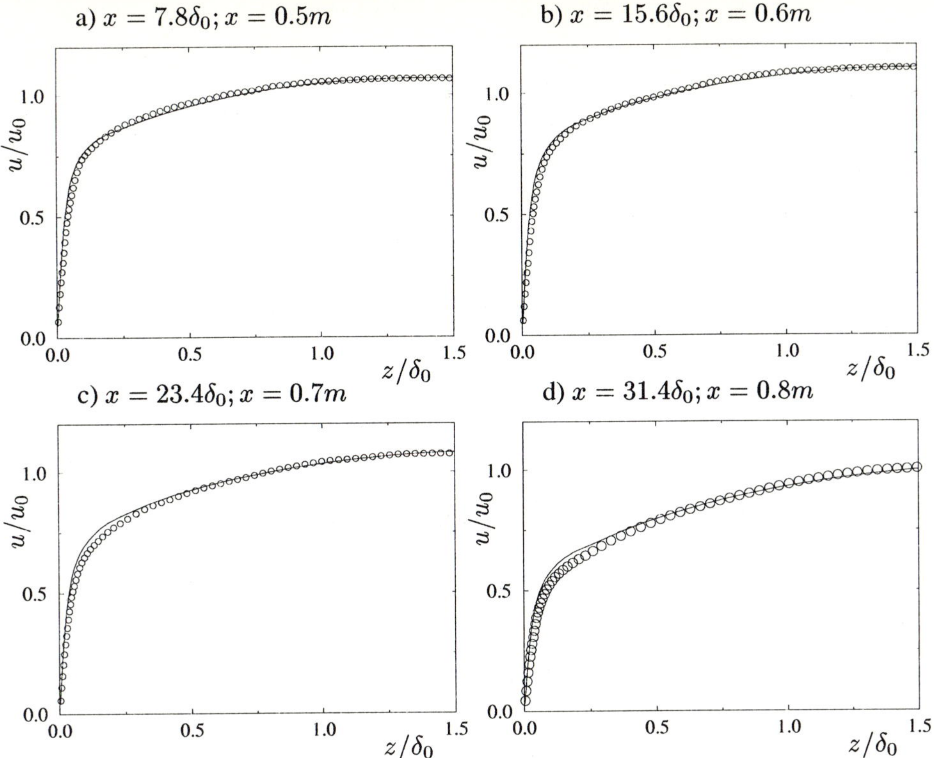

Figure 8: Mean streamwise velocity component at different streamwise positions: ——— APG1, ○ Watmuff [25].

this is not surprising because in the experiment it has been applied for a longer streamwise distance than in the simulation. For the adverse pressure gradient region this is not so obvious and for an explanation of the higher c_f in this region probably a simulation with a longer streamwise domain would be necessary. Nevertheless, the skin friction difference between simulation and experiment is less than 10%.

Profiles of the mean streamwise velocity component obtained by zonal DNS are compared in Figure 8 with experimental results of Watmuff [25] at four streamwise positions. At the same positions, the RMS values of the streamwise velocity fluctuations are shown in Figure 9. Here, the quantities are normalized by the boundary layer thickness $\delta_0 = \delta(x = 0)$ and the freestream velocity $u_0 = u_\infty(x = 0)$.

The behaviour of the streamwise velocity profile during acceleration ($x/\delta_0 = 15.6$) and deceleration ($x/\delta_0 = 31.4$) shows, that both, the acceleration and the deceleration more strongly affect the near wall region than the upper edge of the boundary layer. This effect gives rise to a changing skin friction coefficient, even if it is normalized by the local freestream velocity. The deformation of the time-averaged velocity profile due to this effect is reasonably well represented by the DNS, although it seems that in the buffer layer ($z/\delta_0 \approx 0.1$) the computed deceleration is smaller than the measured one. This observation is consistent with the higher skin friction coefficient in the deceleration region ($x/\delta_0 > 15.0$, Figure 7). The DNS underpredicts the peak RMS value of streamwise velocity fluctuations throughout the whole development of the boundary layer (Figure 9). In order to clarify this observation, we included the RMS values from the zero-

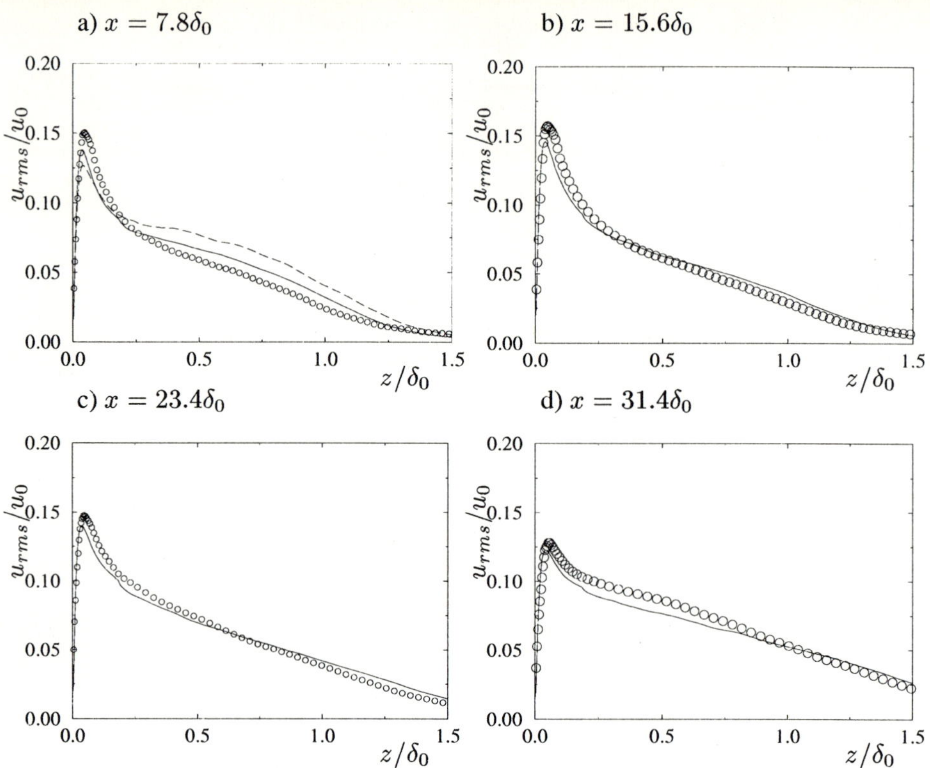

Figure 9: RMS values of the streamwise velocity fluctuations at different streamwise positions: ——— APG1, – – – – ZPGM3, ○ Watmuff [25].

pressure gradient boundary layer (case ZPGM3, see also Figure 5) in the plot for $x/\delta_0 = 7.8$. The DNS results for the accelerated region ($x/\delta_0 = 7.8$) are in between the DNS results for a zero pressure gradient boundary layer and the experiment with an accelerated boundary layer. As one can see in Figure 6 the acceleration that takes place over a long distance in the experiment has just started in the DNS at the position considered. We expect, that the DNS would have needed a longer acceleration zone to achieve the fully developed state of an accelerated turbulent boundary layer. As we can see at the position $x/\delta_0 = 15.6$, the agreement between DNS and experiment increases during the acceleration phase of the boundary layer. It should be noted that in the DNS of Na and Moin [18] the same effect can be observed where the peak values of the DNS are systematically lower than those of the experiment throughout the whole development of the boundary layer, even in the decelerated (adverse pressure gradient) phase. So, the lower RMS values in the decelerated phase do not result from the zonal grid algorithm, although in this region the grid interface at $z/\delta_0 = 0.18$ becomes slightly visible in the RMS values. They rather result from an acceleration zone which is too short in the DNS compared to the experiment.

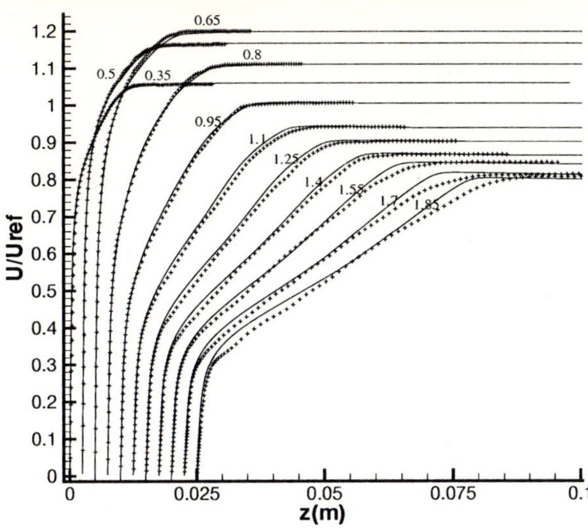

Figure 10: Prediction with the SST model: Symbol: Watmuff, Line: Computation.

6 Computation of adverse pressure gradient boundary-layer flow with turbulence model

In this section, we present numerical results obtained with turbulence models for the turbulent boundary-layer flow under adverse pressure gradient and compare with Watmuff's experiment [25]. Computations have been done mainly with a boundary-layer code. Some RANS computations have also been done for validation. The Reynolds number based on the momentum thickness is about $Re_\theta \approx 500$ at the entrance where the flow becomes turbulent. It increases to $Re_\theta \approx 5000$ at the end of the test section. This Reynolds number is sufficiently high to maintain the turbulence, but still low enough to allow a DNS to be done at the beginning of the test section. The DNS has been performed only in the first half of the test section where Re_θ is smaller than 1500. Computation based on the statistical approach covers the entire test section. For these boundary-layer calculations the inlet profiles are interpolated from the measurement data of Watmuff [25] whenever possible. The streamwise pressure gradient is evaluated from the free-stream velocity by using Bernoulli's equation. A Neumann condition is applied to all other quantities at free-stream.

Three different turbulence models have been tested, i.e. a two-equation model, a nonlinear model and a Reynolds stress transport model. Menter's SST model [17] has been found to perform well in adverse pressure gradient flow. It has been chosen among the two-equation models for the present study. Concerning the nonlinear model, we have chosen the quadratic explicit algebraic stress model (EASM) proposed by Gatski & Speziale [4]. The Reynolds stress transport model tested was originally proposed by Shima [21], but the ϵ-equation which is found to be the source of instability of the model has been replaced by a more stable ω-equation, see Deng & Visonneau [3]. This model will be referred to as Shima-ω model. Since the pressure strain model employed in the Shima model is based on the IP model [21], we use the same IP model for the EASM model. Details about the implementation of the EASM-IP model can be found in Deng & Visonneau [2]. Figure 10 shows the prediction of the mean velocity with

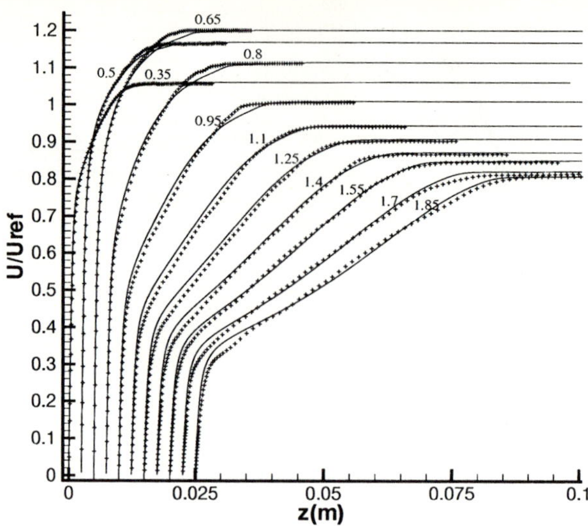

Figure 11: Prediction with the Shima-ω model.

the SST model. The numbers indicate the downstream x-position in meters. Each profile has been shifted in y-direction by $\Delta y = 0.0025$ for better visibility. In the region where the DNS has been performed ($x < 1m$), the agreement between the computation and the measurement is good. Although the adverse pressure gradient beyond $x = 1m$ is less strong, the effect of the adverse pressure gradient becomes more important due to the history effect. Discrepancies are observed, especially near the end of the test section. However, the overall accuracy of the computation remains acceptable. It should be noted that the prediction of a boundary-layer under adverse pressure gradient is a challenging task for a linear two-equation model. The SST model is the only linear two-equation model tested which can give a reasonable prediction. Among the Reynolds stress transport models, the Shima-ω model is found to give reasonable prediction for boundary-layers under adverse pressure gradient. Figure 11 shows the prediction obtained with the Shima-ω model. Overall accuracy is similar to the SST model except at the free-stream under strong adverse pressure gradient (from $x = 0.65m$ to $0.95m$). It should be mentioned here that the success of the Shima-ω model is not due to the fact that it is a Reynolds stress transport model, but only due to the fact that the model coefficients, which are functions of the invariants of the Reynolds stress tensor, are well-tuned for this specific flow. Figure 12 shows the result obtained with the EASM model deduced from the IP model for the pressure strain term. The result is not as good as that one of the previous models. The use of the LRR-model (Launder et al., [8]) or the SSG-model (Speziale et al., [23]) for the pressure strain improves nothing. No attempt has been made in the present studies to improve the prediction of the explicit algebraic stress model. The friction coefficient is shown in Figure 13. Results are consistent with the mean flow prediction. Best prediction is obtained with the SST model, while the EASM model is most unsatisfactory.

Now, we present a comparison with the DNS results of case APG1. To check the accuracy, computations have been done both with a boundary-layer code and a Navier-Stokes code. Necessary boundary conditions are extracted from the DNS data. We use the same computational domain and the same inlet condition as in the DNS. The boundary conditions for the computation with the boundary-layer code are the same as in the previous case. For the Navier-Stokes

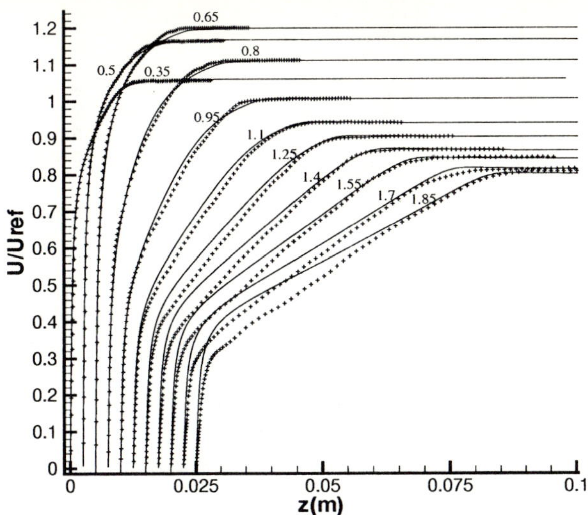

Figure 12: Prediction with the EASM-IP model.

computation, we impose the normal velocity distribution on the top boundary. The pressure field is determined by the code. Figure 14 compares the results obtained with the boundary-layer code and the Navier-Stokes code. The SST model is used in both computations. Although the predictions obtained with the Navier-Stokes code are in better agreement with the DNS data near the free-stream, both computations agree quite well. They are in good agreement with the DNS data of case APG1 except near the zonal boundary. When we compare statistical and DNS results with the experimental data of Watmuff, we observe some discrepancy (Figure 15), especially near the end of the computational domain ($x = 0.85m$). Figure 16 compares computational results obtained with the same SST model but different inflow and upper boundary conditions. Results obtained with the boundary condition extracted from the measurement give good agreement with the experimental data, while those obtained with the boundary condition extracted from the DNS data agree well with the DNS results. This observation suggests that the discrepancy observed between the computation and the measurement are mainly due to the treatment of the free-stream and the inlet boundary conditions rather than the numerical or the model accuracy.

7   Conclusions

Numerical simulation of boundary layer flow with pressure gradient is a challenging task. Due to the huge requirements of computational ressources, the DNS technique can be used for low Reynolds numbers only. At first, three DNS of boundary layer flow at $Re_\theta = 300$, based on the inlet momentum thickness, have been performed in order to test the inflow boundary conditions. They consist in prescribing a mean axial velocity profile and superimposing fluctuations from downstream. Good results can be obtained, if this downstream position is 10 boundary layer thicknesses away from the inlet. Then, three DNS of boundary layer flow at $Re_\theta = 670$ have been performed, in order to test the zonal grid approach. With this approach, grid points and CPU-time can be saved, compared to a full refined DNS. In comparison to a coarse grid simu-

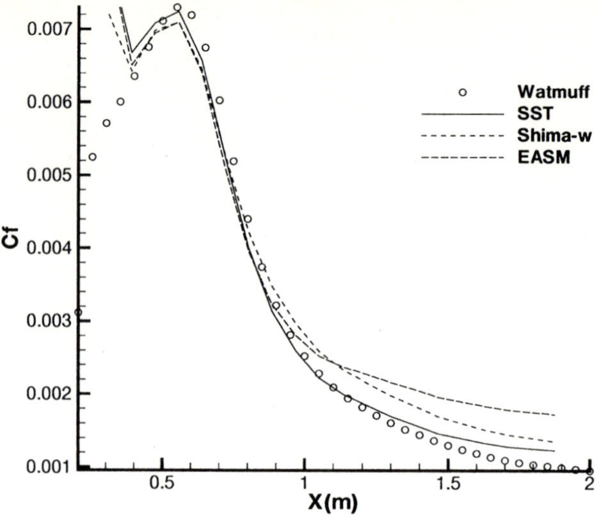

Figure 13: Prediction of c_f.

lation, the accuracy of the results is much higher and reaches the accuracy of a fine grid DNS in the locally refined zone. Direct numerical simulations of adverse pressure gradient boundary layer flow have been performed and compared with experimental results of Watmuff [25]. Due to resolution restrictions, the computational domain cannot be chosen as long as the experimental region. Nevertheless, pressure coefficient, friction coefficient and the mean and RMS values of the streamwise velocity component, achieved by DNS, are in fair agreement with the experiment.

Statistical predictions of Wattmuff's [25] accelerated/decelerated boundary layer have been done with different turbulence models ranging from two-equation linear eddy viscosity models to Reynolds stress transport models. In the first half of the flow domain where the flow is dominated by the pressure gradient, all tested models perform well in predicting the mean flow. In the second half of the domain, where the flow is dominated by the shear stress, only a few well-tuned models such as the SST model and the Shima-ω model predict this decelerated flow reasonably well. The reason for the success of some turbulence models and the failure of others is not well understood. We hope to clarify some of these open issues in the future with a new DNS simulation covering the entire domain. Nevertheless, the comparison between DNS and RANS data has clearly shown the importance of properly specifying inflow conditions in flows with strong spatial development.

Acknowledgment. We gratefully acknowledge the support of the High Performance Computing Centre Stuttgart (HLRS), the Leibniz Computing Centre of the Bavarian Academy of Science (LRZ) in Munich and the Institut du Développement et des Ressources en Informatique Scientifique in Nantes.

References

[1] D. Coles: 'The turbulent boundary layer in a compressible fluid.' In *Report R-403-PR*. The Rand Corporation, Santa Monica, CA, 1962.

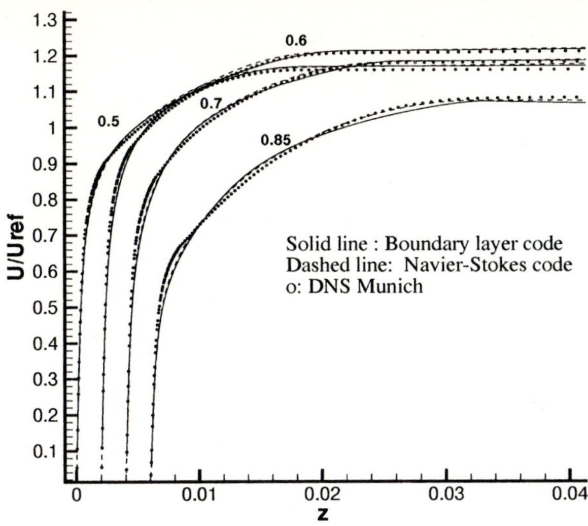

Figure 14: Contrasting results of Navier-Stokes and boundary-layer solvers with those from DNS.

[2] G.B. Deng, M. Visonneau: 'Comparison of explicit algebraic stress models and second-order turbulence closures for steady flows around ships', 7th International Conference on Numerical Ship Hydrodynamics, July 1999, Nantes/France.

[3] G.B. Deng, M. Visonneau: 'Computation of a wing-body junction flow with a new Reynolds-stress turbulence model', 22nd Symposium on Naval Hydrodynamics, 1998.

[4] T.B. Gatski and C.G. Speziale: 'On explicit algebraic stress models for complex turbulent flow', J. Fluid Mech., vol. 254, pp. 59–78, 1993.

[5] C.W. Hirt, B.D. Nichols, and N.C. Romero: 'Sola – a numerical solution algorithm for transient fluid flows.' In *Los Alamos Sci. Lab.*, Los Alamos, 1975.

[6] A.G. Kravchenko, P. Moin, and R. Moser: 'Zonal embedded grids for numerical simulations of wall-bounded turbulent flows.' *J. Comp. Phys.*, 127:412–423, 1996.

[7] Kravchenko, G. and P. Moin: 'B-Sline methods and zonal grids for numerical simulations of turbulent flows'. Report No. TF-73, Flow Physics and Computation Division, Department of mechanical engineering, Stanford University, 1998.

[8] B.E. Launder, G.J. Reece, W. Rodi: 'Progress in the development of a Reynolds stress turbulence closure', Jornal of Fluid Mechanics, vol. 183, pp. 63–63, 1975.

[9] T.S. Lund, X. Wu, and K.D. Squires: 'On the generation of turbulent inflow conditions for boundary layer simulations.' In *Annual Research Briefs - 1996*, pages 281–295. Center for turbulence research, Stanford, 1996.

[10] Lund, T., X. Wu, and K. Squires: 'Generation of turbulent inflow data for spatially-developing boundary layer simulations'. *J. Comp. Phys* **140**, 233–258, 1998.

[11] M. Manhart: 'Zonal direct numerical simulation of turbulent plane channel flow' – In: 'Computation and visualization of three-dimensional vortical and turbulent flows, Proceedings of the Fifth CNRS/DFG Workshop on Numerical Flow Simulation', R. Friedrich and P. Bontoux (Eds.), Notes on Num. Fluid Mech., vol. 64, Vieweg Verlag, 1998.

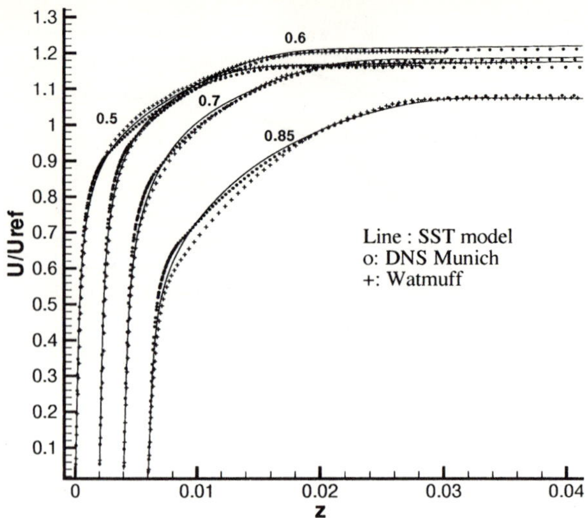

Figure 15: Prediction with the SST model and comparison with DNS data.

[12] M. Manhart: 'Direct Numerical Simulations of an Adverse Pressure Gradient Turbulent Boundary Layer on High Performance Computers', submitted to: Transactions of the High Performance Computing Center Stuttgart (HLRS) 1999, Springer Verlag, 1999.

[13] M. Manhart: 'Direct Numerical Simulations of Turbulent Boundary Layers on High Performance Computers', In: High Performance Computing in Science and Engineering '98, Transactions of the High Performance Computing Center Stuttgart (HLRS) 1998, E. Krause, W. Jäger (Eds.), Springer Verlag, pp. 199-212, 1999.

[14] M. Manhart: 'Using Zonal Grids for Direct Numerical Simulation of Turbulent Boundary Layers with Pressure Gradient', -In: 'New Results in Numerical and Experimental Fluid Mechanics II', Contributions to the 11th AG STAB/DGLR Symposium Berlin 1998, Nitsche W. et al. (Eds.), Notes on Numerical Fluid Mechanics Vol. 72, 299-306, 1999.

[15] M. Manhart, G.B. Deng, T.J. Hüttl, F. Tremblay, A. Segal, R. Friedrich, J. Piquet, P. Wesseling: 'The Minimal Turbulent Flow Unit as a Test Case for Three Different Computer Codes', – In: 'Numerical Flow Simulation I, CNRS-DFG Collaborative Research Programme, Results 1996-1998', E.H. Hirschel, (Ed.), Notes on Numerical Fluid Mechanics, vol. 66, Vieweg Verlag, pp. 365-381, 1998.

[16] M. Manhart and H. Wengle: 'Large-eddy simulation of turbulent boundary layer flow over a hemisphere.' In Voke P.R., L. Kleiser, and J-P. Chollet, editors, *Direct and Large-Eddy Simulation I*, pages 299–310, Dordrecht, March 27-30 1994. ERCOFTAC, Kluwer Academic Publishers, 1994.

[17] F.R. Menter: 'Zonal two-equations $k - \omega$ turbulence models for aerodynamic flows', AIAA 24th Fluid Dynamics Conf., AIAA Paper 93-2906, 1993.

[18] Y. Na and P. Moin: 'Direct numerical simulation of turbulent boundary layers with adverse pressure gradient and separation.' Report No. TF-68, Thermosciences Division, Department of mechanical engineering, Stanford University, 1996.

[19] Y. Na and P. Moin: 'Direct numerical simulation of a separated turbulent boundary layer'. *J. Fluid Mech.* **370**, 175–201, 1998.

[20] K. Richter, R. Friedrich, L. Schmitt: 'Large-eddy simulation of turbulent wall boundary layers with pressure gradient'. In: *6th Symposium on Turbulent Shear Flows, Toulouse.* pp. 22/3/1–22/3/7, 1987.

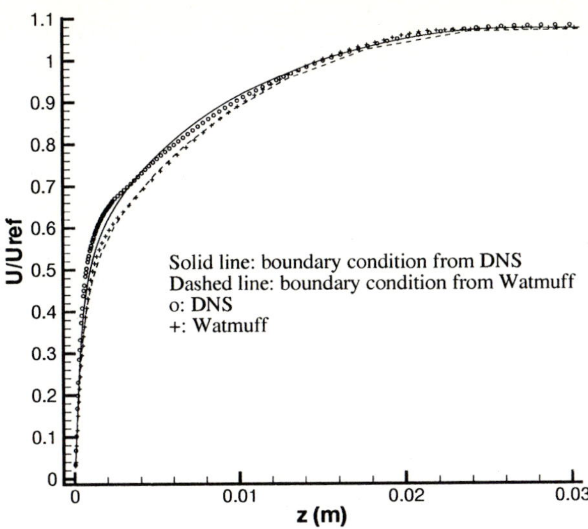

Figure 16: Influence of inflow and upper boundary conditions on predictions with the SST model at $x = 0.85m$.

[21] N. Shima: 'Prediction of Turbulent Boundary Layers With a Second Moment Closure: Part I - Effects of Periodic Pressure Gradient, Wall Transpiration, and Free-Stream Turbulence', ENG, vol. 115, pp. 56–63, 1993.

[22] P.R. Spalart: 'Direct simulation of a turbulent boundary layer up to $Re_\theta = 1410$.' *J. Fluid Mech.*, 187:61–98, 1988.

[23] C.G. Speziale, S. Sarkar, T.B. Gatski: 'Modeling the pressure-strain correlation of turbulence: An invariant dynamical systems approach', Jornal of Fluid Mechanics, vol. 227, pp. 245–272, 1991.

[24] P.P. Sullivan, J.C. McWilliams, and C.-H. Moeng: 'A grid nesting method for large-eddy simulation of planetary boundary-layer flows.' *Boundary-Layer Meteorology*, 80:167–202, 1996.

[25] J.H. Watmuff: 'An experimental investigation of a low Reynolds number turbulent boundary layer subject to an adverse pressure gradient.' In *Ann. Res. Briefs*, pages 37–49. Center for Turbulent Research, 1989.

[26] H. Werner and H. Wengle: 'Large-eddy simulation of turbulent flow over and around a cube in a plate channel.' In F. Durst et al. (eds.), *Turbulent Shear Flows 8*, Springer-Verlag, Berlin, pp. 312-324 1993.

DNS and LES of a backward-facing step flow using 2nd- and 4th-order spatial discretization and LES of the spatial development of mixing of turbulent streams with non-equilibrium inflow conditions.

A. MERI [1], H. WENGLE [1] and R. SCHIESTEL [2]

[1] Institut für Strömungsmechanik u. Aerodynamik, LRT/WE 7,
Universität der Bundeswehr München, D-85577 Neubiberg, Germany

[2] Institut de Recherche sur les Phénomènes Hors d'Equilibre, IRPHE,
UMR 6594 CNRS/Universités d'Aix-Marseille I & II,
IMT - Technopôle de Châteaux-Gombert/La Jetée,
38 rue Frédéric Joliot Curie,F-13451 Marseille Cedex 20, France

Summary

In the first part of this paper, LES results for a backward-facing step flow (with a fully developed channel flow utilized as a time-dependent inflow condition) are evaluated against a corresponding DNS reference data set ($Re_h = 3300$). The two LES cases (using the dynamic subgrid scale model) differ in the spatial discretization of the numerical solution method, using either a second-order (central) or a 4th-order (compact or Hermitian) scheme.

In the second part of the paper, inflow conditions are presented for LES of the spatial development of the mixing of flow streams having different dominant length scales and levels of turbulent kinetic energy. To provide proper inflow conditions the effects of the turbulence producing grids used in the experiments must be realistically modelled. In the case of a co-flowing plane jet with mean shear (experiment of Raddaoui [2]) the effects of different honey comb grids placed in the central and the co-flowing streams must be modelled, and in a shearless turbulence mixing layer (experiment of Veeravalli and Warhaft [3]) the effects of parallel bar grids were simulated to create the desired shearless mean vertical velocity profile (in addition with different length scales and levels of turbulent energy in the upper and lower half of a channel flow).

1 Introduction

The general aim of the current french-german cooperation is the development of advanced tools for the numerical simulation of complex flows involving flow obstacles and non-equilibrium inflow conditions. The basic numerical simulation concepts used are Direct Numerical Simulation (DNS) and Large-Eddy Simulation (LES) to be able to calculate the three-dimensional and time-dependent structure of the turbulent flows. For the numerical solution of the partial differential equations involved, a fourth-order compact (or Hermitian) discretization scheme on non-equidistant grids has been proposed in Meri et al. [1] and results are presented from solving test problems and complex flow cases such as the perturbed channel flow (LES), the transitional backward-facing step flow (DNS) and the turbulent backward-facing step flow (LES), [1].

In the *first part* of this paper we present results from DNS and LES of a backward-facing step flow ($Re_h = 3300$, based on the maximum inflow velocity and on the step height) with a fully developed turbulent channel flow ($Re_\tau = 180$, based on the wall skin friction velocity and the height of the plane channel) as inflow condition. The DNS result serves as a reference data set for the evaluation of different LES results calculated with spatial discretization of second-order and fourth-order accuracy, respectively.

The *second part* of the paper is related to the question of how proper inflow conditions for DNS/LES can be provided, e.g. to study the mixing of different turbulent fields having different dominant length scales and levels of turbulent kinetic energy. We present LES results for two complex flow cases: the spatial development of a confined planar jet with co-flowing streams with mean shear (experiment by Raddaoui [2]), and the spatial development of a plane shearless mixing layer (experiment by Veeravalli and Warhaft [3]). In the first case, different length scales in the central flow stream and in the two co-flowing streams are created by placing honey comb grids of different size into the inflow channels. In the second case, the non-equilibrium inlet turbulence is created by parallel bar grids (different size and distance of bars in the upper and lower half of the channel). To provide sufficiently accurate and detailed inflow conditions, we tried to simulate realistically the effects of the honeycomb grids and bar grids used in the experiments.

2   Numerical solution strategies and subgrid scale models used

Numerical solution method: The numerical code used in this paper is based on a finite-volume formulation of the Navier-Stokes equation for an incompressible fluid on a non-equidistant and staggered Cartesian grid. The discretization in space used here is either the classical second-order (central) discretization, or the 4th-order compact discretization on non-equidistant grids proposed in Meri et al. [1]. For the time advancement of the momentum equations an explicit second-order time step is used (leap-frog with time-lagged diffusion). The solution of the Poisson equation is carried out iteratively. The iterative solver used here is the well known point-by-point velocity-pressure iteration described by Hirt et al. [4]. This iterative pressure-velocity correction cycle is stopped as soon as the dimensionless divergence of the (incompressible) flow reaches levels below 10^{-3} at all grid points after every time step. This iterative method has the advantage that it can easily be applied to quite arbitrary geometry of flow obstacles in the computational domain. Its disadvantage is that it is very costly to reach divergence levels down to machine accuracy (which is not required in our case). Acceleration of the convergence of the iteration cycle is possible by applying smoothing with a multigrid cycle.

Direct numerical simulation (DNS): For a DNS the conservation equations for mass and momentum must be solved without any additional assumptions or modelling related to the effects of the turbulent motion, i.e. the original Navier-Stokes equations must be discretized in space and time such that all the relevant scales in a turbulent flow are resolved. For example, referring to spatial scales, the size of the computational domain must be large enough to accommodate the largest turbulent scales, and the grid spacing must be sufficiently small to enable a resolution of the order of the dissipation length scale, $\eta = (\nu^3/\epsilon)^{1/4}$. Here, ν is the kinematic viscosity of the fluid and ϵ is the dissipation rate in the flow field.

Large-eddy simulation (LES): For a LES the low-pass filtered Navier-Stokes equations are solved, i.e. eddies smaller than the grid spacing are removed and their effect on the resolvable motion is provided by a subgrid-scale (SGS) model. In deriving the resolvable scale equations, see e.g. in Rogallo and Moin [5], the proper choice of a filter and of a filter width is involved (top-hat filter, Gaussian filter) and the result exhibits, in addition to correlations between SGS quantities, the so-called Leonard terms (involving cross-correlations between SGS quantities and resolved quantities). Finally, in numerically solving these resolvable scale equations the numerical errors and the modelling errors are involved. For a strict separation of the modelling physics and the numerical errors a large ratio of filter size to mesh resolution is required. This is certainly attractive from a theoretical point of view. However, in solving practical problems with complex geometry and large Reynolds numbers this would be very costly.

An alternative derivation of the resolvable scale equations has been carried out by Schumann

[6] and leads directly to the integral form of the Navier-Stokes equations in which time derivatives of cell-volume averages of velocities are related to differences of cell-surface averaged stress and momentum flux. Finally, the averages are related to their finite-difference operators (on a staggered computational mesh). In this paper, we follow this approach. From the point of view of the explicit filtering approach we use a filter width equal to the grid spacing and the effects of all the unknown terms are modelled all together by an edddy viscosity model which acts as a sink of energy for the short waves in the flow.

Subgrid scale models: The unknown SGS stresses must be related to the GS velocity via an eddy viscosity model. In this paper, two different SGS models are used. The first one is the classical *Smagorinsky SGS model* (using an empirical model constant $c_s = 0.1$). In grid volumes next to rigid walls we use for the mixing length the smaller value of κx_n and $0.1(\Delta x \Delta y \Delta z)^{1/3}$ (where x_n is the distance normal to a wall). The second SGS model used here is the so-called *dynamic SGS model*, see e.g. in Germano et al. [7]. This model allows to calculate the model constant c_s as a function of time and space. To avoid numerical instabilities, the model parameter must be properly averaged in time and/or space.

Computation of the statistics: The direct results from DNS and LES are the time-dependent and three-dimensional velocity and pressure fields. After an initial transient phase of the calculation (not used for collecting samples for the statistics) the mean fields have been obtained by averaging samples in time and by making use of spatial averaging in the homogeneous direction of the flow problem. As soon as the mean fields have reached stable (i.e. time-independent) values, the fluctuating fields and the second-order statistics can be evaluated.

3 Comparison of DNS and LES of a backward-facing step flow using 2nd-order and 4th-order spatial discretizations

In this chapter, we use a DNS result ($Re_h = 3300$, based on the centerline velocity of the incoming channel flow and on the step height) for the evaluation of LES results calculated with second-order and fourth-order (compact or Hermitian) discretization, respectively. The use of a higher-order numerical solution method seems to be useful also for LES. The improved spatial discretization should give a better representation of the resolved smaller scales and of the steep local gradients in the instantaneous flow field. In addition, the higher-order method should give a better representation in particular of the non-linear convective terms.

For the turbulent backward-facing step there are DNS and LES results available ($Re_h = 5100$) from Le et.al. [8] (DNS) and Akselvoll and Moin [9] (LES), using a turbulent boundary layer as an inflow. In the flow case presented here, a fully developed turbulent channel flow is utilized as inflow condition.

3.1 Computational domain, spatial discretization and boundary conditions

In our case, x is the main flow direction (velocity component u), y is the lateral (velocity component v), z is the vertical direction (velocity component w). Note that in the following all the variables are presented in dimensionless form, i.e. the velocities and the velocity correlations are made dimensionless with the bulk velocity, U_m, lengths are made dimensionless with the step height, h.

Computational domain: The flow configuration selected is a backward-facing step (of height h) in a plate channel (of height 2h before the step and of height 3h after the step), see **figure 1** . The inflow cross-section for the channel flow part is located at X=x/h=-12.6 and its outflow cross-section is located at X=x/h=-3.0 upstream of the position of the step at X=x/h=0.0. The outflow cross-section of the step flow part is located at X=17.3 (DNS) and X=16.2 (LES), respectively. The expansion ratio of this flow configuration, ER=1.5, is defined as the ratio of the channel height after the step to the channel height before the step. The lateral extension of the computational domain was 6.0 step heights. If all length scales are normalized with the step height, h, the dimensions of the computational domain are for the reference DNS $(L_x,L_y,L_z) \approx (30,6,3)$ and for the LES $(L_x,L_y,L_z) \approx (29,6,3)$.

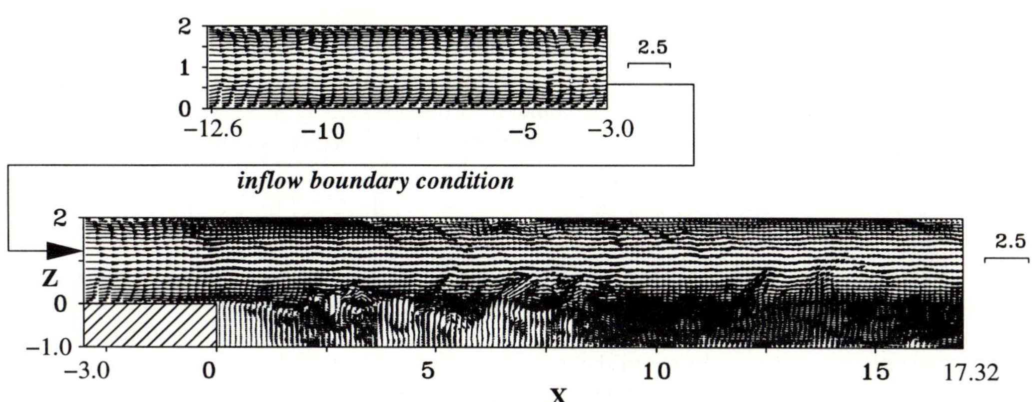

Figure 1 Computational domain for backward-facing step flow,
with turbulent channel flow as inflow condition

Spatial resolution: Different numerical simulations have been carried out for this flow case: a direct numerical simulation (DNS) using about 6 million grid points, with a grid distribution $(N_x,N_y,N_z) = (576,80,132)$, and large-eddy simulations (LES) using about 0.37 million grid points, with a grid distribution $(N_x,N_y,N_z) = (144,40,64)$. In the homogeneous lateral (y-) direction of the flow problem the grid spacing was equidistant, i.e. $\Delta Y = 0.15$ (LES), $\Delta Y = 0.075$ (DNS). In the longitudinal (x-) direction the grid spacing was also equidistant, i.e. for LES $\Delta X = 0.20$, and for the DNS $\Delta X = 0.10$ was used in the channel part $(-12.6 \leq X \leq -3.0)$, a stretched grid was used between $-3.0 \leq X \leq -0.0$, and again an equidistant grid with $\Delta X = 0.04$ was applied downstream of the step. In the wall-normal direction (z) the grid was stretched away from the wall. The velocity grid point closest to a horizontal wall was $\Delta Z_{min} = 0.032$ (LES) and $\Delta Z_{min} = 0.016$ (DNS), respectively. Measured in units of the viscous length (using the wall shear velocity, u_τ, of the incoming channel flow) the wall distance of the first velocity grid point was $\Delta Z_{min}^+ \approx 1.35$ for the reference DNS and $\Delta Z_{min}^+ \approx 2.7$ for the LES cases.

From an *a posteriori* evaluation of the local dissipation rate, ϵ, from the DNS a maximum (dimensionless) dissipation rate of $\epsilon_{max} \approx 0.01$ has been evaluated. With this value of ϵ_{max} the Kolmogorov dissipation length scale could be estimated with $\eta = (\nu^3/\epsilon)^{1/4} = 0.008$. Measured in units of this estimated η the spatial resolution of the DNS was about $(\Delta X_\eta,\Delta Y_\eta,\Delta Z_\eta) = (5\eta,9\eta,2\eta)$. Therefore, the spatial resolution of the DNS was indeed of the order of the Kolmogorov length scale.

Boundary conditions: No slip boundary conditions were used at the rigid walls and periodic boundary conditions were applied in the homogeneous lateral direction. Between X=-12.6 and X=-3.0 the flow was a fully developed channel flow (with $Re_\tau = 180$) with periodic boundary conditions between these two locations. In the same production run, the result from the channel flow at X=-3.0 was simultaneously fed as a time-dependent inflow condition into the step flow problem downstream of this position. As an outflow boundary condition gradients normal to the

271

outflow cross-section have been set to zero in calculations using a second-order numerical solution method, and for the flow calculation using a fourth-order (compact, or Hermitian) method
the outflow boundary condition corresponded to a third-order extrapolation of the velocity components.

3.2   Time-dependent inflow condition (results of turbulent channel flow)

In this flow case, a fully developed turbulent channel flow delivers the time-dependent inflow
conditions at X=-3.0 for the step flow case downstream of this location. **Figure 2a** compares vertical profiles of the rms values from the channel flow part $(-12.6 \leq X \leq -3.0)$ of the DNS runs
(using second-order and fourth-order spatial discretization, respectively) with the DNS data from
Kim, Moin and Moser [10] (using a Fourier-Chebyshev spectral method on a grid with about 4
million grid points). As already shown by Meri et al. [1] there are only significant differences
to be observed for the statistics of higher than second order (e.g. for the skewness of the wall
normal fluctuations). From **Figure 2a** it can be concluded that using a sufficiently large number
of grid points (here the channel flow part contains about 0.5 million grid points) the fourth-order
(compact or Hermitian) discretization and even the second-order (central) discretization delivers sufficiently accurate results up to the second-order statistics. In **Figure 2b** we compare the
rms profiles from LES (with about 60 000 grid points, using the dynamic subgrid scale model)
with the DNS reference data from Kim et al. [10]. Here, the second-order results as well as the
fourth-order results contain the subgrid scale contributions. These subgrid scale contributions
are derived (with the assumption of local isotropy) from the subgrid scale turbulence energy. The
peaks of the u-rms values are higher and the peaks of the v-rms and w-rms values are lower than
those of the reference data. The fourth-order results are closer to the DNS data.

3.3   Discussion of results for the turbulent backward-facing step

In the following we compare results from LES (with about 245000 grid points within the step
flow part of the flow problem) with a corresponding DNS (using about 5 million grid points
within the step flow part). For the DNS and LES cases we used second-order (central) differencing and the fourth-order compact method, respectively. Note, in both LES cases the *dynamic
subgrid scale model* was applied.

Instantaneous flow field: Figure 3 shows two snapshots at arbitrary times from DNS using
second-order and fourth-order discretization, respectively. The second-order result after the step
shows small "wiggles" in the upper part of the flow problem, indicating that the grid spacing
in this part of the flow is not sufficiently small to properly resolve the *instantaneous* velocity
gradients in the flow by the second-order method. In the following we shall always take the
result from the fourth-order DNS as a reference data set.

Mean flow and second-order statistics: In **Figure 4** we first compare vertical profiles of
the first-order statistics (mean flow) and of the second-order statistics from DNS with the LES
also using fourth-order discretization. The agreement is very satisfying. Then, in **Figure 5** a
comparison of LES results for the mean $< U >$ and $< W >$ velocity components from second-
order and fourth-order calculations, respectively, are presented. The differences in the mean flow
are small. The zero-crossing of the downstream $< U >$-velocity distribution through the first grid
point closest to the bottom wall defines the mean reattachment length, x_r/h, of the recirculation
region behind the step. The DNS delivers $x_r/h = 6.0$, the fourth-order LES result gives $x_r/h =
5.75$ and the second-order LES result is about 10% shorter, $x_r/h = 5.4$. There is also a secondary

recirculation zone, with a mean recirculation length, x_{rs}/h, immediately at the step: $x_{rs}/h = 1.9$ (DNS), $x_{rs}/h = 1.7$ (LES, 4th-order), and $x_{rs}/h = 1.2$ (LES, 2nd-order).

From the results for the rms-values for the fluctuations of all three velocity components in **Figure 4** it can be concluded that an assumption of local isotropy is not a bad approximation in parts in this flow field away from the walls. However, for example, in the re-attachment zone (around $x_r/h = 6.0$) there are strong lateral velocity fluctuations (v-rms) created by the re-attaching shear layer.

Grid scale (resolved) and subgrid scale turbulence energy: Figure 6 compares second-order and fourth-order solutions of the grid scale turbulence energy, e_g, and of the (estimated) subgrid scale turbulence energy, e_s, using the two different discretization methods. In general, on an identical grid, and using the same dynamic subgrid scale turbulence model, the higher-order method is capable to resolve a larger part of the grid scale energy and correspondingly, a smaller part of the non-resolved subgrid scale energy, e_s, must be modelled. This conclusion becomes obvious from **Figure 6** (above and middle). In addition, **Figure 6** (below) shows profiles of the ratio of (estimated) subgrid scale energy, e_s, to the resolved grid scale energy, e_g. This ratio is less than about 20% in this flow field. However, note that in the whole recirculation zone, in particular in the so-called dead-water zone close to the step, the percentage of the subgrid scale contributions is relatively large.

3.4   Conclusions:

The backward-facing step flow is a very useful test case if a fully developed turbulent plane channel flow is used as a time-dependent inflow condition (representing the physically correct spatio-temporal structure). A DNS of this combined flow case ($Re_h = 3300$) can be used as a reference data set to evaluate results from LES using 2nd-order and 4th-order (Hermitian) spatial resolution, respectively. Here, the LES cases use about 0.3 million grid points, i.e. a factor of about 20 less grid points than the DNS. In general, the 4th-order method resolves a larger part of the turbulent kinetic energy on the grid scale level and as a consequence, a smaller part of the (non-resolved) subgrid scale energy must be modelled. Nevertheless, in our flow case the non-resolved turbulent energy (in the flow field away from the walls) is between 10% and 20%.

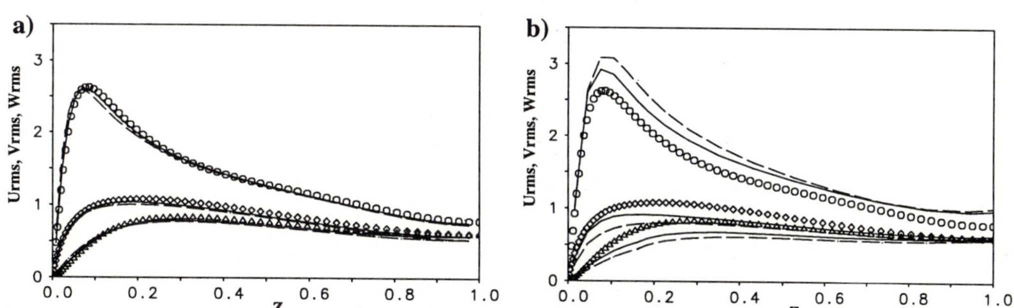

Figure 2   Profiles of u-rms, v-rms and w-rms, using 2nd-order and 4th-order spatial discretization, in the channel flow section ($-12.6 \leq X \leq -3.0$)
a) DNS-4th-order (full line), DNS-2nd-order (dashed line),
b) LES-4th-order (full line), LES-2nd-order (dashed line),
(symbols) DNS of Kim et al. [10]

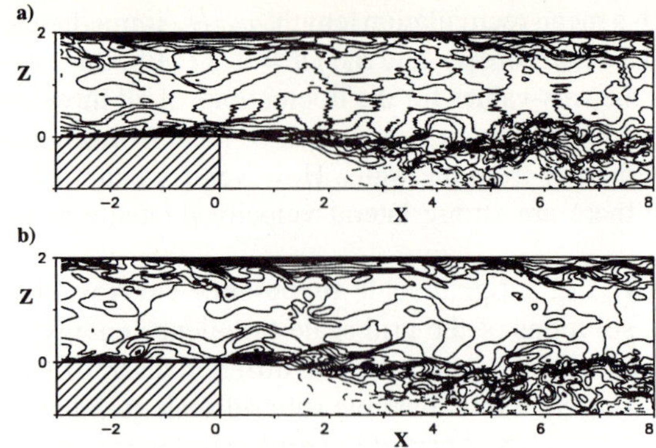

Figure 3 Instantaneous isolines of longitudinal velocity component, U,
DNS-2nd-order (above), DNS-4th-order (below),
positive values (full line), negative values (dashed line).

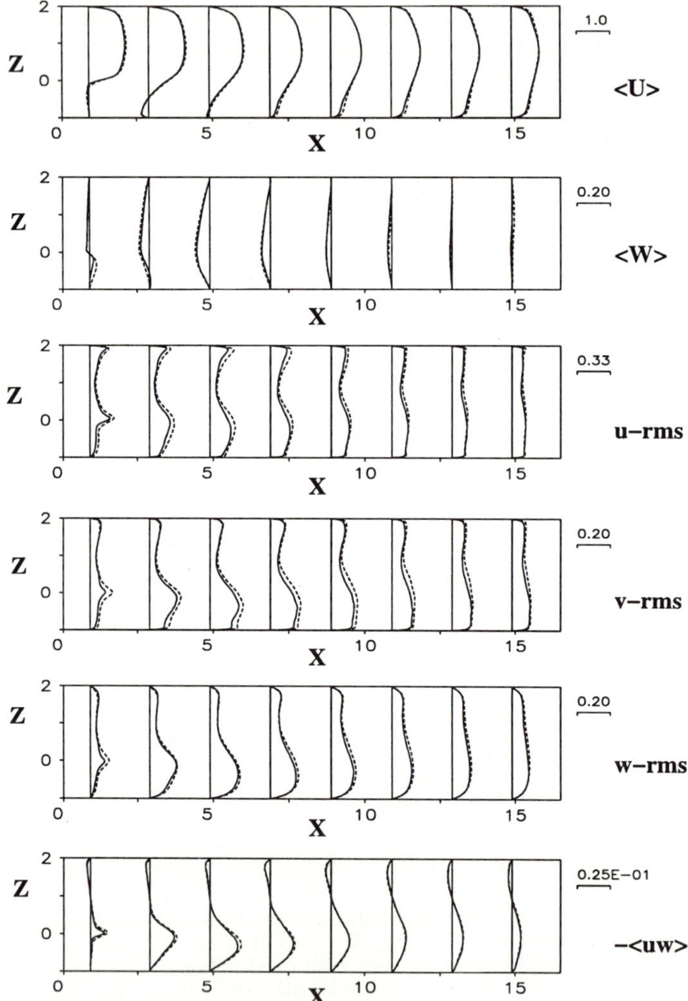

Figure 4 Comparison of DNS and LES results for the mean flow
and for the second-order statistics (containing estimated subgrid scale contributions),
DNS (full line), LES (dashed line), 4th-order solutions.

274

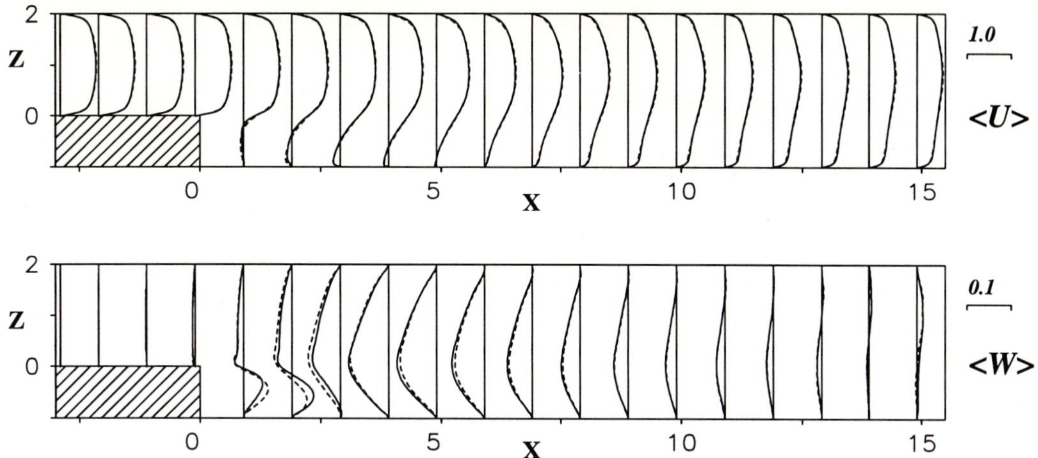

Figure 5 Comparison of mean velocity profiles from LES (dynamic subgrid scale model), LES-4th-order (full line), LES-2nd-order (dashed line).

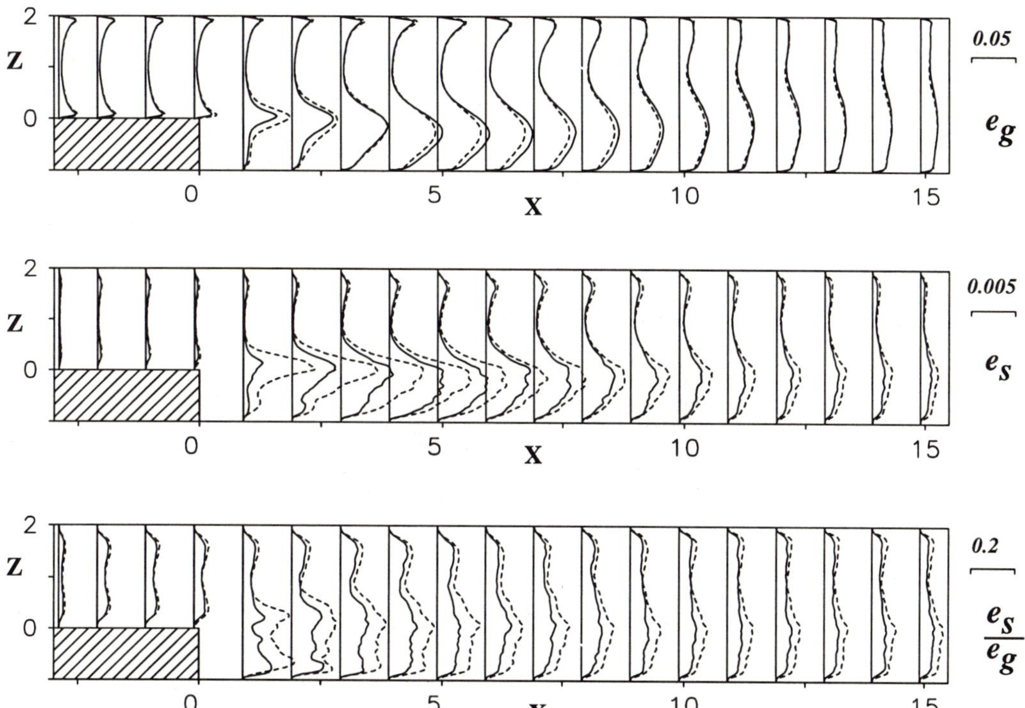

Figure 6 Comparison of grid scale, e_g, and subgrid scale turbulent energy, e_s, from LES (dynamic subgrid scale model), LES-4th-order (full line), LES-2nd-order (dashed line).

4 Spatial development of a confined plane jet with

A very complex non-equilibrium flow situation is created by the interaction of different turbulent fields having different dominant length scales and typical levels of turbulent kinetic energy in the central and the co-flowing streams, Two shear layers are created at the boundaries of the central jet and new turbulent kinetic energy is produced in these shear layers. Finally, after a certain downstream development the two shear layer start to interact. It may be difficult to calculate such non-equilibrium flow situations using classical engineering turbulence modelling methods. Sometimes the required non-equilibrium inflow conditions are also difficult to estimate, or are not completely documented in reports on corresponding experiments.

A complex flow is created in the experimental work on a plane jet with co-flowing streams without and with mean shear by Raddaoui [2], using different honeycomb grids in the incoming flow streams to create different length scales and turbulent kinetic energy levels.

In the following we compare LES results (using second-order discretization and the Smagorinsky-subgridscale model) with data from Raddaoui's experiment for selected flow cases with mean shear, for a Reynolds number of $Re \approx 11000$ (based on the height, h_1, of the central channel and of the maximum velocity, U_1, there). The main objectives here was to provide an equivalent non-equilibrium inflow condition by simulating the detailed effects of the honeycomb grids used in the experiment, to study the details of the spatial development of the flow, and to compare results for the first- and second-order statistics of the flow and of time spectra with available experimental data.

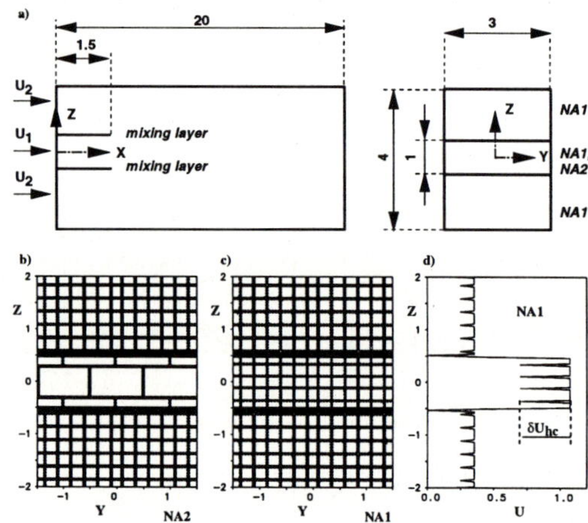

Figure 7 Parameters of the simulation:
    a) geometry of a confined plane jet with co-flowing streams,
    b,c) approximation of the honeycombs (case NA2 and NA1) and
    d) streamwise velocity profile used as inflow condition (case NA1).

4.1 Computational domain and inflow condition

Computational domain and boundary conditions: For the flow cases with shear presented here we used a computational domain of size $(L_x/h_1, L_y/h_1, L_z/h_1) = (20,3,4)$ with (NX, NY, NZ): $(320,120,160)$ grid points. Here, h_1 is the height of the central channel which is separated from the upper and lower side channels by two rigid walls, see **Figure 7** . These walls have a length of $L/h_1 = 1.5$ and a thickness of $\Delta z/h_1 = 0.025$ (which corresponds to the width of just one grid volume in the vertical direction). The boundary conditions were periodic in the lateral direction (y), and rigid-wall condition were applied at the upper and lower boundaries of the computational domain and at the walls of the central channel. At the outflow cross-section we applied (x-gradient=0)-boundary conditions. The use of a relatively large number of grid points in the lateral and the vertical directions was necessary to provide a realistic approximation of the effects of the honeycombs located in the entrance section (X=0.0). The thickness of the walls of the honeycombs used in the experiment, normalized by the reference length h_1, was $d = \frac{d_{exp}}{h_1} = \frac{0.5mm}{20mm} = 0.025$, therefore requiring a minimum number of grid points $NZ = L_z/d = 4.0/0.025 = 160$ for a uniform distribution of the grid points in the vertical direction. In the flow cases with shear we realized that the use of a coarse grid in the streamwise direction (with only 160 grid points) produces too a high level of turbulent kinetic energy, see **Figure 8** . Therefore, we used a fine grid with 320 grid points uniformly distributed in the x-direction between X=0.0 and X=12.0 and then the grid is stretched towards the outflow cross section.

Inflow condition: A realistic approximation of the effects of the honeycombs in the numerical simulation is very important to produce a non-equilibrium flow situation comparable to the experiment. To approximate the effects of the honey combs we simulated only the wakes behind all the walls of the honeycomb grids. In the real world of the experiment they consist of ducts with cross-sections having the shape of a hexagon and in the numerical simulation we approximated this with ducts having a square cross-section.

The vertical and lateral profiles for the streamwise velocity behind a honey comb grid was assumed to be uniform in space, except immediately behind the honeycomb walls where the velocity profile was assumed to be modified by a velocity defect δU_{hc}, see **figure 7d**. Transition to turbulence is provoked by the inherent physical instability of the mean inflow velocity profile. It was not necessary to add random fluctuations to be amplified during the transition process. Small numerical inaccuracies in the flow field immediately behind the inflow section were sufficient to trigger the transition process.

The streamwise velocity integrated over the cross-section of the central channel and of the two side-channels, respectively, had to satisfy the ratio, Q_2/Q_1, of the mass flux in the side channels to the mass flux in the central channel (given by the experiment). Two cases with different mean shear are considered: $Q_2/Q_1 = 0.72$ (with $U_1 = 1.0$ and $U_2 = 0.24$) and $Q_2/Q_1 = 1.00$ (with $U_1 = 1.0$ and $U_2 = 0.33$). The only parameter in our simple approximation of the effects of the honeycombs is the velocity defect, δU_{hc}, behind the honeycomb walls. **Figure 9** shows the influence of this parameter on the downstream development of the rms-value of the streamwise velocity fluctuations, u-rms. There is strong production of turbulent kinetic energy immediately behind the honeycombs and a steep decay further downstream. For all the results presented in the following figures $\delta U_{hc} = 0.3 \cdot U_{max}$ has been chosen.

4.2 Discussion of results

Figure 11 and **Figure 12** compare the downstream development of the mean velocity component, $< U >$, and of profiles of the rms-values of the longitudianal velocity fluctuations, u-rms, with two experimental data sets available for different mean shear ($Q_2/Q_1 = 0.72$ and 1.0). The mean velocity profiles closest to the honeycomb grid (NA1) clearly show the wake effects of the walls of the honeycombs (with similar effects present also in the spanwise direction). This effect is still visible at a downstream position X=6.5. There is also satisfying agreement with the experimental data for the downstream development of the turbulence intensity, u-rms, with the exception of significantly higher experimental values close to the two confining (upper and lower) walls. In the LES the developing boundary layers at these walls could not be resolved properly. High levels of turbulent kinetic energy are created in the two shear layers which still can be distinguished by the two peaks in the vertical u-rms profiles through the whole computational domain (see the center figure in figure 12). In **Figure 13** we compare one-dimensional frequency spectra from experiment and numerical simulation. Data for the u-velocity component only have been available from the experiment. We therefore evaluated corresponding frequency spectra from LES time records which is a very expensive task because of the large computing time and storage space required to cover a sufficiently large window in physical time. For the case NA1 (small honeycomb grids in all flow streams) a typical Strouhal number $Str = fh_1/U_1 \approx 3.2$ (LES) and $Str \approx 3.6$ (experiment), respectively, can be found at X=4.0 on the centerline of the central jet (see figure 13, above). Using Taylor's hypothesis: an eddy with a typical length scale of about 0.25 (created by the small honeycomb grid) and a dimensionless transport velocity on the centerline of the central channel of (less than) 1.0 would just produce this Strouhal number. A similar estimate leads to the conclusion for the case NA2 (see figure 13, below) that the Strouhal number $Str \approx 0.75$ (LES) and $Str \approx 0.7$ (experiment) corresponds to a typical eddy created by the large honeycomb grid NA2. The mixing cabability of the two shear layers bounding the central jet (and possibly the interaction of these shear layers further downstream) may be influenced by pairing events in the shear layers, thereby transferring energy from one dominant frequency to its subharmonics. From the downstream development of the dimensionless frequency spectra along the centerline of the central channel some indirect indication of such pairing events in the shear layers (above and below) is given by these frequency spectra. **Figure 10** presents information about typical dominant length scales in the flow case NA2 (large honeycomb grid placed in the central channel). Considering the wave number spectra for the three velocity components on the centerline of *the central jet* (Z=0.0) it becomes very clear from the result for the streamwise velocity component, u, that (at the selected location X=4.0) most of the energy is contained in the wave number $K_y \approx 6.0$ which corresponds to the typical width $X_{hc} \approx 1.0$ of the large honeycomb grid (NA2). Additional peaks at multiples of this wave number are visible, with decaying energy content for increasing wave number. In the center of *the side channel* (Z=1.25) most of the energy is contained in the wave number $K_y = 25$ which corresponds to the typical width $X_{hc} = 0.25$ of the small honeycombs. *The shear layer* (Z=0.5) developing in between the central and the co-flowing streams is dominated (at X=4.0) by a wave number $K_y = 12$, a value between the dominating wave numbers in the bounding flow streams. In the case NA1 (not shown here) the only dominating wave number in all three flow regimes (central jet, shear layer, co-flowing jets) is $K_y = 25$ corresponding to the small honeycomb grid.

4.3 Conclusions

LES is capable of providing the detailed structure of a complex non-equilibrium turbulent flow with interacting flow streams having different length scales and different levels of turbulent kinetic energy. To reach agreement with available data from an experimental investigation it is necessary to simulate the effects of the honeycomb grids used in the experiment to produce different length scales and energy levels. From our relatively simple approach to simulate these effects realistically, it can be concluded that it seems to be sufficient to use in the inflow cross-section a basically uniform velocity profile together with a velocity defect behind all the horizontal and vertical walls of all the honeycomb ducts of the grids used in the experiment.

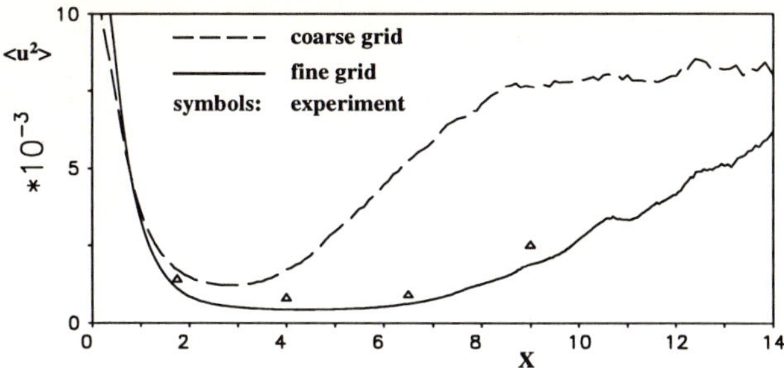

Figure 8 Downstream development of the variance of the longitudinal velocity fluctuations along the centerline of the central jet (case NA1, $Q_2/Q1 = 0.72$), (dashed line) coarse grid (160*120*160), (full line) fine grid (320*120*160), symbols: experiment

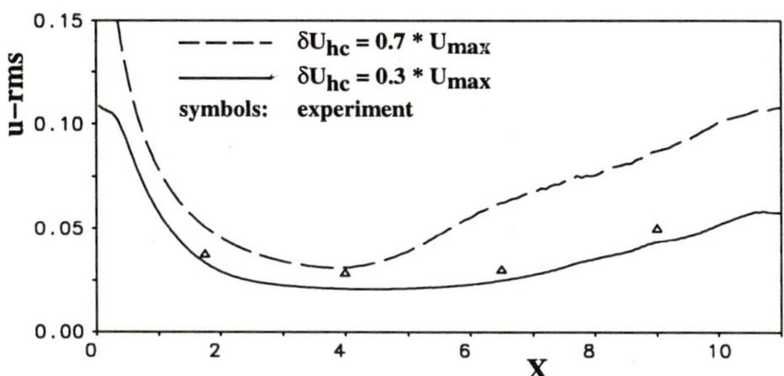

Figure 9 Downstream development of the rms values of the longitudinal velocity fluctuations, u-rms, along the centerline of the central jet, for different velocity defects, δU_{hc}, behind the honeycomb walls, symbols: experiment, lines: LES (case NA1, $Q_2/Q1 = 0.72$).

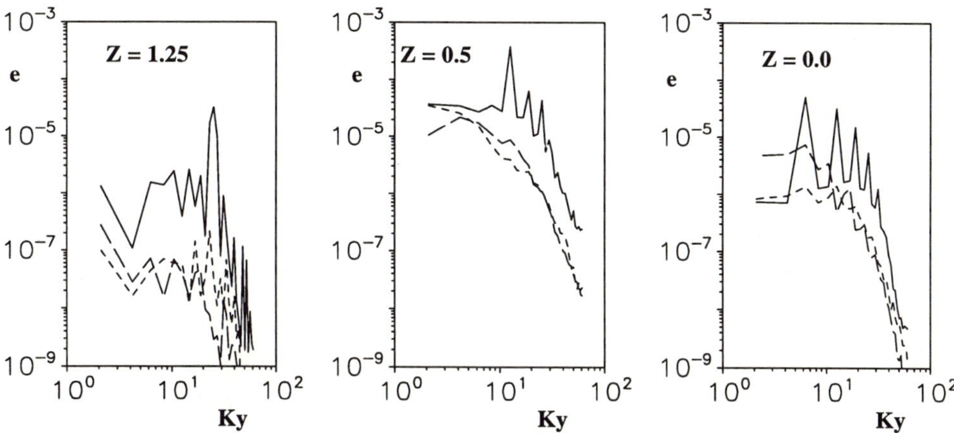

Figure 10 One-dimensional wave number spectra at the center of the co-flowing jet ($Z = 1.25$, left), of the shear layer ($Z = 0.5$, middle) and of the center of the central jet ($Z = 0.0$, right) at a downstream location $X = 4.0$. case NA2, $Q_2/Q1 = 1.0$, —— U, – – – V, - - - W.

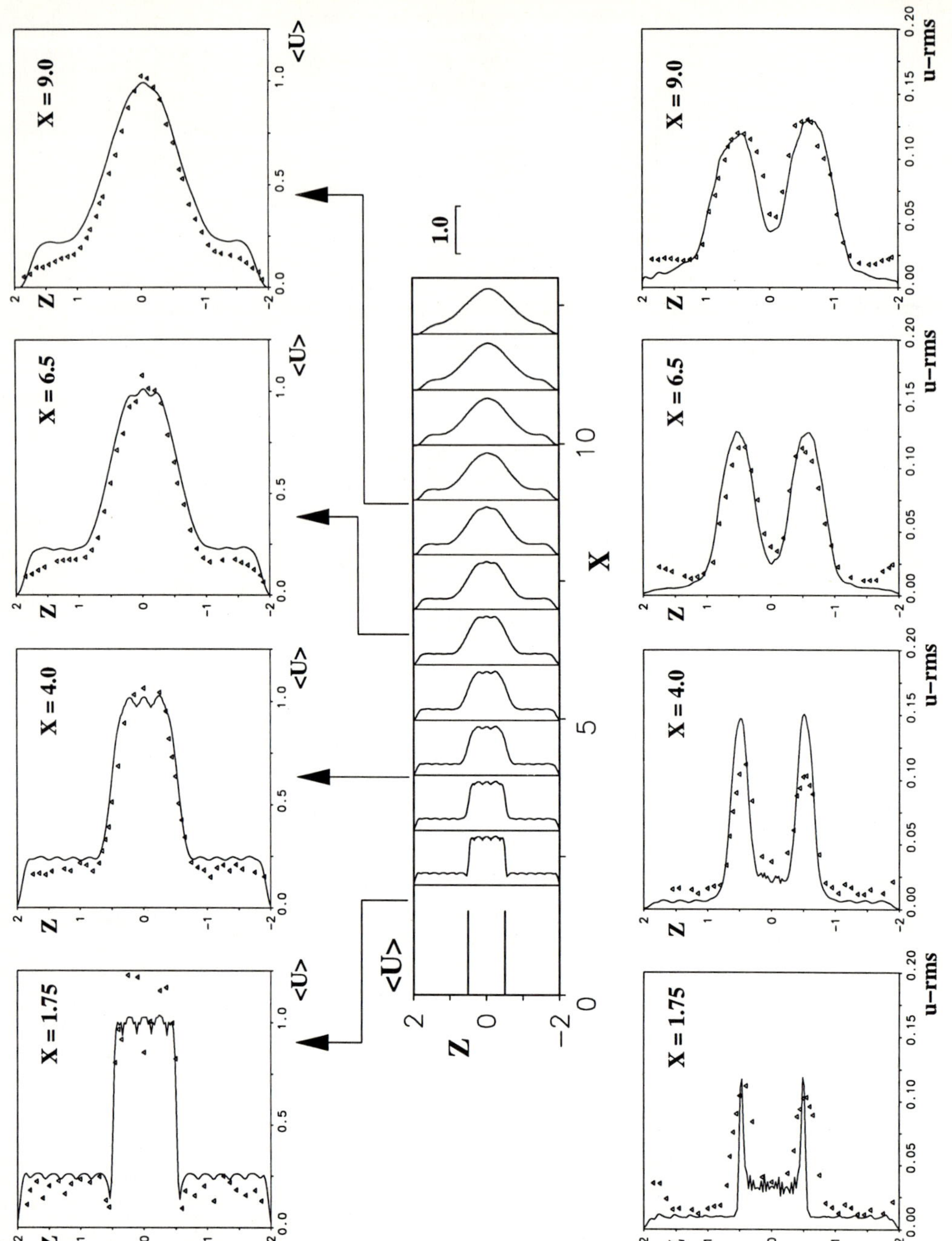

Figure 11 Mean velocity component, $<U>$, and rms values of the longitudinal velocity fluctuations, u-rms, case NA1, $Q_2/Q1 = 0.72$, symbols: experiment, lines: LES.

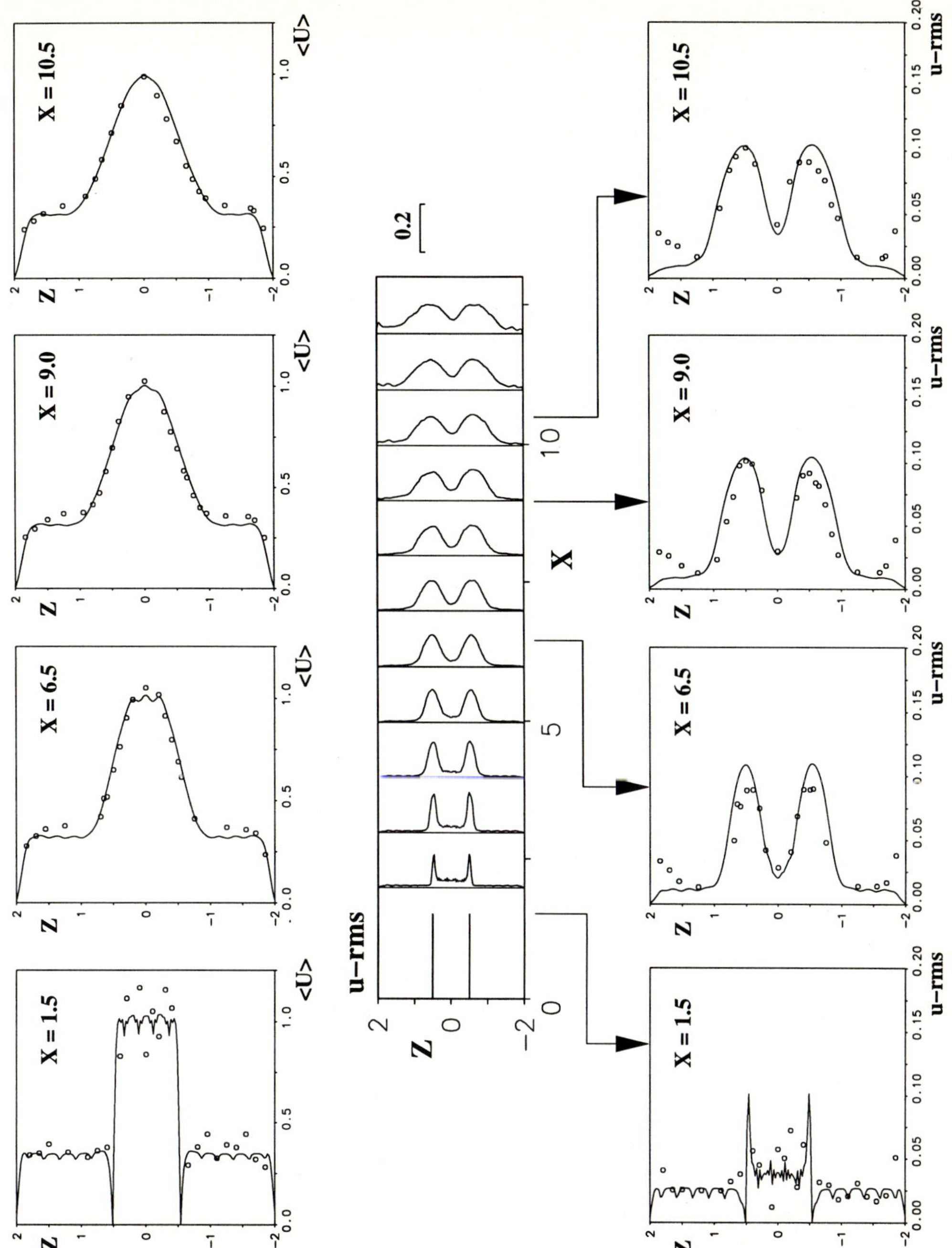

Figure 12 Mean velocity component, $<U>$, and rms values of the longitudinal velocity fluctuations, u-rms, case NA1, $Q_2/Q1 = 1.0$, symbols: experiment, lines: LES.

281

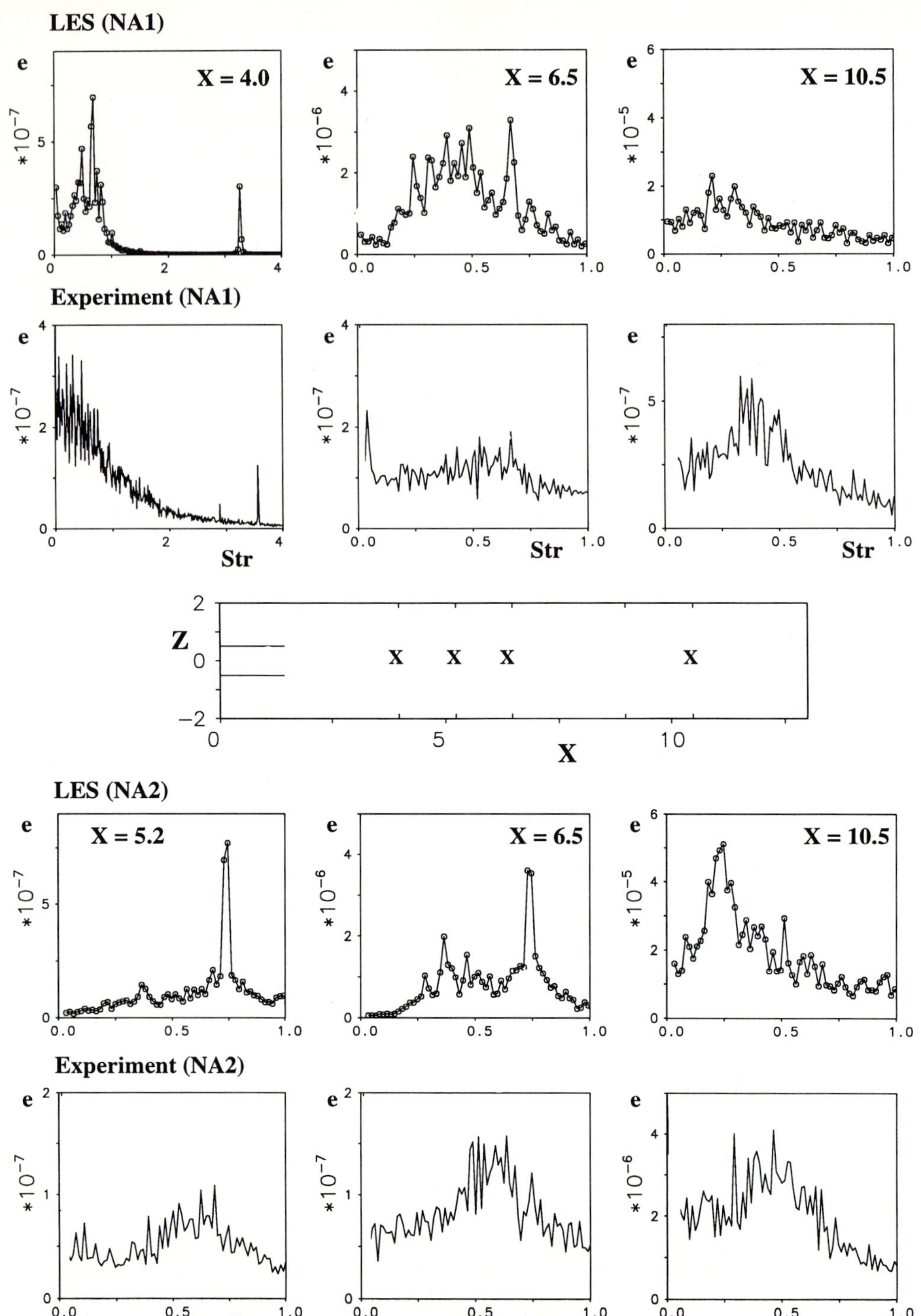

Figure 13 Downstream development of the dimensionless frequency spectrum, along the centerline of the central jet ($Z = 0.0$), $Q_2/Q1 = 1.0$, for the Strouhal number, $Str = f h_1/U_1$, the height of the central channel, h_1, and the maximum velocity, U_1, is used. Case NA1 (above) and NA2 (below).

5 Spatial development of a shearless turbulence mixing klayer

The plane shearless mixing layer of Veeravalli and Warhaft [3] is a relatively simple case of non-equilibrium mixing (excluding the effect of mean shear). The non-equilibrium inlet turbulence is created by two parallel bar grids. In the upper half of the channel there is a bar grid (7 thick bars) and in the lower half there is another bar grid (17 thin bars, with smaller distances between the bars), see **figure 14**. In contrast to a perforated plate with large and small holes there is a chance to numerically simulate the parallel bar grids and the *spatial* development of the turbulence mixing layer downstream of the bars by applying a fine-grid LES. The Reynolds number of this flow case is $Re_{M1} \approx 3700$ based on the distance, M1, between the thin bars, and $Re_{M2} \approx 11000$ based on the distance, M2, between the thick bars, respectively. Comparison with the experimental data is made at the locations X=10.6 and X=15.6.

For a low Reynolds number and very idealized initial conditions results from a DNS (temporal development) are available for a shearless turbulence mixing layer from Briggs et al. [11] and, using LES, from Shao et al. [12], both referring to the experiment of Veeravalli and Warhaft [3].

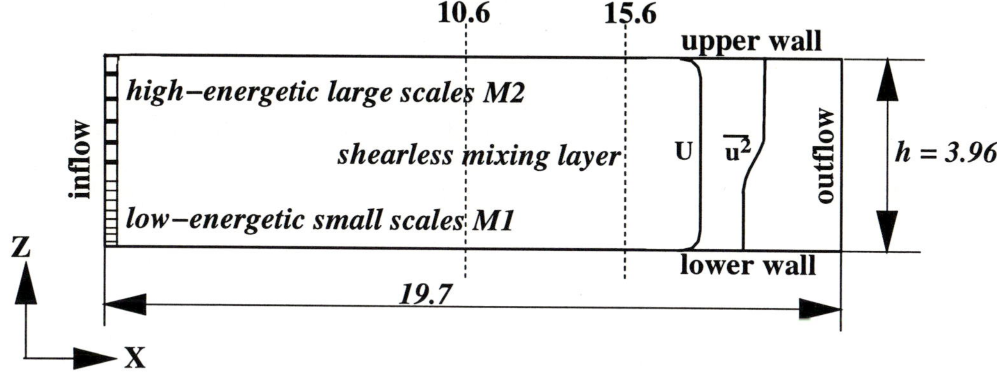

Figure 14 Computational domain for the shearless turbulence mixing layer.

5.1 Computational domain and inflow condition

Computational domain and boundary conditions: The computational domain had the dimensionless size $(L_x, L_y, L_z) = (19.7, 4.0, 3.96)$. The boundary conditions were periodic in the lateral direction (y), and no-slip conditions were applied at the upper and lower walls. At the outflow cross-section the x-gradients of variables were set to zero.

Spatial resolution: In the first phase of the investigation, three grids were used differing in the spatial resolution in the vertical (z-) direction only. The purpose was to improve the spatial resolution of the bar grids:

  coarse grid with $\Delta Z = 0.0300$ and (NX,NY,NZ)=(120,64,132) grid points,

  middle grid with $\Delta Z = 0.0150$ and (NX,NY,NZ)=(120,64,264) grid points, and

  fine grid with $\Delta Z = 0.0075$ and (NX,NY,NZ)=(120,64,528) grid points.

The basic coarse grid uses three control volumes for the vertical width of the thick bars and only one control volume for the vertical width of the thin bars. For all three grids, the spatial resolution in the longitudinal direction was stretched, starting with $\Delta X = 0.1$ at the inflow cross-section and ending with $\Delta X = 0.25$ at the outflow cross-section.

Inflow condition: In the present first stage of the investigations the inflow condition consists in uniform inlet velocity profiles between the bars. This simple inflow condition was already successfully applied for the mixing of different length scales produced by different honey comb grids in a central plane jet and two co-flowing streams, see chapter 4.

5.2   Discussion of results

Figure 15 shows time series of the longitudinal and vertical velocity fluctuations, u' and w', at three different vertical positions: in the high-turbulence region (Z=3.0), the low-turbulence region (Z=0.8), and in the mixing layer (Z=1.4) between these two regions. The position Z=1.4 was chosen with reference to the location of the minimum/maximum of the higher-order statistics, such as the skewness, W_{ske}, and the flatness factor (kurtosis), W_{fla}, of the vertical velocity fluctuations (see figure 18). In comparison to the three series in the experiment [3], the same characteristics can be observed: in particular the vertical velocty fluctuations exhibit intermittent burst (from the large-scale to the small-scale side) in the mixing layer (around Z=1.4).

Figure 16 compares vertical profiles of the mean velocity, $< U >$, and of the turbulent kinetic energy, $< e >$, calculated for the three computational grids. In addition, **Figure 17** shows the three components of the kinetic energy. First, it can be observed that the arrangement of the parallel bar grids (bars placed at the geometrically exact positions) does not lead to the desired shearless mean velocity profile. In the experiment [3] a final fine-tuning of the grid-spacings between the bars (obviously a very difficult and time consuming task) was arranged with the goal of creating a flow situation as shearless as possible. As a consequence of the shear in the vertical mean velocity profile, $\partial < U > /\partial Z$, there is increased turbulent energy production at the location of vertical shear, and local maxima of turbulent energy can be observed. On the low-energy side all three grids produce the correct level of turbulent energy.

In **Figure 18** the higher-order statistics is presented in comparison with the measurements. For a downstream position at X=10.6, the skewness, W_{ske}, and the flatness, W_{fla}, of the vertical fluctuations are shown. Reference values for a Gaussian distribution are $W_{fla} = 3.0$ and $W_{ske} = 0.0$. On the high-energy side these values are reached in agreement with the experiment. On the low-energy side, the deep minimum, $W_{ske} = -1.0$, of the experiment is not reached by our LES results ($W_{ske} = -0.65$), it is also not reached by the DNS result of Briggs et al. [11] ($W_{ske} = -0.70$). In comparison with the maximum of the experimental value for the flatness factor, $W_{fla} = 6.0$, the fine grid simulation gives the closest result ($W_{fla} = 5.0$).

Finally, in **Figure 19** results are presented from an attempt to improve the desired shearless flow situation. An additional small bar was placed between the last small bar and the first big bar. This additional small bar increased the mixing downstream of this vertical location and decreased the mean vertical shear significantly. Figure 19 compares the results from the case with this additional bar (dashed line) with the result on the same fine grid (full line, from figure 16 without this additional bar). The improvement is significant but still not fully satisfying.

5.3   Conclusions

In the first stage of the current investigation of the shearless turbulence mixing layer the creation of a shearless mean vertical velocity profile behind the bars seems to be as difficult as it was in the experiment. The current investigations show significant improvement by placing an additional small bar between the upper channel section with the big bar grid and the lower half with the small bar grid.

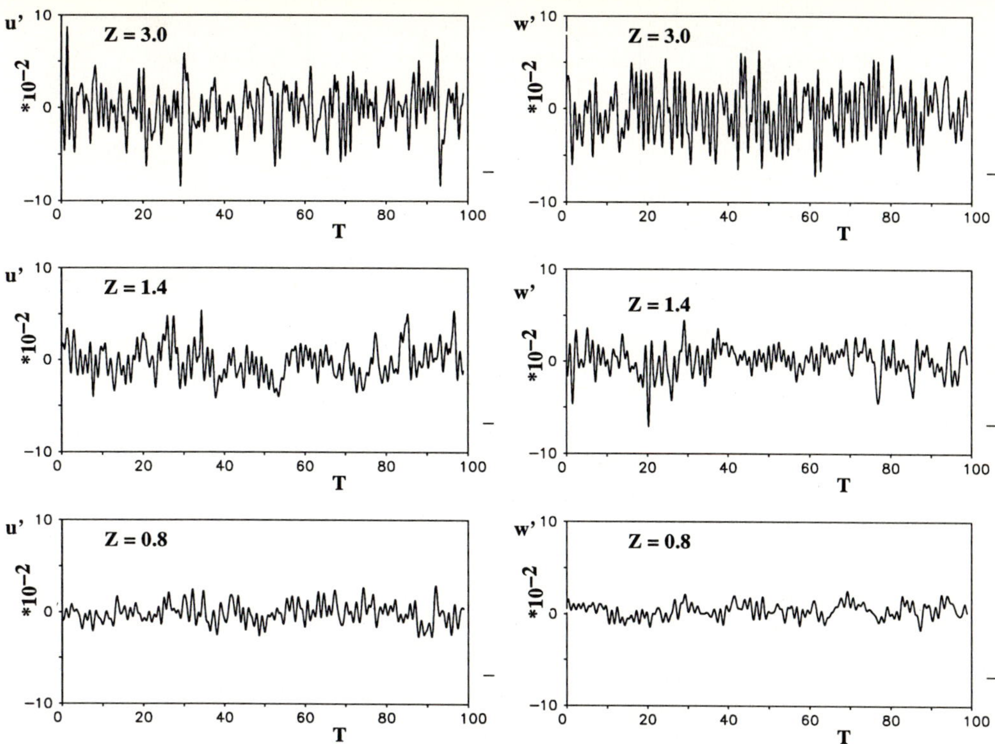

Figure 15 Typical time series of the longitudinal velocity fluctuations u' (left) and the vertical velocity fluctuations w' (right) at three vertical positions and a downstream location $X = 10.6$ (from fine grid calculation)

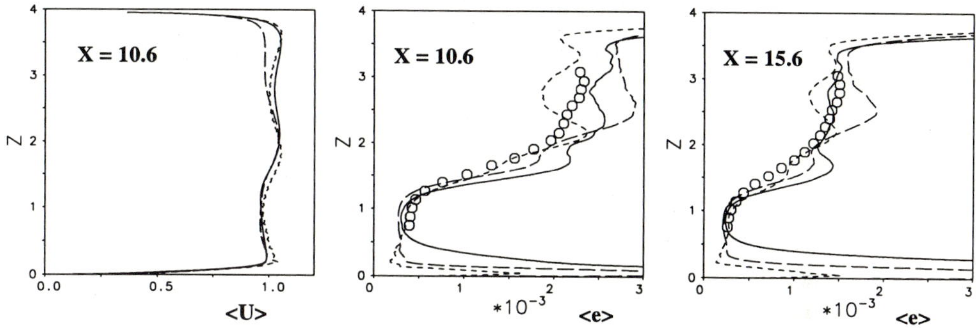

Figure 16 Influence of the spatial resolution of the grids:
left: mean longitudinal velocity component $< U >$ at $X = 10.6$,
middle: turbulence energy $< e >$ at $X = 10.6$, and
right: turbulence energy $< e >$ at $X = 15.6$.
——: fine grid, – – – : middle grid, and - - - : coarse grid, open circles: experiment [3]

285

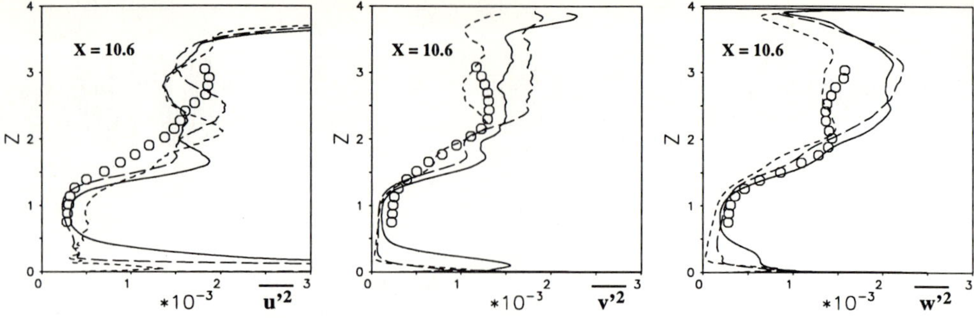

Figure 17 Profiles of the variance $\overline{u'^2}$, $\overline{v'^2}$, and $\overline{w'^2}$ at the downstream location $X = 10.6$.
—: fine grid, – – – : middle grid, and - - - : coarse grid, open circles: experiment [3]

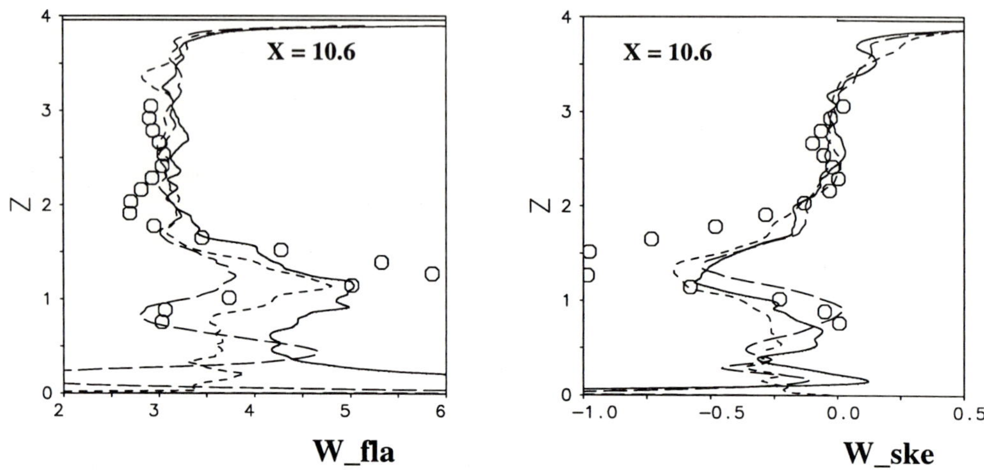

Figure 18 Flatness and skewness of the vertical velocity component at the downstream location $X = 10.6$.
—: fine grid, – – – : middle grid, and - - - : coarse grid, open circles: experiment [3]

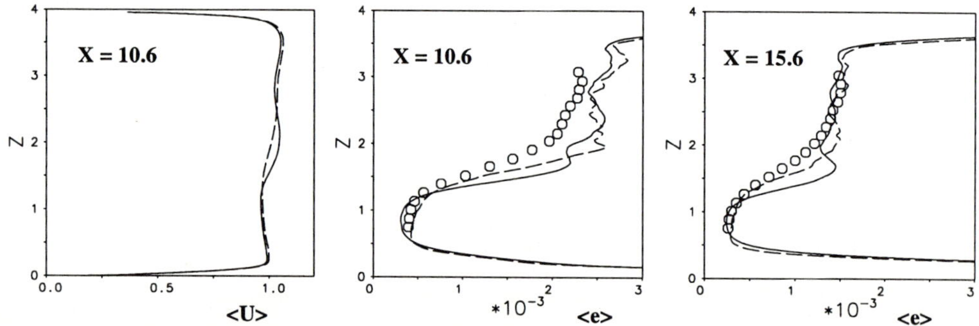

Figure 19 Improvement towards zero mean shear:
left: mean longitudinal velocity component $< U >$ at $X = 10.6$,
middle: turbulence energy $< e >$ at $X = 10.6$, and
right: turbulence energy $< e >$ at $X = 15.6$.
—: fine grid result from figure 16, – – – : fine grid result with additional small bar, open circles: experiment [3]

References

[1] A. Meri, H. Wengle, A. Dejoan, Védy E., and R. Schiestel. Application of a 4th-order Hermitian scheme for non-equidistant grids to LES and DNS of incompressible fluid flow. In E.H. Hirschel, editor, *Numerical Flow Simulation I, Notes on Numerical Fluid Mechanics*, Vol. 66, pp. 382-406, Vieweg, 1998.

[2] M. Raddaoui. Modélisation numérique et étude expérimentale du mélange d'échelles dans un jet turbulent confine. PhD thesis, Univ. de la Méditerranée, Aix-Marseille II, 1997.

[3] S. Veeravalli and Z. Warhaft. The shearless turbulence mixing layer. *J. Fluid Mech.*, 207:191–229, 1989.

[4] C.W. Hirt, B.D. Nichols, and N.C. Romero. Sola – a numerical solution algorithm for transient fluid flows. In *Los Alamos Sci. Lab. Report LA 5852*, Los Alamos, 1975.

[5] R.S. Rogallo and P. Moin. Numerical simulation of turbulent flows. *Ann.Rev.Fluid Mech.*, 16:99–137, 1984.

[6] U. Schumann. Subgrid scale model for finite difference simulations of turbulent flows in plane channels and annuli. *J. Comp. Phys.*, 18:376–404, 1975.

[7] M. Germano, U. Piomelli, P. Moin, and W.H. Cabot. A dynamic subgrid-scale eddy viscosity model. *Phys. Fluids A*, 3:1760–1765, 1991.

[8] H. Le, P. Moin, and J. Kim. Direct numerical simulation of turbulent flow over a backward-facing step. *J. Fluid Mech.*, 330:349–374, 1997.

[9] K. Akselvoll and P. Moin. Large eddy simulation of a backward facing step flow. In W. Rodi and F. Martelli, editors, *Proceedings of Engineering Turbulence Modelling and Experiments 2*, pp. 303–313. Elsevier Science Publishers, 1993.

[10] J. Kim, P. Moin, and R. Moser. Turbulence statistics in fully developed channel flow at low Reynolds number. *J. Fluid Mech.*, 177:133–166, 1987.

[11] D.A. Briggs, J.H. Ferziger, J.R. Koseff, and S.G. Monosmith. Entrainment in a shear-free turbulent mixing layer. 7th Fluid Dynamics Symposium, April 19994, Chania, Greece, in: Application of direct and large-eddy simulation to transition and turbulence, AGARD Conference Proceedings 551, 1995.

[12] L. Shao, J.P. Bertoglio, and M. Michard. Large eddy simulation of the interaction between two distinct turbulent velocity scales. *Advances in Turbulence 3*, 101–112, Springer-Verlag, 1991.

LES of Turbulent Boundary Layers and Wakes

M. Opiela, M. Meinke, and W. Schröder
Aerodynamisches Institut, RWTH-Aachen
Wüllnerstr. zw. 5 u. 7, 52062 Aachen, Germany

P. Comte[1], E. Briand
Institut National Polytechnique de Grenoble, LEGI/IMG
BP 53, 38041 Grenoble, France

Summary

Results of Large-Eddy Simulations (LES) of transitional boundary layers (Grenoble) and wake flows behind flat plates with rectangular and circular trailing edge (Aachen) are presented. With improved inflow conditions that take into account the weakly non-linear development of perturbations upstream of the domain K- and H-type transitions were simulated with good agreement concerning the streamwise evolution of the perturbation amplitudes. The vortex dynamics during the transition process is analyzed using the data of simulations based on 5 million grid points. The influence of spanwise grooves is investigated for the fully turbulent boundary layer. The main effect to partially isotropize the vortical structures in the near wall region is confirmed by statistical results.

So far findings for the wake flow have been obtained using laminar boundary layer profiles as inflow conditions. Solutions with turbulent boundary layers are presently carried out in cooperation with Grenoble. The algorithm applied for the wake flow simulation is a second-order accurate mixed central-upwind scheme for compressible flows. Non-reflecting boundary conditions are used for the lateral and outflow boundaries. The simulations are performed for a free stream Mach number of 0.3 and a Reynolds number of 5.000 based on the trailing edge thickness. The presented results include the visualization of the vortex dynamics and the turbulence statistics of the near wake.

1 Mathematical Model

The Navier-Stokes equations written in tensor notation and in terms of dimensionless conservative variables for an ideal gas read:

$$\frac{\partial \boldsymbol{Q}}{\partial t} + (\boldsymbol{F}_\beta^C - \boldsymbol{F}_\beta^D)_{,\beta} = 0 \quad , \qquad \boldsymbol{Q} = [\varrho,\varrho u_\alpha,\varrho E]^T \quad , \tag{1.1}$$

where $\boldsymbol{Q}$ is the vector of the conservative variables, $\boldsymbol{F}_\beta^C$ denotes the vector of the convective, and $\boldsymbol{F}_\beta^D$ of the diffusive fluxes:

---

[1] new affiliation: Institut de Mécanique des Fluides, Equipe Aérothermodynamique, Université de Strasbourg, 67000 Strasbourg, France

$$\boldsymbol{F}_\beta^C - \boldsymbol{F}_\beta^D = \begin{pmatrix} \varrho u_\beta \\ \varrho u_\alpha u_\beta + p\delta_{\alpha\beta} \\ u_\beta(\varrho E + p) \end{pmatrix} + \frac{1}{Re} \begin{pmatrix} 0 \\ \sigma_{\alpha\beta} \\ u_\alpha \sigma_{\alpha\beta} + q_\beta \end{pmatrix} . \qquad (1.2)$$

The stress tensor $\sigma_{\alpha\beta}$ is written as a function of the strain rate tensor $S_{\alpha\beta}$

$$\sigma_{\alpha\beta} = -2\nu \left(S_{\alpha\beta} - \tfrac{1}{3}S_{\gamma\gamma}\delta_{\alpha\beta}\right) \qquad \text{with} \qquad S_{\alpha\beta} = \tfrac{1}{2}(u_{\alpha,\beta} + u_{\beta,\alpha}) . \qquad (1.3)$$

The dynamic viscosity ν is assumed to be only a function of the temperature. The system is closed by Fourier's law of heat conduction applied to the heat flux q_β, and the equation of state for ideal gases

$$q_\beta = -\frac{k}{Pr(\gamma-1)}T_{,\beta} \quad , \quad p = \frac{1}{\gamma}\varrho T \quad , \quad p = (\gamma-1)\left(\varrho e - \frac{\varrho}{2}u_\beta u_\beta\right) \quad , \qquad (1.4)$$

where γ is the ratio of the specific heat capacities, T the temperature and Pr the Prandtl number. For the discretization of the convective fluxes a low-dissipation second-order scheme is used, which yielded LES results of pipe and channel flows in good agreement with DNS data [7] and LES findings of other authors [13]. It is briefly described in the following section.

2 Method of Solution

The turbulent flows were simulated using different solution methods. A description of the methods used for the turbulent boundary layers can be found in [10]. The turbulent wakes were computed with a second-order finite-volume method based on an AUSM method. The convective fluxes for this scheme read:

$$\boldsymbol{F}_\beta^C = \begin{pmatrix} F_1^C \equiv \frac{M_\beta}{2}((\rho a)^L + (\rho a)^R) + \frac{|M_\beta|}{2}((\rho a)^L - (\rho a)^R) \\ \frac{F_1^C}{2}(u_\alpha^L + u_\alpha^R) + \frac{|F_1^C|}{2}(u_\alpha^L - u_\alpha^R) + (\frac{p^L+p^R}{2} + \chi(p^L u_\alpha^L - p^R u_\alpha^R))\delta_{\alpha\beta} \\ \frac{F_1^C}{2}(\frac{(\rho e)^L+p^L}{\rho^L} + \frac{(\rho e)^R+p^R}{\rho^R}) + \frac{|F_1^C|}{2}(\frac{(\rho e)^L+p^L}{\rho^L}) - \frac{(\rho e)^R+p^R}{\rho^R}) \end{pmatrix} , \qquad (2.5)$$

where superscript ^{L} and ^{R} denote left and right interpolated variables which are obtained using a quadratic MUSCL interpolation of the primitive flow variables. The quantities a and $M_\beta = \frac{u_\beta}{a}$ are the average of the left and right interpolated speed of sound and Mach number. The parameter χ is used to control the amount of numerical dissipation within this scheme. Its influence on the solution is demonstrated in [12, 16]. In all the simulations χ is set to zero, such that a central discretization of the pressure derivative is obtained with a minimal numerical dissipation. The integration in time of the differential equations is carried out using an explicit low storage second-order accurate Runge-Kutta method. The coefficients in the Runge-Kutta steps are chosen as $\alpha_l = (0.25, 0.1667, 0.375, 0.5, 1)$, which are optimized for maximum stability for a central scheme. More details of the solution method can be found in [16, 12].

2.1 Computational Grids

Boundary fitted multiblock structured grids are used for the LES of the wake flow. Since the spatial steps of the computational grid should vary smoothly and the gridlines should be orthogonal and clustered in regions with strong flow gradients the grids are generated using Poisson equations for the grid coordinates. Clustering is achieved via weight functions which control the distance of grid lines in the physical space and the orthogonality at the block boundary. In multiblock arrangements all blocks are generated simultaneously, using a special treatment of non-regular cells. The grid for the circular trailing edge in Figure 1 demonstrates the good properties of this grid generation procedure. The extent of the domain of integration in spanwise direction is about 3 times the trailing edge thickness. It is known that a sufficient resolution of the spanwise flow structures is important for an accurate prediction of the wake flows Therefore 65 grid points are used in this space direction. In Table 1 the flow parameters and the main grid characteristics of the wake simulation are summarized.

Table 1 Geometrical and physical parameters for the simulation of the wakes behind flat plates with rectangular and circular trailing edge.

	rectangular	circular
trailing edge thickness	H=0.05	D=0.05
domain of integration x×y×z	$30 \cdot H \times 11 \cdot H \times 3.2 \cdot H$	$30 \cdot D \times 11 \cdot D \times 3.2 \cdot D$
distance of first grid line from the wall	$2.0 \cdot 10^{-3} \cdot H$	$2.0 \cdot 10^{-3} \cdot D$
spatial step in spanwise direction	$0.05 \cdot H$	$0.05 \cdot D$
number of blocks	3	7
grid points (coarse)	412.203	–
grid points (fine)	$1.9 \cdot 10^{6}$	$2.0 \cdot 10^{6}$
Re_l	100.000	100.000
Re_H and Re_D	5000	5000
Mach number	0.3	0.3
boundary layer thickness	$\frac{\delta}{H} = 0.316$	$\frac{\delta}{D} = 0.316$

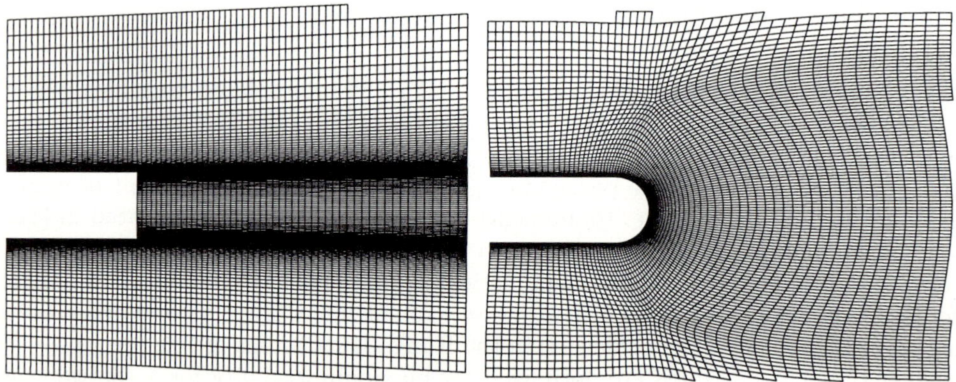

Figure 1 Zoom of the computational grids for the LES of the turbulent wake flows. The meshes consist of 1.9 million grid points for the rectangular (left) and 2.0 million (right) for the circular trailing edge.

2.2 Boundary Conditions

In the wake behind the rectangular and circular trailing edge large scale vortices are generated, that are slowly dissipated by the turbulence generated in the shear layers. Since

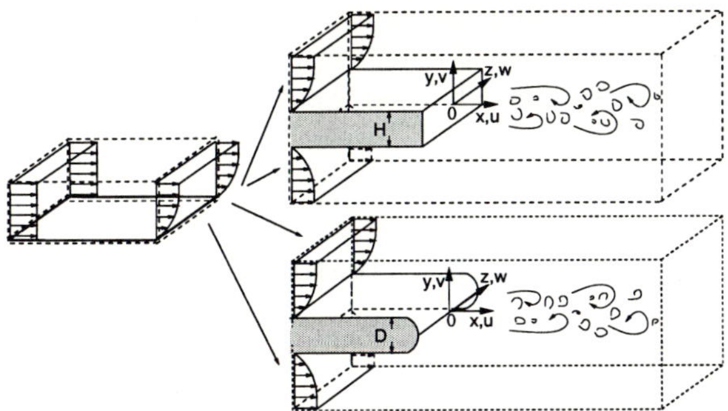

Figure 2 Domain of integration and boundary conditions used for the LES of turbulent wake flows.

the size of the domain of integration is limited, outflow boundary conditions are required that minimize the reflection of numerical waves. Otherwise an unphysical coupling between in- and outflow boundaries may result, since the waves reflected at the outflow boundary travel upstream to the inflow boundary, where they are then reflected as physical waves. Non-reflecting boundary conditions were proposed in [18] and extended to viscous flows in [14]. Their current implementation is described in detail in [15], where these conditions were successfully applied for the simulation of turbulent jets. At the outflow boundary this non-reflecting boundary condition with pressure level relaxation is applied. At the inflow boundary the flow state is transferred from a solution of a flat plate with laminar boundary layer. A sketch of the wake simulation with the boundary layer is given in Figure 2, as well as the coordinate system (x,y,z), the origin and the velocity components which are used for the presentation of results.

2.3 Vortex Identification in Turbulent Flow

To identify vortex structures in turbulent flows the so called λ_2-criterion of [5] is used. The basic idea is that a pressure minimum in a vortical motion corresponds to a vortex core if unsteady straining and viscous effects are neglected. The fundamental equation for the vortex identification is obtained by taking the gradient of the Navier-Stokes equations:

$$a_{i,j} = -\frac{1}{\rho}p_{,ij} + \nu u_{i,jkk}, \tag{2.6}$$

where $a_{i,j}$ is the acceleration gradient, and $p_{,ij}$ is the Hessian of the pressure which is symmetric. Decomposition of $a_{i,j}$ into its symmetric and antisymmetric part gives

$$a_{i,j} = \underbrace{\left[\frac{DS_{ij}}{Dt} + \Omega_{ik}\Omega_{kj} + S_{ik}S_{kj}\right]}_{symmetric} + \underbrace{\left[\frac{D\Omega_{ij}}{Dt} + \Omega_{ik}S_{kj} + S_{ik}\Omega_{kj}\right]}_{antisymmetric}. \tag{2.7}$$

The symmetric part of (2.6) is

$$\frac{DS_{ij}}{Dt} - \nu S_{ij,kk} + \Omega_{ik}\Omega_{kj} + S_{ik}S_{kj} = -\frac{1}{\rho}p,_{ij}. \tag{2.8}$$

If the first two terms of this equation are neglected, see [5], $S^2 + \Omega^2$ determines the existence of a local pressure minimum due to vortical motion and defines a vortex core as a connected region with two negative eigenvalues of $S^2 + \Omega^2$. If λ_1, λ_2 and λ_3 are the eigenvalues and $\lambda_1 \geq \lambda_2 \geq \lambda_3$, the definition is equivalent to the requirement that $\lambda_2 < 0$ within the vortex core.

3 Results

3.1 Transitional Boundary Layers

In Ducros et al. [3] a spatial simulation using a subgrid model (filtered structure function) and a high-order scheme (4th order extension of the Mac Cormack scheme) showed its capacity to describe a complete transition up to developed turbulence.

In the present work, the same method with a special attention to the control of the inflow is used to study the transition and its influence on the downstream turbulence.

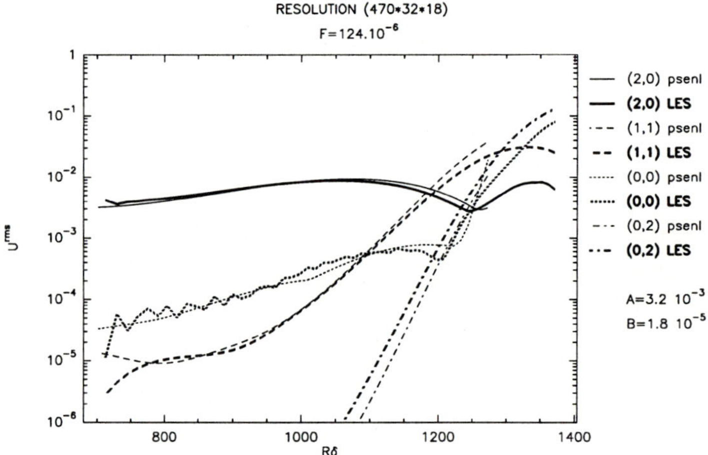

Figure 3   Stream-wise evolution of perturbations for a H-transition. The first two modes are the inflow perturbations. The last two are stationary modes.

First the inflow conditions were improved which should take into account the weakly non-linear development of perturbations upstream of the domain. For that purpose non-linear Parabolized Stability Equations (see Bertolotti et al., [2]) developed by the ONERA-CERT were used. To validate this method, results were compared with non-linear PSE calculations. The physical parameters were chosen to fit the experiment of Kachanov et al. [6] (low Mach and low Reynolds numbers). A very good agreement on the streamwise evolution of the amplitudes of the perturbations was obtained, even for the stationary

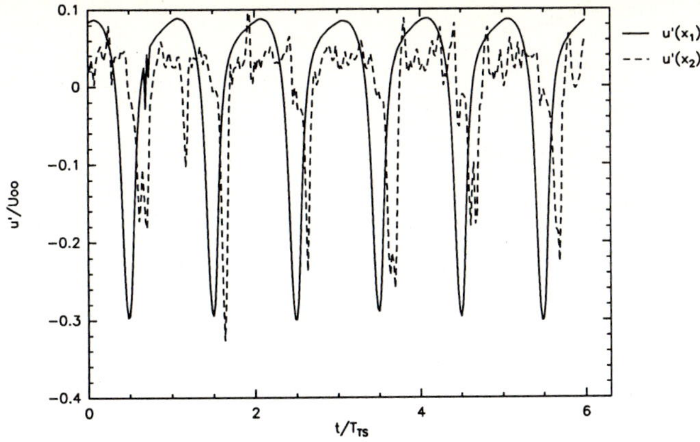

Figure 4   Signal of u' at $x_1 = 415\delta_1, x_2 = 530\delta_1$ for a K transition. Note the periodicity in an almost turbulent zone (downstream the cf peak).

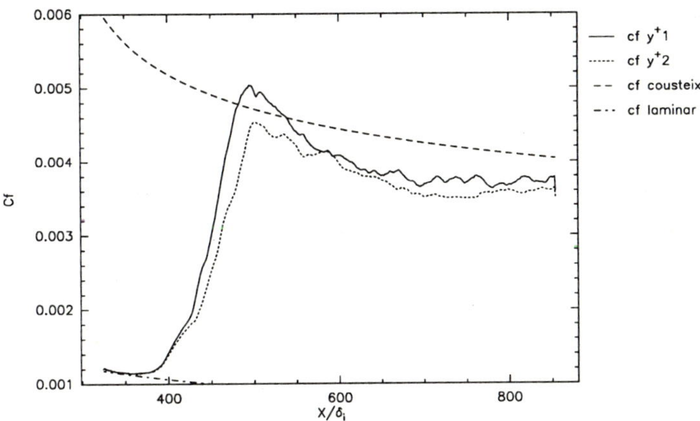

Figure 5   Stream-wise evolution of the friction coefficient for a K-transition (first points at $y^+ = 1$ and $y^+ = 2$) the dashed lines represent the theoretical coefficients.

modes (see Figure 3). With the non-linear PSE calculations as inflow conditions, the two kinds of transition were simulated. The computational domain was large enough to reach a developed turbulent state and the resolution (5 millions points) was sufficient to resolve the waves propagating through the transition region and the large scales of turbulence. First results show the randomization of the flow close to the wall by a stream-wise progressive filling of the spectrum between the harmonics of the Tollmien-Schlichting waves. This contrasts with the violent periodic events which take place in the upper part of the boundary layer, even downstream of the friction coefficient peak (see Figure 4 and 5). In the transition region lambda shaped structures (staggered or aligned) can be seen giving rise to hairpin shaped structures resembling those of the turbulent flow. A striking feature is that the spanwise streak scaling (around 100 wall units) appears at

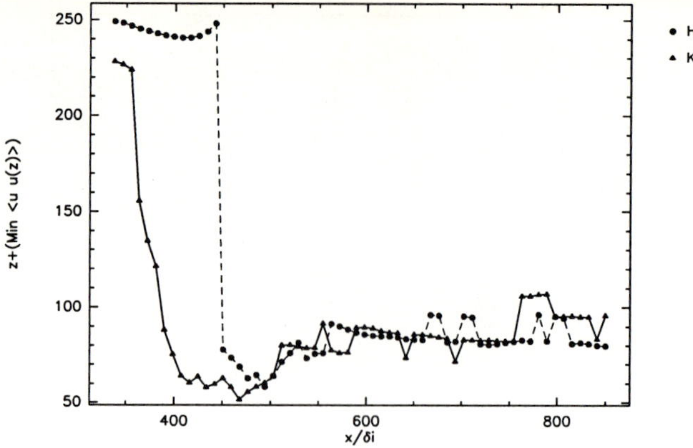

Figure 6 Stream-wise evolution of the first minimum of spanwise correlation on longitudinal velocity for a H and K transition ($y^+ = 15$).

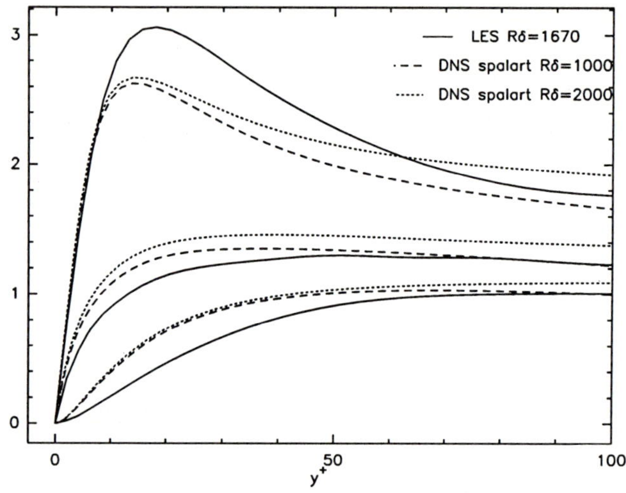

Figure 7 Velocity fluctuations (u^+,v^+,w^+) compared with Spalart (88), $x = 810\delta_1$.

the very beginning of the transition (see Figure 6). The transitional structures interact and evolve, from two distinct situations in our case into a statistically similar turbulent state (see Figure 7). The dynamics of these structures should be investigated in more detail in the next future.

3.2 Turbulent Boundary Layers Perturbed by Spanwise Grooves

Figure 8 shows the effect of a spanwise groove (D-type roughness) the depth d of which is identical to the visual thickness δ_0 of the incoming boundary layer. The Reynolds number

294

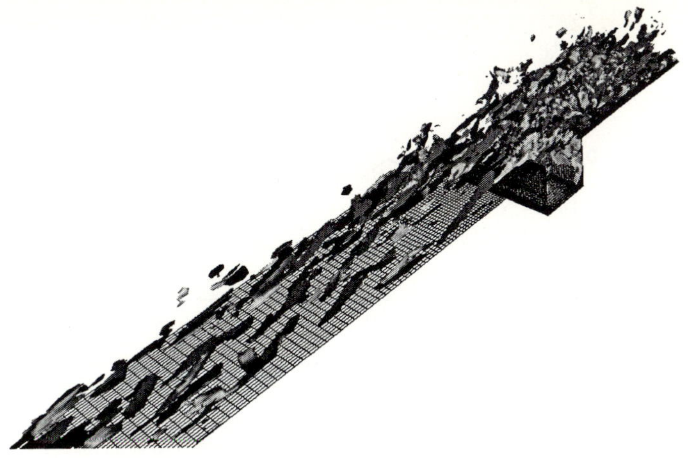

Figure 8 Isosurfaces of vorticity fluctuations filtered by positive $Q = (\Omega_{ij}\Omega_{ij} - S_{ij}S_{ij})/2$, $\omega = 0.3\omega_i$ and shaded/colored by the sign of longitudinal vorticity: black/blue $\omega_x \leq -0.1$, grey/red $\omega_x \geq 0.1$.

based on the free stream velocity U_0 and δ_0 is 5100. It corresponds to the numerical simulation of a turbulent boundary layer performed by Spalart [17] at $R_\theta = 670$ where θ is the momentum thickness of the flow. The Reynolds number is also roughly the same as in the experiment of Elavarasan *et al.* [4]'s experiment ($R_\theta = 700$ for the flow visualization and 1320 for the measurements). The resolution for the groove block is $41 \times 101 \times 40$, that of the two adjacent flat plate blocks is $121 \times 51 \times 40$. The minimum grid spacing at the wall in the vertical direction corresponds to $\Delta y^+ = 1.2$. The streamwise grid spacing goes from $\Delta x^+ = 3.2$ near the groove edges to 20 at the outlet. The spanwise resolution is $\Delta z^+ = 16$. The grid is stretched by an hyperbolic tangent type transformation similar to that used by Le & Moin [9] for a backward-facing step flow. From Figure 8 it is evident that the main effect of the groove is to isotropize partly the vortical structures. This is confirmed by statistical results such as maps of the invariants of the Reynolds-stress anisotropy tensor introduced by Lumley & Newman [11]. Figure 9 shows spanwise correlation distances before and after the groove. The spanwise wavelength of the streaks is estimated from the location of the negative maximum of the correlation. For the streamwise velocity u, this maximum occurs at 50 wall units and the spanwise wavelength is therefore $\lambda_z^+ = 100$ (see Kim *et al.* [8]). Downstream of the groove λ_z^+ drops to approximately 60. This fact evidences that the groove strongly affects the near-wall dynamics. In contrast the outer region is hardly influenced.

3.3 Wakes Behind Plates with Rectangular and Circular Trailing Edge

The wakes behind the flat plates with rectangular and circular trailing edge were simulated for the parameters specified in Table 1. To investigate the influence of the spatial resolution on the solution two different grids were used for the simulation of the flow

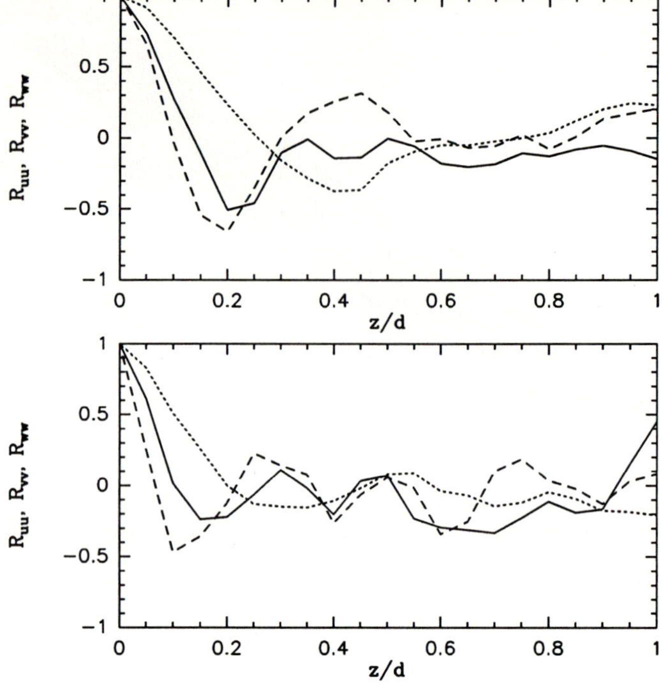

Figure 9 Spanwise correlation of the velocity components, upstream and downstream of the groove (top and bottom plots, respectively): ——, R_{uu} ; - - - -, R_{vv}; · · · · ·, R_{ww};

around the rectangular trailing edge. All simulations were first run in two space dimensions. After the development of a vortex street, the solution was extended in spanwise direction to allow the development of three dimensional turbulent structures. The simulation was then carried out for more than 40 vortex shedding cycles to allow the development of a fully turbulent and a quasi periodic flow field. For the determination of the turbulence statistics the solution of about 80 vortex shedding cycles was then stored to disc with a resolution of 10 time levels per vortex shedding period.

The time and spanwise averaged velocity profiles on the symmetry axis and at different locations in streamwise direction in the near wake flow field are shown in Figure 10. With the coarse grid a shorter separated flow region with a higher negative peak velocity is predicted for the rectangular trailing edge. In the case of the circular trailing edge the negative peak velocity is smaller while the separated region extends further downstream of the trailing edge. The turbulent kinetic energy profiles in Figure 11 show a similar trend. The maximum of the turbulent kinetic energy, which occurs in the vicinity of the rear stagnation point, is over-predicted on the coarse grid for the rectangular trailing edge, whereas further downstream the turbulence energy is smaller than that on the fine grid. The overall differences between the coarse and fine grid solution, however, are relatively small, which indicates that the grid resolution of the fine grid is adequate. For the circular trailing edge the peak value of the turbulence energy is about 30% lower than that for the rectangular trailing edge. The profiles across the wake exhibit the

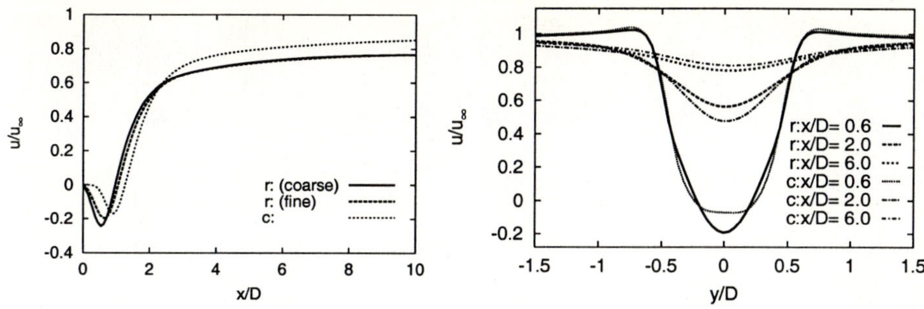

Figure 10   Time and spanwise averaged profiles of the velocity in streamwise direction u in the wake of the rectangular (r) and circular (c) trailing edge. Velocity profile on the symmetry axis $y{=}0$ (left) and velocity profiles in the wake at different locations x (right).

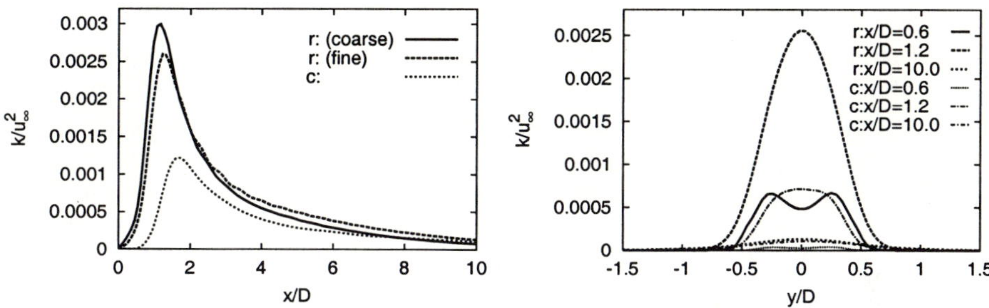

Figure 11   Time and spanwise averaged profiles of the turbulent kinetic energy k in the wake of the rectangular (r) and circular (c) trailing edge. Turbulent kinetic energy profile on the symmetry axis $y{=}0$ (left) and profiles in the wake at different locations x (right).

characteristic double peak in the separated flow region which occurs due to the higher turbulence production in the separated shear layers. Again it can clearly be seen that the turbulence intensity is considerably smaller in the near wake field for the circular trailing edge. The time and spanwise averaged distribution of the Reynolds stress component $u'u'$ in the wake of the flat plate with rectangular and circular trailing edge are shown in Figure 12. In the case of the circular trailing edge the maximum value of $u'u'$ occurs further downstream of the trailing edge. In addition, the high intensity regions of the Reynolds stress component $u'u'$ exhibit different shapes. For the circular trailing edge these regions are closer together and elongate further into the wake. These regions are connected with the formation of the main rollers of the Karman vortex street, so that the initial position and convection of the main rollers must be different for the two different trailing edge shapes.

In Figure 13 the vortex dynamics of the near wake is visualized for one vortex shedding cycle. The surfaces of λ_2 clearly indicate the main rollers of the vortex street and vortex

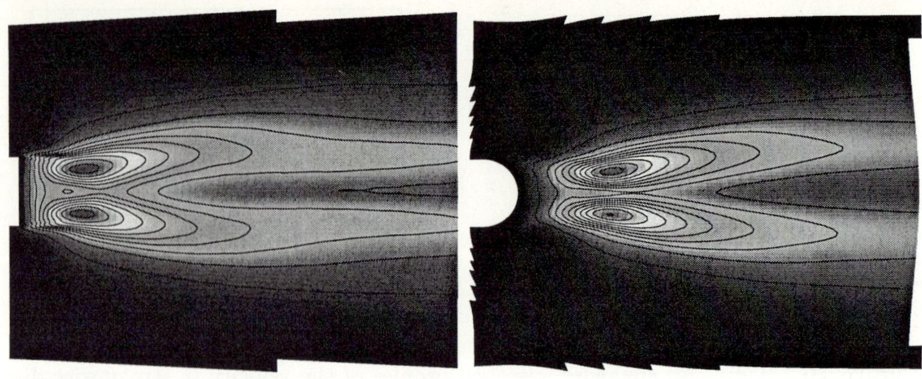

Figure 12 Time and spanwise averaged distribution of the Reynolds stress component $u'u'$ in the wake of the rectangular (left) and circular (right) trailing edge.

structures with an axis in streamwise direction. They appear in pairs with opposite sign of rotation and they connect the main rollers. From the wake behind the circular cylinder it is known that one pair of streamwise vortices appear in a region with an extent in spanwise direction of about one cylinder diameter [1]. Since the extent of the domain of integration in spanwise direction is about 3 times the trailing edge thickness, 3 pairs of such vortices should appear in the present solution. This can clearly be seen in Figure 13 e.g. time level 6. Additionally, smaller scale vortices are visible in the separated shear layer behind the trailing edge. The number of these vortices is influenced by the number of grid points used in spanwise direction. For the coarse grid solution not shown here, considerably less smaller structures are observed.

In Figure 14 surfaces of constant λ_2 in the near wake of the circular trailing edge are presented. The visualization shows a similar vortex dynamics as in the wake behind the rectangular trailing edge. In the separated shear layers less smaller scale structures are visible, although the same grid resolution has been used as for the rectangular trailing edge. The two point correlation in spanwise direction z at two different locations in the x,y-plane is plotted in Figure 15 for the flat plate with rectangular and circular trailing edge. For both locations there is a negative minimum correlation at $z/H{=}0.6$ for the rectangular and at $z/D\approx0.6$ for the circular trailing edge (time averaged results are not fully converged yet). The negative minimum in the two point correlation in y-direction is characteristic for the existence of streamwise vortices. A first indication of the origin of such vortices, can be obtained from the distribution of R_{ww} in a x,y-plane. Such a distribution for the flat plate with rectangular trailing edge is plotted in Figure 16 for the x,y-plane at $z/D{=}0.6$ where the negative minimum occurs. The dark region in the plot near the centerline behind the trailing edge indicates the region, where the streamwise vortices occur most frequently. In addition, a region of smaller negative correlation extends in the upper and lower region of the wake, which corresponds to the location of the vortices which connect the main rollers, which are visible in Figure 13.

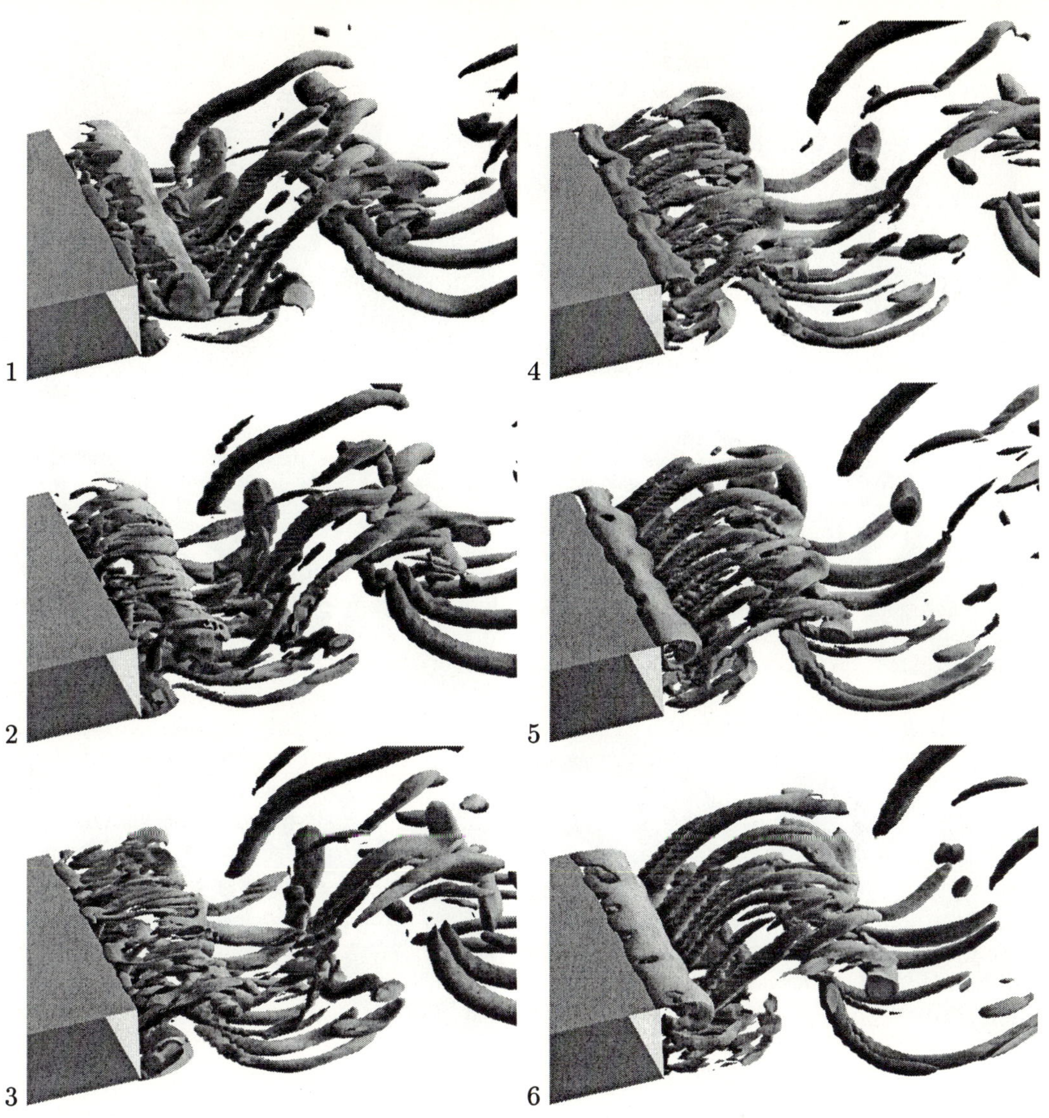

Figure 13 Surface of constant λ_2 color coded with the local Mach number for the wake with rectangular trailing edge (fine grid). Shown are six different time levels in one vortex shedding cycle (Δt=0.740).

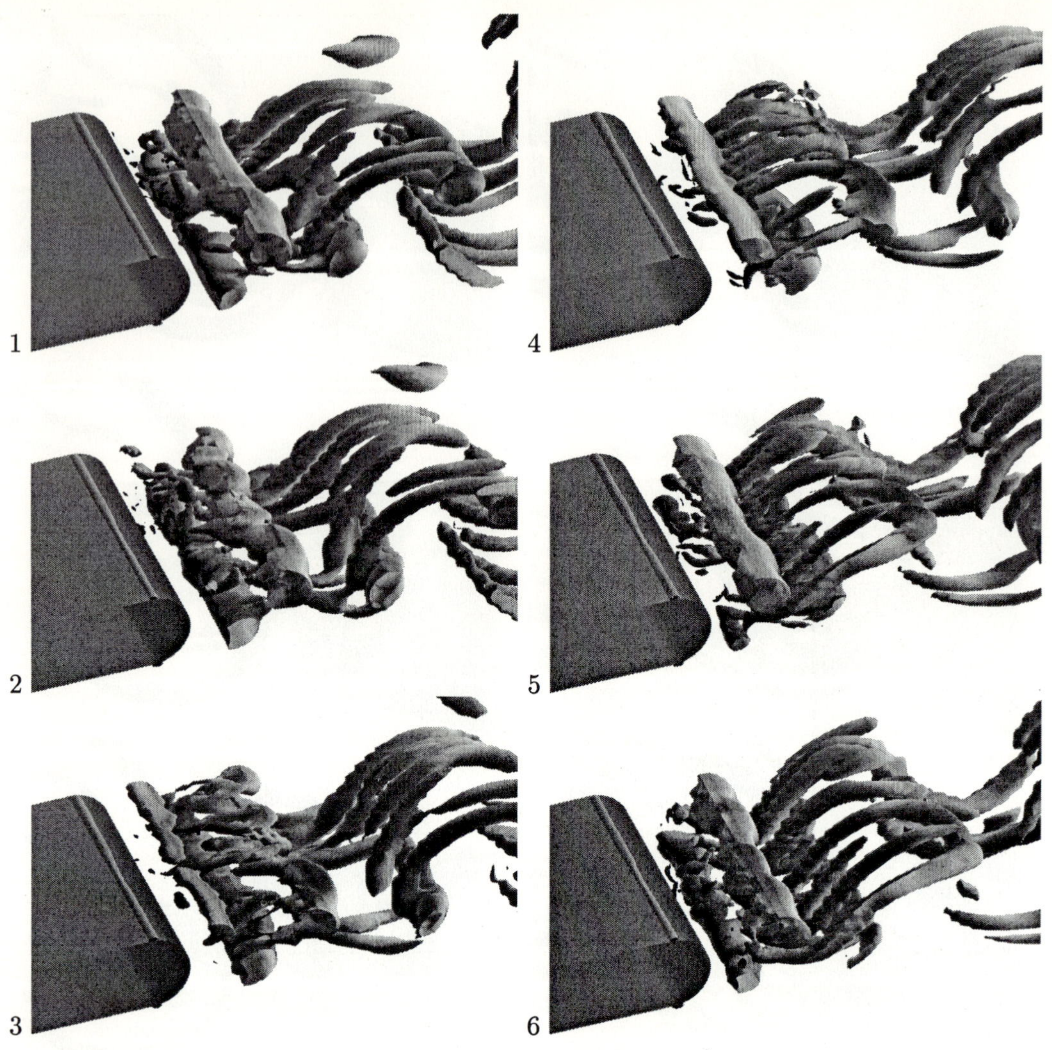

Figure 14 Surface of constant λ_2 color coded with the local Mach number for the wake with circular trailing edge (fine grid). Shown are six different time levels in one vortex shedding cycle ($\Delta t = 0.740$).

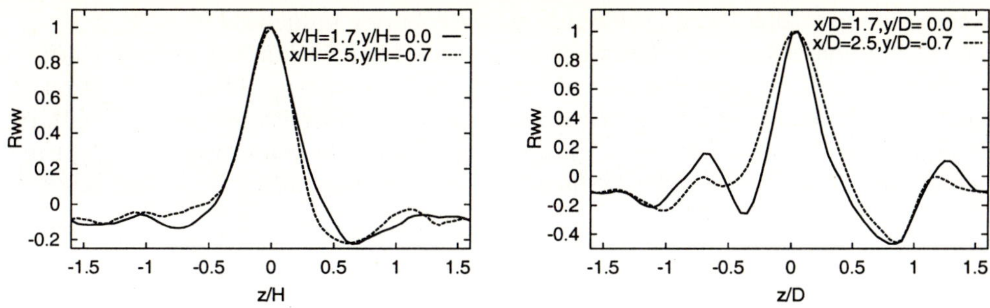

Figure 15   Time averaged profiles of the two point correlation R_{ww} in spanwise direction z for two different locations in the x,y-plane for the rectangular (left) and circular (right) trailing edge.

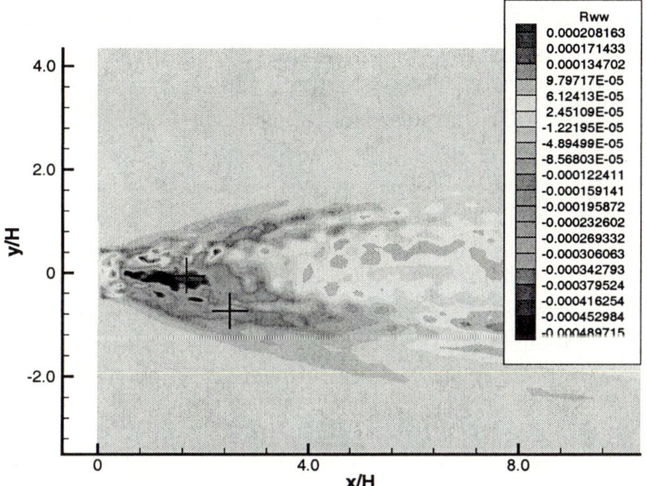

Figure 16   Distribution of the two point correlation R_{ww} in spanwise direction z/D for the flat plate with rectangular trailing edge. Shown is the x,y-plane at z/D=0.7.

4   Conclusion

Solutions of turbulent boundary layers and wakes have been presented. For the simulation of transitional boundary layers improved inflow conditions were formulated using Parabolized Stability Equations. In two kinds of transition the dynamics of the near wall coherent structures were analyzed. One interesting result is that the spanwise streak spacing of the fully developed flow seems to appear already at the very beginning of the transition. First simulations of turbulent boundary layers that are manipulated with spanwise grooves show that shortly after the groove the near wall turbulence statistics becomes more isotropic, while the outer region is barely influenced.

The simulation of the wake flows has been carried out for laminar inflow velocity profiles, different trailing edge shapes and various grid resolutions. The visualization of the vortex dynamics shows characteristic streamwise vortices which connect the main rollers with an axis in spanwise direction. Detailed turbulence statistics have been determined, which are available for the comparison with a turbulent inflowing boundary layer.

References

[1] P. Beaudan and P. Moin. Numerical experiments on the flow past a circular cylinder at sub-critical Reynolds numbers. Technical Report TF-62, Center Turb. Res., 1994.

[2] F. Bertolotti and T. Herbert. Analysis of the linear stability of compressible boundary layer using pse. *Theoret. and Comp. Fluids Dynamics*, 1991.

[3] F. Ducros, P. Comte, and M. Lesieur. Large-eddy simulation of transition to turbulence in a boundary layer spatially developing over a flat plate. *J. Fluid Mech.*, 11:1–36, 1995.

[4] R. Elavarasan, B. R. Pearson, and R. A. Antonia. The response of a turbulent boundary layer to a square groove. *J. Fluid Eng.*, 119:466–, 1997.

[5] J. Jeong and F. Hussain. On the identification of a vortex. *J. Fluid Mech.*, 285:69–94, 1995.

[6] Y. Kachanov and V. Levchenko. The resonant interaction of disturbances at laminar-turbulent transition in a boundary layer. *J. Fluid Mech.*, 1984.

[7] J. Kim, P. Moin, and R. Moser. Turbulence statistics in fully developed channel flow at low Reynolds number. *J. Fluid Mech.*, 177:133–166, 1987.

[8] J. Kim, P. Moin, and R. Moser. Turbulent statistics in fully developed channel flow at low reynolds number. *J. Fluid Mech.*, 177:133–166, 1987.

[9] H. Le and P. Moin. Direct simulation of turbulent flow over a backward-facing step. Technical Report 58, NASA, 1994.

[10] M. Lesieur and O. Métais. New trends in large eddy simulations of turbulence. *Ann. Rev. Fluid Mech.*, 28:45–82, 1996.

[11] J. L. Lumley and G. R. Newman. The return to isotropy of homogeneous turbulence. *J. Fluid Mech.*, 82:161–178, 1977.

[12] M. Meinke, C. Schulz, and T. Rister. LES of Spatially Developing Jets. In R. Friedrich and P. Bontoux, editors, *Computation and Visualization of Three-Dimensional Vortical and Turbulent Flows. Proceedings of the Fifth CNRS/DFG Workshop on Numerical Flow Simulation*, volume NNFM 64, pages 116–131. Vieweg Verlag, 1998.

[13] U. Piomelli. High Reynolds number calculations using the dynamic subgrid-scale stress model. *Phys. Fluids*, A 5(6):1484–1490, June 1993.

[14] T. J. Poinsot and S. K. Lele. Boundary conditions for direct simulations of compressible viscous flows. *J. Comput. Phys.*, 101:104–129, 1992.

[15] T. Rister. *Grobstruktursimulation schwach kompressibler turbulenter Freistrahlen - ein Vergleich zweier Lösungsansätze*. Diss., Aerodyn. Inst. RWTH Aachen, 1998.

[16] C. Schulz. *Grobstruktursimulation turbulenter Freistrahlen*. Diss., Aerodyn. Inst. RWTH Aachen, 1997.

[17] P. R. Spalart. Direct simulation of a turbulent boundary layer up to $R_\theta = 1410$. *J. Fluid Mech.*, 187:61–98, 1988.

[18] K. W. Thompson. Time-dependent boundary conditions for hyperbolic systems, ii. *J. Comput. Phys.*, 89:439–461, 1990.

Adaptive Wavelet Methods for the Navier-Stokes Equations

K. Schneider[1], M. Farge[2], F. Koster[3], M. Griebel[3]

[1] ICT Universität Karlsruhe (TH), Kaiserstraße 12, D-76128 Karlsruhe, Germany

[2] LMD-CNRS Ecole Normale Supérieure, 24 rue Lhomond, F-75231 Paris Cedex 05, France [3] IAM Universität Bonn, Wegelerstr. 6, D-53115 Bonn, Germany

Summary

In this paper we introduce and compare two adaptive wavelet-based Navier–Stokes solvers. The first one uses a Petrov-Galerkin approach for the vorticity-velocity formulation of the Navier–Stokes equations, while the second one is a collocation method for the pressure-velocity formulation. Both codes are applied to the 2D mixing layer test problem and their results are compared to Fourier pseudo–spectral solutions.

Key words. adaptivity, wavelets, vaguelettes, interpolets, nonlinear thresholding

1   Introduction

Turbulent flows are characterized by their large number of active scales of motion increasing with the Reynolds number. Hence, for the numerical simulation of fully developed turbulence the complexity has to be considerably reduced and turbulence models are unavoidable. In current approaches the fine scales of the flow are replaced by a subgrid scale model, e.g. in Large Eddy Simulation (LES), using a linear cut–off filter which does not depend on the actual flow realization.

Wavelets are functions with simultaneous localization in physical and in Fourier space which correspond to filters having a constant relative bandwidth. They allow adaptive filtering of signals and are well suited for investigating unsteady, inhomogeneous or intermittent phenomena like those encountered in turbulence. In the past we have shown that wavelets are an efficient basis to represent turbulent vorticity fields, see *e.g.* [14, 32, 15]. By means of nonlinear filtering of the wavelet coefficients of vorticity, we can separate the dynamical active part of the flow (i.e. the coherent vortices) from the incoherent background flow. This filtering technique is much more efficient than linear low pass filtering employed in LES, because it retains much more enstrophy and energy for the same number of modes.

From a numerical point of view wavelets constitute optimal bases to represent functions with inhomogeneous regularity, such as intermittent turbulent flow fields. The existence of fast pyramidal algorithms (with linear complexity), to transform the computed fields between lacunary wavelet coefficients and structured adaptive grids, allows to design efficient methods for solving nonlinear PDEs [16, 17, 33, 32, 35, 5, 19, 21, 25].

We present two different wavelet schemes to solve the Navier–Stokes equations and compare their performances for computing a two–dimensional temporally growing mixing layer which is a good test–case because it is a typical configuration encountered in many turbulent flows. The results are compared with those obtained with a classical Fourier pseudo–spectral method.

The remainder of this paper is organized as follows: In the next section we give a brief introduction to wavelets and we present the general adaptive time stepping method which is used by both wavelet solvers. In section three and four further details of the two wavelet schemes are given. In particular the differences between them are pointed out. In section 5 we describe the setup of the mixing layer experiment and we present the results of both schemes compared to Fourier spectral simulations.

2　Adaptive Wavelet Methods for Time-Dependent PDEs

In the following, we give a brief introduction to wavelets and the notation used by the Bonn group. The Paris-Karlsruhe group uses the classical notation [8, 14, 32]. We first deal with one–dimensional wavelets and then consider two variants for the generalization to the multivariate case. Finally, we describe how wavelets can be used in the adaptive spatial discretization of time stepping methods for e.g convection–diffusion problems or the Navier Stokes equations.

2.1　Univariate Wavelets

In this subsection, we explain the main features of our approach in a simple notation using wavelets defined on $\mathbb{R}$. We give brief comments on wavelets on the interval later on.

We start with a sequence of spaces $\{V^l\}_{l \geq l_0}$ which are spanned by dilates/translates of a single function ϕ. A simple example for ϕ is the hat function $\phi(x) = \max(0, 1 - |x|)$. We have

$$V^l = \text{span } \Phi^l, \quad \Phi^l := \{\phi^{(l,s)}\}_{s \in \mathbb{Z}}, \quad \phi^{(l,s)}(x) := \phi(2^l x - s) .$$

Here, l denotes the level of refinement and l_0 is a certain given coarsest level. The function ϕ should have the following properties:

(P1)　ϕ has compact support or decays sufficiently fast such that it can be truncated. Hence, $|\text{supp}\phi^{(l,s)}| \sim 2^{-l}$.

(P2)　There is a $P \in \mathbb{N}$ which depends on the particular ϕ such that polynomials of degree less than P can be written as linear combinations of $\{\phi^{(l,s)}\}_{s \in \mathbb{Z}}$ for all $l \geq l_0$. E.g for the hat function we have $P = 2$.

(P3)　ϕ is the solution of a so-called scaling equation

$$\phi(x) = \sum_{s \in \mathbb{Z}} h_s \phi(2x - s) \tag{2.1}$$

with explicitly known coefficients $\{h_s\}$. Therefore, the functions ϕ and $\phi^{(l,s)}$ are called scaling functions. An analytic description of ϕ is often not available, but it is also not needed. All we have to know about ϕ from the practical point of view are its scaling coefficients h_s. An immediate consequence of (2.1) is: $V^l \subset V^{l+1}$.

Now, associated functions, so–called wavelets, can be defined as follows: Wavelets are the basis functions of the complementary spaces W_l:

$$V^l = V^{l-1} \oplus W_l , \quad W_l = \text{span}\Psi_l , \quad \Psi_l = \{\psi_{(l,t)}\}_{t \in \mathbb{Z}} .$$

In our setting the wavelets are also dilates/translates $\psi_{(l,t)} = \psi(2^l x - t)$ of a single function ψ

$$\psi(x) = \sum_{s \in \mathbb{Z}} g_s \phi(2x - s) . \tag{2.2}$$

Again, the coefficients $\{g_s\}$ are all what is needed to use ψ and $\psi_{(l,t)}$ in practice. Furthermore, we can assume that ψ also has compact support or decays quite fast. Repeated decomposition of V^l yields a so-called multi-resolution analysis (MRA) of V^l with the wavelet basis $\{\Psi_k\}_{l_0 \leq k \leq l}$

$$V^l = \oplus_{k=l_0}^{l} W_k , \quad \text{where} \quad W_{l_0} := V^{l_0} \quad \text{and} \quad \Psi_{l_0} := \Phi^{l_0}.$$

Consider successive approximations $u^{l-1} \in V^{l-1}$ and $u^l \in V^l$ of a function u and their difference

$$u_l = u^l - u^{l-1} = \sum_{t \in \mathbb{Z}} u_{(l,t)} \psi_{(l,t)} .$$

Then, a Taylor argument using (P1) and (P2) shows that the coefficients $u_{(l,t)}$ are significant (i.e. 'large') only if their associated wavelets $\psi_{(l,t)}$ lie in the vicinity of a singularity or a quasi-singularity of u. Therefore, for any u which has only a few (quasi-) singularities, the number of active degrees of freedom required to achieve a desired accuracy is substantially less in a wavelet basis compared to a scaling function basis Φ^l.

The most important analytical property of the wavelet basis is the following: Sobolev norms $\|.\|_s$ of e.g. u can be characterized in terms of the wavelet coefficients $u_{(l,t)}$: There are $\gamma^* < \gamma$ such that for $s \in]\gamma^*, \gamma[$ there are $0 < c(s) < C(s)$ with

$$c(s) \sum_{l,t} |u_{(l,t)}|^2 \|\psi_{(l,s)}\|_0^2 4^{ls} \leq \| \sum_{l,t} u_{(l,t)} \psi_{(l,s)} \|_s^2 \leq C(s) \sum_{l,t} |u_{(l,t)}|^2 \|\psi_{(l,s)}\|_0^2 4^{ls}. \tag{2.3}$$

Here $\gamma := \sup\{s \mid \phi \in H^s\}$ and γ^* mainly depends on the particular basis Ψ_l of W_l. Such norm equivalences are at the heart of the wavelet theory and are also the foundation for efficient preconditioning techniques for linear systems [7, 30, 22, 26] or a reliable error control [6].

Furthermore, the so–called dual scaling functions $\tilde{\phi}^{(l,s)}$ and the dual wavelets $\tilde{\psi}_{(l,t)}$ can be considered. They have to fulfill the following biorthogonality relations

$$\langle \phi^{(l,s)} , \tilde{\phi}^{(l,t)} \rangle = \delta(s - t) , \quad \langle \psi^{(k,s)} , \tilde{\psi}^{(l,t)} \rangle = \delta(k - l)\delta(s - t) .$$

Here, $\langle \, , \, \rangle$ is usually the L_2 scalar product; if the dual functions are linear combinations of Dirac functionals, then $\langle \, , \, \rangle$ denotes a dual pairing. From the theoretical point of view the dual functions are required for both, the understanding of (2.3) and for the algorithms later on: They can be used as test functions in a Petrov-Galerkin discretization scheme. There, the biorthogonality relations lead to strong simplifications which make programming easy. Again, the dual functions are in general only known by their mask coefficients in scaling relations similar to (2.1) or (2.2).

Note that everything described in this subsection can be generalized to wavelets on an interval $[a,b]$ as well. There, the wavelets and scaling functions, respectively, are defined by appropriately chosen linear combinations of $\{\phi(2^l x - s)\}_{s \in \mathbb{Z}} \chi_{[a,b]}(x)\}$, see [24] and the references therein. In this way it is also possible to incorporate e.g. homogeneous Dirichlet or Neumann boundary conditions required for the solution of PDEs. In case of wavelets on the interval, the spaces V_l and W_l have a finite dimension. More precisely $\dim V^l \approx 2^l$ and $\dim W_l = 2^{l-1}$ $(l > l_0)$.

2.2 Multivariate Wavelets

The simplest ways to obtain multivariate wavelets are to use the tensor product of 1D multiresolution analyses [8, 5] or to construct a multivariate multiresolution analysis (MRA) [8, 17].

(MRA-d) Here, the multivariate wavelets are defined by tensor products between wavelets in each direction.

$$\psi_{(\mathbf{l},\mathbf{t})}(\mathbf{x}) := \psi_{(l_1,t_1)}(x_1) \cdot \ldots \cdot \psi_{(l_d,t_d)}(x_d) \;, \qquad \mathbf{l} := (l_1,..,l_d) \;, \quad \mathbf{x},\mathbf{t} \text{ analogous.}$$

Since this construction does not yield a multiresolution analysis, for certain choices of $\mathbf{l}$, e.g. $\mathbf{l} = (1,..,1,20)$, the support of $\psi_{(\mathbf{l},\mathbf{t})}$ is anisotropic.

This anisotropy is avoided with the multiresolution analysis. The wavelets are defined for

(MRA) $\mathbf{e} \in \{0,1\} \times .. \times \{0,1\} \backslash \mathbf{0} \subset \mathbb{N}^d$, $l \geq l_0$ and $\mathbf{t} \in \mathbb{Z}^d$ by

$$\psi_{(\mathbf{e},l,\mathbf{t})} := \psi_{(e_1,l,t_1)}(x_1) \cdot \ldots \cdot \psi_{(e_d,l,t_d)}(x_d) \;,$$

where $\psi_{(0,l,t)}(x) := \phi^{(l-1,t)}(x)$ and $\psi_{(1,l,t)}(x) := \psi_{(l,t)}(x)$. The size of the support of $\psi_{(\mathbf{e},l,\mathbf{t})}$ is $\sim 2^{-l}$ in each direction, i.e. the basis functions are quasi isotropic with different directions. The idea behind the above definition is illustrated in the 2D case. The scaling functions are simply the tensor products of the univariate scaling functions. Then,

$$
\begin{aligned}
V^l \otimes V^l &= (V^{l-1} \oplus W_l) \otimes (V^{l-1} \oplus W_l) \\
&= V^{l-1} \otimes V^{l-1} \;\oplus\; (W_l \otimes V^{l-1}) \;\oplus\; (V^{l-1} \otimes W_l) \;\oplus\; (W_l \otimes W_l) \;.
\end{aligned}
$$

Obviously, $V^l \otimes V^l$ plays the role of V^l in the one–dimensional case and $V^{l-1} \otimes V^{l-1}$ that of V^{l-1}. Hence, $(W_l \otimes V^{l-1}) \oplus (V^{l-1} \otimes W_l) \oplus (W_l \otimes W_l)$ plays the role of W_l with three diferent directions: horizontal, vertical and diagonal.

In case of the (MRA)-approach we have a linear order of the approximation spaces $\left(V^{l-1} \otimes .. \otimes V^{l-1} \subset V^l \otimes .. \otimes V^l \right)$ while for the (MRA-d)-approach the underlying approximation spaces $V^{\mathbf{l}} = V^{l_1} \otimes .. \otimes V^{l_d}$ form a d–dimensional array [30]. This explains our naming convention (MRA$\{-d\}$), although stricto sensus it is not an MRA. The constructions for the dual scaling functions/wavelets are the same as for the (primal) scaling functions/wavelets. To unify the notation we denote by ψ_λ and u_λ the wavelets and the wavelet coefficients for both, the (MRA-d)- and the (MRA)-approach. In the first case $\lambda = (\mathbf{l},\mathbf{t})$ and in the second case $\lambda = (\mathbf{e},l,\mathbf{t})$.

A problem closely related to the efficiency of adaptive methods is the best N-term approximation for a given function v. Here, one is interested in the order $\alpha(v,s)$ such that for all $N \in \mathbb{N}$ and a suitable constant C independent of N there holds

$$\inf \left\{ \|v - \sum_{\lambda \in \mathcal{T}} v_\lambda \cdot \psi_\lambda \|_s \;\Big|\; \mathcal{T} \text{ and } \{v_\lambda\}_{\lambda \in \mathcal{T}} \text{ arbitrary, but } \#\mathcal{T} = N \right\} \;\leq\; C N^{-\alpha}. \tag{2.4}$$

Since we can represent each $\phi^{(l,s)}$ by $O(l)$ wavelets $\psi_{(k,t)}$ the order α_{MRA-d} for the (MRA-d)-approach is at least arbitrarily close to the order α_{MRA} for the (MRA)-approach. Of course the constant C_{MRA-d} may be substantially different from C_{MRA}. However, the point is that under some relatively mild assumptions on the smoothness of u, the

(MRA-d)-approach yields a significantly larger order of approximation than the (MRA)-approach: $\alpha_{MRA-d} > \alpha_{MRA}$. Further information on this topic can be found in e.g. [30, 11, 10, 20, 37].

Finally, some comments on the geometry of the considered domain are in order. Due to the intrinsic tensor product construction of both multivariate approaches, rectangular domains are the only geometries for which the (MRA-d)- and the (MRA)-wavelets can be used directly. Thus, complicated geometries pose a serious problem.

With respect to the implementational effort, the simplest solution is to embed a given non-rectangular domain $\Omega \subset \mathbb{R}^d$ into an enclosing rectangular domain and to solve here a modified PDE, where the boundary conditions on $\partial\Omega$ are included in the right hand side. See [13] and the references therein for the special case of the Navier–Stokes equations. However, a drawback of this approach is the additional adaptive refinement needed in the vicinity of $\partial\Omega$ in order to resolve the boundary layer accurately.

Another solution to the geometry problem is the use of parametric mappings, i.e. the computational domain $[0,1]^d$ is mapped to the real domain Ω. Topological complicated situations are handled by structured blocks. This approach is quite common in engineering applications. In context of the Navier–Stokes equations this approach has a further advantage: The underlying curvilinear grid is aligned to the surface of Ω. Now, since the flow is approximately also aligned to the surface of Ω, the flow is approximately aligned to the underlying grid. This is exactly the situation where we can expect the (MRA-d)-approach to work very well as long as the boundary layer is not turbulent.

2.3 Time Dependent Problems and Adaptive Basis Selection

Consider the linear convection diffusion problem

$$
\begin{aligned}
\partial_t a + \nabla \cdot (\mathbf{v} a) &= \nu \Delta a \, , \quad (\mathbf{x},t) \in \Omega \otimes [0,T] \, , \\
a(\mathbf{x},0) &= a^0(\mathbf{x}) \, ,
\end{aligned}
\tag{2.5}
$$

which is a model for the evolution part of the Navier Stokes equations. A common way to discretize (2.5) is to use a FD/E/V or wavelet method for the spatial discretization and a time stepping (i.e. FD) method for the time discretization:

$$
\frac{a^{n+1} - a^n}{\Delta t} + C(a^n, a^{n-1}, ..) = \nu \Delta a^{n+1}.
\tag{2.6}
$$

Here, $C(.)$ is an explicit approximation to the convective term e.g. of Adams-Bashforth or Runge-Kutta type. Note that in case of strongly time-dependent problems like turbulence (our final goal) the additional effort for the implicit treatment of the convective term does not pay off, since the time step size Δt has to be quite small anyway.

An important question for our adaptive wavelet methods is how to select the adaptive basis. A simple, but efficient strategy is the following, as proposed in [28, 17]:
For the initial condition a^0 we use an appropriately chosen initial adaptive basis $\{\psi_\lambda\}_{\lambda \in \mathcal{T}^0}$. Usually $\mathcal{T}^0$ is the set of λ with l or $|\mathbf{l}|_\infty$ less than a given maximum level L. Then, the time evolution of a and $\mathcal{T}$ is calculated as follows:

$$\boxed{\begin{aligned}
&\textbf{(Adaptive Time Stepping Method)}\\
&\text{given are the initial adaptive index set } \mathcal{T}^0 \text{ and } a^0_{\mathcal{T}^0} = \sum_{\lambda \in \mathcal{T}^0} a^0_\lambda \cdot\\
&\psi_\lambda\\
&\texttt{for } n = 0 \texttt{ to } T/\Delta t \texttt{ do}\\
&\quad \texttt{// Time Step:}\\
&\quad \text{Calculate the convective term } C^n_{\mathcal{T}^n}\\
&\quad \text{Solve} \quad (I - \Delta t \nu \Delta) a^{n+1}_{\mathcal{T}^n} = a^n_{\mathcal{T}^n} - \Delta t C^n_{\mathcal{T}^n}\\
&\quad \texttt{// Adaptivity: new adaptive basis}\\
&\quad \text{Determine } \overline{\mathcal{T}}^{n+1} := \{\, \lambda \in \mathcal{T}^n \mid |a^{n+1}_\lambda| > \epsilon \}\\
&\quad \text{Determine } \mathcal{T}^{n+1} := \{\mu \mid \exists\, \lambda \in \overline{\mathcal{T}}^{n+1} \text{ such that } dist(\mu,\lambda) \leq\\
&R\}\\[4pt]
&\quad \texttt{// Adaptivity: prolongation to new basis}\\
&\quad \text{Set} \quad a^{n+1}_{\mathcal{T}^{n+1}} := \sum_{\lambda \in \mathcal{T}^{n+1} \bigcap \mathcal{T}^n} a^{n+1}_\lambda \cdot \psi_\lambda\\
&\texttt{end}
\end{aligned}}$$

The first part of this algorithm is the time stepping method (2.6). Here, we used an Euler scheme for the sake of simplicity. Note however, that other higher order time discretization methods can be plugged in straightforwardly.

In the second part, first we determine the set $\overline{\mathcal{T}}^{n+1}$ of active basis functions. The analog of this step in adaptive FE/V methods would be the selection of elements which should be refined. The refinement criterion we use is whether or not the magnitude of a particular wavelet coefficient is larger than a given threshold, i.e. $|a_\lambda| > \epsilon$. [1]

The use of the magnitude of the wavelet coefficients can also be interpreted as a traditional method of local error estimation. If one takes (P1) and (P2) into account it is quite easy to see that the magnitude of a wavelet coefficient a_λ is a measure of the magnitude of a local finite difference approximation of some higher order derivative of a, i.e. we work with a gradient type error indicator.

Now consider the second step of the adaptivity strategy, i.e. the determination of $\mathcal{T}^{n+1}$ by insertion of basis functions which are in space and scale close to the active basis functions of $\overline{\mathcal{T}}^{n+1}$. We assume that the current solution a^n is quite accurate and we want to preserve this property also for the next time slice a^{n+1}. For the sake of simplicity, let us assume that we calculate a^{n+1} by the following explicit Petrov–Galerkin scheme:

$$\langle a^{n+1} , \tilde{\psi}_\lambda \rangle = \langle (I - \Delta t \nabla \cdot (\mathbf{v}.) + \Delta t \nu \Delta) a^n , \tilde{\psi}_\lambda \rangle \,.$$

The best index set $\mathcal{T}^{n+1}$ with a given number N of degrees of freedom is that with the N largest entries $\langle a^{n+1} , \tilde{\psi}_\lambda \rangle$. Now, the a priori known locality properties of the differential operator $(I - \Delta t \nabla \cdot (\mathbf{v}.) + \Delta t \nu \Delta)$ show that $\langle a^{n+1} , \tilde{\psi}_\lambda \rangle$ can only be significant if λ is near to one of the significant $\mu \in \mathcal{T}^n$. In this sense, the above method is closely related to the works of Cohen, Dahmen and DeVore [6] or the work of Becker and Rannacher [1, 2].

The last issue of the section is the conservation of mass or energy. We may assume that the wavelets ψ_λ have a vanishing mean value, except for the scaling functions on the coarsest level which are always kept in the adaptive basis. Then, the remeshing step does not change the mass budget $\int_\Omega a d\mathbf{x}$. Unfortunately, things are more complicated for the energy $\int a^2 d\mathbf{x}$. For L_2-orthogonal wavelets (which are used in method I described in

---

[1] Another possible strategy would be to choose the N largest wavelet coefficients, where N is a fixed given number. In this case the work count would be almost constant for all time steps.

the next section) there is no problem, but for all other types of wavelet-approaches we usually have only an estimate for the change of energy (see eq. (2.3)) introduced by the remeshing step. Of course this change will be quite small, but we can not say whether the energy defect is positive or negative.

However, in our numerical experiments [26] with non-L_2-orthogonal Interpolet wavelets (see section 4) it turned out that the dominant contribution of the energy defect does not come from remeshing but from the discretization of the convective term which is not energy conservative in the adaptive case, since telescope or partial summation arguments fail. Nevertheless, we observed a very good prediction of the rate of energy dissipation. Figure 1 shows the energy and the number of degrees of freedom (DOF) for the Molenkamp test of a rotating hill (see e.g. [27] pp. 248) with the quite small diffusion of $\nu = 10^{-5}$, i.e $Re = 10^5$, using (MRA-d)-Interpolet wavelets.

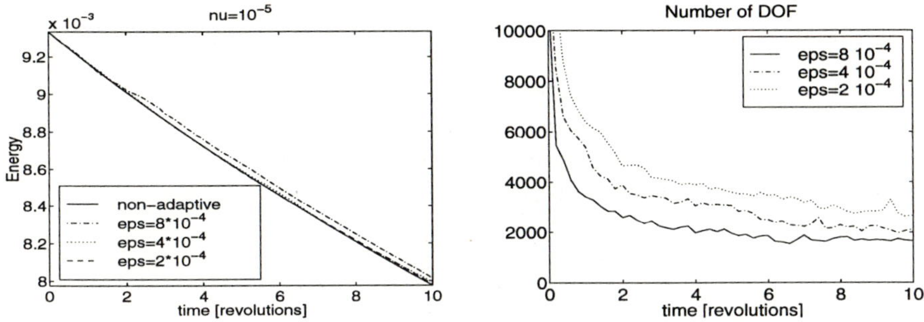

Figure 1 DOF and energy for adaptive Molenkamp test. The maximum #DOF for ($\epsilon = 2_{-4}$) is ≈ 17000.

3 Method I

In the following we briefly describe the adaptive wavelet method used by the Karlsruhe–Paris group. For further details we refer to [33, 32, 17] For the numerical simulation of two–dimensional turbulence we consider the Navier–Stokes equations written in velocity–vorticity formulation

$$\partial_t \omega + \mathbf{v} \cdot \nabla \omega = \nu \nabla^2 \omega \,, \qquad \nabla \cdot \mathbf{v} = 0 \,, \tag{3.7}$$

with the velocity field $\mathbf{v} = (u,v)$, the vorticity $\omega = \nabla \times \mathbf{v}$ and the kinematic viscosity ν. We assume periodic boundary conditions in both directions.

For the time discretization we use finite differences with a semi–implicit scheme, i.e. backward–differences for the viscous term and Adams–Bashforth extrapolation for the nonlinear term, both of second order. We obtain

$$(\gamma I - \nu \nabla^2)\omega^{n+1} = \frac{4}{3}\gamma\omega^n - \frac{1}{3}\gamma\omega^{n-1} - \mathbf{v}^\star \cdot \nabla \omega^\star, \text{ where } \omega^\star = 2\omega^n - \omega^{n-1} \tag{3.8}$$

with time step Δt, $\gamma = 3/(2\Delta t)$ and I representing the identity.

For the spatial discretization we use a Petrov–Galerkin scheme. Therefore the vorticity is expanded into a set of trial functions and the minimization of the weighted residual of (3.8) requires that the projection onto a space of test functions vanishes.

As space of trial functions we employ a multiresolution analysis in two dimensions, and expand ω^n at time step n into an orthonormal wavelet series, from the largest scale $l_{max} = 2^0$ to the smallest scale $l_{min} = 2^{-J}$:

$$\omega^n(x,y) \;=\; \sum_{j=0}^{J-1} \sum_{k_x,k_y=0}^{2^j-1} \sum_{d=1}^{3} \tilde{\omega}_\lambda^n \psi_\lambda(x,y) \;, \quad \text{where } \lambda = (j,k_x,k_y,d) \;. \tag{3.9}$$

One consequence of the orthogonality of the trial functions is that we know exactly what amount of enstrophy is lost in the remeshing step.

The test functions θ_μ used are defined as solutions of the linear part of equation (3.8), i.e.

$$(\gamma I - \nu \nabla^2)\theta_\mu \;=\; \psi_\mu \;. \tag{3.10}$$

Hence,

$$\langle (\gamma I - \nu \nabla^2)\psi_\lambda \,,\, \theta_\mu \rangle = \langle \psi_\lambda \,,\, (\gamma I - \nu \nabla^2)\theta_\mu \rangle = \delta(\lambda - \mu) \;.$$

This avoids the assembly of the stiffness matrix and the solution of a linear equation at each time step. The functions θ_μ, called vaguelettes, are explicitly calculated in Fourier space and have localization properties similar to wavelets [17].

The solution of (3.8) therewith reduces to a simple change of basis,

$$\omega_\lambda^{n+1} \;=\; \langle \omega^{n+1} , \psi_\lambda \rangle \;=\; \langle \frac{4}{3}\gamma\omega^n - \frac{1}{3}\gamma\omega^{n-1} - \mathbf{v}^\star \cdot \nabla\omega^\star , \theta_\lambda \rangle \;. \tag{3.11}$$

An adaptive discretization is obtained by applying at each time step a nonlinear wavelet thresholding technique which retains only wavelet coefficients $\tilde{\omega}_\lambda^n$ with absolute value above a given threshold $\epsilon = \epsilon_0 \sqrt{Z}$, where $Z = \frac{1}{2} \int \omega^2 dx$. For the next time step the index coefficient set (which addresses each coefficient in wavelet space) is determined by adding neighbours to the retained wavelet coefficients, consequently only those coefficients ω_λ in (3.11), belonging to this extrapolated index set are computed using the adaptive vaguelette decomposition [3]. The nonlinear term $\mathbf{v}^\star \cdot \nabla\omega^\star$ is evaluated by partial collocation on a locally refined grid. The vorticity $\omega^\star$ is reconstructed in physical space on an adaptive grid from its wavelet coefficients $\{\tilde{\omega}_\lambda^\star\}$ using the adaptive wavelet reconstruction algorithm [3]. Using the adaptive vaguelette decomposition with $\theta = (\nabla^2)^{-1}\psi$, we solve $\nabla^2\Psi^\star = \omega^\star$ ($\Psi^\star$ being the stream function), get $\{\Psi_\lambda^\star\}$ and finally reconstruct $\Psi^\star$ on the refined grid. By means of centered finite differences of 4th order we finally compute $\nabla\omega^\star$ and $\mathbf{v}^\star = (-\partial_y\Psi^\star, \partial_x\Psi^\star)$ on the adaptive grid and we evaluate the nonlinear term pointwise. Subsequently (3.11) can be solved using the adaptive vaguelette decomposition. A complete description of this algorithm is given in [32, 33, 35]. Finally let us mention that the total complexity of the algorithm is of order $O(N_{ad})$, where N_{ad} denotes the number of wavelet coefficients retained in the adapted basis. As the present implementation has not yet been optimized, the computing time is as effective as a classical spectral method for the resolution used here ($N = 256^2$).

4   Method II

The method of the Bonn group uses Deslaurier-Dubuc Interpolets [9, 25] together with the (MRA-d) construction and their dual counterparts as trial and test functions. The advantage of Interpolets is that not only the scaling functions, but also the wavelets are interpolating. This allows for the efficient evaluation of nonlinear terms. Besides of its potential advantages with respect to an efficient approximation (2.4), the (MRA-d) technique leads to a very simple structure of the algorithms in the multivariate case: All operations we require (adaptive wavelet transform, its inverse, the adaptive evaluation of differential operators) boil down to the corresponding adaptive operations for the 1D case. As an example we consider the evaluation of the ∂_{xx} part of the Laplacian:

$$\langle \partial_{xx} \sum_{\mathbf{k,s}} a_{(\mathbf{k,s})}\psi_{(\mathbf{k,s})} , \tilde{\psi}_{(\mathbf{l,t})}\rangle =$$

$$= \sum_{\mathbf{k,s}} a_{(\mathbf{k,s})}\langle \partial_{xx}\psi_{(k_1,s_1)} , \tilde{\psi}_{(l_1,t_1)}\rangle \cdot \prod_{i=2}^{d} \delta(k_i - l_i)\delta(s_i - t_i)$$

$$= \langle \partial_{xx} \sum_{k,s} a_{((k,l_2,..,l_d),(s,t_2,..,t_d))}\psi_{(k,s)} , \tilde{\psi}_{(l_1,t_1)}\rangle .$$

A description of the complete scheme and of all adaptive algorithms is given in [25]. These algorithms have a work count which is proportional to $\#\mathcal{T}$. An analysis of the consistency order of the Interpolet Petrov-Galerkin scheme is contained in [23].

Now, we consider the Navier-Stokes equations in primitive variable formulation:

$$\partial_t \mathbf{v} + \nabla(\mathbf{v} \otimes \mathbf{v}) = \mathbf{f} - \nabla p + \nu\Delta\mathbf{v}$$
$$\nabla \cdot \mathbf{v} = 0$$
$$\mathbf{v}(\mathbf{x},0) = \mathbf{v}^0(\mathbf{x}) ,$$

with appropriate boundary conditions. The time discretization is achieved by means of a Chorin-type projection method:

$$\text{(Transport step)} \quad \frac{\bar{\mathbf{v}}^{n+1} - \mathbf{v}^n}{\Delta t} + \mathbf{C}(\mathbf{v}^n,\mathbf{v}^{n-1},..) = \mathbf{f}^{n+1} - \nabla p^n + \nu\Delta\bar{\mathbf{v}}^{n+1} \qquad (4.12)$$

$$\text{(Projection step)} \quad \frac{\mathbf{v}^{n+1} - \bar{\mathbf{v}}^n}{\Delta t} = -\nabla(p^{n+1} - p^n) \qquad (4.13)$$
$$\nabla \cdot \mathbf{v}^{n+1} = 0$$

The projection step involves a saddle problem which can be treated by solving

$$\nabla \cdot \nabla(p^{n+1} - p^n) = \frac{1}{\Delta t}\nabla \cdot \bar{\mathbf{v}}^{n+1}$$

with appropriate boundary conditions for the pressure. If periodic boundary conditions are prescribed for the velocity, then $\bar{\mathbf{v}}$ and p have periodic boundary conditions. If Dirichlet boundary conditions $\mathbf{v}_D$ are prescribed for $\mathbf{v}$, then these boundary conditions are also used for $\bar{\mathbf{v}}$ and homogeneous Neumann boundary conditions are used for p to ensure that $\mathbf{v} \cdot \mathbf{n} = \mathbf{v}_D \cdot \mathbf{n}$ on $\partial\Omega$ at least. There is a controversy about Neumann boundary conditions for p in the literature (see e.g. [18] sec. 3.8.2) but at least for boundary layer flows they are the physically correct ones. For all other cases we can use the modified projection method of [31] which overcomes the problems of the Neumann boundary conditions. All boundary conditions are implemented by employing Interpolet–wavelets as trial functions which fulfill the boundary conditions.

The convective term is discretized using a 3rd order Adams-Bashforth scheme.

The operator $\nabla \cdot \nabla$ in the pressure Poisson equation requires special attention. Instead of the usual discrete Laplacian this operator is the nested application of the discrete gradient operator for the pressure and the discrete divergence operator for the velocity. Only this structure ensures that the velocity $\mathbf{v}^{n+1}$ is discretely divergence-free after the projection step. A drawback of this operator is that it is not spectrally equivalent to the continuous Laplacian. Spurious pressure oscillations and the breakdown of the efficiency of the usual diagonal wavelet preconditioner are the result. In particular it can be shown by the techniques of [25] (sec. 6) that the condition number of the preconditioned operator is $O(4^L 2^{(d-1)L/2})$ if L is the finest level appearing in the adaptive index set $\mathcal{T}$ and d the dimension of Ω, i.e. $d = 2,3$. A solution to this problem is to slightly modify (see [25]) both, the discrete gradient operator for the pressure and the discrete divergence operator for the velocity. Then, the spectral equivalence with the continuous Laplacian is established and the diagonal preconditioner yields a condition number of $O(2^{(d-1)L/2})$. This can be further improved to $O(1)$ by employing a more sophisticated technique (see [25] sec. 6) which is based on the lifting scheme. However, at least for the numerical experiments of the next section the latter improvement was not necessary, since the number of iterations of the BiCGStab solver were quite low: between 2-4 in each time step for the pressure Poisson equation and between 1-2 for the transport equations.

5   Numerical Experiments

In [35, 15] the Karlsruhe–Paris group studied a temporally developing mixing layer, schematically sketched in Fig. 2, which has been chosen to compare method I with method II. The initial velocity has a hyperbolic–tangent profile $u(y) = U \tanh(2y/\delta_0)$ which implies a vorticity thickness $\delta_0 = 2U/(du/dy)|_{y=0}$. From linear stability analysis the mixing layer is known to be inviscidly unstable. A perturbation leads to the formation of vortices by Kelvin–Helmholtz instability, where the most amplified mode corresponds to a longitudinal wavelength $\lambda = 7\delta_0$ [29].

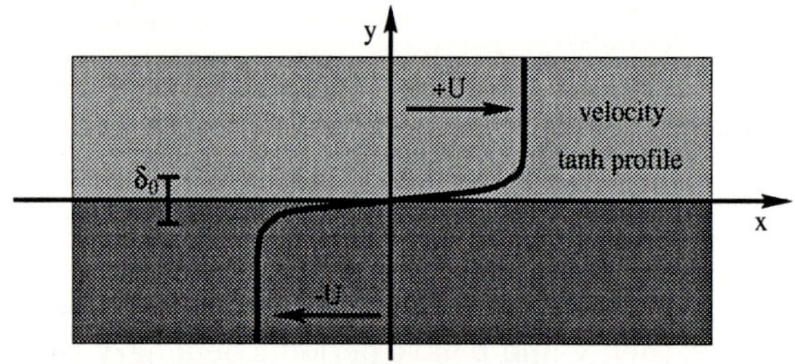

Figure 2   Initial configuration for the mixing layer.

The initial vorticity thickness δ_0 is chosen such that 10 vortices should develop in the numerical domain of size $[0,2\pi]^2$. To trigger the instability we superimposed a weak white noise in the rotational region. The velocity is $U \approx 0.1035$ and the viscosity is $\nu = 5 \cdot 10^{-5}$.

For the reference simulations two Fourier psueod–spectral codes were used. The first code is based on the vorticity formulation of the Navier-Stokes equations and employs a 2nd order Adams-Bashforth (AB2) scheme as in method I; the second code is based on the pressure-velocity formulation and uses a 3rd order Adams-Bashforth (AB3) scheme as in method II.

5.1 Results for method I

For the numerical simulation we employ a maximal resolution of 256×256, which corresponds to $L = 8$ in (3.9), and cubic spline wavelets of Battle–Lemarié type. The time step is $\Delta t = 2.5 \cdot 10^{-3}$. The threshold for the wavelet coefficients was $\epsilon_0 = 10^{-6}$ and 10^{-5}.

In Fig. 3 (left) we compare the energy spectrum at $t = 37.5$ for a reference computation using a classical pseudo-spectral method, and for two wavelet computations using different thresholds ($\epsilon_0 = 10^{-6}$ and 10^{-5}).

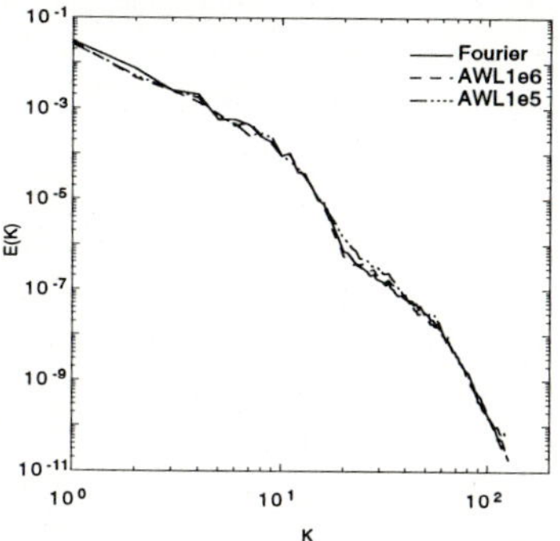
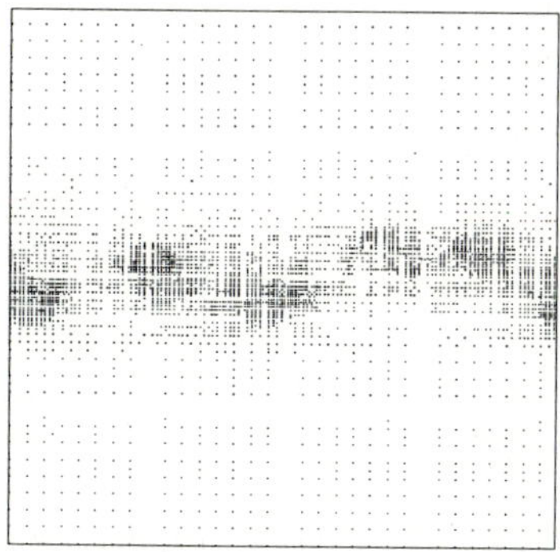

Figure 3   Left: energy spectra for the pseudo–spectral reference run and for the adaptive wavelet simulations with thresholds $\epsilon_0 = 10^{-6}, 10^{-5}$. Right: adaptive grid reconstructed from the index set of the retained wavelet coefficients. Both at time $t = 37.5$.

Fig. 3 (left) shows that all scales of the flow are well–resolved for both thresholds. The underlying grid Fig. 3 (right) which corresponds to the centers of active wavelets for the computation with $\epsilon_0 = 10^{-6}$ at $t = 37.5$ shows a local refinement in regions of strong gradients where dissipation is most active. In Fig. 4 (bottom) we show the evolution of the vorticity field for the adaptive wavelet simulation with threshold $\epsilon_0 = 10^{-6}$ and for the reference pseudo–spectral computation (top). In both simulations, as predicted by the linear theory, 10 vortices are formed, which subsequently undergo successive mergings. In Fig. 5 the active wavelet coefficients (gray entries) are plotted using a logarithmic scale. The coefficients $\tilde{\omega}_\lambda$ are placed at position $(x_1, x_2) = (2^j(1-\delta_{d,1})+k_x, 2^j(1-\delta_{d,2})+k_y$ with the origin in the lower left corner and the y-coordinate oriented upwards, from coarser to finer scales. We observe that the basis dynamically adapts to the flow evolution during the computation with only 8% of the coefficients being used. We observe that in the wavelet simulation the formation and evolution of vortices are well captured, although we find that at later times a slight phase shift appears with respect to the reference run (top). This might be due to the fact that the retained wavelet coefficients contain 94% of the total enstrophy, as observed in Fig. 6 (left) which shows the time evolution of the total enstrophy using the different thresholds. The 6% loss of enstrophy comes from the fact that in the wavelet simulations we have not modelled the effect of the discarded modes onto the retained ones, similarly to the subgrid scale model used in LES. This will be considered in future work, where the enstrophy of the discarded wavelets will be reinjected into the coherent vortices using the wavelet forcing method we have proposed [34].

Finally, we plot the time evolution of the number of degrees of freedom for the two wavelet runs in Fig. 6 (right). First, we observe an initial phase, up to $t = 7$ s, where there

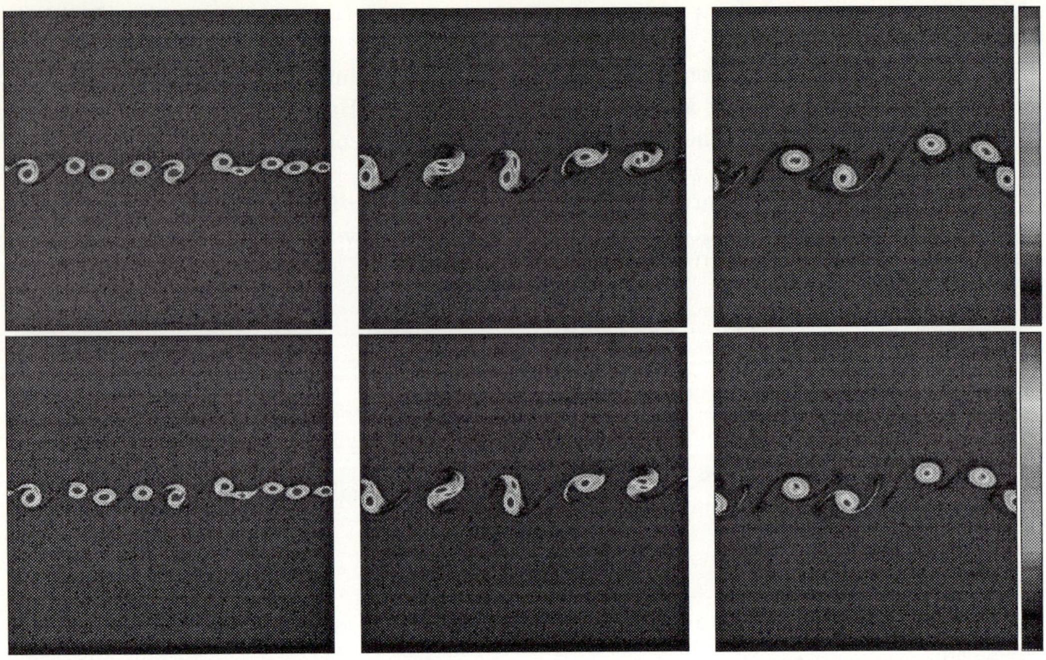

Figure 4 ω at $t = 12.5$, 25, 37.5. Top: pseudo–spectral method (vorticity-based). Bottom: method I ($\epsilon_0 = 10^{-6}$).

Figure 5 Active wavelet coefficients at $t = 12.5$, 25, 37.5. Method I ($\epsilon_0 = 10^{-6}$).

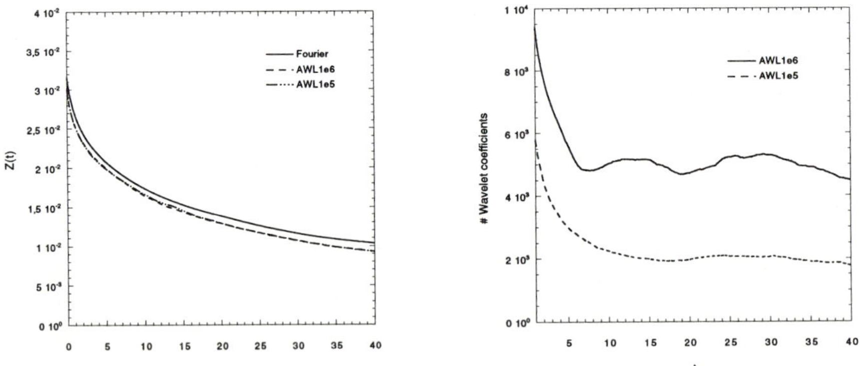

Figure 6 Left: evolution of enstrophy. Right: evolution of #DOF. Method I ($\epsilon_0 = 10^{-6}, 10^{-5}$).

is a strong reduction in the number of active modes, which corresponds to the formation of the coherent vortices. Then the number of active modes remains almost constant, and represent a significant reduction of the number of modes, with $N_{ad} = 5000$ for $\epsilon_0 = 10^{-6}$ and $N_{ad} = 2000$ for $\epsilon_0 = 10^{-5}$ out of $N = 65536$ initial modes.

5.2 Results for Method II

For the numerical simulations we employed tensor products of 6th order Interpolets, i.e. $P = 6$ in (P2), see section 2. The values of the threshold parameter ϵ were set to $\{12, 8, 4\} \cdot 10^{-4}$. Note that these values can not be directly compared to $\epsilon_0 = 10^{-6}$ used in method I, since there, the wavelets are normalized to $\|\psi_{(e,l,\mathbf{t})}\| \sim 1$, while the Interpolet wavelets are normalized to $\|\psi_{(\mathbf{l},\mathbf{t})}\|_0 \sim 2^{-(l_1+..+l_d)/2}$. If we would normalize them to ~ 1 then, the present threshold criterion corresponds to $|u_{(\mathbf{l},\mathbf{t})}| \cdot 2^{(l_1+..+l_d)/2} > \epsilon$, i.e. finer scales get a larger weight. This can be generalized to criteria of type:

$$|u_\lambda| \cdot f(\lambda) > \epsilon, \text{ with e.g. } f(\mathbf{l},\mathbf{t}) = 2^{(l_1+..+l_d)/2}\left(4^{sl_1} + ... + 4^{sl_d}\right).$$

In this case the threshold strategy is adjustable to different Sobolev norms by the parameter s (compare (2.3)). Numerical experiments with respect to the impact of f to the performance of the adaptive scheme will be done in the future.

In the present numerical experiments with method II the finest level $\mathbf{l}$ was (9,10) to take into account the expected anisotropy of the mesh. The time step was $\Delta t = 10^{-2}$ which corresponds to a CFL number of ≈ 0.2 with respect to the maximum velocity encountered during the simulation. The initial velocity has been calculated in Fourier space with 512×1024 modes from the initial vorticity field provided by the Karlsruhe–Paris group and used for method I.

Although, due to algorithmic reasons, the Bonn group preferred to compute a double mixing layer, obtained by imposing a mirror symmetry on the initial condition. This is the reason why the solutions of the Bonn group (Fig. 7) differ from the ones of the Karlsruhe–Paris group (Fig. 6).

In Figure 7 we give the results for the velocity-based spectral code (top) and method II (bottom). Both results agree very well. Only for $t = 37.5$ there is a phase shift for the right most two vortices. Despite of these slight differences in the instantaneous vorticity fields, the agreement between the spectral and the adaptive solutions with respect to the statistical quantity enstrophy $Z = 0.5 \int \omega^2 d\mathbf{x}$ is almost perfect as shown in Figure 9 (left). Note that we calculated the enstrophy only for $t = 0, 2, 4, ...$ This explains the kink at $t = 2$.

The evolution of the number of degrees of freedom (DOF) is shown in Figure 9 (right). The initial velocity is not very smooth because of the white noise added to trigger the instability. Therefore, a large number of DOF is required to resolve the initial velocity sufficiently well. Then, in a first phase the diffusion smooths the velocity very fast which leads to the strong decay of the number of DOF. This process is stopped by the development of the Kelvin-Helmholtz instabilities leading to an increase of the number of DOF ($4 < t < 10$). In the last phase ($t > 10$) the number of coherent vortices constantly decreases by successive merging. (The final state are two counter-rotating vortices which dissipate.) Therefore, in this stage the number of DOF decreases almost constantly.

In Figure 8 adaptive grids are shown which are associated to the adaptive index sets $\mathcal{T}$. Each $(\mathbf{l},\mathbf{t}) \in \mathcal{T}$ corresponds to the node (x_1, x_2), $x_i = 2^{-l_i} t_i$.

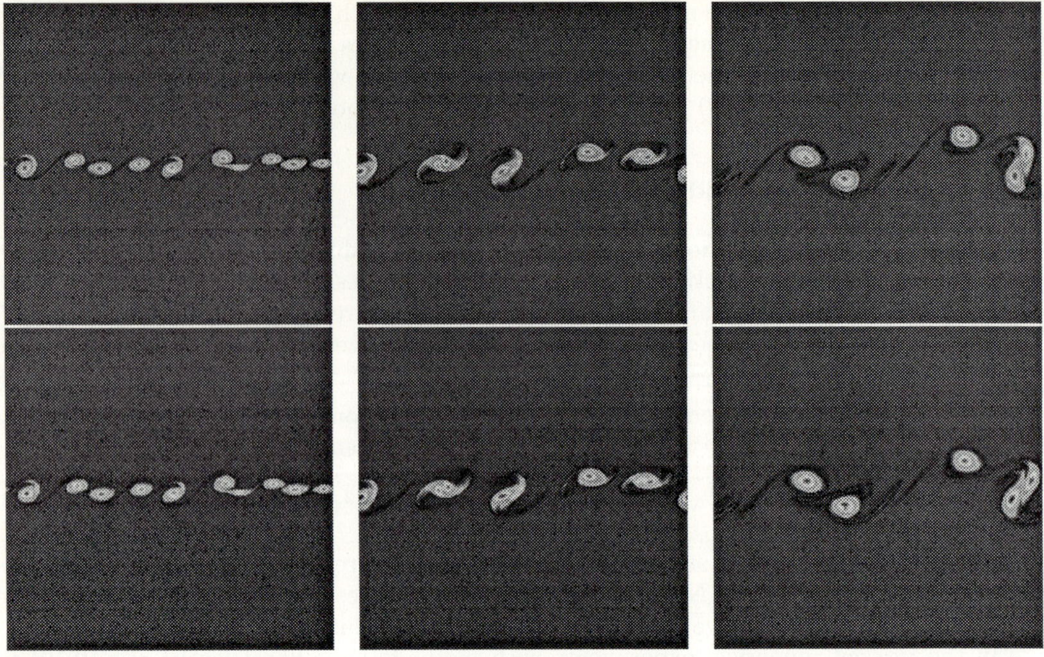

Figure 7 $|\omega|$ at $t = 12.5, 25, 37.5$. Top: spectral (velocity-based). Bottom: method II ($\epsilon = 8_{-4}$).

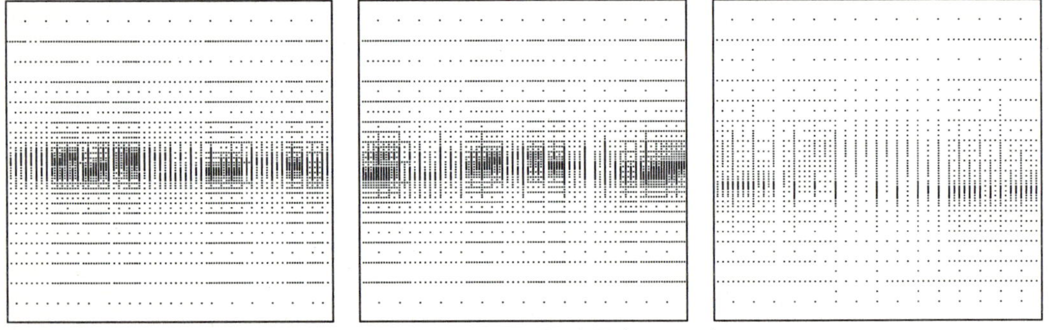

Figure 8   Adaptive grids $t = 12.5, 25, 37.5$. Method II ($\epsilon = 8_{-4}$).

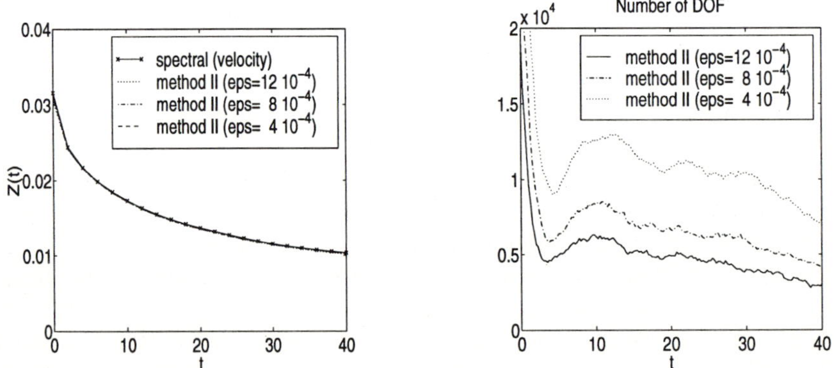

Figure 9   Left: evolution of enstrophy. Right: evolution of #DOF. Method II. The maximum #DOF for ($\epsilon = 4_{-4}$) is ≈ 31000.

316

6   Conclusion

We have shown that the two different adaptive wavelet methods presented here yield reliable and very accurate results with a reduced number of degrees of freedom. Although method I and II are Eulerian and are based on Petrov-Galerkin schemes, their adaptive character in both space and scale allows us to track the displacements and deformations of active flow regions as Lagrangian methods would do. These methods can also be applied to compute other flow configurations, such as jets or wakes, where coherent vortices play an important dynamical role.

Acknowledgements: We acknowledge partial support through PROCPOE (contract PKZ 9822768/99090) and the French–German program on 'Computational Fluid Dynamics' (contract No. Gr1144/7-1).

References

[1] Becker R., Rannacher R., *Weighted a posteriori error control in FE methods*; in ENUMATH-95 in Proc. ENUMATH-97, World Scientific Publ., Singapore (1998).

[2] Becker R., *Weighted error estimators for FE approximations of the incompressible Navier-Stokes equations*; to appear in Comput. Methods Appl. Mech. Engrg. (1998).

[3] Bockhorn, H., Fröhlich, J. and Schneider, K., *An adaptive two–dimensional wavelet–vaguelette algorithm for the computation of flame balls*; Combust. Theory Modelling 3 (1999) 177–198.

[4] Canuto C., Hussaini M., Quarteroni A., Zang T., *Spectral Methods in Fluid Dynamics*; Springer Verlag (1988).

[5] Charton P., Perrier V., *A pseudo–wavelet scheme for the two–dimensional Navier–Stokes equations*; Appl. Comput. Math., 15 (199-) pp.139-160.

[6] Cohen A., Dahmen W., DeVore R., *Adaptive wavelet methods for elliptic operator equations - convergence rates* IGPM-Report, (1998), RWTH Aachen.

[7] Dahmen W., *Wavelet and Multiscale Methods for Operator Equations*; Acta Numerica 6 (1997) pp. 55-228.

[8] Daubechies I., *Ten lectures on wavelets*; Regional Conference Series in Appl. Math. 61 (1992).

[9] Deslaurier G., Dubuc S., *Symmetric iterative interpolation processes*; Constr. Appr. 5 (1989), pp. 49-68.

[10] DeVore R., *Nonlinear Approximation*; Acta Numerica 8 (1999).

[11] DeVore R., Konyagin S.V., Temlyakov V.N., *Hyperbolic wavelet approximation*; Constr. Approx. 14 (1998), pp. 1-26.

[12] Donoho D., *Interpolating wavelet transform*; Stanford University (1992), preprint.

[13] Farge M., Kevlahan N., Bardos C., Schneider K., *Combining deterministic and statistical approaches to compute two-dimensional turbulent flows with walls*; Fundamental Problematic Issues in Turbulence (Eds. A. Gyr, W. Kinzelbach and A. Tsinober), Birkhäuser (1999), pp. 163-174.

[14] Farge M., *Wavelet transforms and their applications to turbulence*; Ann. Rev. Fluid Mech. 24 (1992) pp. 395–457.

[15] Farge M., Schneider K. and Kevlahan N., *Non–Gausianity and Coherent Vortex Simulation for two–dimensional turbulence using an adaptive orthonormal wavelet basis*; Phys. Fluids, 11 (8), (1999) pp. 2187–2201.

[16] Fröhlich J., Schneider K., *An adaptive wavelet Galerkin algorithm for one– and two–dimensional flame computations*; Eur. J. Mech., B/Fluids, 13 (1994), pp. 439–471.

[17] Fröhlich J., Schneider K., *An Adaptive Wavelet-Vaguelette Algorithm for the Solution of PDEs*; J. Compt. Phys. 130 (1997), pp. 174-190.

[18] Gresho P., Sani R., *Incompressible Flow and the Finite Element Method*; Wiley (1998).

[19] Griebel M., *Adaptive sparse grid multilevel methods for elliptic PDEs based on finite differences*; Computing 61 No. 2 (1998), pp. 151-180.

[20] Griebel M., Knapek S., *Optimized tensor-product approximation spaces*; to appear in Constr. Appr.

[21] Griebel M., Schiekofer T., *An adaptive sparse grid Navier-Stokes solver in 3D based on the finite difference method*, Proceedings of ENUMATH97 (1997).

[22] Griebel M., Oswald P., *Tensor product type subspace splitting and multilevel iterative methods for anisotropic problems*; Adv. Comp. Math. 4 (1995), pp. 171-206.

[23] Koster F., *A Proof of the Consistency of the Finite Difference Technique on Sparse Grids*; University Bonn, (2000) preprint.

[24] Koster F. Griebel M., *Orthogonal wavelets on the interval*; University Bonn, (1998) SFB 256 report No. 576.

[25] Griebel M., Koster F., *Adaptive Wavelet Solvers for the Unsteady Incompressible Navier-Stokes Equations*; to appear in J. Malek, M. Rokyta (Eds.), Advances in Mathematical Fluid Mechanics, Springer Verlag, also as Preprint No. 669 (2000), University Bonn.

[26] Koster F. Griebel M., *Efficient Preconditioning of Linear Systems for the Galerkin, Finite Difference and the Collocation Method on Sparse Grids*; University Bonn, (2000) in preparation.

[27] Kröner D., *Numerical Schemes for Conservation Laws*; Wiley-Teubner (1997).

[28] Liandrat J., Tchamitchian, P., *Resolution of the 1D regularized Burgers equation using a spatial wavelet approximation algorithm and numerical results*; ICASE (1990).

[29] Michalke, A., *On the inviscid instability of the hyperbolic tangent velocity profile*; J. Fluid Mech. 19 (1964) pp. 543–556.

[30] Oswald P., *Multilevel Finite Element Approximation*; Teubner Verlag (1994).

[31] Prohl A., *Projection and Quasi Compressibility Methods for Solving the Incompressible Navier Stokes Equations*; Teubner Verlag (1997).

[32] Schneider K. and Farge M., *Wavelet approach for modelling and computing turbulence*; Lecture Series 1998–05 Advances in turbulence modelling, von Karman Institute for Fluid Dynamics, Bruxelles (1998).

[33] Schneider K., Kevlahan N., Farge M., *Comparison of an adaptive wavelet method and non-linearly filtered pseudo-spectral methods for two-dimensional turbulence*; Theoret. Comput. Fluid Dynamics 9 (1997) pp. 191–206.

[34] Schneider K., Farge M., *Wavelet forcing for numerical simulation of two-dimensional turbulence*; C. R. Acad. Sci. Paris, série II b 325 (1997) pp. 263–270.

[35] Schneider K., Farge M., *Numerical simulation of temporally growing mixing layer in an adaptive wavelet basis*; C. R. Acad. Sci. Paris, série II b 328 (2000) pp. 263–269.

[36] Sweldens W., *The lifting scheme: a construction of second generation wavelets*; SIAM J. Math. Anal. 29,2 (1997), pp. 511-546.

[37] Temlyakov V., *Approximation of Periodic Functions*; Nova Science Publishers, N.Y. (1993).

Analysis of Discretization and Modeling Errors in Complex Three-Dimensional Flows

VISONNEAU M. [1], SCHMID M.[2], DENG G. [1], PERIĆ M.[2]

[1] Ecole Centrale de Nantes, Laboratoire de Mécanique des Fluides UMR 6598 CNRS
1, Rue de la Noë, B.P. 92101 44321 Nantes Cedex 3, France

[2] TU Hamburg-Harburg, Arbeitsbereich Fluiddynamik und Schiffstheorie
Lämmersieth 90, D-22305 Hamburg, Germany

1   Summary

The aspects of numerical accuracy in predictions of complex turbulent flows are addressed. Two research groups using different numerical methods and computer codes have computed the same three-dimensional turbulent flows using the same turbulence models. It is demonstrated that under the laminar flow conditions, the results obtained by different codes are identical; however, when turbulent flows are computed, the interaction of numerical (discretization) and modeling errors becomes important and the grid-independent solutions are more difficult to achieve. The agreement between solutions obtained by different codes using nominally the same turbulence model is satisfactory, although not perfect. Finally, the role of the second-moment turbulent closure for some flow types is emphasized.

2   Introduction

Results of turbulent flow predictions should be – on sufficiently fine grids – independent of the numerical method. However, at several workshops in the past it has often been found that the results obtained by different groups using nominally the same turbulence model differ more than when the same group uses different models. The differences often come from different numerics, different grid quality, or different implementation of various model components (e.g. the wall treatment). The difference due to the use of different approximations in the discretization method is unavoidable; however, it becomes smaller as the grid is refined, all other factors remaining the same. It is therefore possible to evaluate the modeling errors by making sure that the discretization errors are small by means of a systematic grid refinement. Different errors can often interact in an unusual manner when the problem is highly non-linear, as is the case with the solution of Reynolds-averaged Navier-Stokes equations. It is not unusual to find out that the results obtained on a coarse grid agree relatively well with experimental data, while the agreement gets worse as the grid is refined. Also, for a given grid, the results of turbulent flow predictions can depend substantially more on the choice of discretization schemes than is the case in predictions of laminar flows. Here the results of comparative studies of two three-dimensional turbulent flows by two research groups using different numerical methods and the same turbulence models will be presented. The emphasis is on the prediction of grid-independent results for a given turbulence model, in order to be able to evaluate the modeling errors by comparison with experimental data.

3 Physical Model

The equations to be solved are the Navier-Stokes equations, which can be written in integral form as:

$$\frac{\partial}{\partial t} \int_V \varrho \underline{v} \, dV + \int_S \varrho \underline{v} \, \underline{v} \cdot \underline{n} \, dS = \int_S \underline{\underline{T}} \cdot \underline{n} \, dS + \int_V \varrho \underline{b} \, dV \,, \tag{3.1}$$

$$\frac{\partial}{\partial t} \int_V \varrho \, dV + \int_S \varrho \underline{v} \cdot \underline{n} \, dS = 0 \,. \tag{3.2}$$

The momentum equation (3.1) contains four unknown variables but supplies only three component equations. The natural choice is to couple it with the mass conservation equation (3.2). In the case of incompressible fluids considered here, the density ϱ is assumed constant in the whole computational domain. The fluid is further regarded as Newtonian with a constant dynamic viscosity μ, depending on the temperature but not on the local deformation rate $\underline{\underline{D}}$,

$$\underline{\underline{D}} = 0.5 \left(\nabla \underline{v} + (\nabla \underline{v})^T \right) \,,$$

which determines the viscous stress tensor

$$\underline{\underline{T}} = 2\mu \underline{\underline{D}} - p\underline{\underline{I}} \,.$$

In turbulent flows the amount of instantaneous data would be too large to handle; besides, at high Reynolds numbers, the direct numerical simulation (DNS) would be impossible anyway. Even for the large-eddy simulation (LES), the Reynolds number in engineering flows like those considered here (of the order of one million) is too high. The widely used method of data reduction is the statistical averaging of the equations proposed by Reynolds, who considered the splitting of instantaneous variables in a mean and a fluctuating part (denoted by $\overline{u}_i$ and u_i'). The Reynolds-averaged Navier-Stokes (RANS) equations read:

$$\frac{\partial}{\partial t} \int_V \varrho \overline{\underline{v}} \, dV + \int_S \varrho (\overline{\underline{v}} \, \overline{\underline{v}} + \overline{\underline{v}'\underline{v}'}) \cdot \underline{n} \, dS = \int_S \overline{\underline{\underline{T}}} \cdot \underline{n} \, dS + \int_V \varrho \overline{\underline{b}} \, dV \,. \tag{3.3}$$

The task of the turbulence model is to provide a good approximation for the elements of the Reynolds-stress tensor $\tau_{ik} = -\varrho \overline{u_i' u_k'}$ in equation (3.3) in terms of the mean quantities, which is discussed in the next section. Different numerical methods use different approximations to discretize the mathematical operators in the equations. For the comparison of methods used in the cooperating groups in Nantes and Hamburg, all important approximations are listed in table 1. The numerical methods are second-order accurate (i.e. the discretization errors are reduced by a factor of 4 when the grid is systematically refined), as demonstrated for laminar flows in [8]. The finite-volume (FV) method used in Hamburg is fully conservative; the convective terms are expressed as a sum of fluxes through CV-faces, which are computed as a product of the mass flux through the face and the interpolated variable value at cell face. Both finite-difference (FD) and FV method used in Nantes are based on the so called "convective" definition of convective terms, i.e. approximated are $\varrho u_j (\partial u_i / \partial x_j)$ instead of $\partial (\varrho u_j u_i)/\partial x_j$. It turned out – as will be discussed below – that this formulation is under some circumstances more accurate than the conservative treatment – contrary to what one might expect.

Table 1 Description of approximations used in the numerical methods from Nantes and Hamburg.

group	Nantes	Nantes	Hamburg
method	finite-difference (FD)	finite-volume (FV)	finite-volume (FV)
internal naming	Horus	Isis	Comet
variable arrangement	colocated	colocated	colocated
discretization	convective	convective	conservative
interpolation	linear	linear	linear
integral approximation	-	midpoint rule	midpoint rule

4 Turbulence Modeling

For the sake of simplicity, the most common turbulence models used today are two-equation models. The assumption of isotropy, which is employed in these models, is not valid for many complex flows, but it holds at very large Reynolds numbers Re=ud/ν and is based upon the Boussinesq approximation:

$$-\varrho\overline{u_i'u_k'} = \mu_t \left(\frac{\partial \overline{u}_i}{\partial x_k} + \frac{\partial \overline{u}_k}{\partial x_i} \right). \tag{4.4}$$

The eddy viscosity $\mu_t = \varrho C_\mu k/\varepsilon^2$ is a function of the turbulent kinetic energy k and the dissipation rate ε, which physically represent the velocity and the length scale in turbulent flows, respectively. Some models use besides the velocity scale a time scale and solve the model transport equations for k and the specific dissipation rate $\omega = C_\mu k/\varepsilon$:

$$\varrho\frac{\partial k}{\partial t} + \varrho\overline{u}_k\frac{\partial k}{\partial x_k} = \frac{\partial}{\partial x_k}\left((\mu + \mu_t/\sigma_k)\frac{\partial k}{\partial x_k} \right) + \varrho\overline{u_i'u_k'}\frac{\partial \overline{u}_k}{\partial x_i} - C_\mu\varrho k\,\omega \tag{4.5}$$

$$\frac{\partial \omega}{\partial t} + \varrho\overline{u}_k\frac{\partial \omega}{\partial x_k} = \frac{\partial}{\partial x_k}\left((\mu + \mu_t/\sigma_\omega)\frac{\partial \omega}{\partial x_k} \right) + C_1\frac{\omega}{k}\varrho\overline{u_i'u_k'}\frac{\partial \overline{u}_k}{\partial x_i} - C_2\varrho\,\omega^2. \tag{4.6}$$

The performance of the model strongly depends on the chosen set of constants; these are calibrated to reproduce well some typical flows, but may need to be adjusted to the particular flow problem, as no universal turbulence model or set of constants exists [12]. In this study, the k-ω model with standard constants [11] has been used: $C_\mu = 0.09$, $\sigma_k = 0.5$, $C_1 = 5/9$, $C_2 = 3/40$, $\sigma_\omega = 0.5$. In addition, the Reynolds-stress model will be used, which provides transport equations for the six components of the Reynolds-stress tensor $-\varrho\overline{u_i'u_k'}$ and the specific dissipation rate ω. Details on this model and the parameters used can be found in [2].

5 Three-Dimensional Turbulent Flows

Two complex flow problems were computed on a fine grid by both groups, using the same turbulence model. The cases considered are real flow problems encountered in hydrodynamics, which are relevant to many other fields of fluid dynamics. The first test-case represents a wing-body junction; similar configurations can be found on airplanes, submarines, and in turbo-machines. The second test-case is a model of a tanker ship, which has never been built but whose geometry is close to built ships. For both test cases there are extensive experimental data that can be used for evaluating modeling errors. In both cases the boundary layers are three-dimensional and the separation occurs close to the end of the body due to the adverse pressure gradient. The aim was to achieve grid-independent solutions of the steady RANS equations and compare them to experimental data. The correctness of the implementation of turbulence models was tested in two-dimensional cases [7],[8],[9], which involved large recirculation zones and separated boundary layers. The grids were generated and optimized in Nantes, where these two cases are studied in detail with many different turbulence models. The same grids were then used in Hamburg.

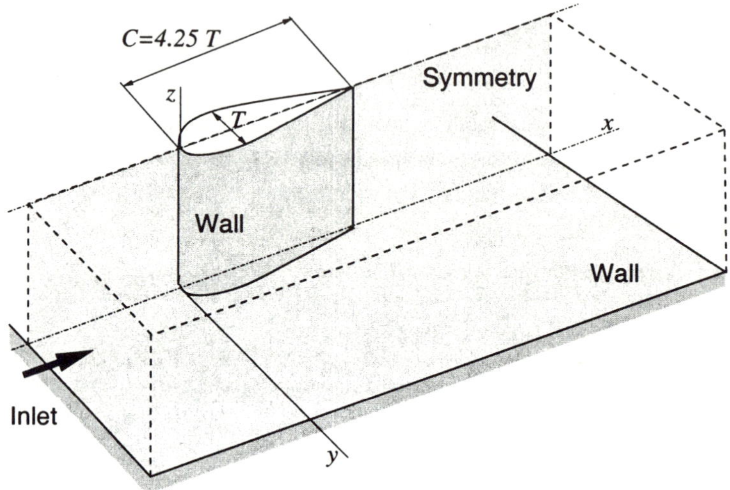

Figure 1 Configuration of the wing-body-junction: maximum wing thickness $T = 71.7$ mm, chord length $C = 305$ mm.

5.1 Wing-Body junction

The wing consists of a 3:2 half-elliptic nose and a NACA 0020 tail, joined at the cross-section of maximum thickness and mounted at 90°on a solid plate. The configuration as used in the experiment [5], [6] is outlined in figure 1. The finest grid used for comparisons had 563 200 control volumes; it was fine in wall vicinity to allow the use of a low Reynolds number version of the turbulence model, the normalized wall distance of the first grid point being less than unity. It was not possible to match exactly the boundary conditions of the experiment in the computation, but nearly-equivalent conditions have been specified. In both computations the same turbulent boundary-layer profile was specified as inlet condition; the profile has been computed in a separate turbulent boundary-layer computation. It closely corresponds to the velocity profile at the inlet cross-section observed in the experiment.

At the wing and plate walls, the no-slip boundary condition ($\underline{v} = 0$) was applied. The boundary at the upper end of the wing was treated as a symmetry boundary, which is close to experimental conditions where the wing was free at one end in order to minimize the influence of the upper channel wall on the secondary vortex at the junction. The outlet boundary was placed far away

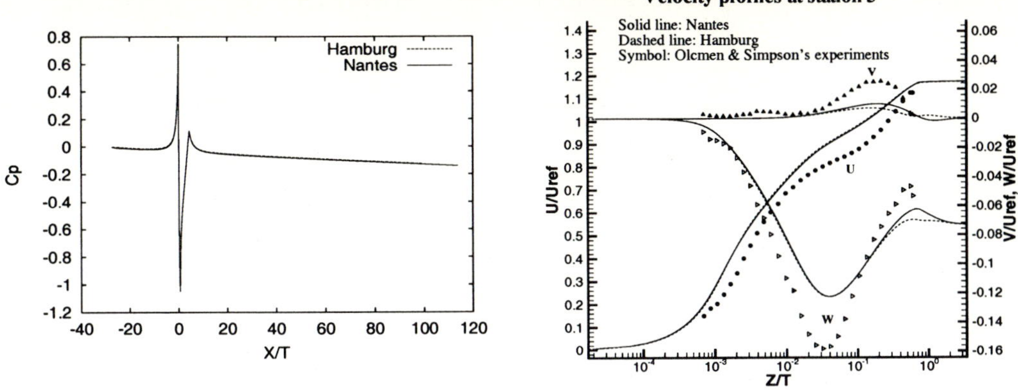

Figure 2 Comparison of results from computations in Nantes and Hamburg using the k-ω turbulence model and a 563 200 CV grid for the wing-body junction flow: Pressure coefficient along the line of symmetry on bottom plate and around the wing (left) and velocity profiles at one cross-section ($x/T = 0.092$, $y/T = 1.041$) compared to experimental data [6].

downstream from the wing and no influence on the flow around the junction by the zero-gradient boundary condition used in Nantes and Hamburg was detected. In front of the stagnation point a secondary, oscillating vortex can be observed, according to the double-peaked histograms of the measured velocity fluctuations. The statistical turbulence models are not able to predict this unsteadiness of the flow and the Reynolds number is too high for using the large-eddy simulation technique. The result of solving the RANS equations is thus a steady solution. The comparison of performance of different turbulence models for this flow has already been done in Nantes [2]; it showed a superior performance of the Reynolds stress model (RSM) compared to two-equations models used in comparisons.

Here, the comparison of *different discretization methods* with *the same turbulence model* is performed, in order to evaluate the numerical uncertainties. The comparison of computed results obtained by using the same grid and the same turbulence model (k-ω) but different codes is shown in figure 2 for the dimensionless pressure coefficient and for the velocity profile at one cross-section ($x/T = 0.092$, $y/T = 1.041$). The agreement is very good: the profiles of the x-component of the velocity are indistinguishable, while only small differences can be seen in profiles of the y- and z-components (note that the scale is an order of magnitude smaller for these two components). This suggests that both the model implementation is correct in both codes and that the grid used is fine enough so that the discretization errors can be considered negligible. The modeling errors of the low-Re k-ω turbulence model can be estimated by comparison with experimental data in figure 2. The difference between computed and measured profiles is much larger than the difference between the two computations, indicating that the modeling errors are indeed much larger than the discretization errors. The former are of the order of 10% of the mean streamwise velocity, why the latter are below 1%. The modeling errors are partly due to the assumption of isotropy in the model; anisotropic higher-order models result in appreciably smaller modeling errors, as demonstrated in [2].

Table 2 Description of meshes used for calculation of HSVA tanker.

Grid Level	coarse	medium	fine
grid	structured	structured	zonal
hull mesh	85 × 60	119 × 80	170 × 120
outer mesh	85 × 60	119 × 80	85 × 60
Number of CV	127 500	314 160	748 000

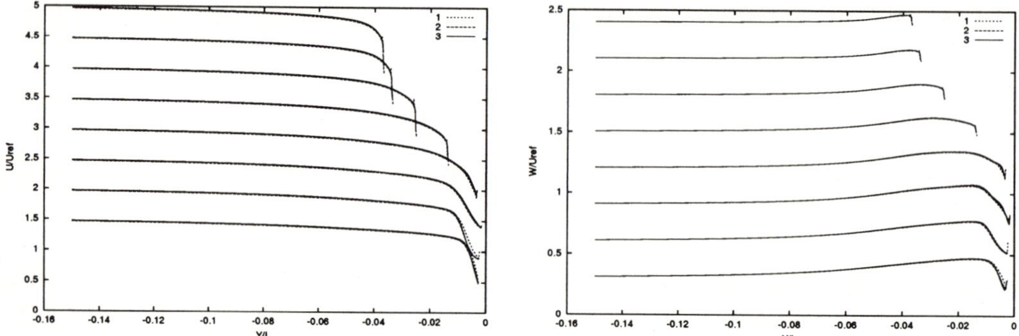

Figure 3 Comparison of computed velocity profiles on three grid levels at $x/L = 0.962$: x-component (left) and z-component (right) in different depths.

5.2 HSVA Tanker

HSVA tanker is the name for a ship hull designed by the HSVA (Hamburgische Schiffbau-Versuchsanstalt). The flow around this hull has been extensively studied at model scale experimentally in a wind tunnel [10]. The LDA-measurements of velocities and fluctuations have been performed using double model of the ship hull (i.e. neglecting the free-surface effects) at Re = $5 \cdot 10^6$, based on the length $L = 2.74$ m and the undisturbed velocity $v = 27$ m/s. The most interesting flow region from the ship design point of view is that around the propeller location (not present in experiments and simulations). Here, the lines of constant streamwise velocity component curve to form a famous hook-like shape – a feature which is normally not well predicted by two-equation turbulence models. Similar as in the wing-body junction case, a longitudinal vortex of positive vorticity forms along the ship and continues into the wake, where it is dissipated. Separation due to inverse pressure gradient takes place at the stern but is restricted to a small region on the hull. Computations were performed for a quarter of the model (half hull) on three grids (see Tab. 2), using the k-ω turbulence model. The third grid was systematically refined from coarse to fine grid in the vicinity of the hull by halving cell size in each direction, while in the outer region the coarse-level grid was retained. This is justified by the fact that the variation of variables is rather small further away from the hull, so that the discretization errors are also small there already on the coarsest grid. The results from the medium grid also showed little difference in this region. In the plane at $x/L = 0.962$, the velocity profiles on three grids are compared at different depths in figure 3 for the FV method used in Hamburg. Far from the hull the profiles are identical; close to the solid wall, differences are visible. They are smaller between medium and fine grid than between coarse and medium grid. Discretization errors are especially visible in regions of separation and of poor grid quality, while in planes at midship no differences can be seen between results from different grids (not shown here). It is difficult to obtain grid-independent solution for flows with separation as the results are very sensitive to the resolution of the separated boundary layer. In the two-dimensional calculation of the flow around a sphere [9], a fully grid-independent solution has not been achieved, although the grid contained around 175 000 CV. It is thus not surprising that the discretization errors are not negligible in the computations of a three-dimensional flow with separation on a smooth, curved surface. The comparison of results on three grid levels allows to estimate the error of the used scheme. On the second grid, which is used for comparisons with different methods, the error appears to be small as can be judged by the difference between solutions on medium and fine grid. The effort of computing the flow on a finer grid is disproportionally high (the computation lasts about one week on 4 Pentium III CPUs) compared to the expected change of results.

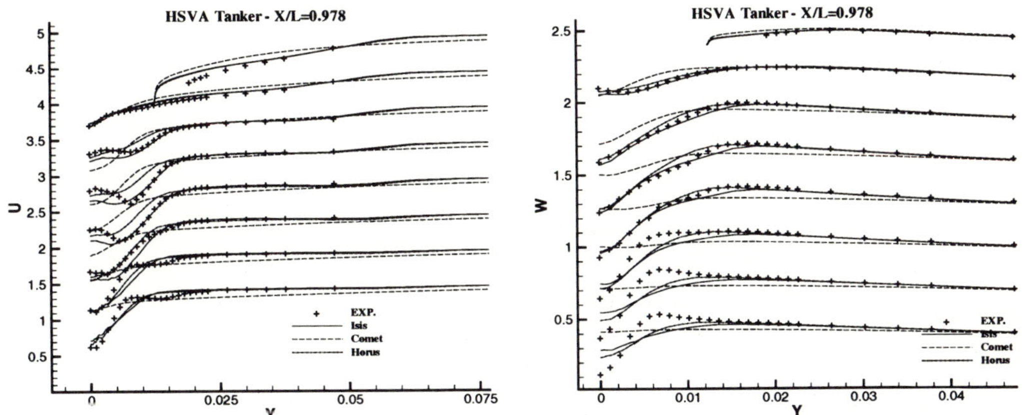

Figure 4 Comparison of streamwise and spanwise velocity contours at $x/L = 0.978$: computations in Nantes and Hamburg using the k-ω turbulence model and experimental data [10].

The two-equation turbulence model fails to predict accurately the typical hook-shaped contours of the streamwise velocity component in the propeller plane (see fig. 5) of the experiment; the hook-shaped contours indicate the presence of a longitudinal vortex which forms along the side of the ship. The results on three grids do not show any hook-like structure, which is unlikely to be due to discretization errors as shown by comparison of solutions on three grids. Partly the poor resolution in lateral direction (from the side to the bottom of the ship where the curvature is large only 4 CVs are located) is responsible for the damping of the vortex. On the medium grid, a small hook-like structure is visible if the computation is not fully converged; however by continuing the computation, the residuals first grow substantially and the flow changes to the state which looks different. In figure 4 results of computations on the medium grid level, obtained with the same turbulence model but different numerical methods, are compared to experimental data. There are noticeable differences between the computations as well as between computations and experiment. The group in Nantes uses two different numerical methods: a finite-volume one with a colocated arrangement of variables (Isis), which is similar to that used in Hamburg (Comet), and a finite-difference method with the same variable arrangement (Horus). Although the FV solution obtained in Hamburg seems to be grid independent, there is still a substantial difference between this solution and those obtained in Nantes. Initially large discrepancy between solutions obtained bt the FV-code and FD-code using the same grid has been observed in Nantes as well; the solutions obtained by the FV-code was close to the one obtained in Hamburg. This prompted

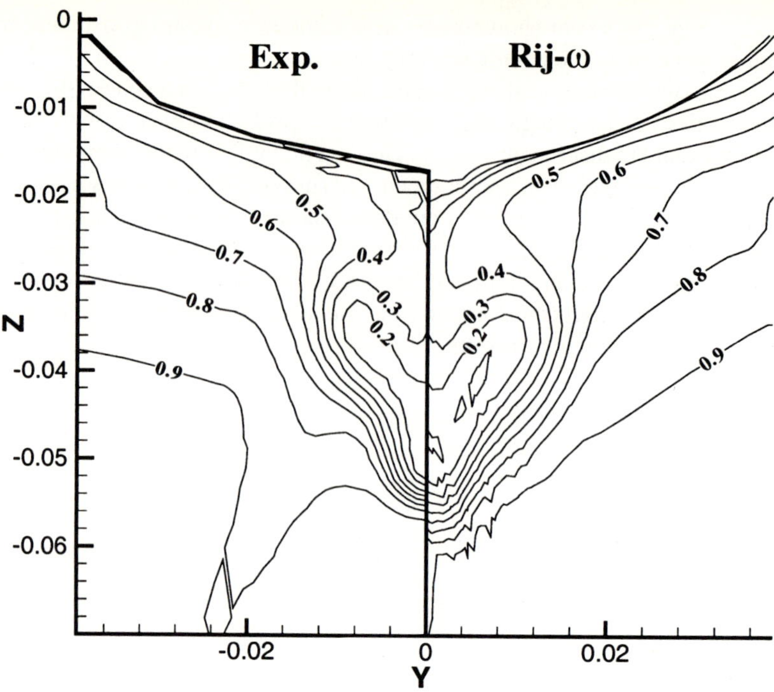

Figure 5   Conturs of the streamwise velocity component at x/L=0.978: Results from computations using the R_{ij}-ω turbulence model, compared to reference data [10].

the group in Nantes to change the discretization of convective terms in their FV-code to be conforming with that used in the FD-code, i.e. from fully-conservative to semi-conservative. This resulted in a much better agreement between their FV and FD codes, as can bee seen in figure 4. It appears that the difference between the two discretizations becomes important when the grid is highly non-orthogonal, as is the case for the HSVA-tanker geometry. Similar observations were already made at two workshops, where the HSVA-tanker was one of the test cases. This issue will be further analyzed in Hamburg.

The remaining difference between the computed velocity of the FD method and experimental data is due to the modeling errors; the most likely source of error lies in the assumption of isotropic turbulence in the two-equation turbulence model used here. This conclusion can be drawn by referring to the results of computations obtained by using the second-moment closure [3]; an example is shown in figure 5 as obtained on the medium grid.

5.3   Discussion

Three different methods predict for the wing-body junction flow solutions which are nearly indistinguishable from each other and are thus considered grid-independent; for the HSVA tanker, predictions differ notably. The comparison of ship-flow results on three grids show relatively

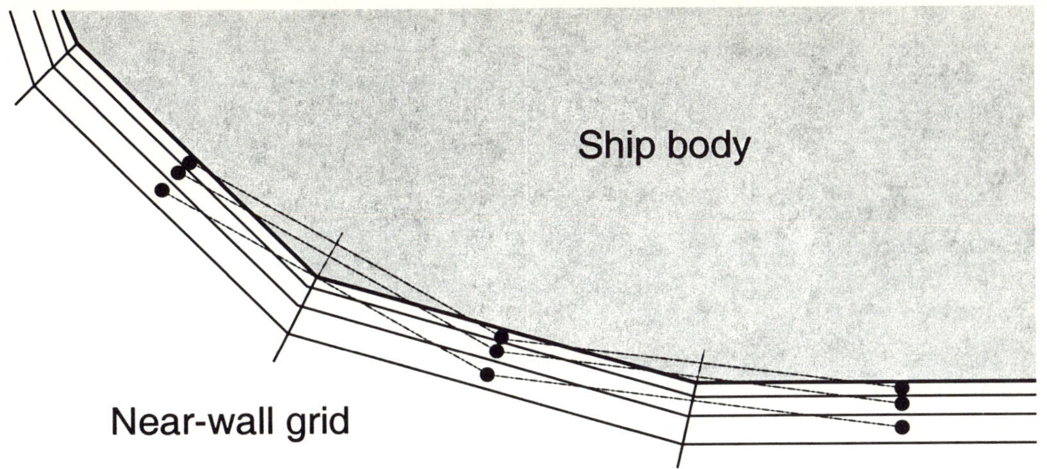

Figure 6 Typical grid deformation on curved surface with turbulent boundary layer resolution (the aspect ratio is much higher in the real grid - spacing normal to wall has been enlarged in this figure for clarity).

small changes, while the difference between results obtained by two different codes on the same grid is relatively large. The reason for this appears to lie in the sensitivity of the FV-discretization used in Hamburg to grid quality. For the wing-body junction geometry a high-quality grid could be generated and all codes produced nearly identical results. In the case of the HSVA-tanker geometry, the grid exhibits strong non-orthogonality and in some regions also strong warpage; these seem to introduce discretization errors of very low order and thus don't reduce with grid refinement as expected of a nominally second-order method. The interpolation may be substantially in error when the cells are thin and not aligned, as is the case near curved wall surfaces. The cells have to be very thin to allow the use of a low Reynolds number model; the lateral resolution may thus be critical in the case of a colocated arrangement, since the line connecting two cell centers not only does not pass through the center of the cell face, but it does not pass through the face at all, see figure 6. These interpolation errors affect also the computation of gradients at cell centers and thus make the approximation of diffusive fluxes less accurate. Since the grids used in engineering applications are often of poor quality, the ways of increasing the accuracy of approximations on such grids will be further studied. Especially the issues related to the semi-conservative treatment of convective terms will be evaluated in more detail after the experience from Nantes suggests that it may have some important advantages over the more common fully-conservative treatment when the grid is not optimal.

6   Conclusions

Complex three-dimensional flows in engineering practice are usually computed using isotropic turbulence models. In order to evaluate the performance of the turbulence model for a particular flow, it is necessary to make sure that the numerical errors are small enough and that the model has been properly implemented in the computer code. The former requirement can be met by repeating computations on a sequence of systematically refined grids; the latter has been met here by comparing results produced by different codes in Hamburg and Nantes using the same grids and turbulence models. It has been found that the low Reynolds number version of the k-ω model, which does not use wall functions, can not predict all the features of complex three-dimensional flows like those around wing-body junctions and ship-like bodies, since the assumptions on which the model is based are violated when strong secondary flows are induced. On the other hand, it was shown that the Reynolds-stress model is much more suited for this kind of flows.

7 Acknowledgments

Attribution of CPU on the Cray C98, the Cray T3E and the VPP machines by the Scientific Committee of IDRIS and the DS/SPI made the french part of these comparisons possible. The financial support by the Deutsche Forschungsgemeinschaft and CNRS is greatly appreciated.

References

[1] G. B. Deng, J. Piquet, P. Queutey, M. Visonneau, Incompressible Flow Calculations with a Consistent Physical Interpolation using the CPI Method, *Computers & Fluids*, Vol. 23,**8**, 1020–1047 (1994).

[2] G. B. Deng, M. Visonneau, Computation of a Wing-Body Junction Flow with a New Reynolds-Stress Transport Model. 22nd Symposium on Naval Hydrodynamics, Washington DC, August 9-14 (1998).

[3] G. B. Deng, M. Visonneau, Comparison of explicit algebraic stress models and second-order turbulence closures for steady flows around ships. 7th International Conference on Numerical Ship Hydrodynamics, Nantes, July 19-22 (1999).

[4] J. H. Ferziger and M. Perić, *Computational Methods for Fluid Dynamics*, Springer, Berlin (1996).

[5] W. J. Devenport, R. L. Simpson, Time-dependent and time-averaged turbulence structure ner the nose of a wing-body junction, *Journal of Fluid Mechanics*, **210**, 23–55 (1990).

[6] S. Ölçmen, R. L. Simpson, An experimental study of a three-dimensional pressure-driven turbulent boundary layer. *Journal of Fluid Mechanics*, **290**, 225–262 (1995).

[7] J. Piquet, M. Visonneau, M. Perić, Computation of Turbulent Flows with Separation by Coherent Structure Capturing, Progress Report Joint CNRS-DFG Research Programm (1997).

[8] M. Schmid, G. Deng, V. Seidl, M. Visonneau, M. Perić, Computation of Complex Turbulent Flows, Numerical Flow Simulation I (ed. E. H. Hirschel), Vieweg Verlag (1998).

[9] J. Piquet, M. Visonneau, M. Perić, Computation of Turbulent Flows with Separation by Coherent Structure Capturing, Progress Report Joint CNRS-DFG Research Programm (1999).

[10] K. Wieghart, J. Kux, Nomineller Nachstrom auf Grund von Windkanalversuchen, Jahrbuch der Schiffbautechnischen Gesellschaft (STG), 1980.

[11] D. C. Wilcox, Reassesment of the Scale-Determing Equation for Advanced Turbulence Models, *AIAA Journal*, **26**, 1299–1310 (1988).

[12] D. C. Wilcox, *Turbulence Modeling for CFD*, DCW Industries Inc. (1993).

Notes on Numerical Fluid Mechanics

Available volumes

Volume 74: D. Henry, A. Bergeon (eds.): Continuation Methods in Fluid Dynamics. Contributions to the ERCOFTAC/EUROMECH Colloquium 383, Aussois, France, 6.–9. September 1998. ISBN 3-540-41556-4

Volume 73: M. Griebel, S. Margenov, P. Yalamov (eds.): Large-Scale Scientific Computations of Engineering and Environmental Problems II. Proceedings of the Second Workshop on "Large-Scale Scientific Computations" Sozopol, Bulgaria, June 2–6, 1999. ISBN 3-540-41551-3

Volume 72: W. Nitsche, H.-J. Heinemann, R. Hilbig (eds.): New Results in Numerical and Experimental Fluid Mechanics II. Contributions to the 11th AG STAB/DGLR Symposium Berlin, Germany 1998. ISBN 3-540-41552-1

Volume 70: W. Hackbusch, G. Wittum (eds.): Numerical Treatment of Multi-Scale Problems. Proceedings of the 13th GAMM-Seminar, Kiel, January 24–26, 1997. ISBN 3-540-41562-9

Volume 69: D. P. Hills, M. R. Morris, M. J. Marchant, P. Guillen (eds.): Computational Mesh Adaptation. ECARP – European Computational Aerodynamics Research Project. ISBN 3-540-41548-3

Volume 67: C. H. Sieverding, G. Cicatelli, J. M. Desse, M. Meinke, P. Zunino: Experimental and Numerical Investigation of Time Varying Wakes behind Turbine Blades. Results of the Industrial and Material Technology Aeronautics Research Project AER2-CT92-0048, 1992–1996. ISBN 3-540-41546-7

Volume 66: E. H. Hirschel (ed.): Numerical Flow Simulation I. CNRS-DFG Collaborative Research Programme. Results 1996–1998. ISBN 3-540-41540-8

Volume 64: R. Friedrich, P. Bontoux (eds.): Computation and Visualization of Three-Dimensional Vortical and Turbulent Flows. Proceedings of the Fifth CNRS-DFG Workshop on Numerical Flow Simulation, München, Germany, December 6 and 7, 1996. ISBN 3-540-41557-2

Volume 63: M. Fiebig, N. K. Mitra (eds.): Vortices and Heat Transfer. Results of a DFG-Supported Research Group. ISBN 3-540-41558-0

Volume 62: M. Griebel, O. Iliev, S. D. Margenov, P. S. Vassilevski (eds.): Large-Scale Scientific Computations of Engineering and Environmental Problems. Proceedings of the First Workshop on "Large-Scale Scientific Computations" Varna, Bulgaria, June 7–11, 1997. ISBN 3-540-41560-2

Volume 57: H. Deconinck, B. Koren (eds.): Euler and Navier-Stokes Solvers Using Multi-Dimensional Upwind Schemes and Multigrid Acceleration. Results of the BRITE/EURAM Projects AERO-CT89-0003 and AER2-CT92-00040, 1989–1995. ISBN 3-540-41535-1

Printing: Mercedes-Druck, Berlin
Binding: Buchbinderei Lüderitz & Bauer, Berlin